ENVIRONMENTAL RESEARCH ADVANCES

STEPPE ECOSYSTEMS

BIOLOGICAL DIVERSITY, MANAGEMENT AND RESTORATION

ENVIRONMENTAL RESEARCH ADVANCES

ENVIRONMENTAL RESEARCH ADVANCES

STEPPE ECOSYSTEMS

BIOLOGICAL DIVERSITY, MANAGEMENT AND RESTORATION

MANUEL B. MORALES PRIETO
AND
JUAN TRABA DIAZ
EDITORS

nova publishers
New York

For permission to use material from this book please contact us:
Telephone 631-231-7269; Fax 631-231-8175
Web Site: http://www.novapublishers.com

NOTICE TO THE READER

Additional color graphics may be available in the e-book version of this book.

Library of Congress Cataloging-in-Publication Data

Steppe ecosystems : biological diversity, management and restoration / editors, Manuel B. Morales Prieto and Juan Traba Diaz.
pages cm
Includes bibliographical references and index.
ISBN 978-1-62808-298-2 (hardcover)
1. Steppe ecology. 2. Biotic communities. 3. Biodiversity. 4. Ecosystem management. 5. Restoration ecology. I. Morales Prieto, Manuel B. II. Traba Diaz, Juan.
QH541.5.P7S738 2013
639.9--dc23

2013021111

Published by Nova Science Publishers, Inc. † New York

CONTENTS

Preface vii

Chapter 1 Steppes across the World: An Overview with Emphasis on the Iberian Peninsula 1
Helios Sainz Ollero

Patterns and Processes in Steppe Ecosystems 27

Chapter 2 Factors Determining Species Richness and Composition of Steppe Bird Communities in Peninsular Spain: Grass-Steppe vs. Shrub-Steppe Bird Species 29
Juan Traba, Pablo Sastre and Manuel Borja Morales

Chapter 3 Species Richness and Niche Differentiation of Darkling Beetles (Coleoptera: Tenebrionidae) in Mongolian Steppe Ecosystems 47
Omid Paknia, Michael Grundler and Martin Pfeiffer

Chapter 4 Diversity Patterns in the Steppe of Argentinean Southern Patagonia: Environmental Drivers and Impact of Grazing 73
Pablo Luis Peri, María Vanessa Lencinas, Guillermo Martínez Pastur, Grant W. Wardell-Johnson and Romina Lasagno

Chapter 5 The Vegetation Structure of the Mongolian Altai Mountain Steppes 97
Galina N. Ogureeva, Inessa M. Mikljaeva and Maxim V. Bocharnikov

Chapter 6 Resilience or Vulnerability? Vegetation Patterns of a Central Tibetan Pastoral Ecotone 111
Georg Miehe, Sabine Miehe, Kerstin Bach, Karsten Wesche, Elke Seeber, Lena Behrendes, Knut Kaiser, Christoph Reudenbach, Jasmin Nölling, Jan Hanspach, Mark Herrmann, Ma Yaoming and Volker Mosbrugger

Chapter 7 Estimation of Carbon Dioxide Exchange of Cascade Geochemically Conjugated Steppe Ecosystems in Salinity Condition 153
Aleksei Naumov and Elena Smolentseva

Chapter 8	The Late Quaternary Paleoecology and Environmental History of Hortobágy, a Unique Mosaic Alkaline Steppe from the Heart of the Carpathian Basin *Pál Sümegi, Gábor Szilágyi, Sándor Gulyás, Gusztáv Jakab and Attila Molnár*	**165**
Conservation and Management of Steppe Ecosystems		**195**
Chapter 9	The Conservation of High-Interest Plant Species Offers the Chance to Preserve Unique and Vulnerable Representatives of Gypsum Steppes *Màrcia Eugenio, Carlos Molina and Gonzalo Montamarta*	**197**
Chapter 10	The Cover of Steppe Vascular Plants by the Protected Area Network of the Republic of Mordovia (Russia) *Anatoliy A. Khapugin and Tatyana B. Silaeva*	**211**
Chapter 11	Recovery of Degraded Steppe Soils of Northern China after Grazing Exclusion *Martin Wiesmeier*	**221**
Chapter 12	Effects of Ungulate Grazing on Soil Microbial Diversities in Inner Mongolian Steppes of Northern China *Xiaoqi Zhou and Yanfen Wang*	**253**
Chapter 13	The Pseudo-Steppe of La Crau (Southeastern France): Origin, Management and Restoration of a Mediterranean Rangeland *Thierry Dutoit, Elise Buisson, Sylvain Fadda, Frédéric Henry, Clémentine Coiffait-Gombault, Renaud Jaunatre, Jean-François Alignan, Solène Masson and Adeline Bulot*	**287**
Chapter 14	Critical Review of the Algerian Experience to Manage the Land Degradation *Aziz Hirche, Mostefa Salamani, Boughani Abdelmadjid, Dalila Nedjraoui, Ratiba Hourizi and Youcef Oukil*	**303**
Chapter 15	Sustainable Development in the Caldenal (Argentina) *Hugo Laborde, Osvaldo A. Fernández, María N. Fioretti, Sandra S. Baioni and Roberto E. Brevedan*	**321**
Index		**335**

PREFACE

Steppes are among the most fragile and human-impacted ecosystems of the world. Their mainly flat and open physiography, along with their character of areas transitional from forested regions to deserts has favored their use and exploitation by humans since the early Neolithic, when the first sedentary populations began plowing steppe soils and feeding their herds of domesticated grazers on steppe vegetation. This process has continued for millennia transforming large parts of original steppes and favoring their ecological impoverishment and desertification, frequently beyond their capacity of resilience. In parallel, formerly wooded areas have been cleared to expand agriculture and livestock. When this clearing occurred in regions where climate and soil characteristics hardly allowed vegetation recovery after agriculture or grazing cessation, as it is frequently the case under Mediterranean climate conditions, landscapes became steppic, acquiring many of the ecological features of natural steppes. Therefore, diverse typologies of steppe ecosystems can be found in different parts of the world as a result of both natural and anthropic processes.

Steppe environments impose strong selection pressures on organisms due to limiting climate conditions related with the length and severity of the frost period in some cases, and/or of the summer drought in others, as well as with other factors such as constant drying wind in many of them. Biotic factors such as herbivory (by either wild or domesticated grazers) or competition for limiting resources also exert important pressures on steppe organisms, which display notable and interesting adaptations such as tussocky or spiny cushion-like growth in the case of plants, or very strict microhabitat selection patterns in that of some animal groups like birds.

As a result of these ecological and conservation features, steppe ecosystems and organisms have long deserved the attention of ecologists, biogeographers and conservation biologists. In spite of that, some basic questions still require unambiguous answers and continue to be the subject of open scientific debate. How did present-day steppe ecosystems arise? Is the steppe a perfectly identifiable and homogeneous biome, or are there different types of steppes? If the latter is true, what are the abiotic and biotic factors that define steppe ecosystems? Do they function differently? In short, what is a steppe? Can we identify clearly steppe-specific taxa? Are their functional traits homogeneous across steppe ecosystems? How are their main biodiversity patterns?

On the other hand the current and increasing use of steppe ecosystems by human populations brings out the need of their sustainable management on the basis of adequate scientific knowledge. The result of steppe land-use intensification is, in many cases, desertification (particularly in the ecotones with arid environments), and therefore the loss of valuable resources that may guarantee the way of life of human inhabitants, such as agriculturally suitable soil, pastures, or primary materials for local manufactures. Many steppe organisms are endangered or show worrying population declines. Their extinction would not only imply the irreversible loss of a common natural heritage, but the impoverishment of biological communities that are particularly original from a biogeographical and ecological perspective. Therefore, the evidence-based assessment of these ecosystems' resilience and their response to different management models is critical to prevent further degradation.

The present book aims to address these issues through the contribution of steppe ecology experts from different regions of the world, which also means that different types of steppe ecosystems are encompassed in the volume. Some contributions are focused on steppe biodiversity patterns and ecosystem functioning. Others mainly address management and conservation problems. Consequently, the book's chapters are arranged in two main sections, a first one dedicated to steppe ecosystem patterns and processes, and a second one devoted to their conservation and management. However, and with the purpose of answering the fundamental questions of what is a steppe and which are the main types of steppe ecosystems in the world, the very first chapter, written by *Helios Sainz Ollero*, describes steppe-like ecosystems across the world from a geobotanical perspective, from Central Asia to Patagonia. This author pays particular attention to the steppes of the Iberian Peninsula, often neglected in the international context as marginal or resulting from the anthropic degradation of former forested areas. Sainz Ollero highlights their character of genuine Mediterranean steppes, as well as their physiographical and biogeographical affinities with other Palearctic steppe regions. Such affinity is patent even for the man-extended cereal steppes, whose bird communities, for example, are comprised by species typical of Eurasian natural steppes.

The Iberian steppe bird community is, precisely, the subject of the first chapter in the patterns and processes section, contributed by *Juan Traba*, *Pablo Sastre* and *Manuel B. Morales*. Using data from 5000 100-km^2 quadrats of the Spanish breeding bird atlas, these authors analyze the relative importance of geographical position and environment (climate, topography and land use) in steppe bird richness and composition across continental Spain and the Balearic Islands. Their results highlight the importance of low-intensity land use practices on the richness of steppe birds, as well as the dependence of community composition on topography and land use. Consistently with the two main steppe environments present in Iberia, two steppe bird assemblages are identified: the grass-steppe bird species and the shrub-steppe bird species.

Another faunal group closely linked to steppe ecosystems is that comprised by the darkling beetles or beetles of the tenebrionid family, on which the next chapter is focused. In this contribution, *Omid Paknia*, *Michael Grundler* and *Martin Pfeiffer* analyze data from extensive tenebrionid museum records and own samplings covering all steppe ecosystems of Mongolia, finding a clear north to south increasing gradient in the species richness of this group. Moreover, tenebrionid richness consistently increased with maximum and minimum temperatures, but decreased with rainfall, providing an example of a community that enriches as the environment becomes less productive due to aridity, which reveals the high degree of adaptation of these beetles to arid and semiarid environments.

The role of beetles as indicators of steppe ecosystem state is evaluated, along with vegetation composition, by *Pablo Luis Peri*, *María Vanessa Lencinas*, *Guillermo Martínez*

Pastur, *Grant W. Wardell-Johnson* and *Romina Lasagno*, who studied coleopteran and vegetation diversity patterns of the steppes of southern Patagonia in relation to different environmental and disturbance factors. The association of changes in those communities with various sources of disturbance (sheep grazing, fire, beaver activity) and their interaction with environmental variables are shown to serve as adequate indicators of the Patagonian steppe ecological health.

High altitude intensifies the extreme conditions of steppe environments favoring the evolution of specific adaptations of plants and therefore the formation of singular but extremely fragile communities. Mountain steppes of Central Asia still remain insufficiently surveyed and inventoried. The chapter written by *Galina Ogureeva*, *Inessa M. Mikljaeva* and *Maxim V. Bocharnikov* contributes to fill this gap by providing a comprehensive description of the steppe vegetation formations of the Mongolian Altai Mountains, highlighting the need to improve this type of information in order to adequately manage their use by local populations while keeping them from anthropic degradation.

Another high altitude region of Asia where steppe communities need both adequate surveying and proper assessment of their ecological state is the Tibetan Plateau. These issues are successfully addressed by *Georg Miehe, Sabine Miehe, Kerstin Bach, Karsten Wesche, Elke Seeber, Lena Behrendes, Knut Kaiser, Christoph Reudenbach, Jasmin Nölling, Jan Hanspach, Mark Herrmann, Ma Yaoming* and *Volker Mosbrugger* in their complete chapter on the Tibetan plant communities, whose composition and potential for resilience against land-use intensification is evaluated combining both field survey and remote sensing data. These authors discuss how Tibetan steppes have changed since first recorded human settlements and how the more humidity-dependent formations like the turf pastures of *Kobresia pygmea* are particularly vulnerable to increasing grazing pressure, turf extraction and mining by small mammals, whose effects may be magnified by current climate trends.

The combination of low precipitation and high summer temperatures found in steppe environments favors salt precipitation in soils and the formation of saline or salty shallow lakes. Differences in soil or water salt concentration influence different ecosystem processes such as soil or water CO_2 exchange. The field experiment reported in the chapter by *Aleksei Naumov* and *Elena Smolentseva* shows how CO_2 exchange varies along a catena of different vegetation formations neighboring the salty Bagan Lake in the steppes of southern Siberia. Understanding the carbon biogeochemical cycle in these extreme environments is important to evaluate how current climate change may affect the productivity and general function of steppe ecosystems.

Salt and alkali precipitation, along with hydrogeological dynamics, seems to have played a central role in the origin of the steppes of the Carpathian basin, limiting the development of closed forest formations. In the last chapter of the patterns and processes section of the volume, *Pál Sümegi, Gábor Szilágyi, Sándor Gulyás, Gusztáv Jakab* and *Attila Molnár* combine hydrogeological information and palinological data to describe the evolution of vegetation communities living on the alkali and salt-rich soils of the Hungarian plain of Hortobágy, inferring the dominance in this European region of steppe landscapes since the end of the last glacial period. The main conclusion of this chapter is that this discontinuity in the current temperate forest biome expected in the latitudinal band ranging from the Atlantic Ocean to the Urals has not been produced by anthropic action, but has remained largely unchanged throughout the Holocene.

The conservation and management section of the book also begins with a chapter on Iberian steppes. More precisely, its authors, *Màrcia Eugenio*, *Carlos Molina* and *Gonzalo Montamarta*, evaluate the role of flora micro-reserves aimed at protecting local populations of highly endangered plant species in preserving fragments of gypsophilous steppes classified as high-priority habitats after the European Union Habitats Directive, discussing the implications of conserving small and isolated steppe patches for the species of interest.

The role of protected areas in preserving steppe vascular flora is also evaluated in the next contribution to the volume by *Anatoliy A. Khapugin* and *Tatyana B. Silaeva*, who examine the capacity of the protected area network of the Russian republic of Mordovia to include populations of steppe plants included in that republic's Vascular Plant Red Data Book. Evaluating the efficiency of protected area networks is critical to establish an adequate large scale strategy to conserve highly transformed and fragmented ecosystems like the Mordovian steppe.

The next chapters are more focused on ecosystem management than on the conservation of particular species or habitats. The contribution by *Martin Wiesmeier* describe the results of a long-term field experiment involving exclusion of grazing after a long period of intense use by livestock of steppe areas of Inner Mongolia (northern China). More precisely, the author focuses on the changes of soil characteristics (organic matter-mineral aggregates) after grazing exclusion in relation to grazed areas, and uses geo-statistical techniques to analyze the spatial pattern of good quality soil recovery and its effect on pasture plant species such as *Leymus chinensis* and *Stipa spp.* This chapter illustrates the capacity of these steppe soils to recover from high intensity use, but also their sensitivity to grazing resumption and questions the sustainability of long-term land-use intensification.

The effect of different grazing intensities in steppes of Inner Mongolia is also analyzed in the chapter contributed by *Xiaoqi Zhou* and *Yanfen Wang*. However, this study focuses on the effect on the structure and diversity of soil methanotrophic bacteria, reporting the results of a field experiment in which the bacterial community under different grazing load treatments was sampled and analyzed using molecular genotyping techniques. Authors show that intermediate grazing intensities enrich the bacterial community, as well as temporary but not long-term grazing exclusion. Changes in the community correlate with some easily measured environmental variables such as soil moisture and pH, and biomass of some indicator plants like *Artemisia frigida*. These results are useful for pasture management aiming at restoring degraded soil properties and provide cues for their monitoring.

The long history of human use of Mediterranean steppes has profoundly modified those landscapes. Agriculture has contributed to the fragmentation and reduction of original steppe ranges of the Mediterranean basin such as the steppe of La Crau, in south-eastern France. In the complete chapter contributed by *Thierry Dutoit*, *Elise Buisson*, *Sylvain Fadda*, *Frédéric Henry*, *Clémentine Coiffait-Gombault*, *Renaud Jaunatre*, *Jean-François Alignan*, *Solène Masson* and *Adeline Bulot*, authors present the results of a long term field and experimental study aimed at evaluating the capacity of the cultivated land of La Crau to revert to the plant and insect (coleopterans) biodiversity levels of the original steppe. This chapter highlights the important limitations of current ecological engineering techniques to restore degraded steppe ecosystems.

Steppe degradation seems to be the combined effect of enduring drought and overgrazing in the Algerian High Plateaus as reported by Aziz *Hirche, Mostefa Salamani, Boughani Abdelmadjid, Dalila Nedjraoui, Ratiba Hourizi* and *Youcef Oukil* in the volume's last chapter.

According to these authors, the strategy to restore the steppe and its natural resources followed by the Algerian administration during the last forty years has not been fruitful and only those actions regarding livestock reduction and grazing exclusion have proved effective.

Another steppe ecosystem strongly transformed by human use is the *Caldenal*, a vast steppe rangeland in the western rim of the Argentinean Pampa devoted to livestock grazing during the last hundred years. In the final chapter of this volume, *Hugo Laborde*, *Osvaldo A. Fernández*, *María N. Fioretti*, *Sandra S. Baioni* and *Roberto E. Brevedan* report on the ongoing attempts to promote an economically sustainable livestock production compatible with the conservation and even expansion of native grassland plant species that can improve the quality of those pastures. According to these authors, there is an urgent need to develop that kind of range exploitation models to prevent the Caldenal's further desertification and abandonment by local settlers.

These last chapters bear a clear message to researchers and, above all, to managers: steppes are highly vulnerable ecosystems that 'live dangerously' close to that ecological and environmental frontier beyond which resilience is no longer possible. Different threats related with global change, and particularly land use intensification, may be, and actually are, taking them across that line and pushing them down the slope of depletion and desertification. However, this preface must end with an optimistic message. The chapters comprising this book not only diffuse current scientific knowledge on steppe ecosystems, which is not a minor aim, but provide cues and tools to evaluate their state and to scientifically inform and help their management. Let us hope these messages reach the adequate ears.

ACKNOWLEDGMENTS

We wish to thank NOVA Science Publishers for giving us the opportunity to edit this volume, and particularly to Carra Feagaiga for her help and patience. We are indebted to John Muddeman for translating some of the chapters into English, and to José Antonio Martínez and Juan E. Malo for kindly and freely give us permission to use some of their photographs. Of course, we are also indebted to Marina Otero for her engagement in the chapter formatting work and to the REMEDINAL2 network of the Comunidad Autónoma de Madrid (S-2009/AMB/1783) for funding it. Finally, but not least, we wish to thank all authors for their valuable contributions to the book.

Manuel B. Morales and Juan Traba
Cantoblanco, Madrid. April 12th, 2013

In: Steppe Ecosystems
Editors: M. B. Morales Prieto and J. Traba Diaz

ISBN: 978-1-62808-298-2

Chapter 1

STEPPES ACROSS THE WORLD: AN OVERVIEW WITH EMPHASIS ON THE IBERIAN PENINSULA

Helios Sainz Ollero*

Botany Unit. Department of Biology. Universidad Autónoma de Madrid, Madrid, Spain

ABSTRACT

Steppes are treeless semiarid landscapes of diverse origin which have persisted almost without change since the last ice age. Their post-glacial stability is related to both climatic (continental climate, "rain shadow" effect) and anthropomorphic (pastoralism) aspects. The concept of steppe nowadays is rather lax and is used almost worldwide simply when referring to herbaceous or low shrub formations which are related to the cold arid regions of the planet. In this chapter, the climatic and eco-physiological factors conditioning the presence of steppe regions across the world are reviewed, their principal differentiating characters are defined and the peculiarities of the Mediterranean steppes and their diversity within the Iberian Peninsula are described.

Keywords: Aridity gradient, biomes, biome ecotones, Central Asia, Mediterranean basin, steppe diversity

INTRODUCTION

What Is a Steppe?

Various ecological factors, more complementary than exclusive, can be referred to in order to explain the presence of the world's steppe areas. These are principally climatic in nature, but also edaphic, historic and anthropic in origin.

---

* Corresponding author: Helios Sainz Ollero. Botany Unit. Department of Biology. Universidad Autónoma de Madrid, C/ Darwin 2, 28049 - Madrid, Spain. E-mail: helios.sainz@uam.es.

Steppe is currently a lax and very widely used geobotanical concept, despite being somewhat controversial as it encompasses rather different formations in very varied areas across the planet. Without doubt the common character of all steppe landscapes is essentially their appearance. These are open, extensive, treeless habitats, with a simple horizontal structure to the landscape. The vegetation is usually sparse, frequently with little ground cover and dominated by herbaceous plants or low shrubs. They show wide horizons, are generally very flat and located at diverse altitudes, generally in rather extreme climates.

In climate terms steppe formations almost invariably occur in semiarid and principally temperate environments, but also appear in Mediterranean and even subtropical regions. They constitute transitional areas towards more arid zones and frequently appear in the periphery of deserts.

Another characteristic feature of steppes is paleogeographic in origin and related to ecological dynamics. In the majority of cases they are relatively stable, resilient landscapes, adapted to extreme, more-or-less cold and dry climates, which have changed little since the end of the last ice age (Berglund, 1986; González-Sampériz, 2004). Palynology studies have highlighted the structural and floristic similarity of the steppe formations which developed widely during the glacial periods with those still persisting today in some cold arid regions of the world (Tibet, Pamir, Utah, Nevada, Patagonia, Tierra de Fuego, Birks and Birks, 1980; Lowe and Walker, 1997). These surviving steppe formations, which coincide with those detected during paleoecological studies of the Quaternary glacial periods carried out in currently forested areas, are known as "modern analogies" (Overpeck et al., 1985). This method requires using existing pollen system models (in this case of steppe zones), for them to be used as analogies in the interpretation of fossil pollen spectra. It is based on the principle of actualism and the search for relict landscapes which can aid in paleoenvironmental interpretation (Jackson and Williams, 2004).

Permanent shrubby or grassland types of vegetation formations predominate in steppes, conditioned by climatic, geomorphological, edaphic and anthropic factors. These aspects made their natural evolution towards woodlands during the Holocene difficult and have contributed to their persistence with few changes.

Frequently, herds of wild herbivores exploiting these ecosystems have collaborated in their conservation (Walter, 1984). Historically, nomadic human groups settled in the steppes and the herds of mammoths, bison, horses and antelopes were substituted, sooner or later, by other domesticated animals which have continued to keep the trees "at bay" (Walter and Breckle, 1986).

Civilizations have flourished in many semiarid areas of the world, thanks to grazing and the cultivation of cereal crops of domesticated species. In consequence, steppes have been widely transformed over extensive areas of the planet.

However, man, through his artificial practices which have enormously simplified the ecosystems, continues to cultivate grasses and shrubby species (grape vines, olives, pistachios, etc.) so maintaining the almost treeless nature of these areas. Consequently, some steppe animal groups like the birds and some rodents continue to use them as long as anthropic factors (hunting, poisons, agricultural intensification, Santos and Suárez 2005) permit. The term "pseudosteppes" has frequently been used to refer to these transformed areas which still retain some of their landscape structure and original fauna (Suárez et al. 1997). And from the structural and faunal points of view their similarity to natural steppes is very high.

Diversity and Controversy in Terminology

The term steppe originally derives from the Russian *stepj*, but numerous ecosystems of relatively varied characteristics are currently united under this concept, though tend to be treeless, more-or-less flat and located in arid or semiarid environments, which relates them to cold or hot deserts. Very often the continentality and the rain shadow phenomena produced by high mountain ranges are determining for understanding the presence of steppe ecosystems in different world regions.

From a functional and ecophysiological viewpoint the steppe territories reflect two contrasting situations in which the plants either experience a long, very low quality growth period, or a very brief period of high quality (Table 1). In both cases the development of woodland is particularly difficult. However, trees can grow in conditions where competition with the herbaceous substrate is reduced or where browsing is reduced. This latter aspect is also linked to another of the characteristic features of steppes, their glacial relict nature in balance with herbivory by large herds of wild ungulates (Walter, 1970).

Although the steppes in some regions are predominantly herbaceous in structure, forming prairies or grasslands (swards of *Stipa, Poa* or *Festuca* spp., esparto grass *Stipa tenacissima*, feather grass -*Stipa* spp., or albardine *Lygeum spartum*), woody shrubs predominate in others (thymes -*Thymus* spp.-, saltworts -*Salsola* spp.-, wormwoods -*Artemisia* spp.-, whins and gorses -*Genista* spp. *Ulex* spp.-, etc.). In these semiarid environments the gradual transitions which exist due to subtle climatic variations are of great importance. Transitions towards cold deserts, deciduous broad-leaved woodlands or boreal conifer forests, evergreen Mediterranean woodlands or the savannahs and hot deserts are all seen. And all intermediate situations are recognizable across the world in steppe environments (Figure 1).

The inadequacy of the term "steppe" under a strict ecosystem point of view, has been highlighted on numerous occasions (Huguet del Villar, 1925; Walter, 1970; Arroyo, 1988), despite the fact that it is a widely used term in geobotanical circles and is well established in general language.

The broad nature of the concept means that extraordinarily diverse ecosystems are frequently included within it, ranging from cold temperate, even subpolar, climates to subtropical savannah environments. Treeless steppe formations are found just as widely in cold climate transition areas, such as Mongolia, Kazakhstan, Tibet, Nevada, Utah or Patagonia, as they are in hot regions, e.g. Australia, Arizona, Sudan or the Sahel.

The high cold plateaus (*páramos*) and Mediterranean steppes of the Iberian Peninsula occupy an intermediate position between these two extremes (Arroyo, 1988), with floristic influences which can be considered as related to both (Irano-Turanian and Saharo-Sindian steppes, Sainz Ollero, 1988).

Table 1. Types of steppes according to growth periods and substrates

Growth period	Steppe type	Soil types
Relatively long and of poor quality	WOODY (chamaephytes) Dispersed shrubs	Sandy, poor stony
Brief (c. 2 months) optimum	HERBACEOUS (tufted hemicryptophytes). Prairies (predominance of grasses)	Clays, gypsum/marls, Loess or Chernozem (A/C, Superficial organic material)

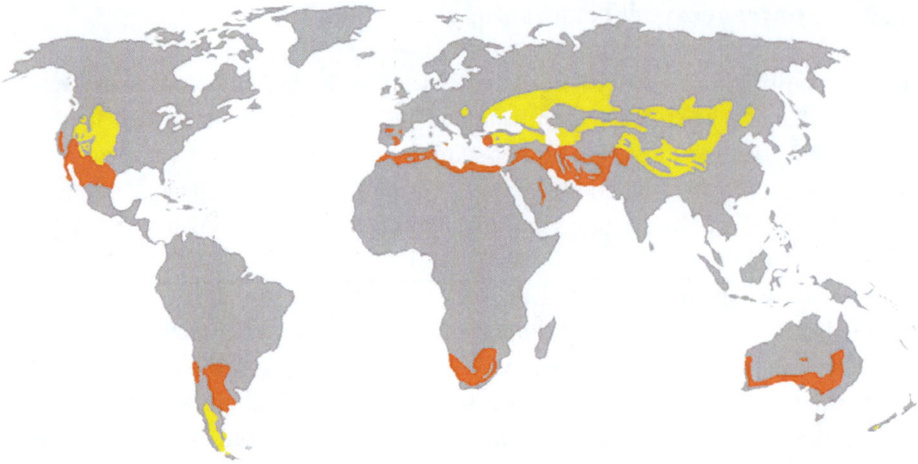

Figure 1. The world's steppes: http://commons.wikimedia.org/wiki/File:Steppe_world.png. Temperate steppes (in yellow) and Mediterranean and subtropical steppes (in orange); the more tropical formations such as the Orinoco Plains or the sub-Saharan Sudanese and Sahelian grasslands are missing.

While woody chamaephytes predominate in the Iberian steppes, with the exception of esparto grass and albardine, and occupy poor and degraded rocky soils (rendzinas or lithosols), frequently with salts or gypsum, the Russian steppes are established over Chernozem-type A/C substrates. Functionally, these have nothing in common and there is no doubt that these are very different ecological formations. However, and despite the correct arguments of the purist geobotanists (Huguet del Villar, 1929), the term steppe has continued to be used in both cases successfully amongst other naturalists (botanists and zoologists).

The greatest paradox is that following the evolution of the steppe concept, given its generalization and widespread use in geobotany, the environment which least concurs with the most widespread use of the term today are the true *stepjs* of southern Russia, Ukraine and Kazakhstan, which are unique landscapes with exceptional soils, and most similar in structure and function to the prairies.

The breadth of the concept of steppe, which principally designates a more-or-less superficial physiognomic aspect to the landscape, recommends the use of adjectives to more precisely define them, either related to the macroclimate in which they are found, to the conditioning substrate or to the geographic area where they are located. Consequently, we consider that we ought to refer from now on to Russian, Mediterranean, halophile, Asiatic, Tibetan, Patagonian, etc., steppes.

Diversity of the World's Steppe Landscapes

Steppes of Southern Russia, Ukraine and Kazakhstan

Considered by purists as the most genuine steppes on the planet, it is no surprise that the term steppe derived from the Russian *stepj*. They are without doubt the most singular, given their originality and the richness of their soils and grasslands.

In Eastern Europe and Central Asia the continentality and xericity reach impressive levels given the vast extensions present within the continental landmass. In winter, the temperatures drop dramatically with the establishment of the Siberian anticyclone, while in summer its aridity increases due to isolation from the influence of the Indian Ocean by the huge natural barrier of the Himalaya Range. In the same latitudinal belt where deciduous broadleaved forests grow in Western Europe, in southern Russia, the Ukraine, Kazakhstan and Mongolia, steppe ecosystems appear. The winters here are severe and dry, but the short summers are mild and humid. These conditions are especially favorable for the famous Russian, Ukrainian and Kazakh *stepjs*, pastures dominated by perennial hemicryptophytes, whose buds are protected in winter and develop quickly after thawing during the optimal, albeit short (three–month, May - July), growth conditions. These are botanically very rich communities which results in a spectacular display of spring flowers; these plants belong to the Holarctic cold-temperate floral communities (Walter and Breckle, 1986).

The soils of these steppes are very fertile (see also Chapter 7 of this book), exhibiting a deep surface layer of organic material containing a high percentage of humus (6 – 10%), which derives from the summer biological mineralization of organic matter. This very darkly colored soil horizon lies directly on the bedrock, which is comprised by a loess, rich in calcium and magnesium which was wind deposited during the last ice age. In the "Chernozem" black soils, the organic horizon can reach depths of over one and a half meters, in marked contrast to most of the world's other steppes where the soils are only poorly developed and contain very little organic material. The surface of this layer is decarbonised through washing and this effect occurs in spring when the spring snow melt saturates the ground. The surface water, rich in CO_2 given the intense summer biological activity (plants, worms, insects and rodents), converts the calcium carbonate into soluble bicarbonate which soaks away through gaps and cavities in the soil, giving rise to somewhat diffuse ochre colored calcium horizon. This thick organic layer is also cut through by numerous ancient abandoned rodent tunnels (*Krotowinas*), testament to the stability of these environments since the last ice age (Crespo i Veigas, 1988).

Given their fertility, these steppe areas offer very favorable conditions for the settlement of man and for agricultural development, given their combination of moderate summer temperatures, adequate sunlight levels, edaphic development and humidity. Cultivated cereals produce high yields and in consequence the great majority of the steppes have been transformed. Even here however, the agricultural productivity is affected by the erratic nature of the rains, including on occasions, catastrophic droughts.

In the really genuine prairie steppes, the vegetative growth period is marked by an extraordinary exuberance of flowers. The timing of flowering is also sequential. After the May snowmelt, a few geophytes (*Hyacinthus, Iris*), Ranunculaceae (*Pulsatilla patens, Anemone sylvestris, Adonis vernalis*) and sedges (*Carex humilis*) appear. The grasses begin to grow afterwards (numerous species of *Stipa, Festuca, Bromus, Koeleria*) and among them appear scattered Campanulaceae, composites and Leguminoseae. Clovers (*Trifolium montanum*), forget-me-nots (*Myosotis sylvatica*) and daisies (*Chrysanthemum leucanthemum*) are abundant. The colors of the steppe consequently change quickly over time with substitution of different species. Die-off begins in mid July and from August the steppe turns golden yellow until covered by the autumn snows (Walter, 1970). The cause for the substitution of forest by steppe in these areas is a result of the limited period for which water is available during the growing season, which in these areas is too short for trees.

Vegetative activity starts with snowmelt and water is available in the surface layers from the end of April. This continues until the end of July, when water becomes scarce and the steppe dries. Clear competition for water resources occurs between woody and herbaceous plant species, which is related to their life strategies. Trees and woody shrubs require a relatively long period of water availability and have an extensive and deep-reaching root system. In contrast, grasses and other herbaceous plants can adapt to much shorter growth periods with their denser and more superficial root systems. Their fasciculate and abundantly branched roots are very effective in fine, sandy or organic soils.

These extract water intensively from the upper soil horizons from the first moments following spring snowmelt. For these reasons while woody species are specially adapted to pump water from deep, fractured rocky soils in winter, they are uncompetitive in areas where the water is only accessible for a short summer growth period.

Soils which are dense, clayey or with a high organic content, in sites where water penetrates only through gravity during the summer, are far more favorable for grasses and herbaceous species in general. The most favored biotype in these circumstances is the tussock hemicryptophyte, with superficial fasciculate roots and buds protected among an interwoven mass of leaves.

In these Central Asian regions, where conditions are not too arid and genuine pasture-type steppes occur on Chernozem soils, an interaction between forests and steppe appears (Walter and Breckle, 1986). Above all this is related to the quantity of water available for the different ecosystems in function with their edaphic conditions. The forests, needle-leaved or deciduous, establish on the areas of greater permeability and better drainage, and principally occupy stony areas on slopes of varying angles.

In contrast, the steppe is found in the more-or-less flat areas of the plains or in the valley bottoms, on soils which are denser, badly draining and clayey or rich in organic content. A consequence of this preferential localization is that mosaic, forest-steppe landscapes predominate. As aridity increases towards desert areas, steppe landscapes increase and are increasingly independent of soil types. Typical steppes are located in more arid climates, on black soils with lower humus content (4-6%, Walter, 1984). The vegetation is more open and shows less diversity. Grasses dominate, particularly of the genera *Stipa* (*S. pennata, S. capillata, S. joannis, S. laessiagiana*), *Festuca* (*F.* gr. *ovina*) and *Koeleria* (*K. gracilis, K. cristata*). The vegetation cover of the steppes progressively decreases in parallel with rainfall, as does the diversity and the soil's organic content. These are progressively more carbonated and frequently show saline endorheic depressions. Mat-plants and shrubs begin to appear among the grasses, principally wormwoods (*Artemisia incana, A. austriaca, A. sieberi*) and Chenopods (*Kochia, Salicornia, Haloxylon, Camphorosma*), marking the transition towards the Central Asian deserts, which are very clearly marked by the presence of white saksaul (*Haloxylon persicum*).

Another member of the Chenopodiaceae is *Eurotia* (= *Krascheninnikovia*) *ceratoides*, which shows a wide distribution in Central Asia and Tibet and also appears in isolated sites around the Mediterranean (Pannonian Basin, Anatolia, semiarid regions of the Iberian Peninsula, Morocco; Figure 2). Interpreted as a glacial relict, its presence establishes a connection between the Mediterranean and Central Asian steppes (Domínguez et al., 2001).

Nowadays the great majority of the Russian and Kazakh steppes have been converted into cereal-growing steppes which make contact with increasingly arid formations in the surroundings of the Asian deserts and the high plains of Mongolia, Manchuria and Tibet.

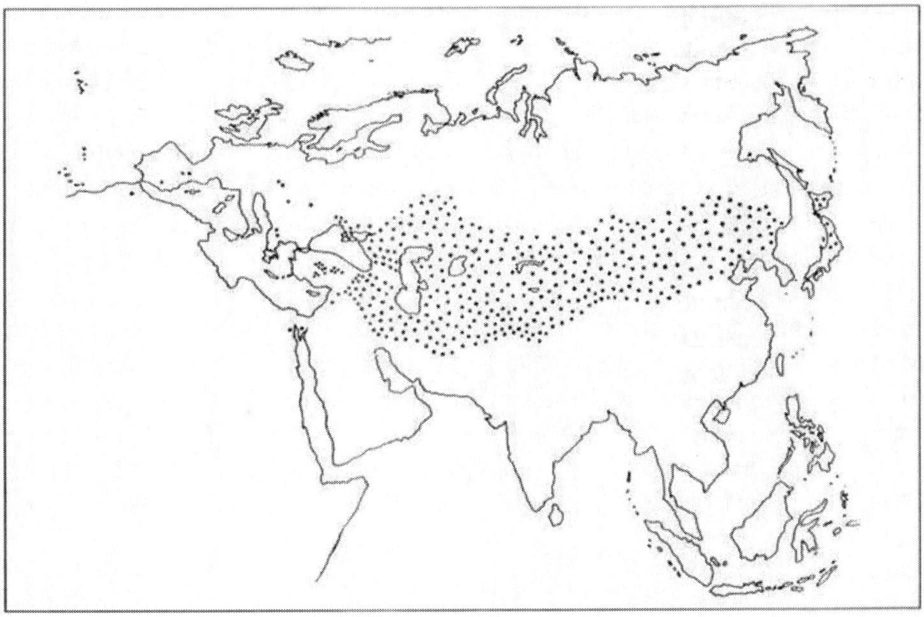

Figure 2. World distribution of *Eurotia ceratoides* (Domínguez et al., 2001).

The steppes of Tibet and the Pamir (see Chapter 6 in this volume) are now montane variants in which the environmental conditions are truly extreme (the air is very dry, frosts are only absent 1 month a year and rainfall is <100mm per year and concentrated in the summer months). The vegetation is flat, very dispersed and comprised by woody short shrubs with more radial than vertical growth. Composites, Polygonaceae and Chenopod species dominate, alongside a few *Ephedra*.

North American Prairies and Steppes

The outstanding features of the American prairies are their complexity, diversity and gradual variations. These features were probably in part responsible for the success of the ideas related to the continuum expressed by North American scientists dedicated to the intensive study of this vegetation type (Curtis and McIntosh, 1951; Whittaker, 1970). Without doubt they are also a response to the opposing climatic and floristic influences registered on the boundary of the Holarctic and Neotropical kingdoms. In the centre of the North American continent, in an extensive continental zone located in the rain shadow of the Rocky Mountains, prairies and shrubby steppes predominate. An East-West zoning is seen, produced by the progressive decrease in Atlantic precipitation, and a North-South gradation with temperate (circumboreal) or subtropical floral influences, related to temperature is observed (Coupland, 1979). From the north derive cold-temperate plants, in particular northern grasses with a short growth period, regulated by the photoperiod and with relatively shallow roots. These are species of the genera *Festuca, Poa, Bromus, Avenula, Dactylis, Deschampsia, Stipa*, etc., many of them present in other temperate grasslands widely distributed across the planet in both Hemispheres: Europe and Asia in the north and Patagonia and the South Island of New Zealand in the south.

In contrast, via the Mexican deserts have come xerothermic grasses, of the genera *Bouteloua, Spartina, Andropogon* and *Panicum*. These are subtropical species with a long growth period, regulated by temperature and xericity, and deep roots which in general lead to big tussock biotypes. (Archibold, 1995).

The driest habitats are found in the high plains at the foot of the Rocky Mountains (Great Basin province), where the steppes are influenced by the mountain biomes and show some similarities with the Tibetan steppes. Chenopods (*Atriplex*) and composites, mainly *Chrysothamnus* and *Artemisia* are abundant, especially *A. tridentata*, the widespread sagebrush, which reaches as far as the Mediterranean steppes of California, Mohave and Sonora, all of which are related to the deserts of Mexico. Of particular note are the large number of endemic species of the legume genus *Astragalus* and also *Larrea divaricata,* a member of the Zygophyllaceae which establishes links between the Californian steppes and the most arid areas of Argentina. Towards the south the influence of the hot deserts of the Madrean region becomes progressively more patent, and succulent plants of the Cactaceae, Agavaceae and Crassulaceae families appear, especially of the *Opuntia, Carnegia, Dasylirion, Agave*, etc. genera.

The endemic *Fouquieria* (11 spp.) genus is also worth highlighting, the only one of an extraordinarily isolated and singular family, the Fouquieriaceae.

Mediterranean Steppes

In the areas of Mediterranean climate, the most modern climate on the planet, and characterized by its dual thermal and water stresses in summer, any factor exacerbating the environmental conditions makes the continued existence of needle-leaved and sclerophyllous woodlands difficult; in consequence, appreciable areas of the Mediterranean regions lack forest formations and show steppe characteristics (Quézel and Médail, 2003). It is above all in the transition areas between the Mediterranean biomes and hot deserts (Biome III of Walter, 1970) or in pockets scattered across the rain shadow of mountain ranges, or in singular edaphic habitats, where Mediterranean steppe communities appear. Despite their relevant common landscape features, the environmental and biological characteristics of the Mediterranean steppe areas are quite diverse.

The location of the regions with a Mediterranean climate in the Northern Hemisphere, lying on the ecotone between the Holarctic and Paleotropical kingdoms gives rise to a unique convergence of floral elements in these steppe areas (Raven, 1973; Suc, 1984). On one hand plants related to the cold continental deserts are present (Biome VII, Walter, 1970) and on the other, a thermophile, arid or semiarid flora of the subtropics (Biome III) as well (Walter, 1977, 1984).

The major factor responsible for the coexistence of two seemingly disparate strategies in the Mediterranean steppes appears to be the continentality, understood as a heightening of the seasonal and daily variations in climatic components. The main Mediterranean steppe habitats are found in the interior regions of the Iberian Peninsula, Turkey and North African countries. In the Mediterranean region the thermal continentality factor is found above all in interior valleys and in upland flat areas, plains or plateaus. In these zones this effect is further reinforced by their isolated nature and the barrier effects of adjacent or surrounding mountain ranges which produce interior areas of higher aridity.

The Mediterranean Basin (Iberia, Turkey and North Africa)

On the boundary between Europe and Africa a diverse sample of steppe types occurs within the Mediterranean basin. This richness of nuances in the steppe formations located on the Euro-Siberian – Mediterranean and Palearctic/Paleotropical) ecotones, along a relatively reduced boundary area, is probably the most relevant feature of these treeless spaces. This variability is a response to subtle climatic differences and floral influences, from the cold-temperate Holarctic (Irano-Turanian) element which extends along the Danube valley (the Hungarian *puszta* of the Pannonian region), to the Paleosubtropical Mediterraneo-Saharo-Sindian formations of North Africa. Succulent biotypes predominate in the halophilous Tertiary depressions, while in the upland interior continental plains of Iberia and Anatolia mountain-loving, spiny cushion plants, adapted to the winds and herbivory appear. These formations show many common traits with those in North Africa and their adaptations remind us of the Patagonian steppes or of the North American and Central Asian deserts. Several formations of tussock grasses occur, characteristic of subtropical habitats, such as *Stipa gigantea, S. tenacissima, Ampelodesma mauritanica* and in the coastal ravines, *Saccharum* (=*Erianthus*) *ravennae*.

The main differentiating feature among the Mediterranean steppes relates to their position in the ecotone between both cold and subtropical arid regions. This is revealed by the presence of plants which belong to the Mediterraneo-Irano-Turanian and also Mediterraneo-Saharo-Sindian groups (*sensu* Pignatti, 1982), which are related to the deserts mentioned before. Together with these Asiatic and African species of great biogeographical interest, a unique Mediterranean element also occurs, comprising endemic taxa or species of restricted distribution in the xeric regions of Iberia and Northern Africa.

The Mediterraneo-Irano-Turanian element is comprised of eastern plant species which expanded across the Mediterranean basin at the end of the Tertiary era during the drying-out period of the Mediterranean Sea (Jäger, 1971, Súarez et al., 1992). A whole series of transcendental paleogeographic events took place in the Mediterranean basin at that time as a consequence of the convergence of the lithospheric plates in the region of the ancient Tethys and Paratethys Seas (García Antón et al., 2002). This process was the principal cause of the uplifting of the Alps, the changes in oceanic currents and the origin of the Mediterranean climates. The Tethys Sea progressively disappeared and at the end of the Miocene (Messinian period) important drying of the Mediterranean Sea occurred. The entire basin was reduced to a series of salt lakes and pools, surrounded by extensive halophile steppe and a halophilic steppe floral element extended across from the east and consolidated across the basin, with extensions radiating across the Paratethys area, now the Pannonian Basin of the Danube (Bocquet et al., 1978). The Alpine orogeny permitted the appearance of a unique high mountain circum-Mediterranean steppe environment above the cloud condensation level. The colonization of this habitat appears to have induced an extremely important adaptive evolutionary process. Numerous genera exhibited convergence towards thorny-cushion forms ('tragacanthic' biotype), which was well adapted to the continental aridity and herbivory. Many of these genera are also highly diversified in the lowland Mediterranean steppes (Sainz Ollero and Domínguez Lozano, 2011). Various examples of floral vicariance at the generic level (*Salsola*, *Suaeda*, *Astragalus*, *Artemisia*, *Stipa*, *Microcnemum*, *Gypsophila*, *Thymelaea*, *Juniperus* sect. *sabina*, etc.), and some notable coincidences (*Krascheninnikovia ceratoides*, *Microcnemum coralloides*), highlight the undoubted existence of a common floral community, of pre-glacial origin, in this element.

The Mediterraneo-Saharo-Sindian element is comprised of plants which derived from the subtropical flora of the open African regions of the Tertiary era, where savannah-like landscapes existed (*Rand Flora* after Lebrun, 1947; Quézel, 1978). These are consequently Paleotropical plants related to the hot deserts.

One of the best examples of these types of plants of the open regions north of the Sahara and one which is well represented in the Maghreb is the esparto grass (see, for example, Chapter 14 in this volume). The formations in which it dominates (the so called "Alfa Sea" in the Maghreb) exhibit tussock biotypes and neatly mark the boundary of the thermophilous Mediterranean steppes in both Spain and North Africa. Other taxa characteristic of this element and landscapes are *Artemisia herba-alba, Astragalus* spp., *Thymus* spp., *Tephrosia* spp., *Withania* spp., *Peganum harmala, Zygophyllum* spp., *Fagonia* spp. *Periploca laevigata, Salsola vermiculata*, and *Anabasis articulata.*

Endemism levels are of notable importance in the Mediterranean steppes. In his classic monograph, Willkomm (1852) already highlighted the existence of 126 (42%) Iberian endemic species among the 302 species estimated for the peninsula's steppe flora. In addition in these areas, and particularly in the semiarid south-eastern Iberian sector, the Ibero-Mauritanian estenocorous plants are well represented, increasing this singularity even further.

One of the most significant examples within this endemic Mediterranean steppe element is given by three ancient and evolutionarily isolated woody crucifers (paleoendemics) with fragmented distributions within the Iberian Peninsula, and which occupy small and clearly relict areas (Sainz Ollero and Domínguez Lozano, 2011): *Vella pseudocytisus* (present in the Tagus valley, the Hoya de Baza and the Teruel steppes), *Boleum asperum* (present in the Ebro Valley) and *Euzomodendron bourgaeanum* (only present in the Tabernas desert, south-eastern Spain). The woody biotype is interpreted as primitive in this family which has evolved towards annual herbaceous forms occupying ruderal habitats (Gómez Campo, 1981).

In the Mediterranean region and principally in the Iberian Peninsula and in Turkey (Anatolian plain), steppes occur relatively frequently on gypsum soils and an abundance of endemic plants is found. Willkomm in 1852 referred to 32 endemics of the gypsophilous steppes in Spain; Mota et al., (2009) mention 69 species which behave more or less as strict gypsophiles, from a total of 140 species analyzed from areas where tree growth is naturally difficult due to substrate characteristics, and so where this favors the continued presence of singular steppe landscapes.

The Diversity of the Iberian Steppes

Given the breadth of the steppe concept, it is fundamental to specify their distribution, characteristics and diversity within the Iberian Peninsula. Very varied formations exist, including some which occur in particular edaphic or microclimatic locations, or as mosaics in fragmented landscapes –such as the *páramos*, principally cold high plains in the surroundings of the Iberian System mountain range, where they coexist with juniper 'woodlands' and some open pinewood habitats.

In the first maps of the Iberian steppes by Willkomm (1896) and Reyes Prósper (1915, Figure 3), a series of steppe nuclei are shown in the interior depressions containing Tertiary (Oligocene and Miocene) substrates, in the Duero, Tagus (La Mancha steppes) and Ebro catchments, plus in the semiarid south-eastern sector, the most arid part of the Iberian Peninsula where the North African influence is patent. Reyes Prósper (1915) delimited areas of greater size, but which fundamentally coincide with the sectors proposed by Willkomm.

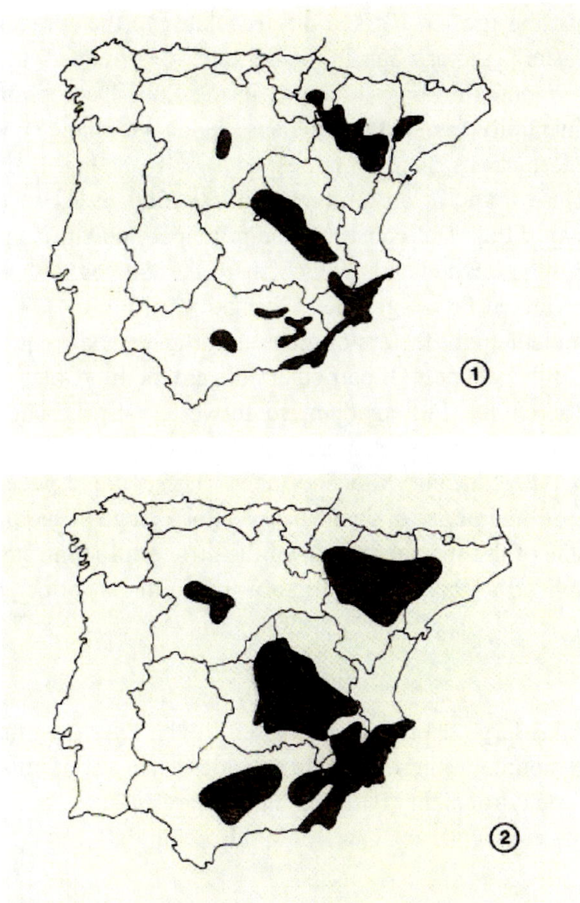

Figure 3. Distribution of Iberian steppes after (1) Willkomm (1896) and (2) Reyes Prósper (1915).

Ornithologists (De Juana *et. al*., 1988), in order to reflect the distribution of steppe birds, subsequently proposed an increase in these areas and also included the *páramos* of the Iberian System, the short pastures of the SW of the peninsula and some cereal steppes. Extensive areas of cereal fields in the Iberian Peninsula are still cultivated on a two-year rotation basis. This practice has produced abundant fallow aeras, which are cultivated in alternate years ("Iberian dry-farming" after the terminology proposed by Dantín Cereceda, 1916). These extensive territories, which have frequently been coined pseudosteppes are very favorable for the continued existence of typically steppe birds such as great bustard (*Otis tarda*), little bustard (*Tetrax tetrax*) or the sandgrouse (*Pterocles alchata, P. orientalis,* see Santos and Suárez, 2005 and Chapter 2 in this volume). The great diversity of the Iberian steppes is related to the heterogeneous nature of the landscape which in turn derives from the highly complex relief and geomorphology, a marked geographical and lithological compartmentalization and contrasting climates (Sainz Ollero, 2003, Sainz Ollero et al. 2010). All this is further modulated by an original history marked by its privileged position in the SW of Europe, in an area of transition between the cold-temperate Circumboreal (Euro-Siberian) and Mediterranean regions (García-Antón et al., 2002).

The principal differences between the various Iberian steppes lie in their substrates. As noted before, all these formations show the common characteristic of being more arid than the average which restricts tree growth. This aridity is related to the general climate, such as in the semiarid south-eastern sector, but in many others is related to local rain shadow phenomena caused by mountain ranges, such as in the Ebro Depression and other, interior catchments. The continentality frequently heightens the local climatic conditions, favoring the stability of the steppe formations, such as in the *páramos*. These highly characteristic areas of the Iberian Peninsula are found at an altitude of 1000-1200 m asl in the periphery of the Iberian System mountain range and constitute the principal habitat of the *Juniperus* (juniper spp.) and *Pinus nigra* steppe woodlands, which share these areas with spiny shrubs (whins, gorses, buckthorn) and mixed thyme-grassland formations (Figure 4). The limitations related to soils tend to be complementary factors of great importance, and frequently determinant for the presence of many of the steppes. Of particular note are the lithosols poor in nutrients, such as in the region of La Serena (Extremadura, south-western Spain) and other spots on the Paleozoic plate of W Spain, and the evaporate substrates rich in gypsum, marls and saline deposits deriving from the ancient Miocene lakes. These are especially frequent in the endorheic depressions of the interior drainage basins (Ebro, Tagus, Duero, Hoyas de Baza and Guadix, Figure 5). Some of the special features of the different Iberian steppes are highlighted below. Figure 6 is provided to facilitate location of the particular principal steppe area in the Iberian Peninsula.

1. The Ebro Valley

The landscape of this large depression, enclosed by the Pyrenees, the Iberia System and the coastal Catalonian mountain ranges contains a wide visual variation and is constituted by a high variety of microenvironments (small plains, hillocks, steep rocky banks, salt pools, brackish stream gulleys, etc. See also Chapter 9 in this volume).

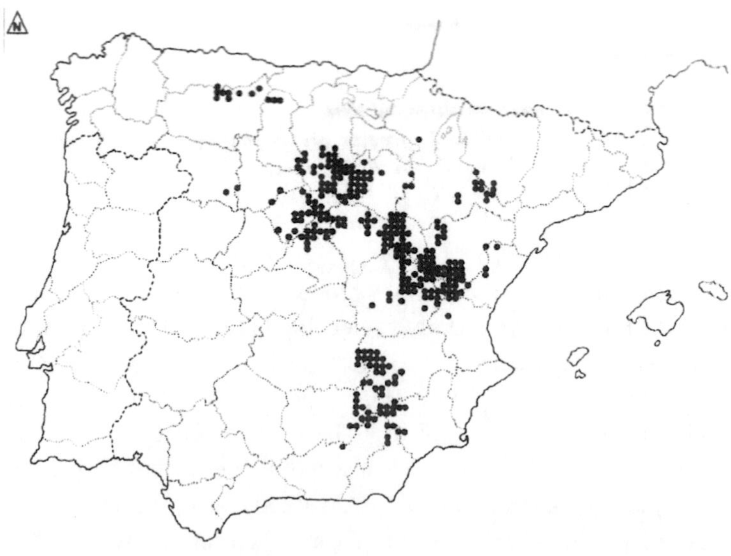

Figure 4. Distribution area of the Spanish juniper (*Juniperus thurifera*) which coincides notably with the steppe areas of the Spanish *páramos*. (Gómez Manzaneque et al., 2000).

Figure 5. Distribution map of *Ononis tridentata*, a gypsophilous legume which fairly precisely delimits the distribution of edaphoxerophilous steppes.

This is one of the most characteristic and extensive steppe territories in the Iberian Peninsula, although it is very compartmentalized given its extensive transformation for cultivation. The principal factors responsible for its steppe character are its continentality, the rain shadow produced by the surrounding mountain chains and the dispersed halophilous environments present deriving from Holocene lagoons. The flora includes species of the Irano-Turanian cold Asian steppe elements (*Krascheninnikovia ceratoides, Microcnemum coralloides, Gypsophila perfoliata*).

Among the best conserved areas are the Bardenas Reales of Navarra, the Monegros badlands, which are generally small areas between cultivated fields, the Site of Community Interest (SCI) "Planas y estepas de la margen derecha del Ebro" and "Tomillar de Alfés y Mas de Melons", one of the few steppe areas of Catalonia, and the sector containing the saline endorheic lagoons of Bujaraloz-Sástago and of Chiprana.

Figure 7 shows two of these steppe landscapes typical of the Ebro Valley. Different irrigation projects using water from the Ebro cathment and promoted by the regional governments, have denaturalized a large part of the best steppe areas of some of these regions.

The landscape of a "Val" (small valley) of the Monegros, characteristic of the Ebro Valley as a whole, is comprised of a mosaic of dense woody shrubs and herbaceous formations and halophyte tussocks: rosemary (*Rosmarinus officinalis*) bushes on gypsum soils with a scattering of Spanish and Phoenician junipers (*Juniperus thurifera* and *J. phoenicea*), Kermes oak (*Quercus coccifera*) and Aleppo pine (*Pinus halepensis*) on the ridges; thyme shrubs, saltwort scrub (*Salsola vermiculata*) and white wormwood stands (*Artemisia herba-alba*) with *Sideritis* spp. in the rockier soils, and albardine (*Lygeum spartum*) on the finer materials which have accumulated in the bottom of the "val" (Figure 8).

Figure 6. Principal steppe areas in the Iberian Peninsula (numbers correspond to those in the text). 1: Ebro Valley; 2: Semiarid Southeast; 3: Hoya de Baza, Hoya de Guadix and the high plains of Granada; 4: La Mancha and the Tagus catchment basin; 5: Extremadura (the plains of Cáceres, Trujillo and La Serena and the Alcudia valley); 6: Duero catchment basin (Villafáfila, Valladolid and Palencia gypsum soils); 7: *Páramos* of the Iberian and Betic Systems, and the Duero catchment basin (Cerrato, páramos of La Lora, Masa and Bureba).

2. Semiarid Southeast

The steppes in the Murcia-Almería region (Figure 6) are the most typical of the Iberian Peninsula in terms of their general climate, and are located over a wide variety of soils. These include semidesert areas of the interior (Tabernas), gypsum and salt outcrops, quartzitic or slaty terrains, marine sediments (today submerged under intensive polytunnel agriculture as in the *campo de Dalias* known by the Romans as *campus spartarius* for its extensive esparto grass formations), continental detritic loams (on the southern edge of the southern Iberian plateau), Dolomitic and other limestones or volcanic areas. A marked level of endemism is characteristic of the flora (*Euzomodendron bourgeanum, Antirrhinum charidemi*, *Teucrium intricatum*, *T. turredanum*) plus notable North African and Saharo-Sindian influences amongst the shrubs and mat-plants (*Periploca laevigata, Maytenus senegalensis, Ziziphus lotus, Stipa tenacissima, Withania frutescens, Launaea arborescens*, and esparto grass) and herbaceous or succulent species (*Androcymbium gramineum, Fagonia cretica, Caralluma europea, Mesembryanthemum cristallinum*).

The steppes of Murcia extend inland along the Segura catchment basin (Cieza, Yecla), coming into contact with the esparto grass and gypsum substrate formations of the southern edge of the southern plateau.

Photos by José A. Martínez.

Figure 7. Two different steppe landscapes of the Ebro Valley (Spain). Above, the semiarid shrub-steppe of Bardenas Reales. Below, mosaic of cereal stubbles and 'bad-lands' in the Belchite steppes.

These are wooded plains where the esparto grass steppe patches appear between stands of Aleppo pine. *Sideritis serrata* is one of the most notable endemics of this area, which extends to inner plateau areas, where the steppes become cooler and come into contact with the formations of the Iberian System *páramos* with their spiny hummock species such as *Genista pumila.*

3. The Hoya de Baza, Hoya de Guadix and the High Plains of Granada

These are interior steppe zones, connected to a greater or lesser degree with the arid regions of the south-east noted before.

The arid character of these corridors between mountain ranges derives from the orographic isolation imposed by the complexity of the Betic mountain ranges or by the continental *páramos* which encircle these ranges. Marine and continental sediments, with a predominance of gypsum-loam materials, and endorheic saline areas, with well developed halophilous formations take on importance in the lower interior regions.

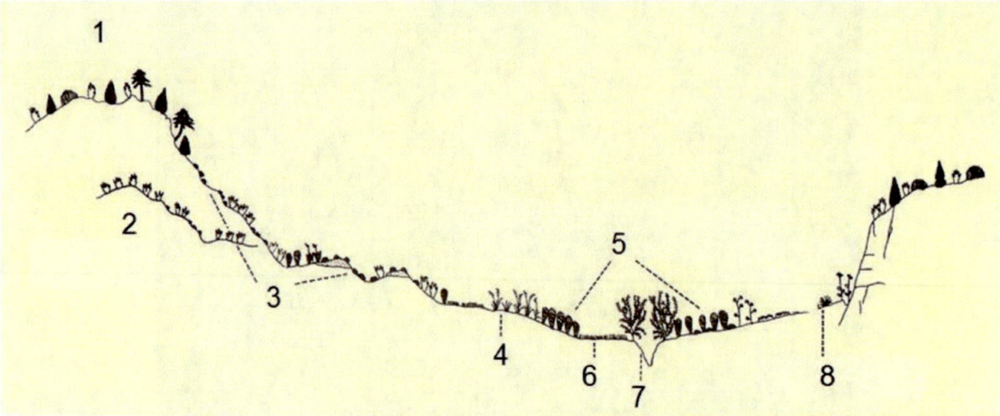

Figure 8. Zonation of the plant communities in the steppes of the Ebro Depression. 1: gypsophilous rosemary (*Rosmarinus officinalis*) formations with juniper (*Juniperus phoenicia* and *J. thurifera*) and Aleppo pine (*Pinus halepensis*); 2: gypsophilous rosemary formations; 3: thyme formations on steep bare gypsum soil; 4: albardine formations; 5: white wormwood-saltwort scrub; 6: cereal cultures; 7: Tamarisk (*Tamarix* spp.); 8: gypsophilous shrubs.

The presence of shrubs of great biogeographic significiance, such as *Vella pseudocytisus* or *Krascheninnikovia ceratoides*, is particularly notable, as are the extensions of esparto grass which in ancient times were the object of traditional exploitation for the production of baskets, hoods and espadrilles.

4. La Mancha and the Tagus Catchment Basin

The areas with natural vegetation in La Mancha, on Iberia's southern plateau, are now greatly reduced due to extensive cultivation. Steppe landscapes only persist in the areas of gypsum soils of low agricultural interest, on steep slopes, on the incised flanks of the Tagus river channel, or in the surroundings of some of the relatively numerous salty endorheic lakes present in this region filled with Miocene materials.

The steppes of this region in general are interpreted as secondary (Huguet del Villar, 1925), given that on the ridge tops a few enclaves with western holm oak woodlands survive, which appears to be the principal potential vegetation (Figure 9), except in the Campo de Montiel (*Juniperus thurifera* stands) or on the Quaternary sands of Cuenca or Ciudad Real (stone pine woodlands, *Pinus pinea*). It is very likely that the steppes in La Mancha were always rather reduced in size, conditioned to the edaphic enclaves (gypsum and salty soils) which were unfavorable for the development of woodland.

5. Extremadura

In the interior of Extremadura and at the foot of the Sierra Morena range, a few singular steppe areas occur. These are principally located in the region of La Serena (Badajoz), and the Alcudia Valley (Ciudad Real) and smaller areas in around Trujillo and Cáceres.

These steppe areas are very different from the others in Iberia, being based on a Paleozoic bedrock platform (Hercinian or Hesperian blocks) and being formed by pastures poor in therophytes or hemicryptophytes and where woody species are rare (Figure 10).

The region is generally covered by oak-dominated *dehesas* (wood-pasture) and forests (western holm oak *Quercus ilex* subsp. *ballota*, cork oak *Q. suber*, Portuguese oak *Q. faginea* subsp. *broteroi* and Hercinian oak *Q. pyrenaica*), and where the steppes cover relatively small extensions. This is a response to the poor soils (lithosols over rocky substrates in flat or nearly flat areas) and their location in colder, higher altitude plains affected by drying winds.

Photo by Manuel B. Morales.

Figure 9. Cereal steppes of the Tagus catchment (central Spain) where scattered holm oak (*Quercus ilex* subsp. *ballota*) bushes indicate the area's potential vegetation.

Photo by José A. Martínez.

Figure 10. Extensive therophyte pastures of La Serena, in Extremadura (southwestern Spain).

6. The Duero Catchment Basin

The northern Iberian plain is not a favorable territory for steppe formation given an average rainfall of more than 500-600 mm, and the vegetation potential is mainly for marcescent subsclerophyllous woodlands of Iberian (*Q. faginea* subsp. *Faginea*) and Hercinian oaks, and locally, western holm oak. Consequently the steppes in the Duero basin are very small, located on areas with detritic or evaporated Tertiary substrates (gypsum/ marls and salt-rich soils), unfavorable for woodlands, especially in the Tierra de Campos and the surroundings of the Laguna de Villafáfila. Their singularity was already highlighted by Willkomm (1852) and is supported by the presence of a few species characteristic of steppe woodlands, and very rare in the region, such as Spanish juniper (*Juniperus thurifera*). However, these cover very small areas within an area dominated by cereal crops. These same steppes however, support the densest world population of great bustard, a highly characteristic species of the herbaceous steppes of Russia and Central Asia (Morales and Martín, 2002).

7. Páramos of the Iberian System and the Duero Catchment Basin

The steppes of the *páramos* are one of the most original landscapes and best represented examples in the Iberian Peninsula. These are woody shrub and thyme-grassland mixed steppes, studded with tree formations (forest-steppes), mainly with *Juniperus, Pinus* and *Quercus* species. These occupy more-or-less flat high altitude areas (900 to 1200 m), on carbonated substrates from the Tertiary or Secondary eras. Messinian limestones, formed some 6 million years ago during the filling-up of the interior lagoons of the Iberian depressions, or Dolomitic, Cretaceous or Jurassic limestones which edge the Iberian System and the eastern part of the Spanish Central Range. These steppe formations appear principally in the periphery of the Iberian System and in a few *páramos* of the northern plateau.

The vegetation of these areas is comprised of a mosaic of spiny shrubs (whins, *Genista scorpius, G. pumila. Erinacea anthyllis*, see Figure 11) or not (*Thymus vulgaris, Salvia lavandulifolia, Lavandula latifolia*) and thyme-grassland mixed formations (*Thymus praecox, Helianthemum* spp. *Teucrium polium, Poa ligulata, Festuca hystrix, Koeleria vallesiana*) which appear beside wooded patches of Spanish juniper, and occasionally European black pine (*Pinus nigra*), Iberian oak or western holm oak. This is a territory with a landscape fossilized since the last ice age in which the cold-Mediterranean steppes and juniper forests have experienced few changes. The harshness of the climatic conditions (continentality, cold, drying winds), even today, together with the karstic geology which exhibits poor humidity retention, explain the scarce competition existing on the part of the tree species. These appear on the periphery of the *páramos*, in either warmer or less xeric locations.

California, Mohave, Sonora

The Californian steppes and those appearing in the surroundings of the Mohave, Sonora and Chihuahua deserts constitute a prolongation of the semiarid continental formations of the Great Basin, east of the Californian Sierra Nevada, which acts as an impassible barrier for Pacific rains (Archibold, 1995). These are woody steppes, dominated by the 'creosote bush' (actually different species of the genus *Larrea* of the Zygophyllaceae) and a few composites (*Franseria, Artemisia, Flourensia*, etc.) and legumes (*Zuccagnia punctata*) (Walter and

Breckle, 1986). Halophilous steppes dominated by Chenopods (*Atriplex* spp. *Ceratocarpus lanata*) also abound.

Photo by Juan E. Malo.

Figure 11. Woody steppes dominated by spiny and cushion-like short shrubs typical of the Iberian *páramos*.

Their more southerly distribution determines the greater weight shown in the landscape by Neotropical species of the hot Mexican deserts, especially cacti (*Opuntia, Carnegia, Encelia, Ferocactus*), agaves (*Yucca* spp., including the Joshua tree *Yucca brevifolia*) and members of the Fouquieriaceae (*Fouquieria* spp.).

South Africa: The Veldt

Veldt is a generic term used in South Africa, Botswana and Namibia referring to the open rural spaces where trees are scarce. These are generally flat areas covered by herbaceous plants or shrubs that fall within the lax concept of steppes. This is a term equivalent to the Australian outback which unites various different semiarid vegetation types in transition towards more-or-less continental warm deserts (Biome III according to Walter 1970).

Australia and New Zealand: The "Outback" and Tussock Pastures

The centre of Australia is desertic and various semiarid steppe zones appear around this, showing intermediate characteristics between the subtropical arid and the Mediterranean climates. The rainfall in these areas is between 200 and 300 or 350 mm annually and is frequently relatively evenly distributed throughout the year. The plant formations which can be considered as steppes are herbaceous, with a predominance of tussock biotypes on sandy or stony semidesert soils, or woody, dominated by Chenopods in saline areas. The increase in this biotype appears to be related to the abundance of natural or man-induced fires which these types of plants resist well (Walter, 1970).

A characteristic monospecific pasture –the "Mitchell grasslands"– of robust tussocks of the genus *Astrebla*, where half of the soil is left exposed, is found in the northeast (Northern territories).

A similar but more diverse formation predominates in the central north (central Australia), dominated by xeromorphic grasses of the Australian endemic genus *Triodia* (45 spp.). These exhibit tightly rolled, toughened and often spiny leaves, with the stoma protected on the inside. A few of the smaller *Plectrachne* species also appear. These are known locally as porcupine grass or *Triodia spinifex* desert. Between the tussocks a few shrubs of the *Hakea* (Proteaceae) genus appear. These grassy steppes frequently come into contact with open Mediterranean or savannah-like *Acacia* woodlands (mainly "mulga" *Acacia aneura* and "malle" or xerophytic *Eucalyptus* species, Archibold, 1995).

In the catchments discharging into the Gulf of Adelaide and in the Nullabor Plains of southern Australia, extensive steppes of succulent halophytes proliferate, including *Atriplex vesicaria, Kochia sedifolia, Arthrocnemum* spp., *Salicornia* spp. (Chenopods) and *Carpobrotus* (Aizoaceae).

In New Zealand a tussock grassland also appears which recalls those in Australia, and given its floral composition, also those in Patagonia. Grasses of cold-temperate genera appear such as *Festuca novae-zelandiae* and *Poa caespitosa* which inhabit the gaps between the big tussocks of *Chionochloa* (21 spp. in Australia and 19 spp. in New Zealand), close to *Cortaderia* and *Danthonia.*

The Chilean Steppes

In the Mediterranean region in Chile there are virtually no steppe areas. The existing steppe formations appear to have originated as a consequence of the degradation of the original sclerophyllous woodlands. Xerophytic and spiny shrubs form an important component, with such species as *Colliguaja odorifera* (Euphorbiaceae), *Trevoa trinervis* (Rhamnaceae) and numerous composites, mainly of the genus *Baccharis.* Roman cassie (*Acacia caven*) and Chilean mesquite (*Prosopis chilensis*) are typical accompanying small trees in these sub-Andean steppe formations.

The Pampa

The grasslands of the pampa constitute the most extensive steppe territory in the southern hemisphere, and covering around half a million km^2 are very closely followed in extension by the Patagonian steppes. The pampa is an extensive flat region extending down the Atlantic coast of southern Brasil (Rio Grande do Sul state), through Uruguay, to northern Argentina, between 30° and 39° south. The centre of this region coincides with the mouth of the rivers Paraguay and Paraná in the Rio de la Plata (Crespo i Veigas, 1988).

The steppe character of this region of the world is somewhat surprising, since it does not appear to be clearly justified by its largely subtropical climate, nor by its annual rainfall of between 600 and 1100 mm, nor also by its relatively mild temperatures of between 14° and 19°C. However, frosts are frequent in winter and spring. Various authors have suggested that the treeless nature of the region is due to ancient deforestation by man (through logging and burning) in order to favor cattle grazing which constitutes the principal use of these grasslands.

In consequence, the interpretation of these landscapes has long generated controversy. Initially the impression was that a relatively humid and warm-temperate climate could not justify such an absolute dominance by grasses, and the fact that trees here are restricted to more-or-less linear riparian formations, formed by stands of ombú (*Phytolacca dioica*), tala (*Celtis tala*), and southern elder (*Sambucus australis*). But considering the figures separately shows that the climate is rather drier than it appears. The high temperatures are responsible for high evapotranspiration potential, producing physiological drought during various moments in the annual cycle. In the most humid areas of the pampa the evapotranspiration potential exceeds the rainfall by some 100 mm, but in the driest regions to the south of the Río de la Plata, the water deficit can reach up to 700 mm. This occurs mainly in summer, in January and February, coinciding with the period of lowest rainfall. In consequence, the biotypes of the plants which develop in the region exhibit numerous xerophytic traits such as spinescence, leafless or assimilating stalks, minuscule, needle-like, coriaceous, narrow or rolled leaves, which are frequently impermeable thanks to waxy or resinous secretions,.

One characteristic which is not frequent in colder steppes appears in the pampa, which is the local importance of ephemeral therophyte communities, such as those typical of hot deserts. These are formed by generally non-grass herbaceous plants, which develop by exploiting the optimal spring and autumn periods, with a certain similarity to Mediterranean steppes.

Another important factor to comprehend the presence of grassy steppes in the pampa is the dense, clayey, poorly draining soils of these flat areas. Numerous endorheic areas with shallow lakes or winter-wet marshlands which dry in summer exist. These are very alkaline halophilous areas, with solonetz type soils containing abundant sodium carbonate. A salt-tolerant grass of the genus *Distichlis* (*D. spicata*) dominates in these areas.

The classic grasslands of the pampa were tussocky. This growth type is characteristic of the grasses growing in subtropical environments with mild winters. These are large clumps, over a meter high, or over two including the flowering spikes. The leaves are rigid and often rolled, which persist within the tussock for a relatively long time when dry; the new green leaves proliferate within these. This biotype is currently very familiar across the planet, following use of pampas grass (*Cortaderia selloana*) as a garden plant for its attractive whitish inflorescences. The presence of other typically subtropical grasses such as *Paspalum, Aristida, Panicum, Bothriochloa* or *Axonopus* is characteristic of the pampa.

The dry pampa, to the south of Buenos Aires, was formerly a region dominated by the large tussocks of *Stipa brachycheata* and *S. trichotoma*. The ground between these was bare or covered by ephemeral communities of annuals. Southern Neoptropical genera abounded, including *Bothriochloa, Piptochaetium* and *Aristida*. Since all these large grasses have little forage value, they have tended to be replaced by more productive, finer, and more readily cattle-grazed European pasture species. Consequently, the majority of the pampa is currently a cattle-rearing pseudosteppe (see Chapter 15 of this book).

In the most humid part of the pampa, towards the north, a transition towards temperate oceanic laurel forest of the Missiones region can be observed. There, the pampa grasslands are more delicate and show great diversity. The soils are relatively deep and rich in humus. Apart from the geographic and climatic differences, there is a certain similarity to the Russian steppes. These contain around twenty grass species and among them are important genera of typical temperate climates such as *Festuca, Poa, Bromus* or *Melica,* plus others of wider distribution such as *Stipa.*

The Patagonian Steppe

The Patagonian steppe extends across the southernmost part of South America, south of the 38° S parallel, from the foothills of the Andes to the Atlantic coast. Most is located in Argentina, although it also appears locally in Chile, especially near the Magellan Strait and the Torres del Paine National Park. The climatic conditions are so harsh in the Tierra del Fuego that the Patagonian steppes occurring there present similarities with the polar tundra (see also Chapter 4 in this volume). In general terms these steppes occupy extensive glacis at the foot of the Andean Range and a gigantic alluvial plain which slopes down gently to the east. The soils are sandy or stony with materials of glacial origin. The climate is cold and dry and conditioned by the mountainous barrier of the Andes. The average temperature varies between 5° and 10°C and progressively decreases towards the south, where frosts occur almost throughout the year. On the Pacific slope of the Andes high rainfall occurs (over 2000-3000 mm annually) which allows the presence of exuberant laurel or deciduous-leaved forests, whereas on the opposite side in Patagonia, annual rainfall is generally between 150 – 300 mm and their occurrence is rather irregular. The most relevant meteorological factor is the almost constant westerly wind, which exacerbate the dry nature of the treeless landscapes. The vegetation is formed by tough grasses or hummock shrubs. The more-or-less compact pillow-like hemispherical shapes are very favorable against the constant winds. The microclimatic conditions created within them are far less extreme than outside and on occasions the mass of twigs can retain sand, soil and a little humidity. The small, sometimes scale-like leaves also contribute my minimizing losses through transpiration. Plants with anostomous twigs, covered with tiny densely adpressed leaves frequently occur, such as the "colapiche" (*Nassauvia glomerulosa*) or the "mata negra" (*Verbena tridens*). Spinescence is also very frequent, being useful for minimizing water loss as well as being an effective defense against herbivory. The grasses develop best in the sandy soils or in the finer loess soils and in the more humid depressions with a little organic material. In the region they are known as "coironales". They are dominated by species of cosmopolitan cold-temperate genera such as *Festuca* (*F. pallescens, F. argentina, F. monticola, F. gracillima* in Tierra de Fuego), *Stipa* (*S. icho, S. neaei, S. speciosa, S. humilis, S. duriuscula*, etc.) and *Poa* (*P. huecu, P. ligularis*). The shrubs appear in the stonier areas and exhibit very variable cover efficiently and precisely reflecting available water resources. Diversity is high (with some 20 principal species) and among the most characteristic species are the "neneo" (*Mulinum spinosum*), the baccharises (*Baccharis* spp.), Magellan barberry (*Berberis buxifolia* and others), *espino negro* (*Colletia spinosissima*), "mata negra" (*Verbena tridens*), "quilenbais" (*Chuquiraga* spp.), "colapiche" (*Nassauvia glomerulosa*), "duraznillo patagónico" (*Colliguaya integerrima*), "malaspina" (*Trevoa patagonica*), "colimamuil" (*Anarthrophyllum rigidum*) and the Patagonian ephedra (*Ephedra frustillata*).

ACKNOWLEDGMENTS

John Muddeman translated the text of this chapter into English. José Antonio Martínez, Juan E. Malo and Manuel B. Morales kindly gave permission for the use of some of their photographs as chapter figures.

REFERENCES

Archibold, O. W. (1995). *Ecology of World Vegetation*. Chapman and Hall. London.

Arroyo, B. (1988). *Páramos y Estepas*. Enciclopedia de la Naturaleza de España, Vol. I. Debate/Círculo, Barcelona.

Berglund, B. E. (ed.). (1986). *Handbook of Holocene Palaeoecology and Palaeohydrology*, Wiley, New York.

Birks, H. J. B., Birks, H. H. (1980). *Quaternary Palaeoecology*. Edward Arnold, London.

Bocquet, G., Wilder, B., Kiefer, H. (1978). The Messinian model. A new outlook for the floristic and systematics of the Mediterranean area. *Candollea* 33, 269-287.

Bolòs, O. de. (1951). Algunas consideraciones sobre las especies esteparias de la península Ibérica. *Anales del Real Jardín Botánico de Madrid* 10, 1-9.

Costa, M., Morla, C., Sainz Ollero, H. (2000). Fitoecología y corología de *Krascheninnikovia ceratoides* en el valle del Ebro. *Actas del Congreso de Botánica en homenaje a Francisco Loscos. Alcañiz,* 1986, pp. 497-511. Instituto de Estudios Turolenses. Teruel, Spain.

Coupland, R. T. (1979). The nature of grassland. In: Coupland, R. T. (ed.), *Grassland Ecosystems of the World: Analysis of Grasslands and their Uses*, pp. 23-31. Cambridge University Press, Cambridge.

Curtis, J. T., McIntosh, R. P. (1951). An upland forest continuum in the prairies-forest border region of Wisconsin. *Ecology* 32, 476-496.

Crespo i Veigas, M. (1988). *Atlas de Botànica. Flora i Vegetació*. Edibook, Barcelona.

Dantín Cereceda, J. (1916). *Dry-Farming ibérico. Cultivo de las tierras de secano en las comarcas áridas de España*. Imprenta Gutemberg. Guadalajara, Spain.

De Juana, E., Santos, T., Súarez, F., Tellería, J. L. (1988). Status and conservation of steppe birds and their habitats in Spain. In: Goriup, P. D. (ed.), *Ecology and conservation of grassland birds,* pp. 113-123. ICBP Technical Publication nº7. ICBP, Cambridge.

Domínguez, F., Franco Múgica, F., Galicia, D. Moreno Saiz, J. C., Orueta, D., Sainz Ollero, H., Blasco, J. (2001). *Krascheninnikovia ceratoides* (L.) Gueldenst. (*Chenopodiaceae*) en Aragón (España): Algunos resultados de su plan de conservación. *Boletín de la Real Sociedad Española de Historia Natural (Sección Biología)* 96, 15-26.

García Antón, M., Maldonado, J., Morla, C., Sainz Ollero, H. (2002). Fitogeografía histórica de la península Ibérica. In: Pineda, F. D., de Miguel, J. M., Casado, M. A., Montalvo, J. M. (eds.), *La Diversidad Biológica de España,* pp. 45-64. Pearson Educación. Madrid.

Gómez Campo, C. (1981). Taxonomic and evolutionary relationships in the genus *Vella*. *Botanical Journal of the Linnean Society* 82, 165-179.

Gómez Manzaneque, F., Costa, M., Morla, C., Sainz Ollero, H. (2000). Elementos para una interpretación paleogeográfica de los sabinares albares de la península Ibérica. *Les Dossiers Forestiers* 6, 171-178.

González-Sampériz, P. (2004). *Evolución paleoambiental del sector central de la Cuenca del Ebro durante el Pleistoceno Superior y Holoceno.* Instituto Pirenaico de Ecología – CSIC-Departamento de Ciencias de la Antigüedad. Universidad de Zaragoza, Zaragoza, Spain.

Huguet del Villar, E. (1925). Avance geobotánico sobre la pretendida estepa central de España. *Ibérica* 23, nº 570, 381-393; nº 577, 293-502; nº 579, 328-333; nº 580, 347-350.

Huguet del Villar, E. (1929). *Geobotánica.* Editorial Labor, Barcelona.

Jackson, S. T., Williams, J. W. (2004). Modern analogs in quaternary paleoecology: here today, gone yesterday, gone tomorrow? *Annual Review of Earth and Planetary Sciences* 32, 495-537.

Jäger, E. J. (1971). Die Pflanzengeographische Stellung der «Steppen» der Iberian Halbinsel. *Flora* 160, 217-256.

Lebrun, J. (1947). *La végétation de la plaine alluviale au sud du Lac Edouard. Exploration du Parc National Albert.* Institute des Parcs Nationaux du Congo Belge, Brussels.

Lowe, J. J., Walker, M. J. C. (1997). *Reconstructing Quaternary Environments.* Longman. Edinburgh.

Menéndez Amor, J., Florschutz, F. (1963). Sur les éléments steppiques dans la végétation quaternaire de l'Espagne. *Boletín de la Real Sociedad Española de Historia Natural (Geología)* 61, 121-133.

Morla, C. (2003). El paisaje vegetal ibérico durante el Cuaternario. *Monografías del Jardín Botánico de Córdoba* 11, 75-93.

Morales, M. B., Martín, C. A. (2002). The Great Bustard *Otis tarda. The Journal of Birds of the Western Paleartic* 4, 217-232.

Mota, J. F., Sánchez Gómez, P., Merlo, M. E., Catalán, P., Laguna E., de la Cruz, M., Navarro, F. B., Marchal, F., Bartolomé, C., Martínez Labarga, J. M., Sainz Ollero, H., Francisco Valle, F., Serra L., Martínez Hernández, F., Garrido J. A., Pérez García, F. J. (2009). Aproximación a la checklist de los gipsófitos ibéricos. *Anales de Biología* 31, 71-80.

Overpeck, J. T., Webb, T., III, Prentice, I. C. (1985). Quantitative interpretation of fossil pollen spectra: Dissimilarity coefficients and the method of modern analogs. *Quaternary Research* 23, 87-108.

Pignatti, S. (1982). *Flora d'Italia.* Edagricole, Bologna.

Quézel, P. (1978). Analysis of the flora of Mediterranean and Saharan Africa. *Annals of the Missouri Botanical Garden* 65, 479-534.

Quézel, P., Médail, F. (2003). *Écologie et Biogéographie des Forêts du Basin Méditerranéen.* Elsevier, Paris.

Raven, P. H. (1973). The evolution of Mediterranean floras. *Abstracts of the Symposium "Plant life of SW Asia"*, pp. 213-224. Edinburgh.

Reyes Prósper, E. (1915). *Las Estepas de España y su Vegetación.* Sucesores de Rivadeneira, Madrid.

Riera, S. (2006). Cambios vegetales holocenos en la región mediterránea de la península Ibérica: ensayo de síntesis. *Ecosistemas* 15, 17-30.

Río, J. del, Peñas, J. (2006). *Krascheninnikovia ceratoides* (l.) Gueldenst., redescubierta en el Marquesado del Zenete (Granada). *Acta Botanica Malacitana* 31, 200-202.

Sainz Ollero, H. (1988). Las estepas ibéricas. Su importancia biogeográfica. *La Garcilla* 71/72, 8-11.

Sainz Ollero, H. (2003) Introducción al paisaje vegetal español. In: Martí, R., del Moral, J. C. (eds.), *Atlas de las Aves reproductoras de España*, pp. 49-53. Dirección General de Conservación de la Naturaleza-Sociedad Española de Ornitologia, Madrid.

Sainz Ollero, H., Domínguez Lozano, F. (2011). El contexto geobotánico. In: Domínguez Lozano, F., Guzmán Otano, D., Moreno Saiz, J. C. (eds.) *Biología de la Conservación de* Vella pseudocytisus *subespecie* paui, *una planta amenazada en Aragón*, pp. 19-42. Consejo de Protección de la Naturaleza de Aragón, Zaragoza, Spain.

Sainz Ollero, H., Sánchez de Dios, R., García-Cervigón Morales, A. (2010). La cartografía sintética de los paisajes vegetales españoles: una asignatura pendiente en geobotánica. *Ecología* 23, 249-272.

Santos, T., Suárez, F. (2005). Biogeography and population trends of Iberian steppe birds. In: Bota, G., Morales, M. B., Mañosa, S., Camprodon, J. (eds), *Ecology and Conservation of Steppe-Land Birds*, pp. 69-102. Lynx Edicions, Barcelona.

Suárez, F., Sainz Ollero, H., Santos, T., González Bernáldez, F. (1992). *Las estepas ibéricas*. Unidades temáticas ambientales, MOPT, Madrid.

Suc, J. P. (1984). Origin and evolution of the Mediterranean vegetation and climate in Europe. *Nature* 307, 429–432.

Walter, H. (1970). *Vegetationzonen und Klima*. Ulmer, Stuttgart.

Walter, H. (1984). *Vegetation of the Earth*. Springer Verlag, New York.

Walter, H., Breckle, S. W. (1986). *Ecological Systems of the Geobiosphere*. Springer Verlag, Berlin.

Whittaker, R. H. (1970). *Communities and Ecosystems*. Collier-Macmillan, London.

Willkomm, M. (1852). *Die Strand-und Steppengebeite der Iberischen Halbinsel und deren Vegetation*. Friedrich Fleischer, Leipzig.

Willkomm, M. (1896). Grundzuge der Pflanzenverbritung auf der Iberischen Halbinsel un deren Vegetation. In: Engler, A., Drude, O. (eds.) *Die Vegetation der Erde* (I), pp. 1-395. Friedrich Fleischer, Leipzig.

WWF-España (2009). *Bosques españoles. Los bosques que nos quedan y propuestas de WWF para su restauración.* WWF-ADENA. Madrid.

Patterns and Processes in Steppe Ecosystems

In: Steppe Ecosystems
Editors: M. B. Morales Prieto and J. Traba Diaz
ISBN: 978-1-62808-298-2

Chapter 2

FACTORS DETERMINING SPECIES RICHNESS AND COMPOSITION OF STEPPE BIRD COMMUNITIES IN PENINSULAR SPAIN: GRASS-STEPPE VS. SHRUB-STEPPE BIRD SPECIES

Juan Traba,[1,*] Pablo Sastre[2,†] and Manuel Borja Morales[1,‡]
[1]Department of Ecology, Universidad Autónoma de Madrid, Madrid, Spain
[2]Instituto Geológico y Minero de España, Madrid, Spain

Dedicated to the memory of Francisco Suárez Cardona (1953-2010)

ABSTRACT

Within the European Union, Spain supports the most abundant and richest community of steppe birds, i.e. those linked to open and treeless habitats, which is the most threatened group of birds in the European context. Here we analyse the relative importance of two sets of variables (spatial and environmental, including climate, topographic and land use variables in the latter) on the geographic variation in steppe bird richness in UTM 10x10 km grid squares using GLM. These same combinations of variables were used to characterize and estimate their relative contribution in the composition of the community using partial CCA. Our results highlight the importance of rainfall, but principally the presence of low-intensity land use practices on the richness of steppe birds in each square. Land use covers of less than 10% for fallow fields or more than 50% of cereal crops produce marked reductions in the steppe bird communities present. In terms of community composition, our results indicate the low relevance of the geographic factors compared to environmental variables. Among the latter, the climatic factors show relatively little influence in comparison to the topographic and land use

* Corresponding autor: Juan Traba. Department of Ecology, Universidad Autónoma de Madrid, c/ Darwin, 2, 28049, Madrid, Spain. E-mail: juan.traba@uam.es.
† Pablo Sastre: Instituto Geológico y Minero de España, c/ Ríos Rosas, 23, 28003 Madrid, Spain.
‡ Department of Ecology, Universidad Autónoma de Madrid, c/ Darwin, 2, 28049, Madrid, Spain.

variables. Two main steppe bird assemblages emerge from this work: the grass-steppe bird species (previously named pseudosteppe birds) and the shrub-steppe bird species.

Keywords: Cereal crops, fallow, grass-steppe, shrub-steppe, Spain, species richness, steppe birds

INTRODUCTION

The European steppes are home to the most threatened assemblage of birds in the continent (Santos and Suárez, 2005; Sanderson et al., 2005), the so-called steppe birds. These birds, characterized by their use of open and treeless habitats have suffered a clear decline over at least the last 20 years. This has enabled important sums of money to be invested over this period to, at least in part, improve their conservation status.

The Iberian Peninsula is crucial in a European context for its diversity of steppe habitats and the richness of the birds bird communities associated with these (Traba et al., 2007). However, until now the factors influencing this richness have not been described, nor if the steppe bird assemblage responds in relation to habitat composition and climate. In an earlier work, Valverde (1958) described two different steppe bird communities, one characteristic of grassy habitats product of the implantation of cultivated crops, especially cereals, and another of shrubby habitats, mainly those on salty soils. Latterly, in a seminal work for the understanding of these steppe bird communities in Spain and Europe, Suárez et al. (1997) also considered a community associated with pseudosteppes (mosaic of cereal crops, dry grasslands and winter grazed fallows) different from others linked to natural steppes. This classification has frequently been used since (see, for example, Santos and Suárez, 2005), although there have been no recent studies which have supported these groupings with data at the regional or national scales.

Species richness tends to be associated with spatial and environmental factors (Gaston, 1996; Lobo et al., 2001). Given that species richness of a particular community, such as birds, may show heterogeneous variation, understanding the determining factors which explain this variation may be crucial for the correct application of management and conservation measures for these species. For steppe birds, the factors most explaining their presence at the continental scale (Delgado et al., 2011) are related with productivity (e.g. great bustard *Otis tarda*) or sunlight/temperature (e.g. little bustard *Tetrax tetrax*), although no studies to date have evaluated the importance of abiotic variables on the species richness' patterns.

One of the typical problems of describing trends in species richness or community composition is the reliability of the data regarding species distributions (Gaston, 1996), combined with the difficulty of differentiating between causal relationships and their consequences (Gaston and Williams, 1996). However, and given the absence of complete reliable inventories, at least over large scales, the use of data from atlases and other similar sources has been habitual (see, for example, Lobo et al., 2001; Traba et al., 2007), based on the assumption that the general patterns are maintained despite the potential absence of complete or balanced data across the finer scale.

The analysis of relationships between species richness and environmental factors requires the identification not only of the environmental factors which show relation with richness, but also evaluation of the potential spatial autocorrelation of the dependent variable, given that its variation may be due to historic factors or intra- or interspecific relationships, which mask the true relationships with the environmental variables (Legrendre and Fortin, 1989; Legendre, 1993). This study has used a comprehensive database of the breeding steppe birds in peninsular Spain, collected at the UTM 10x10km grid square level, as well as explanatory variables estimated at the same scale, with the objective of determining the relative importance of the different groups of variables on the richness of steppe bird communities and their composition. Additionally we examined the definition of groups within the steppe bird communities in function to their response to different environmental variables.

METHODS

Study Area

The study area comprised the whole of peninsular Spain and the Balearic Islands, a total of over 500,000 km^2. We used data on breeding bird presence within the Universal Transverse Mercator (UTM) grid of 10x10 km grid cells. Quadrats with more than 50% falling within the sea or in extralimital countries were excluded, so that the final analysis was limited to 5,035 grid cells. Grid cell size was always 100 km^2 except for those cells joining different UTM coordinate zones.

Species

The species included in the analysis (Appendix 1) were selected in accordance with Suárez et al. (1997) and Traba et al. (2007); in summary, the species were selected on the basis of four associated criteria: i) Species typical of, or very frequent in, the Mediterranean region of the Iberian Peninsula; ii) ground-nesting species; iii) species exclusive to treeless, principally flat zones; and iv) species whose principal European population is in Spain. To this list we have added the lesser kestrel (*Falco naumanni*), which is not a ground-nester but is clearly tied to steppe habitats through their preferential use as foraging habitat (Bustamante, 1997), and the Eurasian skylark (*Alauda arvensis*), which breeds in many steppe ecosystems within the Iberian Peninsula (Suárez et al., 2003).

From the original list we have excluded those species that were present in less than 4% of the 5,035 grid cells analyzed for the Iberian Peninsula: short-eared owl (*Asio flammeus*), cream-coloured courser (*Cursorius cursor*) and collared pratincole (*Glareola pratincola*). The data used here are derived from the Spanish breeding bird atlas, the *Atlas of Spanish Breeding Birds* (Martí and del Moral, 2003), which covers the whole of the Spanish state using 10 x 10 km grid cells.

Environmental Variables

We used two groups of explanatory variables (Appendix 2): i) geographical position, ii) environmental variables, including climate, topography, and land cover and habitats.

Climate variables (temperature, precipitation and seasonality) were selected from the WorldClim database (www.worldclim.org) at a resolution of 30 arc seconds (approximately 1 kilometer). Data for Europe were downloaded from the European Distributed Institute of Taxonomy (http://edit.csic.es/GISdownloads.html).

Topography (mean elevation and slope) variables were derived from GTOPO30 (http://www1.gsi.go.jp/geowww/globalmap-gsi/gtopo30/gtopo30.html), a digital elevation model with a horizontal grid spacing of 30 arc seconds (approximately 1 kilometer). Data for Spain were downloaded from the Spanish National Museum of Natural Sciences (http://www.mncn.csic.es/LBI/Index.htm).

Land cover variables were obtained from two different sources: Areas of cereal crops and fallows were obtained at county level from the agricultural census of 1999 provided by the Spanish Statistics Institute (http://www.ine.es/inebmenu/mnu_agricultura.htm). Areas of natural grasslands were selected from the Corine Land Cover 2000 database (http://www.eea.europa.eu/themes/landuse/clc-download). We have included as a special group of variables the area of steppe habitats corresponding to EU Habitat Directive classes, extracted from the Spanish Natural Habitats map. Five steppe-habitats were defined in accordance with floristic and ornithological criteria (Appendix 2), once we suppressed those areas not belonging to EU-steppe habitats (sensu Suárez et al., 1992; Santos and Suárez, 2005), such as coastal habitats.

Climate, elevation and land cover data for each 10x10 km grid cell were extracted and mean values estimated per UTM 10x10 km grid cell with GIS software gvSIG (http://www.gvsig.gva.es) and IDRISI (http://www.clarklabs.org).

Data Analysis

Species Richness

The relationship between species richness of steppe birds in Peninsular Spain and the spatial and environmental variables (Appendix 2) was estimated by Generalized Linear Models (GLM, McCullagh and Nelder, 1989) assuming a normal error distribution for the response variable, after log-transformation of the data. To estimate the importance of geographical position on the species richness, the nine-terms of the third-degree polynomial of the central latitude (LAT) and longitude (LON) of each grid cell (Legendre and Legendre, 1998) were included in the model. To facilitate the interpretation of the results, we initially regressed the species richness against these nine terms by a backward selection method, in order to remove the non-significant spatial terms (see Lobo et al., 2001 for a similar procedure). In addition, to account for the possibility of non-linear relationships, the linear and the quadratic function of each environmental variable were included in the set of explanatory variables tested, and the scatterplots of residuals of the regression model against the predictor continuous variables were explored.

We reduced the number of environmental variables to be included in the final models via a forward selection method, and repeated the process until no more statistically significant variables remained ($P < 0.05$). We have included only pure effects, avoiding interaction terms between variables due to the difficulty of biologically interpreting many of them. The model goodness-of-fit was measured by the deviance statistic, and compared with deviance from a cross-validated model (McCullagh and Nelder, 1989).

We have transformed independent variables (arc-sin transformation for percentages and log-transformations for the rest) to reduce the effect of differences in measurement scales. All these statistical analyses were performed with STATISTICA 8.0 (Statsoft, 2007).

Community Analysis

We used cluster analysis to determine species clustering, over the absence-presence matrix for the whole data grid. We selected complete linkage as the amalgamation rule, and percent disagreement as the distance measure between species.

Canonical Correspondence Analysis (CCA; CANOCO 4.5, ter Braak and Smilauer, 2002) was used to describe the relationships between bird species presence per grid square and the environmental variables. CCA is a forced (constrained) ordination method which directly and simultaneously relates the species composition with the environmental variables, in contrast to non-forced ordination methods (e.g. Detrended Correspondence Analysis) which produce sequential ordination. CCA is an extension of multiple regression and is robust against moderate deviations from normal distribution of the data (Leps and Smilauer, 2003) and to skewed distributions.

The significance of the canonical axes in explaining the variation in species composition (abundance) was tested using Monte Carlo (n=500) permutations (n= 500).

The relative importance of each environmental variable in the species composition (abundance) was also determined by a forward selection procedure with non-restricted Monte Carlo (n=500) permutations (n= 500).

The overall results of the ordination were additionally confirmed through visual inspection of the species-versus-environmental variables biplots.

Finally, the variance was partitioned to evaluate the relative weight of the spatial and environmental variables on the community composition. This was undertaken performed using two additional partial CCAs, including the geographic coordinates as the environmental variables in the former and as covariates in the second (Leps and Smilauer, 2003).

RESULTS

Factors Determining Species Richness

The GLM forward stepwise selection analysis using the environmental and geographic variables, only selected four variables: percentage cover of fallow land per quadrat (fallow), percentage of non-irrigated cereal croplands (NICC), and annual precipitation (AnnPp) in both its linear and quadratic forms.

The final GLM constructed on these four variables showed a high goodness of fit (scaled deviance = 1.001; cross-validated scaled deviance = 0.927). The model indicated the positive effect of the percentage of fallow and NICC, in addition to an effect in optimum of the annual precipitation, being positive with the linear and negative with the quadratic factor (Table 1, Figure 1). The species richness of steppe birds stabilised above the minimum value of 10% fallow (Figure 1a).

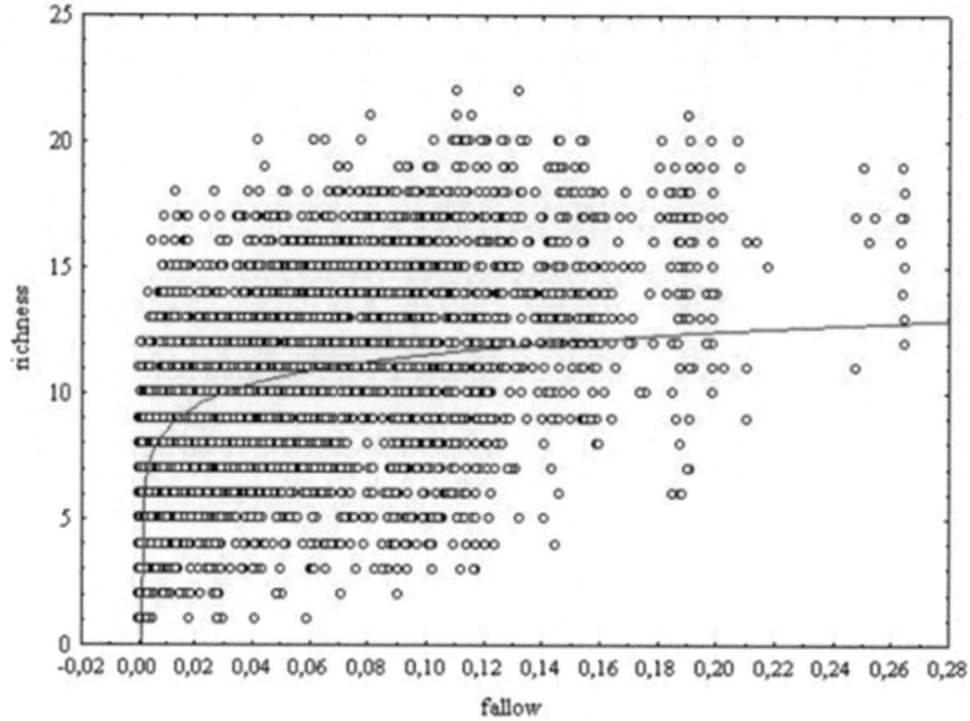

a

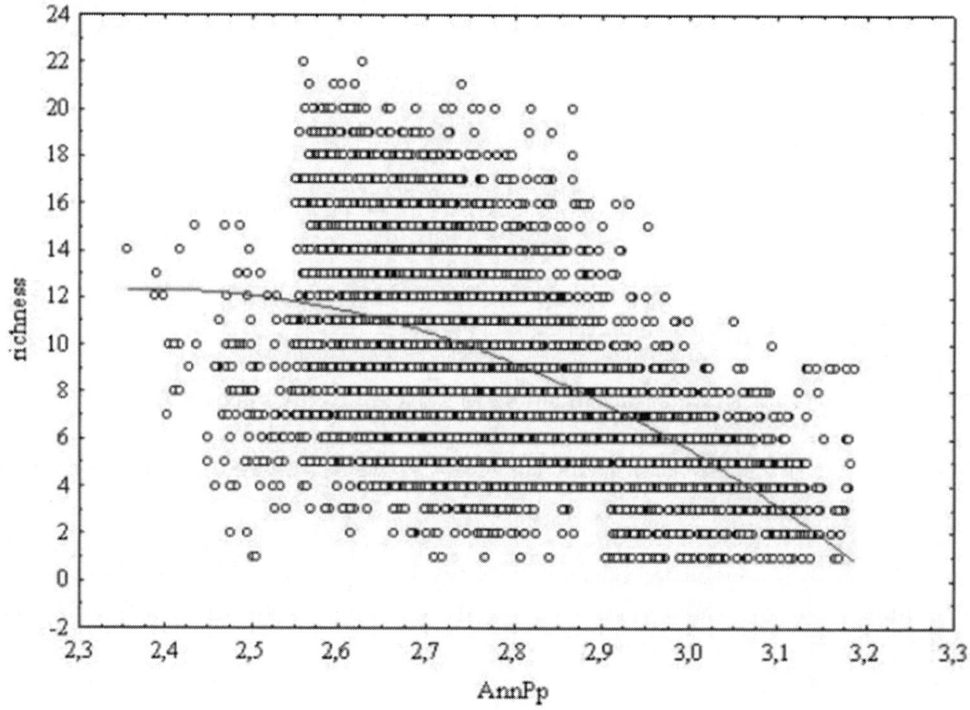

b

Figure 1. (Continued)

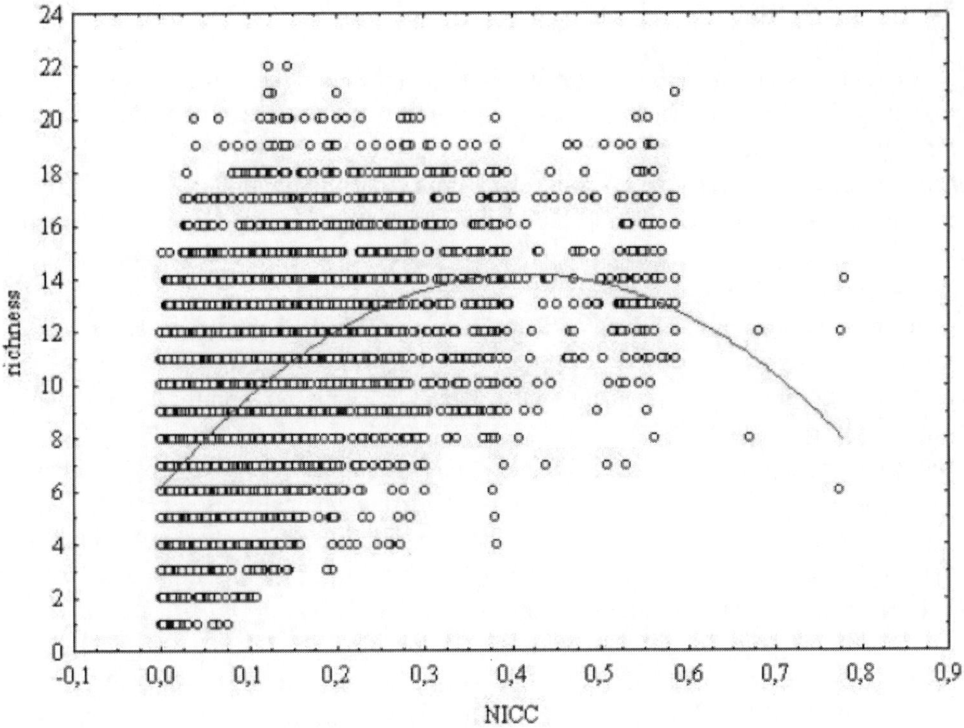

c

Figure 1. a) Relationship between percentage (%/100) of fallow land and steppe bird species richness in UTM 10x10km grid cells. Above the 10% of fallow land threshold, steppe bird species richness stabilizes. b) Relationship between annual precipitation (log-transformed) and steppe bird species richness. Actual values of the optimum are around 400-500 mm/year. c) Relationship between percentage (%/100) of non-irrigated cereal croplands (NICC) and steppe bird species richness. Above the 50% of NICC threshold, the steppe bird richness per cell declines steeply.

Table 1. Results of the GLM performed on the reduced set of environmental and climate variables

	Estimate	Standard Error	Wald Stat.	p
Intercept	-12.8326	0.951264	181.983	0.000
NICC	0.3524	0.023059	233.591	0.000
Fallow	0.884	0.067466	171.698	0.000
AnnPp	9.6388	0.681767	199.884	0.000
$AnnPp^2$	-1.8194	0.122037	222.262	0.000
Scale	0.1527	0.00182	7036	0.000

The relation of bird species richness with rainfall showed a maximum around the 400-500 mm/year value (after back-transformation of the original values), with clearly lower figures as this value increased (Figure1b).

The relationship with NICC showed that richness fell rapidly when NICC exceeded 50% in a quadrat (Figure 1c).

Community Composition

Cluster analysis revealed no clear differentiation in the combined presence of the species. The resulting dendrogram showed two main groups (or four, after reduction of the dissimilarity level between them).

Analysis of the gradient length indicated a unimodal distribution of species presence, recommending the use of a CCA (Leps and Smilauer, 2003). The CCA showed a significant relationship between the frequency of presence of all steppe bird species and habitat and climate variables (Table 2).

Once the number of original variables was reduced, habitat variables with no geographical control described 32.3% of the variation in frequency of presence of the different bird species, whereas geographical variables with no control by environment explained 12.2% of the variance.

The CCA partial models, controlling for the reciprocal effect of geographical and environmental variables, were significant for the first canonical axis as well as for the combined canonical axes (Table 3). Variance partitioning, in spite of the great proportion of unexplained variance (69.0%), showed a higher importance of the environmental variables (19.7%) versus the very low importance of the geographical position (1.7%, Table 4).

Table 2. Environmental variables introduced in the final canonical correspondence analysis model after prior screening by Monte Carlo permutation test (n=9,999)

Variable	Explained Variance	p
Annual Mean Temperature (AMT)	0.11	0.001
Annual Precipitation (AnnPp)	0.07	0.001
Mean slope (slope_mean)	0.04	0.001
Temperature Annual Range (TAR)	0.02	0.001
Percentage of Fallow (fallow)	0.02	0.001
Percentage of Non-irrigated cereal crops (NICC)	0.01	0.001
Seasonality of precipitation (PpSeas)	0.01	0.001
Natural Grasslands (NG)	0.00	0.001
Mean altitude (h_mean)	0.00	0.001
Percentage of steppe habitat 1520	0.01	0.001
Percentage of steppe habitat 4090	0.00	0.001
Percentage of steppe habitat 1430	0.00	0.001
X-coordinate	0.03	0.001
Y-coordinate	0.08	0.001

Table 3. Results of the Monte Carlo permutation for the validation of the partial CCA models

	First canonical axis		All the canonical axis	
	F	p	F	p
CCA only geographical variables	357.02	0.001	252.17	0.001
CCA only environmental variables	658.77	0.001	110.42	0.001
Shared: geographical + environmental	85.58	0.001	45.55	0.001
Shared: environmental + geographical	463.15	0.001	75.40	0.001

Table 4. Variance partitioning estimated after partial CCA with covariable matrix

Variance partitioning	
Variable	Variance explained (%)
Only geographic	1.7
Geographic + environmental	9.6
Only environmental	19.7
Residual	69.0

In general, species typical from grassy habitats were strongly associated with each other and independent (orthogonal) from species more frequent in shrubby steppes (Figure 2). Species like the great and little bustards, pin-tailed (*Pterocles alchata*) and black-bellied (*P. orientalis*) sandgrouse, stone curlew (*Burhinus oedicnemus*), greater short-toed (*Calandrella brachydactyla*) and lesser short-toed larks, (*C. rufescens*), lesser kestrel or the calandra lark (*Melanocorypha calandra*) appeared in association with plains and with the presence of grassy non-irrigated crops and fallow fields.

On the opposite side of the axis the Eurasian skylark showed preference for sites with steep slopes, and to a lesser degree areas of higher rainfall, as did hen harrier (*Circus cyaneus*). Other species in this group were red-legged partridge (*Alectoris rufa*), common quail (*Coturnix coturnix*) and corn bunting (*Emberiza calandra*) (Figure 2).

The second axis of the CCA principally separated species typical of high, cold and shrubby areas, such as the Dupont's lark (*Chersophilus duponti*, Figure 2). Other species which also appeared in this group include tawny pipit (*Anthus campestris*), Northern wheatear (*Oenanthe oenanthe*) and the spectacled warbler (*Sylvia conspicillata*). Other species, such as crested (*Galerida cristata*) and thekla (*G. theklae*) larks, black-eared wheatear (*Oenanthe hispanica*) and Montagu's harrier (*Circus pygargus*) are difficult to assign to either one of these groups.

The importance of the climatic variables, such as the rainfall seasonality (PpSeas) and the annual mean temperature (AMT) were much lower, given that almost no species appeared clearly associated with them. In general, all of the climatic variables, including the average annual rainfall (AnnPp) and the temperature annual range (TAR) showed very little importance in describing the community composition in comparison with the topographic or land use types variables (Figure 2).

DISCUSSION

Models which relate species richness with climatic, topographic or land use type factors have been shown to be especially adequate for describing large scale variation patterns (Guisan and Zimmermann, 2000). In this sense our results have also shown the association existing between the richness of steppe birds and land use and climatic factors.

However, the climatic factors have far less importance when considering variations in the communities of these birds, and topographic factors and land use types are what allow grouping of these species into diverse assemblages.

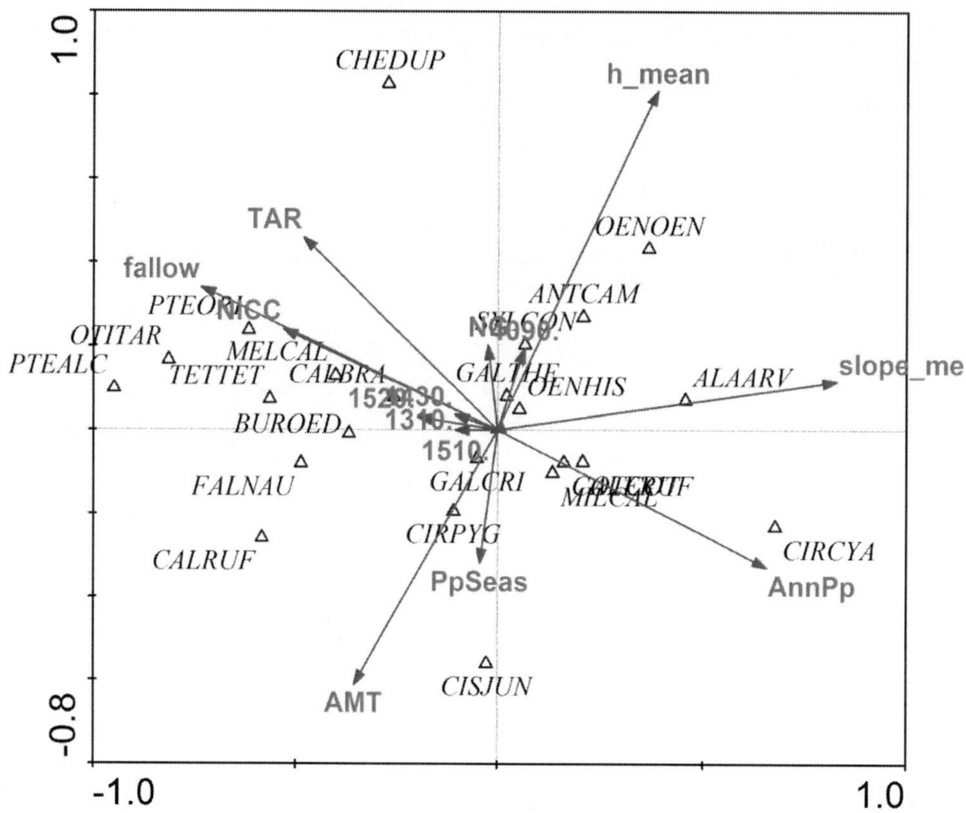

See abbreviations in Appendix 1 and 2.

Figure 2. Biplot from CCA showing the relationship between environmental and climate variables and steppe bird species in peninsular Spain.

Consequently a steppe bird assemblage reflecting the bird community linked to agricultural habitats, the grass-steppe birds has been identified, while the shrub-steppe birds appear linked to habitats with a dominance of woody vegetation, in more continental climatic conditions and with a more limited distribution. Species richness showed a positive association with the extent of fallow land and non-irrigated crops. The inclusion of the quadratic terms did not reveal linear relationships, although examination of the scatterplots appears to show a sharp decline in species richness when the proportion of fallow falls below 10% and the non-irrigated cereal crops exceed 50%. The association of species richness with minimum levels of fallow appears to be related with the close dependence on this type of agricultural substrate described for many of these species typical of steppe areas. For example, little bustards, red-legged partridges, black-bellied and pin-tailed sandgrouse, stone curlews, and to a lesser degree great bustards and small passerines appear to show preference for fallow fields, or landscapes with a significant proportion of fallow (Suárez et al., 1999; Morales et al., 2005; 2013a; Moreira et al., 2007; Buenestado et al. 2008; López-Jamar et al. 2011; Traba et al. 2013). The decline in species richness when the proportion of fallow falls below 10%, together with that observed when grid cells also increase their non-irrigated cereal crop surface area to over 50%, should draw attention to the serious risk resulting from changes in agriculture which suppose an increase in intensification (see a recent revision in Morales et al., 2013b), to the already threatened steppe bird community.

In this sense, the decision adopted by the European Commission in 2007, eliminating the obligation to maintain a minimum of 10% of the agricultural surface area as fallow, with the objective of increasing production, may be having important consequences for the conservation of steppe birds, which are highly dependent on non-intensive agricultural systems and mosaic landscapes (Santos and Suárez, 2005; Concepción and Díaz, 2011; Guerrero et al., 2011). However, to date there have been no studies over large scales to evaluate the possible decline in species richness as a result of the decline in fallow.

The richness of steppe bird species also showed a non-linear relationship with rainfall, with an optimum of around 400-500 mm/year. In reality these average figures are frequent across virtually the entire Iberian Peninsula, highlighting the generally ideal situation in climate terms for the steppe bird community. Only the most humid environments showed a more marked reduction, though even here a few species of wider Euro-Asiatic distribution such as the hen harrier (*Circus cyaneus*) or the Eurasian skylark (see below) can be found. Previous studies have found associations between species richness and climatic factors, more specifically rainfall (De Juana and García, 2005). In fact, rainfall during the pre-breeding period (January-April) explained 27.1% of the variance in steppe bird species richness in Iberian steppes (De Juana and García, 2005).

Only less than 1% of the 5000 UTM 10x10 km quadrats analysed showed a complete absence of steppe birds (Traba et al., 2007). Indeed, the high average values described by these authors suggest a marked tendency for co-occurrence at this spatial scale, despite their differences in habitat selection at smaller scales.

The groups arising from the classification tree do not appear to represent real classifications as judged from expert knowledge, given that the birds analysed form an eclectic group, occupying broad niches and wide distributions, which can be expected to show high levels of coincidence over large scales. Indeed, the more specialist species of the woody shrub steppes and with more limited distributions, such as Dupont's lark or the lesser short-toed lark, fall within a much larger group, together with numerous species of the grassy steppes, such as great and little bustards.

The results of the canonical correspondence analysis however, where the relationships between the species and the environmental factors are explored, far better explained the differences between species, in relation to their habitat and climatic requirements. Firstly, the low importance of the geographic factors stands out, indicating that the observed processes respond directly to environmental variables, and not to population processes or interactions between species within the community. Such dependence on the environmental variables is supported by the high importance of the latter shown by the CCA. The high levels of unexplained variance may be due to various reasons, among them historic factors which contribute to the current distribution of the species, in addition to factors which operate a smaller scale than that analysed, such as agriculture intensification processes or urban expansion at the local scale, which may lead to loss of species from areas where they were previously present.

In community configuration terms, axis 1 of the graph in Figure 2 structures the species in function to the topographical variable of slope. At one extreme are species of broad Eurasian distribution, including Eurasian skylark and hen harrier or to a lesser degree corn bunting (*Emberiza calandra*), common quail (*Coturnix coturnix*) and the red-legged partridge (*Alectoris rufa*, although this last species shows markedly Mediterranean affinities, Cramp and Simmons 1980).

Indeed, some of these species tend to show an increased probability of presence towards the north, even in mid mountain levels (Suárez et al., 2003; Arroyo and García, 2007). The hen harrier also shows a preference for areas with higher rainfall (García and Arroyo, 2001; Arroyo and García, 2007). At the other extreme of the gradient are the species characteristic of the flat plains, associated in the Iberian Peninsula to areas of grassy crops (NICC) and fallow, as the GLM had already indicated: great and little bustards, pin-tailed and black-bellied sandgrouse, stone curlew, greater short-toed, lesser short-toed and calandra larks, and the lesser kestrel. Axis 1, in consequence defines the grass-steppe birds group, linked to grassy agricultural areas, although the proportion of fallow versus crops and the area's slope may favour some species over others. Axis 2 of the CCA structures the community as a function of the average altitude on one hand and to the seasonal rainfall (PpSeas) and the average mean temperature (AMT) on the other. This appears to reflect a second community of steppe birds, the shrub-steppe birds, which are species tied to cold, high, continental regions, to the steppes of the Iberian System and other similar areas (see also chapter Chapter 1 of this volume), and among which the Dupont's lark stands out, but also includes tawny pipit, Northern wheatear and the spectacled warbler. Other species, including both *Galerida* larks, black-eared wheatear and Montagu's harrier are difficult to assign to either of these two large groups, in some cases given their cosmopolitan nature, and in other cases for possible errors in identification (the case of the *Galerida* larks). Finally, a more widespread species such as the zitting cisticola (*Cisticola juncidis*), frequently tied to ditches and small streams in the steppe areas, has become unlinked from the other groups and so perhaps should not be considered in the list of steppe birds, as it was in some other previous studies (see e.g. Suárez et al., 1996; Traba et al., 2007). Finally, it should also be mentioned that the axis of annual rainfall (AnnPp) and annual temperature range (TAR), as for the distribution of pure steppe habitats as defined by the Habitats Directive, have had very little impact in determining community composition in comparison to the relevance of topographic and land use variables. Our results demonstrate that the steppe bird community of mainland Spain appears to be highly dependent on the structural factors at the landscape level, in the same way as that previously described for microscale factors (Morales et al., 2008; Morales and Traba, 2009). This dependence on environmental landscape factors at the macroscale contrasts with the scarce importance attributed to climatic factors, suggesting that we are dealing with a community relatively resistant to climate change, as long as the land use types remain little changed.

The traditional agricultural mosaic systems of the Iberian Peninsula, the so-called pseudosteppes (Suárez et al., 1997) in which grassy crops coexist with non-cultivated areas of varying age, harbour the most typically steppic bird community, including such emblematic species as the great and little bustards, the pin-tailed and black-bellied sandgrouse, the stone curlew and the calandra lark. These agricultural systems are also the most similar environment from the physiognomic, bird community structure and even floral (at least the dominant higher level taxa: e.g. grasses versus woody species, see Chapter 1 of this volume) point of view, with the 'original' steppes of Central Asia, from where the name steppe derives. The degree of similarity with the true steppe, albeit as a result of anthropogenic change of the original vegetation cover, suggests that at least in the Iberian Peninsula, the denomination of "pseudosteppe" should be changed, and we propose "grass steppe" instead.

This is descriptively analogous to shrub steppe for those flat and treeless areas covered with shrubby vegetation, which differ in terms of horizontal and vertical structure, as well as of floristic and bird species composition.

APPENDIX 1

Species list, and number of 10x10 km grid cells occupied by each species in Spain (Canary Islands not included)

Species	N° Cells
Circus cyaneus Hen Harrier	700
Circus pygargus Montagu's Harrier	2168
Falco naumanni Lesser Kestrel	1027
Alectoris rufa Red-legged Partridge	4644
Coturnix coturnix Common Quail	3686
Tetrax tetrax Little Bustard	1288
Otis tarda Great Bustard	527
Burhinus oedicnemus Stone Curlew	2180
Cursorius cursor Cream-coloured Courser*	1
Glareola pratincola Collared Pratincole*	191
Pterocles orientalis Black-bellied Sandgrouse	837
Pterocles alchata Pin-tailed Sandgrouse	437
Asio flammeus Short-eared Owl*	73
Alauda arvensis Eurasian Skylark	2815
Chersophilus duponti Dupont's Lark	231
Galerida cristata Crested Lark	3876
Galerida theklae Thekla Lark	2925
Melanocorypha calandra Calandra Lark	2083
Calandrella brachydactyla Greater Short-toed Lark	2127
Calandrella rufescens Lesser Short–toed Lark	243
Anthus campestris Tawny Pipit	2001
Oenanthe oenanthe Northern Wheatear	2332
Oenanthe hispanica Black-eared Wheatear	2930
Cisticola juncidis Zitting Cisticola	2369
Sylvia conspicillata Spectacled Warbler	1273
Emberiza calandra Corn Bunting	4491

*Species present in less than 4% of the 5,035 grid cells and excluded from the analyses.

APPENDIX 2

Environmental variables used in analyses, and abbreviations (see Methods)

Climate (Worldclim database): mean (temperature in tenths of degree, precipitation in mm)
Annual Mean Temperature -BIO1 (AMT)
Temperature Annual Range (P5-P6) -BIO7 (TAR)
Annual Precipitation -BIO12 (AnnPp)
Precipitation Seasonality (Coefficient of Variation) -BIO15 (PpSeas)

Environmental variables used in analyses (see Methods) (Continued)

Elevation (GTOPO30)
Mean elevation (m) (h_mean) Mean slope (%) (slope_mean)
Climate (Worldclim database): mean (temperature in tenths of degree, precipitation in mm)
Land cover I (agricultural census 1999): area-weighted average of county data on areas (%)
Non-irrigated cereal croplands (NICC) Fallow (Fallow)
Land cover II (Corine 2000 database): area of classes (%)
Natural grasslands (NG)–code 321
Steppe-Habitats (per EU Habitats Directive): area of habitats (%)
Salicornia and other annuals colonizing mud and sand (H1310) -code 1310 Halo-nitrophilous scrubs (*Pegano-Salsoletea*) (H1430) –code 1430 Mediterranean salt steppes (*Limonietalia*) (H1510) –code 1510 Iberian gypsum vegetation (*Gypsophyletalia*) (H1520)–code 1520 Endemic oro-Mediterranean heath with gorse (H4090) –code 4090-subclass 309084

1 Annual Mean Temperature The mean of all the weekly mean temperatures. Each weekly mean temperature is the mean of that week's maximum and minimum temperature.

2 Mean Diurnal Range(Mean(period max-min)) The mean of all the weekly diurnal temperature ranges. Each weekly diurnal range is the difference between that week's maximum and minimum temperature.

3 Isothermality 2/7 The mean diurnal range (parameter 2) divided by the Annual Temperature Range (parameter 7).

4 Temperature Seasonality ANUCLIM (cov=TRUE) returns the temperature Coefficient of Variation (C of V) as the standard deviation of the weekly mean temperatures expressed as a percentage of the mean of those temperatures (i.e. the annual mean). For this calculation, the mean in degrees Kelvin is used. This avoids the possibility of having to divide by zero, but does mean that the values are usually quite small. Worldclim (cov=FALSE) returns the standard deviation of the weekly mean temperatures.

5 Max Temperature of Warmest Period The highest temperature of any weekly maximum temperature.

6 Min Temperature of Coldest Period The lowest temperature of any weekly minimum temperature.

7 Temperature Annual Range (5-6) The difference between the Max Temperature of Warmest Period and the Min Temperature of Coldest Period.

8 Mean Temperature of Wettest Quarter The wettest quarter of the year is determined (to the nearest week), and the mean temperature of this period is calculated.

9 Mean Temperature of Driest Quarter The driest quarter of the year is determined (to the nearest week), and the mean temperature of this period is calculated.

10 Mean Temperature of Warmest Quarter The warmest quarter of the year is determined (to the nearest week), and the mean temperature of this period is calculated.
11 Mean Temperature of Coldest Quarter The coldest quarter of the year is determined (to the nearest week), and the mean temperature of this period is calculated.
12 Annual Precipitation The sum of all the monthly precipitation estimates.
13 Precipitation of Wettest Period The precipitation of the wettest week or month, depending on the time step.
14 Precipitation of Driest Period The precipitation of the driest week or month, depending on the time step.
15 Precipitation Seasonality(C of V) The Coefficient of Variation (C of V) is the standard deviation of the weekly precipitation estimates expressed as a percentage of the mean of those estimates (i.e. the annual mean).
16 Precipitation of Wettest Quarter The wettest quarter of the year is determined (to the nearest week), and the total precipitation over this period is calculated.
17 Precipitation of Driest Quarter The driest quarter of the year is determined (to the nearest week), and the total precipitation over this period is calculated.
18 Precipitation of Warmest Quarter The warmest quarter of the year is determined (to the nearest week), and the total precipitation over this period is calculated.
19 Precipitation of Coldest Quarter The coldest quarter of the year is determined (to the nearest week), and the total precipitation over this period is calculated.

ACKNOWLEDGMENTS

We are grateful to SEO-BirdLife for providing breeding bird atlas data. John Muddeman translated the chapter into English. This work is a contribution to project CGL2009-13029/ BOS of the Spanish Ministry of Science, as well as to the REMEDINAL2 network of the CAM (S-2009/AMB/1783). We wish to dedicate this book chapter to the memory of our beloved friend Francisco 'Quico' Suárez, who devoted his scientific career to brilliant and passionately widen the knowledge and appreciation of steppes and steppe birds.

REFERENCES

Arroyo, B., García, J. T. (2007). *El aguilucho cenizo y el aguilucho pálido en España.* Población en 2006 y método de censo. SEO-BirdLife, Madrid.

Buenestado, F. J., Ferreras, P., Delibes-Mateos, M., Tortosa, F. S., Blanco-Aguiar, J. A., Villafuerte, R. (2008). Habitat selection and home range of red-legged partridge in Spain. *Agriculture Ecosystems and Environments* 126, 158-162.

Bustamante, J. (1997). Predictive models for lesser kestrel Falco naumanni distribution, abundance and extinction in southern Spain. *Biological Conservation* 80, 153-160.

Concepción, E. D., Díaz, M. (2011). Field, landscape and regional effects of farmland management on specialist open-land birds: Does body size matter? *Agriculture, Ecosystems and Environment* 142, 303-310.

De Juana, E., García, A. M. (2005). Fluctuations related to rainfall in richness and abundance of birds of steppic Mediterranean habitats. *Ardeola* 52, 53-66.

Delgado, M. P., Traba, J., Morales, M. B. (2011). Climate niche constraints in two coexisting steppe birds: the little and the great bustards. *Ardeola* 58, 223-238.

García, J. T., Arroyo, B. (2001). Effect of abiotic factors on reproduction in the centre and periphery of breeding ranges: a comparative analysis in sympatric harriers. *Ecography* 240, 393-402.

Gaston, K. J. (1996). Species richness: measure and measure- ment. In: Gaston, K. J., (ed.), Biodiversity. *A Biology of Numbers and Difference*, pp. 77– 113. Blackwell Science, Oxford.

Gaston, K. J. and Williams, P. H. (1996). Spatial patterns in taxonomic diversity. In: Gaston, K. J. (ed.), Biodiversity. *A Biology of Numbers and Difference*, pp. 202–229. Blackwell Science, Oxford.

Guerrero, I., Morales, M. B., Oñate, J. J., Aavik, T., Bengtsson, J., Berendse, F., Clement. L. W., Dennis, C., Eggers, S., Emmerson, M., Fischer, C., Flohre, A., Geiger, F., Hawro, V., Inchausti, P., Kalamees, A., Kinks, R., Liira, J., Meléndez, L., Pärt, T., Thies, C., Tscharntke, T., Olszewski, A., Weisser, W. W. (2011). Taxonomic and functional diversity of farmland bird communities across Europe: effects of biogeography and agricultural intensification. *Biodiversity and Conservation* 20, 3663-36813.

Guisan, A., Zimmermann, N. E. (2000). Predictive habitat distribution models in ecology. *Ecological Modelling* 135, 147–186.

Legendre, P. (1993). Spatial autocorrelation: trouble or new paradigm? *Ecology* 74, 1659–1673.

Legendre, P., Fortin, M. J. (1989). Spatial pattern and ecological analysis. *Vegetatio* 80, 107–138.

Legendre, P., Legendre, L. (1998). *Numerical Ecology*, 2nd ed. Elsevier, Amsterdam.

Lobo, J. M., Castro, I., Moreno, J. C. (2001). Spatial and environmental determinants of vascular species richness distribution in the Iberian Peninsula and Balearic Islands. *Biological Journal of the Linnean Society* 73, 233-253.

López-Jamar, J., Casas, F., Díaz, M., Morales, M. B. (2011). Local differences in habitat selection by Great Bustards Otis tarda in changing agricultural landscapes: implications for farmland bird conservation. *Bird Conservation International* 21, 328-341.

Martí, R., del Moral, J. C. (eds.). (2003). *Atlas de las Aves Nidificantes de España*. Dirección General de la Naturaleza-Sociedad Española de Ornitología, Madrid.

McCullagh, P., Nelder, J. A. (1989). *Generalized Linear Models*. Chapman and Hall, London.

Morales, M. B., J. T. García, Arroyo, B. (2005). Can landscape composition changes predict spatial and annual variation of little bustard male abundance? *Animal Conservation* 8, 167-174.

Morales, M. B., Traba, J., Carriles, E., Delgado, M. P., García de la Morena, E. L. (2008). Sexual differences in microhabitat selection of breeding Little Bustards Tetrax tetrax: ecological segregation based on vegetation structure. *Acta Oecologica* 34, 345-353.

Morales, M. B. and Traba, J. 2009. Compromisos adaptativos en la selección de hábitat de aves esteparias. In: Dopazo, H., Navarro, A. (eds.), *Adaptación y Evolución: 150 años después del Origen de las Especies.* Eds. Obrapropia, Valencia.

Morales, M. B., Traba, J., Delgado, M. P., García de la Morena, E. (2013). The use of fallows by nesting Little bustard Tetrax tetrax females: implications for conservation in mosaic cereal farmland. *Aredola* 60, in press3-15.

Morales, M. B., Guerrero, I., Oñate, J. 2013. Efectos de la gestión agraria en las aves de los cultivos cerealistas: un proceso multiescalar. *Ecosistemas* 22, in press25-29.

Moreira, F., Leitão, P. J., Morgado, R., Alcazar, R., Cardoso, A., Carrapato, C., Delgado, A., Geraldes, P., Gordinho, L., Henriques, I., Lecoq, M., Leitão, D., Marques A. T., Pedroso, R., Prego, I., Reino, L. Rocha, P., Tomé, R., Osborne, P. E. (2007). Spatial distribution patterns, habitat correlates and population estimates of steppe birds in Castro Verde. *Airo* 17, 50-30.

Snow, D. W., Perrins, C. M. (1998). *The birds of the western Paleartic.* Oxford University Press, Oxford.

Sanderson, F. J., Donald, P. F., Burfield, I. (2005). Farmland birds in Europe: from policy change to population decline and back again. In: Bota, G., Morales, M. B., Mañosa, S., Camprodon, J. (eds.), *Ecology and conservation of steppe-land birds*, pp. 211-236. Lynx Edicions, Barcelona.

Santos, T., Suárez, F. (2005) Biogeography and population trends of iberian steppe bird. In: Bota, G., Morales, M. B., Mañosa, S., Camprodon, J. (eds.), *Ecology and conservation of steppe-land birds*, pp. 69–102. Lynx Edicions, Barcelona.

Statsoft (2007). Statistica 8.0. Statsoft, Inc, http://www.statsoft.com/

Suárez, F., Sainz Ollero, H., Santos, T., González Bernáldez, F. (1992). *Las estepas ibéricas.* Unidades temáticas ambientales, MOPT. Madrid.

Suárez, F., Naveso, M. A., De Juana, E. (1997). Farming in the drylands of Spain: birds of pseudosteppes. In: Pain, D. J., Pienkowski, N. W. (eds.). Farming and birds in Europe. *The common agricultural policy and its implications for bird conservation*, pp. 79-116. Academic Press, San Diego.

Suárez, F., Garza, V., Morales, M. B. (2003). Skylark Alauda arvensis densities and habitat selection in Mediterranean Iberia: the role of shrub-steppes and extensive dry-pastures. *Agriculture Ecosystems and Environment* 95, 551-557.

Suárez, F., Herranz, J., Martínez, C., Manrique, J., Astrain, C., Echeverrría, A., Curcó, A., Estrada, J., Yanes, M. (1999). Utilización y selección de hábitat de las gangas ibérica y ortega en la Península Ibérica. In: Herranz, J., Suárez, F. (eds.), La Ganga Ibérica (Pterocles alchata) y la Ganga Ortega (P. orientalis) en España. *Distribución, Abundancia, Biología y Conservación,* pp. 127-156. Ministerio de Medio Ambiente, Madrid.

Traba, J., García de la Morena, E. L., Morales, M. B., Suárez, F. (2007). Determining high value areas for steppe birds in Spain: hot spots, complementarity and the efficiency of protected areas. *Biodiversity and Conservation* 16, 3255-3275.

Traba, J., Delgado, M. P., Albendea, D. J., Morales, M. B. (2013). Selección de macro y microhábitat del Alcaraván común (Burhinus oedicnemus) en dos localidades madrileñas durante la estación reproductiva. *Anuario Ornitológico de Madrid* 13, in press

Valverde, J. A. (1958). *Las aves esteparias de la Península Ibérica.* Instituto de Biología Aplicada 27, 41-48.

In: Steppe Ecosystems
Editors: M. B. Morales Prieto and J. Traba Diaz
ISBN: 978-1-62808-298-2

Chapter 3

SPECIES RICHNESS AND NICHE DIFFERENTIATION OF DARKLING BEETLES (COLEOPTERA: TENEBRIONIDAE) IN MONGOLIAN STEPPE ECOSYSTEMS

Omid Paknia,[1,*] ***Michael Grundler***[2,†] ***and Martin Pfeiffer***[2,‡]
[1]Institute of Animal Ecology and Cell Biology (ITZ), Hannover, Germany
[2]Department of Ecology, National University of Mongolia, Ulaanbaatar, Mongolia

Dedicated to the memory of Gleb Sergeevich Medvedev (1931- 2009)

ABSTRACT

Located in the center of Asia, Mongolia lies in an ecological transition zone where the Central Asian Steppe meets the Siberian taiga forest and the Gobi desert. These unique ecosystems support significant invertebrate biodiversity that plays crucial roles in ecosystem functioning and services.

Darkling beetles of the family Tenebrionidae (Insecta, Coleoptera, Polyphaga) form the most conspicuous group of detrito- and phytodetritophagous arthropods, which dominate - together with ants - the vast Mongolian steppe habitats in terms of biomass and abundance.

The spatial pattern of their species richness, however, is hardly known. We used a comprehensive geo-referenced museum record (ca. 3500 records) of 187 beetle species to quantify habitat associations and estimate species richness of darkling beetles in different ecoregions of Mongolia.

[*] Corresponding author: Omid Paknia. Institute of Animal Ecology and Cell Biology (ITZ), Hannover, Germany. Phone: +49 511 953 8489. E-mail: omid.paknia@ecolevol.de.

[†] Michael Grundler: Department of Ecology, National University of Mongolia, Bagatoiruu 47, P.O. Box 377, Ulaanbaatar 210646, Mongolia.

[‡] Martin Pfeiffer: Department of Ecology, National University of Mongolia, Bagatoiruu 47, P.O. Box 377, Ulaanbaatar 210646, Mongolia.

Darkling beetle species richness was negatively associated with latitude, but positively correlated with maximum and minimum temperature, demonstrating that tenebrionid species richness is greater in Mongolian regions of pronounced aridity and low productivity. High beta diversity between ecoregions was due mainly to species replacement rather than species loss.

Keywords: Community composition, community similarity, latitudinal climate gradient, species turnover, species range

INTRODUCTION

Mongolia lies in the center of the Asian continent, where the wide Asian steppes stretch out between the Sibirian taiga forests and the Gobi desert. In the north and west, highly forested mountain ranges alternate with dry, lake-dotted basins. Mongolia includes some of Earth's most pristine lakes. Its remarkable variety of nature is characterized by a steep climatic gradient from north to south that shapes the vegetation sequence of forest, steppe, semi-desert, and desert. Most of the sparsely populated country comprises pasture land, in part degraded by overgrazing, but less impacted by roads and human settlements. Mining is beginning to influence wide areas, but apart from livestock grazing most locations are unaffected by major human disturbance.

Much of the country is mountainous and there are three major mountain ranges: the alpine Altai Mountains in western Mongolia reach more than 4000 m above sea level (a.s.l.) and many peaks are snow capped year round. The Khangai Mountains are broad dome-shaped mountains that extend 800km from west to east in central Mongolia. Finally, the Khentii Mountains are smaller and lower than the other mountain blocks and extend from northeast Mongolia into Siberia. At the foot of these mountains vast steppes extend into the "Gobi desert", which comprises a mosaic pattern of arid vegetation types.

Beetles, especially three groups of ground-dwelling beetles (Carabidae, Scarabidae, and Tenebrionidae), are diverse, abundant and have numerous roles in steppe ecosystems (Crist 2008). Beetles of the family Tenebrionidae (Coleoptera, Polyphaga) comprise one of the most conspicuous and dominant detrito- and phytodetritophagous groups of arthropods in the Asian steppes and deserts (Konstantinov et al., 2009). Mongolia has a diverse tenebrionid fauna of 215 species from 50 genera (Medvedev, 1990). Dominant genera include *Anatolica*, *Blaps*, *Cyphogenia*, *Microdera*, and *Sternoplax*, but many of these species are restricted to arid areas, where they reach high abundances (Pfeiffer and Enkhmunkh, in press). Mongolian tenebrionids have been studied in detail by Kaszab (1965; 1967; 1977) and later by the great taxonomic expert G. S. Medvedev (Richter and Korotyaev, 2011), who described more than 250 taxa, mainly of darkling beetles, and developed a key to the Mongolian Tenebrionidae (Medvedev, 1990). In collaboration with A. L. Lobanov he published a faunistic list of tenebrionids of the Mongolian People's Republic with coordinates of the localities, which provides a large part of the database for this book chapter (Medvedev and Lobanov, 1990).

Different groups of beetles are increasingly used as biological indicators by insect conservationists. Selection of the taxa depends mainly on the study regions. For example, rove beetles (Staphylinidae) are dominant in agro-ecosystems and are strongly influenced by structure of landscapes and intensity of land-use (Da Silva et al., 2009).

Darkling beetles are abundant and diverse in steppe ecosystems across the world and are considered an appropriate biological indicator for these ecosystems (Carrara and Flores, 2012).

Community ecology of darkling beetles has been studied in the Mediterranean (Fattorini, 2002; 2006), the US (Parmenter et al., 1989; Stapp, 1997; McIntyre, 2000), Israel (Krasnov and Shenbrot, 1996; Krasnov and Shenbrot, 1998), and Patagonian steppe (Carrara and Flores, 2012).

However, little is known about the steppe-dominated central Asian region (Sagdy, 1996; Liu et al., 2012; Pfeiffer and Enkhmunkh, in press). Here we provide a comprehensive assessment of the biodiversity and community ecology of the Mongolian Tenebrionidae.

METHODS

Climate

The climate of land-locked Mongolia is continental with hot summers and extremely harsh winters. Winters are long, lasting from early November until April, and temperatures between -30° to -40°C are not uncommon in some regions.

In the northwest it may even reach – 50°C. In summer, temperatures rise strongly and may reach over +40°C in some desert regions. Rain is scarce and most of the precipitation occurs in summer.

Mean annual precipitation does not exceed 200 mm and ranges between 35 mm and 680 mm. The northern part of the country is nearer to the arctic sea and receives more rainfall than the southern areas. As a result there is a strong temperature and precipitation gradient from north to south.

Vegetation Zones and Ecoregions

Vegetation reflects the climate situation and the major vegetation zones are arranged from north to south in the following order, with the names slightly differing according to author (Grubov, 1982; Hilbig 1995; corresponding WWF ecoregions given in brackets; WWF 2012):

- Mountain taiga forest (Trans-Baikal Conifer Forests, Altai-Sayan Montane Conifer Forests, Khangai Mountains Conifer Forests)
- Forest steppe (Altai Montane Forest and Forest Steppe, Selenge-Orkhon Forest Steppe, Daurian Forest Steppe)
- Steppe (Sayan Intermontane Steppe, Mongolian-Manchurian Grassland)
- Semi desert with desert steppe (Eastern Gobi Desert Steppe, Gobi Lakes Valley Desert Steppe, Great Lakes Basin Desert Steppe)
- Desert and steppe desert (Junggar Basin Semi-Desert, Alashan Plateau Semi-Desert)
- Extrazonal vegetation including alpine vegetation (Khangai Mountains Alpine Meadow, Altai Alpine Meadow and Tundra).

A detailed description of the WWF ecoregions treated in this book chapter is listed below (WWF, 2012), and Table 1 includes the basic climate and plant characteristics of the given ecoregions. The steppe ecoregions can be divided into three groups: (i) forest steppe, (ii) steppe, and (iii) desert steppe (Figure 1 and 2).

Forest Steppe

Generally, in this ecosystem, coniferous forest is restricted to moist northern slopes and steppe grows on other slopes. In summer, Mongolian forests are prone to forest fires, which can devastate large areas.

Altai Montane Forest and Forest Steppe: The Altai stretches out from the Beluha Mountains (up to 4200 m) at the Russian Kazakhstan border southwest into central Mongolia. The Altai region is extremely cold and dry.

Figure 1. General view of (from above) forest steppe, steppe, desert steppe and desert habitats in Mongolia where tenebrionid beetles were sampled.

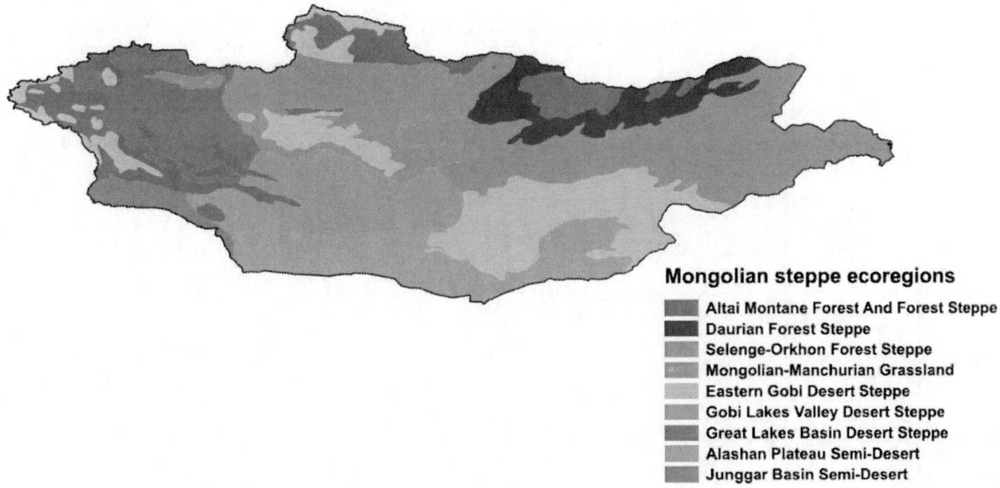

Figure 2. Map of the Mongolian ecoregions; only the steppe ecoregions, including forest steppe, steppe and desert steppe, are shown in colour. Forest steppe, steppe and desert steppe have been displayed in different tones of green, olive, and orange, respectively.

It is floristically and faunally diverse, and there are many endemic species. Forests are mostly larch (*Larix* spp.), larch-cedar and fir (*Picea* spp.). Desert-steppe vegetation grows in parts of the southern regions (see also Ogureeva et al. in this volume).

Selenge-Orkhon Forest Steppe: Valley floors of the Selenge and Orkhon river basins are covered with dry steppe. Mountain tops (1500-2000 m) support pine (*Pinus silvestris*) and aspen forests (*Populus tremula*).

Daurian Forest Steppe: The Daurian forest steppe follows the margin of the Khentii Mountains in northeast Mongolia. Rivers Onon and Ulz originate here. The northern slopes are covered with Sibirian larch and mixed forests of birch, pine or larch trees. Steppe dominates on southern slopes and at lower elevation.

Steppe

Mongolian steppes are boundless and sparsely inhabited. Only a few asphalt roads cross these vast areas. Human influence is visible in the large herds of livestock that graze the steppe.

Mongolian-Manchurian Grassland: This is a large ecoregion that stretches westward from China into Mongolia along the Kherlen River and includes vast undulating tracts of feathergrass steppe (mean altitude 1,000 to 1,300 m). Low annual rainfall falls during summer. A diversity of grasses composes the vegetation. Steppe fires are a frequent danger in this ecoregion.

Desert Steppe

Desert-steppes are moderately dry semi-deserts that support short grass cover and shrubs. The Gobi desert comprises the three desert steppe ecoregions (*Eastern Gobi Desert Steppe, Gobi Lakes Valley Desert Steppe, Gobi Lakes Valley Desert Steppe*) and the two desert ecoregions of Mongolia.

Table 1. Description of ten steppe ecoregions within Mongolia's territory their area size, climate and plant characteristics

Ecoregion name	Area (km^2)	Annual Mean Temp.	Mean Diurnal Temp. Range	Max * Temp.	Min * Temp.	Temp. Annual Range	Annual Precipitation	Mean Temp. Coldest Quarter	Mean Temp. Warmest Quarter	Plant characteristic
Alashan Plateau Semi-Desert	413534	5.1	14.1	28.6	-20.3	49	85	-11.7	20.6	*Haloxylon ammodendron, Reaumuria soongolica, Artemisia salsoloides*
Altai Montane Forest And Forest Steppe	196619	-1.8	13.1	20.5	-26.8	47.3	148	-18.5	13.4	*Larix sibiria, Picea spp. Betula spp., Populus pilosa, Stipa pennata, Allium polyrrhizum*
Daurian Forest Steppe	216834	-1.5	13.9	23.7	-29.1	52.9	306	-21	16	*Larix sibiria, Picea obovata, Carex-poaceae community, Festuca lenensis, Stipa spp.*
Mongolian-Manchurian Grassland	666894	0.3	13.6	25.4	-26.4	51.8	224	-18.7	17.6	*Stipa baicalensis, S. grandis, S. Capillata, Cleistogenes sqarrosa, Festuca ovina, Aneurolepidium chinense*
Sayan Intermontane Steppe	684	-7.9	14	19.2	-37.5	56.7	327	-29.8	10.9	*Betula pendula, Pinus silvestris, Stipa spp.*
Selenge-Orkhon Forest Steppe	464425	-3.2	14.3	20.6	-29.7	50.3	277	-21.4	12.9	*Larix sibiria, Pinus silvestris, Stipa cleistogenes*
Eastern Gobi Desert Steppe	351937	3.3	13.4	27.6	-22.5	50.1	130	-14.7	19.8	*Caragana bungei, Stipa gobica, Cleistogenes soongorica*
Gobi Lakes Valley Desert Steppe	283650	0.7	14.6	23.8	-24.3	48.1	141	-15.5	15.9	*Stipa gobica, Allium polyrrhizum*
Great Lakes Basin Desert Steppe	308381	-1.6	13.5	24.2	-31.7	55.9	147	-23.1	16.6	*Caragana leucophloea, Phragmites communis, Stipa spp.*
Junggar Basin Semi-Desert	68303	3.9	14	27.4	-23	50.4	91	-13.9	19.6	*Haloxylon ammodendron, Stipa glareosa, Anabasis brevifolia*

*Maximum and minimum temperatures represent the maximum temperature of warmest and coldest month, respectively.

Eastern Gobi Desert Steppe: The Eastern Gobi desert steppe is an ecotonal region between the Mongolian-Manchurian grassland and the Alashan desert. Shrubby semi-desert or grassy desert steppes cover the northern parts. Xerophytic communities live on the sandy, cobbled substrates of the south.

Gobi Lakes Valley Desert Steppe: The Gobi Lakes Valley divides the foothills of the Gobi-Altai and the Khangai ranges for 500 kilometers. Intermittent rivers runoff the surrounding mountains into valley floor lakes. Shrub communities surround the lakes, but most of the area is covered by desert steppe vegetation.

Great Lakes Basin Desert Steppe: The Great Lakes Basin includes a series of large freshwater and saltwater lakes that are fed by rivers originating in the Khangai and Altai mountains. Reed beds of *Phragmites communis* may cover the shores of the lakes. Most of the basin consists of desert steppe with grass and shrubs. Halophytes dominate the lower, salty parts of these areas.

Junggar Basin Semi-Desert: The Mongolian part of the Junggar Basin Semi-Desert supports a great variety of vegetation types. Desert steppes are widespread, but shrub forest communities of *Tamarix* or *Haloxylon* (Saxaul) grow in other areas. Mountain steppes include many species of grasses (von Wehrden et al., 2006).

Alashan Plateau Semi-Desert: The Alashan desert is an exceptionally arid region of mountains (up to 2500m) and depressions. Saxaul forests grow on mountain slopes and thinly vegetated desert steppe grows at high elevations (von Wehrden et al., 2006). Semi-desert and semi-shrub communities grow in some of the less arid lands (Hilbig, 1995). Stony deserts nearly barren of plant cover occur in southern regions. Open water sources are extremely scarce.

Species Data Collection

Most of the data are based on the faunistic study of Medvedev and Lobanov (Medvedev and Lobanov, 1990). MP received the raw data of the publication, including a species list and coordinates, from A. Lobanov almost 10 years ago. Additional data have been sampled during various expeditions in 1997, 1999 and 2003 (Pfeiffer and Enkhmunkh, in press). We validated the data by verifying accepted taxonomic names of species and removing specious records.

Species Data Preparation

The data set consists of 3,521 tenebrionid beetle records from a total of 187 species from Mongolia. Each record is associated with latitude and longitude. Many species are represented by one or only a few records (median = 6, standard deviation = 31.3). As for most insect groups, fine-grained distribution data are not available for Mongolian tenebrionids. Previous studies at the country scale, however, demonstrate that coarse-grained data are enough to identify environmental variables shaping species diversity and species composition of insects (Baselga, 2008; Fattorini and Baselga, 2012).

For community analyses, we extracted records of species across grid cells of equal area (1° x 1°) to avoid sampling errors associated with varying area sizes. We transformed the resulting grid cells into a presence-absence matrix where rows represent grid cells and columns represent species.

We excluded grid cells containing fewer than five species and species with fewer than five records because such small sample size can potentially cause misrepresentation in community ecology analyses such as similarity and range analyses.

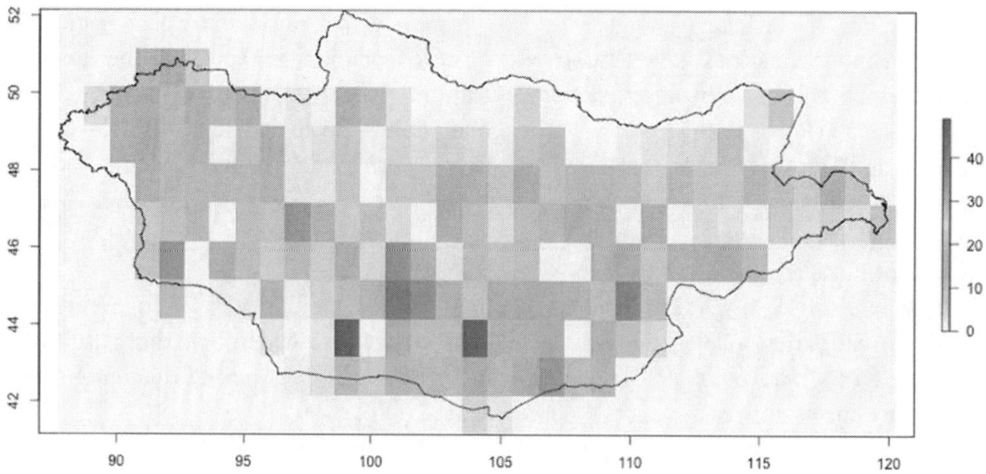

Figure 3. Map of study region (Mongolia) and distribution of 1° x 1° grid cell records; colors represent the species richness within grid cells.

These exclusion criteria left us a dataset including 130 grid cells and a total of 103 species belonging to 28 genera (Figure 3). Most analyses were restricted to this dataset.

We constructed two additional presence-absence matrices where rows represent Mongolian ecoregions and columns represent species: one for all Mongolian ecoregions (12 ecoregions) for which there was any tenebrionid record, and a subset only for steppe ecoregions (including forest steppe, steppe, and desert steppe: nine ecoregions).

Environmental Variables

We used the 19 WorldClim variables at 30 arc-seconds resolution as climate variables (Hijmans et al., 2005). We merged WorldClim raster data for global tiles 18 and 19 and extracted the result to a longitude and latitude extent of 87 to 120 degrees and 41 to 53 degrees, respectively.

The extent of Mongolia's administrative boundary is 87.7 to 119.9 degrees and 41.6 to 52.2 degrees (GADM, 2012). In addition, we extracted the altitude variable from Shuttle Radar Topography Mission (SRTM) images at 90-meter resolution (USGS 2004), and the Normalized Difference Vegetation Index (NDVI) variables from MODIS images at 250-meter resolution (Carroll et al., 2004). NDVI is strongly correlated with Net Primary Productivity and is used as surrogate of productivity in ecological research (Gschweng et al., 2012; Paknia and Pfeiffer, 2012).

We extracted values of these 21 environmental variables for all records of the 96 species and calculated their average values within each 1° X 1° grid cell (Table 2). Mean latitude and longitude of all beetle records within each cell were calculated and used as geographical variables.

Table 2. List of the environmental and geographical parameters, which have been entered into the random forest variable selection analysis. The selected variables were later regressed against species richness of darkling beetles. Note that the values in the table are for the whole Mongolia

	Abiotic parameters	Maximum	Mean	Minimum	Standard Deviation	Reference
1	Annual Mean Temperature (C°)	9.2	-0.1	-12.6	3.7	Hijmans et al., 2005
2	Mean Diurnal Range (C°)	16.2	13.7	10.6	0.79	
3	Isothermality[1]	3.2	2.7	2	0.23	
4	Temperature Seasonality[2]	18.2	13.5	9.5	1.4	
5	Max Temperature of Warmest Month (C°)	33.8	24.1	9.8	3.8	
6	Min Temperature of Coldest Month (C°)	-16.8	-26.5	-39.4	4.6	
7	Temperature Annual Range (C°)	62.3	50.6	38.7	3.7	
8	Mean Temperature of Wettest Quarter (C°)	25.3	16.3	-19.8	3.7	
9	Mean Temperature of Driest Quarter (C°)	10.3	-16.8	-31.9	4.8	
10	Mean Temperature of Warmest Quarter (C°)	26.3	17.4	3.3	3.6	
11	Mean Temperature of Coldest Quarter (C°)	-8.4	-18.3	-31.9	4.7	
12	Annual Precipitation (mm)	679	196	34	95	
13	Precipitation of Wettest Month (mm)	175	56	10	26	
14	Precipitation of Driest Month (mm)	12	1.6	0	1.3	
15	Precipitation Seasonality[3]	139	105	24	10	
16	Precipitation of Wettest Quarter (mm)	423	134	21	65	
17	Precipitation of Driest Quarter (mm)	39	6.2	0	4.2	
18	Precipitation of Warmest Quarter (mm)	423	134	21	65	
19	Precipitation of Coldest Quarter (mm)	49	6	0	4.43	
20	NDVI[4]	0.99	0.34	0.07	0.25	Carroll et al 2004
21	Altitude (m)	4202	1487	534	539	USGS 2004

[1](Mean diurnal range / temperature annual range) x 100.

[2]Standard deviation x 100.

[3]Coefficient of Variation.

[4]The Normalized Difference Vegetation index (ranges between 0 and 250).

Data Analyses

We plotted species accumulation curves (SACs) to determine our sampling coverage of tenebrionid beetles using grid cells as sampling units. We used the "random" method of the "specaccum" function in the R package *vegan* to calculate SACs (Oksanen et al., 2009). This method finds the mean SAC and its standard deviation from random permutations of data.

We used environmental variables in explanatory correlation analyses to determine variability in species richness across the environmental gradients. To test the significance of correlations of the response variable and environmental variables, we employed a modified t-test (Dutilleul, 1993) for partialling out the effect of spatial autocorrelation,, using SAM (Spatial Analyses in Macroecology) v. 4.0 software (Rangel et al., 2010). In the results, "corrected *P*" indicates the correction of the *P* value after factoring out the effect of spatial autocorrelation.

As distributional data usually show high degrees of spatial autocorrelation, which can affect multi-regression analyses of the data, we applied spatial autoregressive (SAR) models to account for spatial autocorrelation. To do this, we applied an estimation of Ordinary Least Square regression using SAM. SAR incorporates spatial autocorrelation using neighbourhood matrices, which specify the relationship between the residuals at each location and those at neighbouring locations (Dormann et al., 2007). Prior to model construction, sets of the most important variables were selected using random forest. Random forest is an efficient algorithm, which can be used for selecting important variables for regression analyses. This algorithm is effective even in the presence of complex interactions among many variables (Genuer et al., 2010). To do this, the "cforest" function of the R package party was used (Hothorn et al., 2012). Then the "varimp" function of the same package was used to select the most important environmental and geographical variables as predictor variables for regressing against darkling beetle species richness as response. All selected variables entered into the SAR models.

Nonmetric Multidimensional Scaling (NMDS) was used to examine the differences in beetle assemblage composition between steppe ecoregions. We conducted this analysis on the ecoregion community matrix. Ordination of beetle assemblages at the ecoregional scale was based on the Bray-Curtis dissimilarity measure and was computed with the *vegan* package. We used the Bray-Curtis dissimilarity as it often performs better than other dissimilarity indices (Oksanen et al., 2009).

We used two methods to calculate beta diversity among gridded samples. First, we used the relationship between geographic distance and similarity in species composition (gridded samples) as a measure of turnover. This relationship was calculated as a correlation coefficient using the Mantel test (Oksanen et al., 2009). Similarity among grid cells was measured by subtracting the Jaccard dissimilarity index from 1 (Koleff et al., 2003). Community similarity decreases smoothly with distance if the environment is completely homogenous, but abruptly if the environment is heterogeneous or there are major dispersal barriers. Second, we calculated the pairwise beta diversity among all gridded samples. To do this we also used the Jaccard dissimilarity index. This index can be partitioned into two components: spatial turnover and nestedness. Spatial turnover implies the replacement of some species by other species. Nestedness occurs when species in habitats with a lower number of species are subsets of the species set in habitats with a higher number of species (Baselga, 2010).

We calculated (i) turnover-fraction and (ii) nestedness-fraction of Jaccard pairwise dissimilarity. The pairwise measure of diversity was also conducted among ecoregions. Jaccard beta diversity index was used for both methods because it has been rated highly among 39 binary similarity indices tested by Shi (1993) and is one of the most commonly used in macroecology studies (e.g. Qian and Ricklefs, 2012).

Like the Mongolian ecoregions, the shapes of tenebrionid species ranges are characteristically elongate. We identified the two most distant sample points for each species and determined the great circle distance and initial bearing of the chord joining these two points. We call this chord the major axis of a species range. Great circle distance between points *i* and *j* was determined with the formula (Williams, 2012):

$$d_{ij} = \cos^{-1}(\sin(lat_i) \times \sin(lat_j) + \cos(lat_i) \times \cos(lat_j) \times \cos(long_i - long_j)) \times R$$

The mean radius of Earth, *R*, was set to 6,371 kilometers (Moritz, 2000). We determined the initial bearing of the chord joining points *i* and *j* with the formula (Williams, 2012):

$$b_{ij} = \tan^{-1}\left(\frac{\sin(long_i - long_j) \times \cos(lat_j)}{\cos(lat_i) \times \sin(lat_j) - \sin(lat_i) \times \cos(lat_j) \times \cos(long_i - long_j)}\right) \mod 2\pi$$

Once we determined the initial bearings for the major axis of species ranges, we compared those directions to the directions of maximum change in climate values. Using WorldClim extracted data, for each non-edge cell in our raster (1° x 1°) we identified its immediate neighboring cell with the largest absolute difference in mean climate value. We constructed a chord between the centers of these two cells and determined the initial bearing of that chord using the above formula (Figure S1). We performed this analysis for mean annual temperature and mean annual precipitation. To compare the directions of species ranges to the directions of maximum temperature and precipitation change we binned directions in intervals of 10 degrees and overlaid the resulting histograms. We performed the analysis a second time by enlarging the square annulus used to identify neighbors around the focal cell by one cell width (sixteen neighbors instead of eight). All calculations were performed in the R programming environment (R Development Core Team 2007). Raster manipulation and visualization were aided by the R package *raster* (Hijmans and van Etten, 2012).

RESULTS

We used 3280 records from 130 grid cells on Mongolian territory to produce sample-based richness estimates. The species richness curve is clearly asymptotic, which indicates that sampling has reached the maximum number of species observable under the conducted sampling effort and methodology (Figure 4). Two tested estimators (Chao and Jackknife 1; Figure S2 in appendix) achieved the same number of observed species richness (103 species), suggesting that the observed number of species is close to the complete fauna of sampled grids.

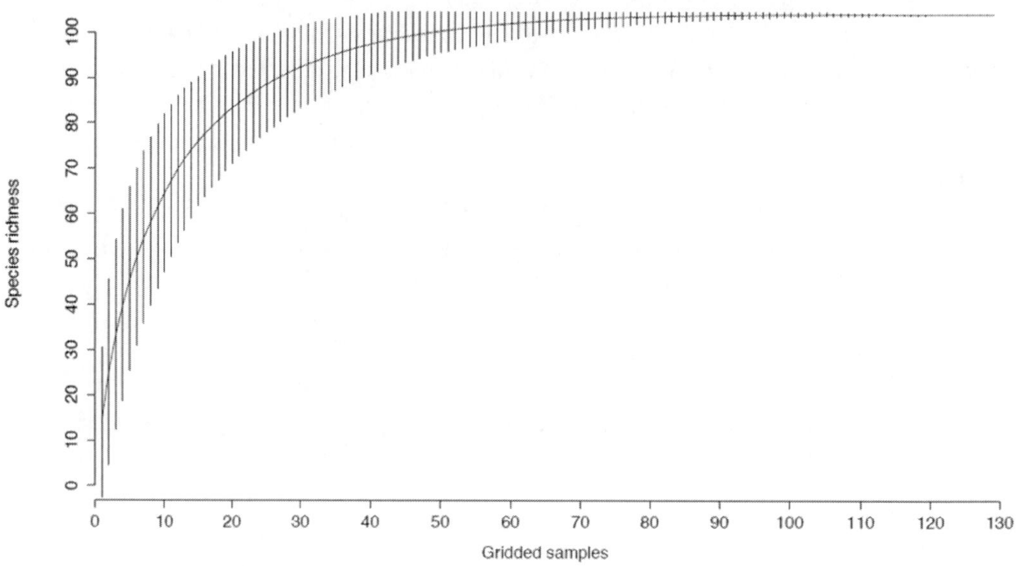

Figure 4. Sample based species accumulation curve of darkling beetles collected from Mongolia; axes X and Y represent number of gridded samples and number of species after the minimum 5 records/species criterion, respectively.

The five most recorded species were *Microdera kraatzi* (177 records), *Scytosoma pygmaeum* (161 records), *Platyscelis brevis* (142 records), *Blaps medusula* (124 records), and *Anatolica gobiensis* (122 records).

Darkling beetle species richness was highly significantly associated with latitude (r = -0.36, t = -4.409, df= 128, corrected P = 0.01; Figure 5), but did not show any association with longitude (r = -0.01, t = -0.155, df = 128, corrected P = 0.9). Among environmental variables, species richness showed positive and strong correlation with maximum (r =0.31, t = 3.687, df = 128, corrected P < 0.018; Figure 5) and minimum temperature (r= 0.29, t = 3.479, df = 128, corrected P = 0.05; Figure 5). Species richness also showed a significant negative association with annual precipitation (r = -0.30, t= -4.024, df = 128, corrected P = 0.006; Figure 5). There is a significant correlation between annual precipitation and minimum temperature (r = -0.45, t = -5.723, df = 128, corrected P = 0.06), indicating that the colder regions receive higher precipitation.

Random forests analysis ranked the environmental variables based on their importance for the multivariate regression analysis as the following (four most important variables are listed): mean temperature of coldest quarter, temperature seasonality, minimum temperature, and maximum temperature. We entered these four predictor variables in the spatial autoregressive analysis. The constructed model was significant (n= 129, F = 6.118, P < 0.001). Three out of four predictor variables were significant in the SAR model: Temperature seasonality, maximum temperature and mean temperature of the coldest quarter (Table 3). In total all four environmental variables explained only 16.5% of variation. Environmental variables and spatial variation together explained almost the same, 16.6%.

Faunistic similarity between sites declined significantly as geographic distance increased (r = -0.37, t = -27.547, df = 4752, P < 0.0001; Figure 6). The average community similarity measured by the Jaccard index was 0.12 (range = 0.0 – 1.0).

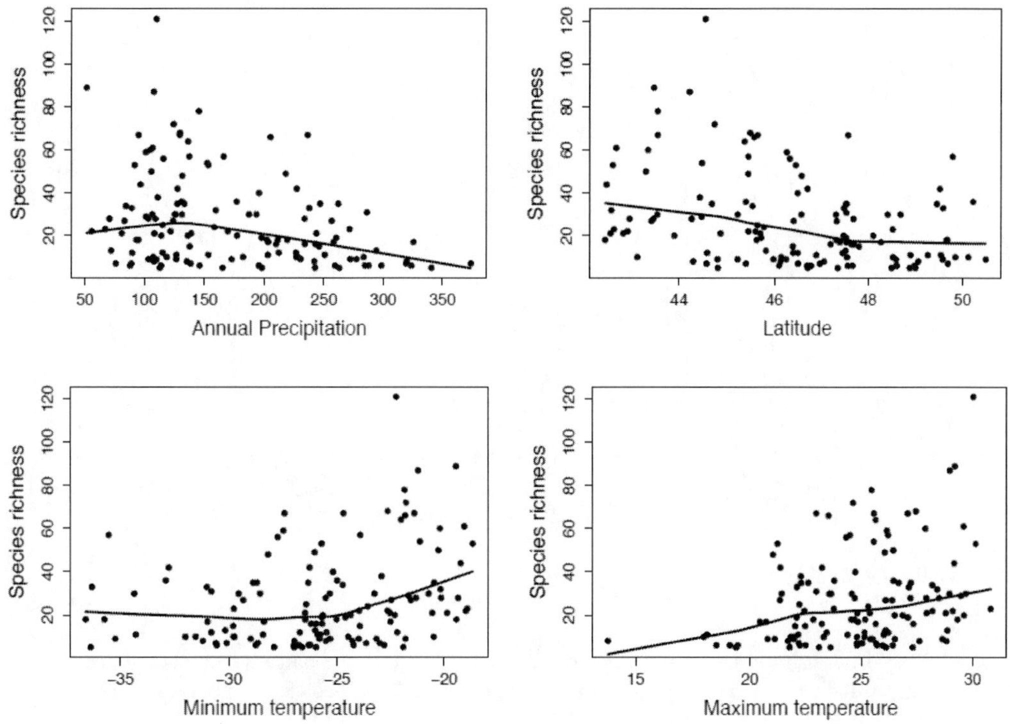

Figure 5. Correlation of tenebrionid beetle species richness with important environmental and geographical variables: annual precipitation (above left), latitude (above right), minimum temperature (below left) and maximum temperature (below right). A loess smoothing line was fitted to each correlation graph to show the detailed trend of species richness across environmental and geographical gradients.

Table 3. Ordinary least square (OLS) and SAR parameter values, standard errors and associated probability levels of the complete spatial autoregressive model for darkling beetle species richness with four environmental variables selected by random forests

Variable	OLS Coefficient	SAR Coefficient	Standard Coefficient	Standard error	t	*P*
Temp. seasonality	0.043	0.043	2.885	0.019	2.258	0.026
Minimum temp.	-4.154	-4.043	-0.8	2.982	-1.356	0.178
Maximum temp.	-13.903	-13.861	-1.93	6.869	-2.018	0.046
Mean temp. cold quarter	21.741	21.539	4.259	8.989	2.396	0.018

The Mantel test also indicated a significant negative correlation between beetle assemblage similarity and distance between grid cells (Mantel test: $r = -0.22$, $P < 0.001$).

Average pairwise beta diversity, and one of its components (spatial turnover) was not associated with latititude. Average nestedness increased with latitude, although the result was only marginally significant ($r = 0.16$, t = 1.841, df = 129, $P = 0.067$; Figure 7). All three components (beta diversity, spatial turnover and nestedness) showed a strong and significant correlation with longitude.

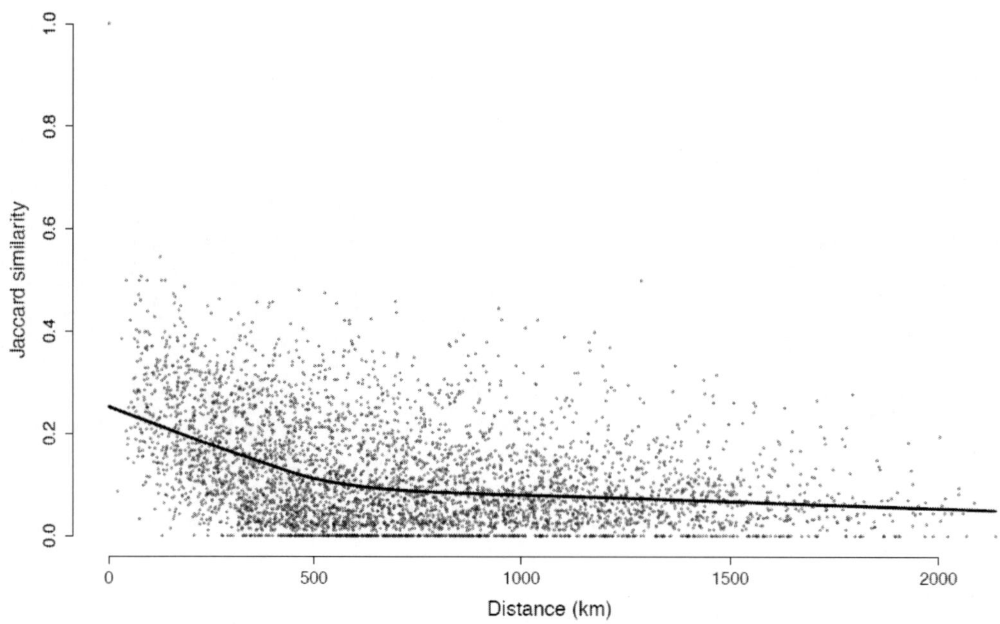

Figure 6. The pairwise similarity of all sampled grids plotted against the pairwise Euclidean distance between them. A distance-decay relationship for Mongolian darkling beetle assemblages.

Beta diversity and spatial turnover showed negative correlations (beta diversity: $r = -0.43$, $t = -5.414$, $df = 129$, $P < 0.0001$; spatial turnover; $r = -0.39$, $t = -4.847$, $df = 129$, $P < 0.0001$; Figure 7) and nestedness showed a positive correlation ($r = 0.24$, $t = 2.774$, $df = 129$, $P < 0.001$; Figure 7).

Species richness of darkling beetles in steppe ecoregions of Mongolia ranged from 26 species in the Daurian Forest Steppe ecoregion to 95 species in the Alashan Plateau Semi-desert ecoregion (Figure 8). Only two species (*Blaps rugosa*, and *Platyscelis brevis*) occurred in all nine steppe ecoregions. Eight more species appeared to be generalists and occurred in eight of nine steppe ecoregions. In contrast 63 species occurred only in one of the steppe ecoregions of Mongolia. NMDS biplot ordination (stress = 0.21) showed a cluster of Selenge Orkhon and Daurian Forest steppe ecoregions. Altai Montane Forest and Forest Steppe clustered with Junggar Basin semi-desert. Three desert steppe areas (Eastern Gobi and Gobi Lakes Valley and Alashan Plateau) clustered in one group. Great Lakes Basin Desert Steppe and Mongolian Manchurian Grassland did not cluster with any other ecoregion (Figure 9).

Beta diversity and its components (spatial turnover and nestedness) differed noticeably among the Mongolian steppe ecoregions. Nevertheless, a general pattern was obvious: between-ecoregion pairwise beta diversity was high (mean $\beta_j = 0.72$; range = 0.64 – 0.82), due mainly to spatial turnover (mean $\beta_t = 0.59$; range = 0.50 – 0.71) rather than nestedness (mean $\beta_n = 0.13$; range = 0.1 – 0.19).

The lowest pairwise beta diversity between ecoregions occurred between the adjacent Gobi Lake Valley Desert Steppe and the Eastern Gobi Desert Steppe ($\beta_j = 0.45$) and the highest pairwise beta diversity occurred between the disjoint Daurian Forest Steppe and the Alashan Plateau Semi Desert ($\beta_j = 0.92$). Junggar Basin Semi Desert showed the highest mean value of pairwise beta diversity and Gobi Lake Valley Desert Steppe showed the lowest value of mean pairwise beta diversity (Figure 10).

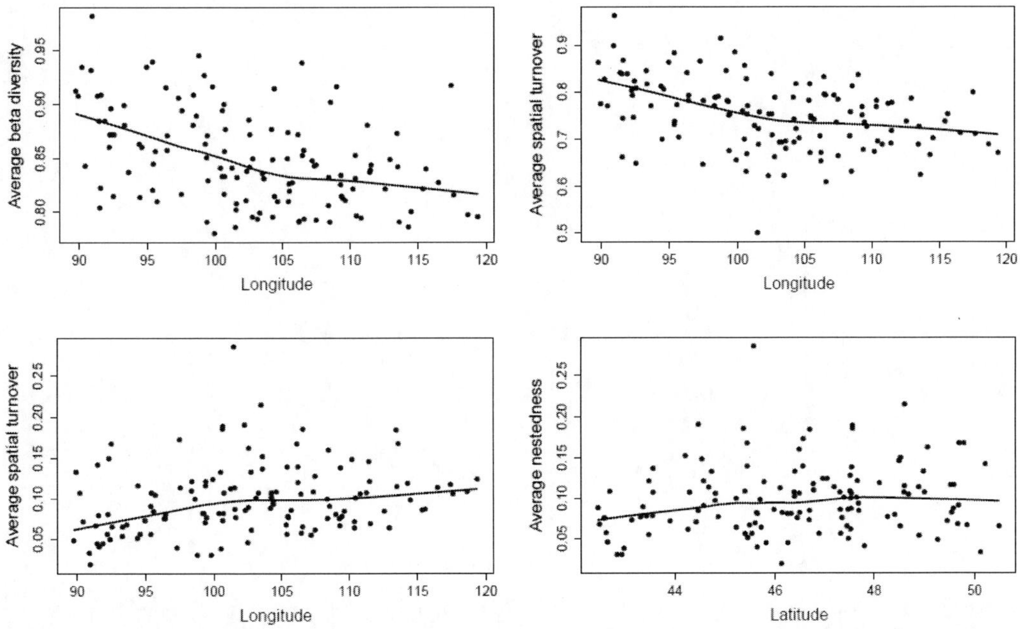

Figure 7. Correlation of average beta diversity (above left), spatial turnover (above right) and nestedness (below left) of tenebrionid beetles with longitude and correlation of average nestedness of tenbrionid beetles with latitude (below right). A loess smoothing was fitted on each correlation graph to show the detailed trend of species richness across environmental and geographical gradients.

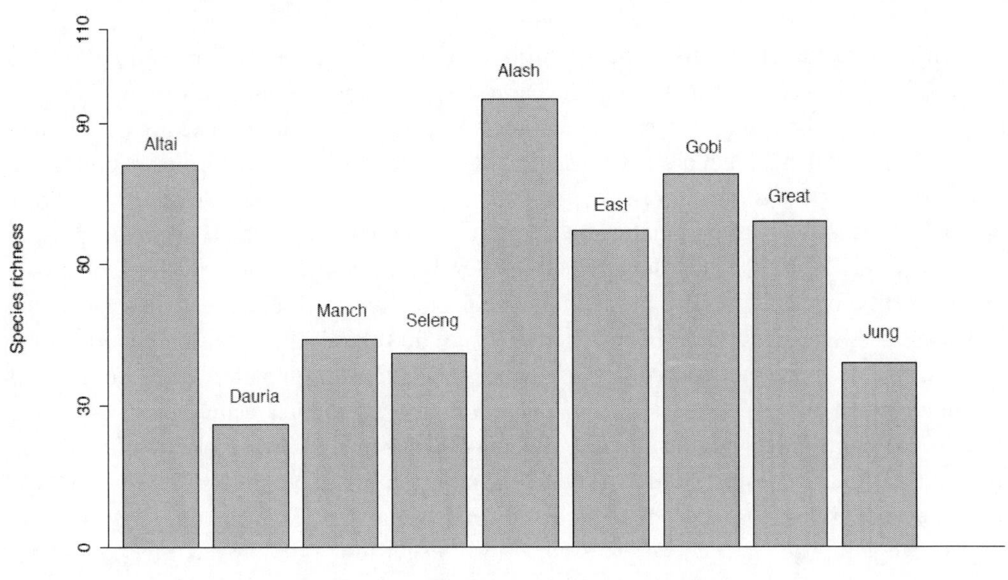

Figure 8. Inter-steppe ecoregion differences in species richness of tenebrionid beetles in Mongolia; ecoregion abbreviations are as follows: Altai: Altai Montane Forest, Dauria: Daurian Forest Steppe, Manch: Mongolian Manchurian Grassland, Seleng: Selenge Orkhon Forest Steppe, Alash: Alashan Plateau Semi Desert, East: Eastern Gobi Desert Steppe, Gobi: Gobi Lake Valley Desert Steppe, Great: Great Lakes Basin Desert Steppe, Jung: Junggar Basin Semi Desert.

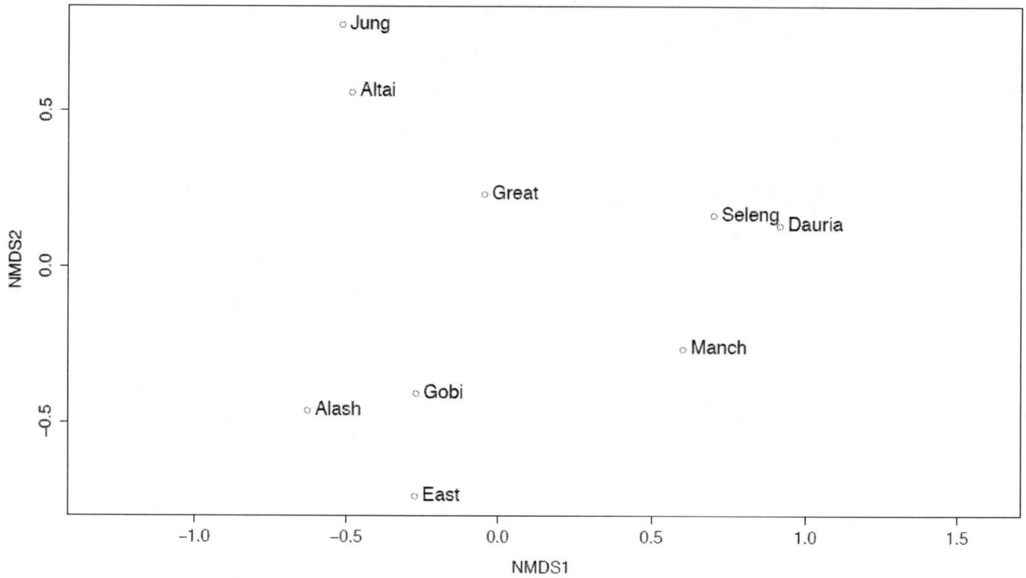

Figure 9. Non-metric multidimensional scaling (NMDS) ordination (first two axes) of nine steppe ecoregions of Mongolia according to their dissimilarity in beetle species assemblage (Bray Curtis index for presence-absence data; stress = 0.21); Ecoregion abbreviations are as follows: Altai: Altai Montane Forest, Dauria: Daurian Forest Steppe, Manch: Mongolian Manchurian Grassland, Seleng: Selenge Orkhon Forest Steppe, Alash: Alashan Plateau Semi Desert, East: Eastern Gobi Desert Steppe, Gobi: Gobi Lake Valley Desert Steppe, Great: Great Lakes Basin Desert Steppe, Jung: Junggar Basin Semi Desert.

When the overall beta diversity was partitioned into its components, it became clear that patterns of species replacement made a large contribution to beta diversity between ecoregions. The lowest species replacement occurred between Gobi Lake Valley Desert Steppe and Mongolian Manchurian Grassland (β_t= 0.31). The highest species replacement occurred between the Daurian Forest Steppe and the Junggar Basin Semi Desert (β_t= 0.87). Similar to the beta diversity pattern, the mean value of the pairwise spatial turnover (species replacement) was highest in the Junggar Basin Semi Desert and lowest in the Gobi Lake Valley Desert Steppe (Figure 10). Nestedness of species was lowest between Junggar Basin Semi Desert and Selenge Orkhon Forest Steppe (β_n= 0.004), indicating that the beta diversity between these two steppe ecoregions is generated almost completely due to species replacement rather than species loss from one ecoregion to another ecoregion. Nestedness (species loss) contributed largely to beta diversity between the Gobi Lake Valley Desert Steppe and the Daurian Forest Steppe (β_n= 0.32), indicating that beta diversity between these two steppe ecoregions has been generated due to loss of species from one ecoregion to another ecoregion. The Daurian Forest Steppe and the Eastern Gobi Desert Steppe had the highest and the lowest values of mean nestedness (Figure 10), indicating that the greatest species loss occurs between Daurian Forest Steppe and other steppe ecoregions and the lowest species loss occurs between Eastern Gobi Desert Steppe ecoregion and other ecoregions, respectively.

The initial bearings of the major axes of species ranges were bimodal with modes near 90 and 290 degrees. In contrast, the direction of maximum temperature and precipitation change was multimodal.

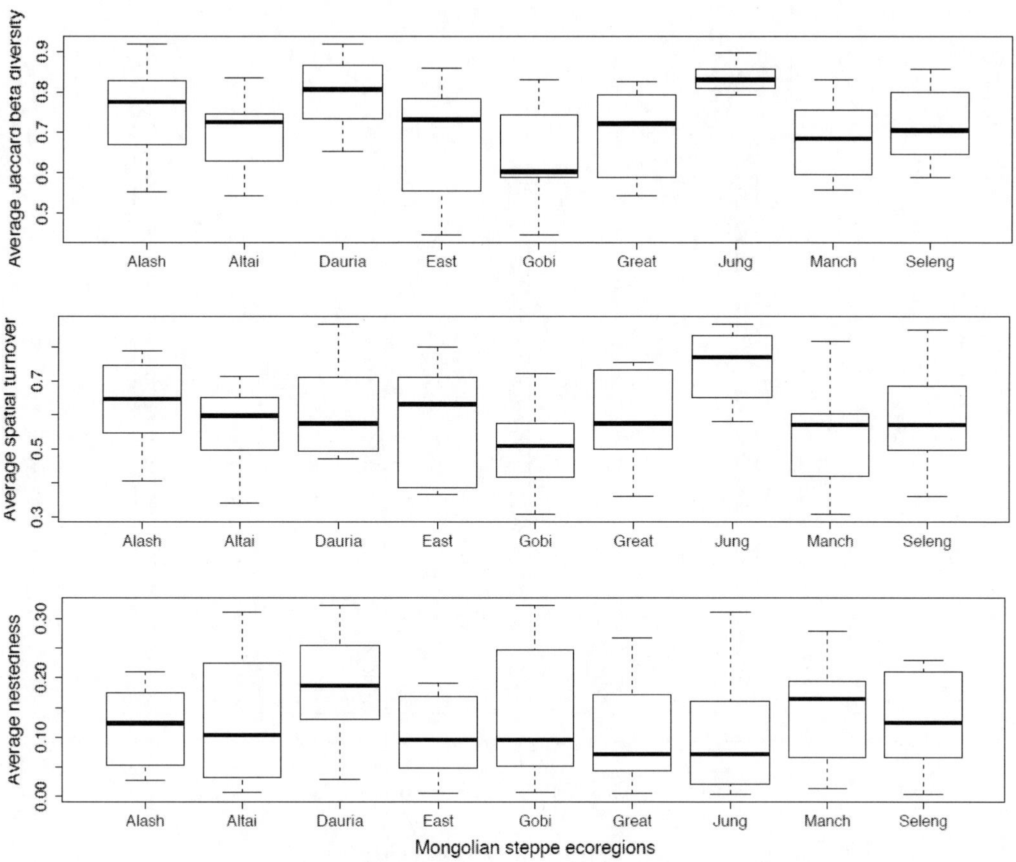

Figure 10. Inter-ecoregional pairwise Jaccard beta diversity (above), spatial turnover (middle) and nestedness (below) mean values for the Mongolian steppe ecoregions; Ecoregion abbreviations are as follow: Alash: Alashan Plateau Semi Desert, Altai: Altai Montane Forest, Dauria: Daurian Forest Steppe, East: Eastern Gobi Desert Steppe, Gobi: Gobi Lake Valley Desert Steppe, Great: Great Lakes Basin Desert Steppe, Jung: Junggar Basin Semi Desert, Manch: Mongolian Manchurian Grassland, Seleng: Selenge Orkhon Forest Steppe.

Results were qualitatively identical for mean annual temperature and precipitation at both window sizes, and we illustrate only mean annual temperature at the smallest window size. The two modes of species ranges are out of phase with the modes of maximum temperature change, indicating that most species distributions follow the axes of minimal environmental variation (Figure 11).

Discussion

In this chapter we examined how species richness of Mongolian tenebrionid beetles varies across environmental and geographical gradients and how diversity is generated among a wide range of steppe ecoregions. Specimen records were concentrated along main roads (Figure S3 in Appendix) and were unevenly distributed among ecoregions. Although most specimens were collected from the Alashan Plateau semi-desert ecoregion, vast areas of this ecoregion have remained unsampled.

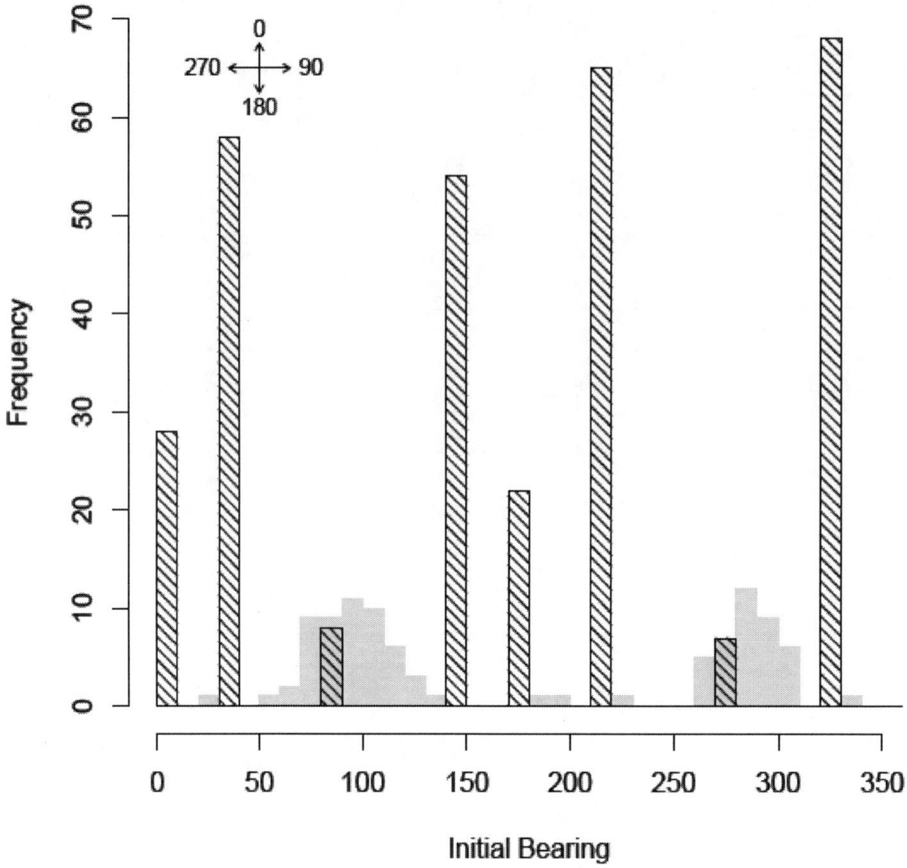

Figure 11. Histograms depict the direction of maximum change in mean annual temperature, hatched bars, and the direction of the major axis of sampled species distributions, gray bars. Compass rose arrows point in the four cardinal directions, which are shown as their numeric values on the X-axis.

The same is true for the Mongolian Manchurian grassland ecoregion, which is the largest Mongolian ecoregion. Using a criterion of "five records per cell – five records per species", however, increased the quality of the data significantly. Such a cut-off criterion has been used to increase the quality of museum data in ecological research (Kreft and Jetz, 2010; Hof et al., 2012). Our sampled-based species accumulation curve analysis supported the quality of our dataset. Furthermore, there was no difference between observed species richness and estimated species richness calculated with two independent estimators (Figure S2 in Appendix). Tenebrionid beetle species richness decreased as latitude increased, the most commonly observed distributional pattern among many insect taxa in different habitats and at different spatial scales (Baselga, 2008; Fattorini and Ulrich, 2012; Labidi et al., 2012). Pfeiffer and colleagues (2003) used a systematic sampling approach and did not observe a significant decrease in ant species richness at higher latitudes in Mongolia, but did observe a decrease in tenebrionid species richness (Pfeiffer and Enkhmunkh, in press). In Mongolia, latitude is strongly correlated with low temperatures (the mean temperature of the coldest quarter and the minimum temperature of the coldest month), which suggests that the correlation between species richness of tenebrionid beetles and latitude is in part a reflection of the temperature decrease.

The major axes of species ranges are roughly perpendicular to the directions of steepest change in temperature. This suggests that temperature has played an important role in the evolutionary history of tenebrionid species and communities and that most species occupy a small subset of the total temperature range experienced by the entire set of species. In fact, the range in maximum and minimum temperatures of the warmest and coldest months experienced by all species is 19.6 and 21.2 degrees Celsius, respectively. The median range for individual species, however, is, respectively, only 8.4 and 9.4 degrees Celsius. Regression analysis also stresses the role of temperature in determining species richness of darkling beetles in our study system.

All four predictor variables selected by the random forests algorithm were temperature attributes. Three of them entered the model significantly: temperature seasonality expresses the seasonal fluctuation in temperature, and maximum temperature and mean temperature of the coldest quarter represent the extremes that could positively or negatively affect beetle species richness (Figure 5). Our regression model, however, explained only a small portion of the variation in beetle species richness, indicating that there are unmeasured and unaccounted factors, which could have a large influence on the observed pattern.

Other environmental factors such as soil chemistry and structure may significantly affect composition and richness of tenebrionid beetle assemblages (Kasanov and Shenbrot, 1996). Spatial variation in resource quality and its effect on reproduction and larval survival is important in other insect groups (Price, 2003). Environmental variables explained little variation in tenebrionid species richness in South America as well (Carrara et al., 2011).

These species-temperature and temperature-latitude correlations imply, and our results demonstrate, that tenebrionid species richness is greater in Mongolian regions of pronounced aridity and low productivity. In Patagonia, Sackmann and Flores (2009) found greater tenebrionid species richness in steppe habitat than in the more productive forest and scrub habitat. In our study system annual precipitation and primary productivity peak in the coldest regions of northern Mongolia.

Although it appears that these regions offer more resources and food, the extremely cold winters most likely prevent the colonization of many beetle species.

This strong co-variation between low temperatures, precipitation and latitude also prevented Pfeiffer et al. (2003) from reporting any correlation between ant species richness and precipitation.

Beetles of the family Tenebrionidae are behaviorally and physiologically well adapted to arid environments (reviewed by Zachariasson, 1996; Cloudsley-Thompson, 2001). They are abundant macrodetritivores and prey to many ectothermic and endothermic predators in low productivity environments (Ayal, 2007). They can escape harsh conditions by entering diapause or torpor (Cloudsley-Thompson, 2001). A comparable study to ours in South America demonstrated a negative relationship between actual evapotranspiration (water availability) and tenebrionid species richness (Carrara et al., 2011).

However, in contrast to our findings they found a negative association between potential evapotranspiration (temperature) and tenebrionid beetles. Overall, these findings are opposite to those documented in most plant, vertebrate and invertebrate groups (reviewed in Hawkins et al., 2003, although that review does not include Tenebrionidae). What factors are responsible for these broad patterns of species richness?

In the case of Patagonia steppe Tenebrionidae, Sackmann and Flores (2009) suggested that higher species richness of some species groups in steppe habitat corresponded to the evolutionary origin of those groups in steppe habitat.

This mechanism is also reported to explain many latitudinal diversity gradients (reviewed in Wiens and Graham, 2005). Using a larger data set from the same region (Patagonia, South America), Carrara et al. (2011) stress that the adaptation of Tenebrionidae to arid habitats confounds predictions resulting from species energy theory (Wright, 1983).

Colder and more productive environments may be difficult for tenebrionid lineages to invade because of ancestral physiological adaptation to aridity or lack of adaptation to extreme cold. Desert lineages of beetles, rodents, and birds all have lower metabolic rates than non-desert lineages (Zachariasson, 1996; Mueller and Diamond, 2001; Williams and Tieleman, 2002). Investigating potential trade-offs between physiological adaptations for living in hot, arid habitats and physiological adaptations for cold-hardiness would help reveal the ecological limits that restrict tenebrionid diversity in cold, wet environments compared to hot, dry environments.

Beetle assemblage similarity among gridded samples was low and decreased strongly over the first 500 km, indicating a large amount of turnover with increasing distance. Three processes are suggested to explain distance decay in community similarity: (i) species sorting due to decreasing similarity in environmental features, (ii) dispersal limitation due to the nature and configuration of the landscape, and (iii) random dispersal and random speciation and extinction in a completely homogenous environment (neutrality) (Soininen and Hillebrand, 2007; Soininen et al., 2007b).

Our distance decay pattern does not reveal explicitly the underlying mechanism or mechanisms for decreasing similarity with distance. However, we sampled beetles across very heterogeneous environments, so a neutrality-based explanation is probably of little importance. Further analyses are needed to disentangle the roles of the other two processes because tenebrionid beetles show strong environmental preferences and are considered a strongly dispersal-limited group (Fattorini and Baselga, 2012). Latitudinal gradients in beta diversity are common (Soininen et al., 2007a). We partitioned beta diversity into spatial turnover and nestedness. Latitude showed a weak positive correlation with nestedness, indicating that species communities at higher latitudes are subsets of species communities at lower latitudes.

A similar pattern was documented in European longhorn beetles (Baselga, 2008). Mean values of beta diversity declined from the west towards the east. Interestingly, spatial turnover and nestedness showed opposing patterns. Spatial turnover (species replacement) decreased from the west towards the east.

In contrast, species loss increased from the west towards the east. Thus, although Mongolian tenebrionid beetle species richness (alpha diversity) stays almost constant from the west towards the east, eastern assemblages are subsets of the western assemblages and show lower mean values of beta diversity. This underscores the high ground beetle diversity of the Mongolian Altai region, which may be a historic centre of speciation (Dudko et al., 2010).

Large values of inter-ecoregional Jaccard beta diversity indicate that steppe ecoregions contain exclusive assemblages. Many species are restricted to only one ecoregion. As a result, a large fraction of species richness and diversity of tenebrionid beetles is generated at the ecoregional scale.

Similar patterns have been found in tenebrionid beetles of South America and Europe (Carrara and Flores, 2012; Fattorini and Baselga, 2012) and in other dispersal limited insects, such as ants in Central Persian desert steppe regions (Paknia and Pfeiffer, 2011). Steppe ecoregions clearly segregate in the NMDS ordination space. To a certain extent, neighbouring ecoregions grouped together, suggesting that the similarity of ecoregional assemblages is correlated with distance.

However, this was not the case for the Junggar Basin Semi Desert and the Alashan Plateau Semi Desert, which were separated from each other along the second ordination axis. The southern foothills of the Altai Mountains separate these two semi deserts, and in the ordination space the Altai Forest and Forest Steppe sits between them. The Mongolian Altai may constitute a historical border between both ecoregions. Climatic parameters were not correlated with the ordination axes, although our other results demonstrate their importance at the scale of grid cells.

Conclusion

Beetles are a keystone insect in steppe ecosystems because of their role in nutrient cycling and decomposition. This chapter is a contribution to the diverse but poorly known Mongolian tenebrionid beetle fauna.

We demonstrate unique patterns of species richness and high species turnover among sample sites and ecoregions. Temperature variables emerge as limiting factors of species distributions and of species richness. Future research should address the mechanisms that can generate the patterns we observed.

Appendix

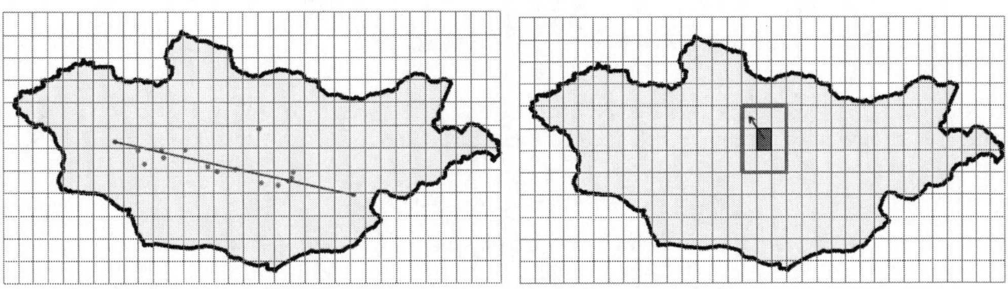

Figure S1. The left panel depicts the calculation of a distribution chord for a single species. Sample records are shown as red points and the blue segment connects the two points separated by the farthest great circle distance. We calculated the initial bearing of this chord and repeated the procedure for each species. The right panel depicts the calculation of a single environmental chord. The focal grid cell is colored in red and its 8 nearest neighbors are circumscribed by the red polygon. The blue vector points from the center of the focal cell to the center of the neighbor cell that experiences the greatest absolute difference in climate value (mean annual precipitation or mean annual temperature). We calculated the initial bearing of this vector and repeated the procedure for each grid cell.

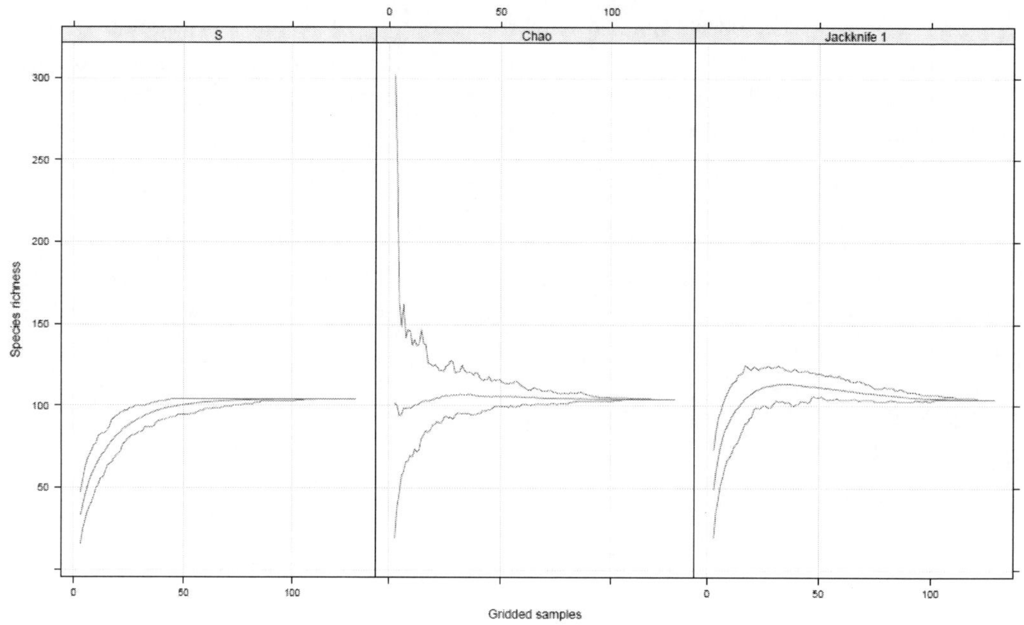

Figure S2. Extrapolated estimation of darkling beetle species richness using two (Chao and Jackknife) estimators.

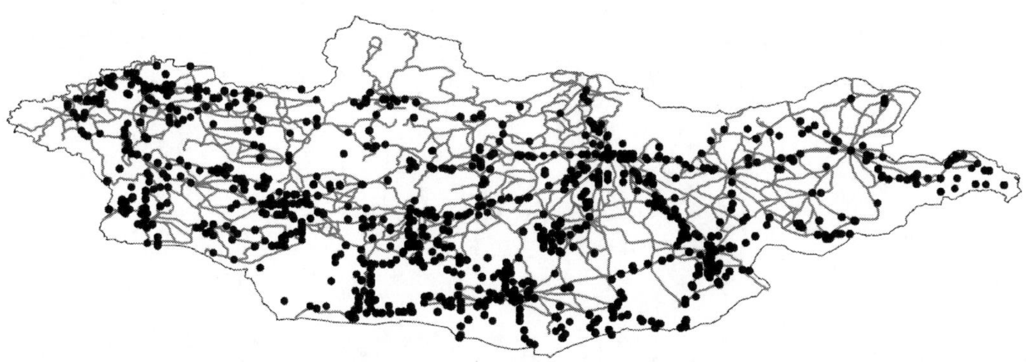

Figure S3. Distribution of darkling beetle records (black dots) and roads (grey lines) in Mongolia.

ACKNOWLEDGMENTS

We are grateful to Andrei Lobanov for sharing the data from the publication Medvedev and Lobanov (1990).

REFERENCES

Ayal, Y. (2007). Trophic structure and role of predation in shaping hot desert communities. *Journal of Arid Environments* 68, 171-187.

Baselga, A. (2008). Determinants of species richness, endemism and turnover in European longhorn beetles. *Ecography* 31, 263-271.

Baselga, A. (2010). Partitioning the turnover and nestedness components of beta diversity. Global *Ecology and Biogeography* 19, 134-143.

Carrara, R., Flores, G. E. (2012). Endemic tenebrionids (Coleoptera: Tenebrionidae) from the Patagonian steppe: a preliminary identification of areas of micro-endemism and richness hotspots. *Entomological Science.*

Carrara, R. D. P. Vázquez, Flores, G. E. (2011). Habitat specificity can blur the predictions of species–energy theory: A case study of tenebrionid beetles adapted to aridity. *Journal of Arid Environments* 75, 703-710.

Carroll, M. L., DiMiceli, C. M. R., Sohlberg, A., Townshend, J. R. G. (2004). *250m MODIS Normalized Difference Vegetation Index*, 250ndvi28920033435, Collection 4. University of Maryland, College Park, Maryland.

Cloudsley-Thompson, J. L. (2001). Thermal and water relations of desert beetles. *Naturwissenschaften* 88, 447-460.

Crist, T. O. (2008). Insect populations, community interactions and ecosystem processes in the shortgrass steppe. In: Lauenroth, W. K., Burke, I. C. (eds.) *Ecology of the Shortgrass Steppe: A Long-Term Perspective*, pp. 215-247. Oxford University Press, Oxford.

Da Silva, P. M., Aguiar, C. A. S., Niemelä, J., Sousa, J. P., Serrano, A. R. M. (2009). Cork-oak woodlands as key-habitats for biodiversity conservation in Mediterranean landscapes: a case study using rove and ground beetles (Coleoptera: Staphylinidae, Carabidae). *Biodiversity and Conservation* 18, 605-619.

Dormann, C. F., McPherson, J. M., Araujo, M. B., Bivand, R., Bolliger, J., Carl, G., Davies, R. G., Hirzel, A., Jetz, W., Kissling, W. D., Kühn, I., Ohlemüller, R., Peres-Neto, P. R., Reineking, B., Schröder, B., Schurr, F. M., Wilson., R. (2007). Methods to account for spatial autocorrelation in the analysis of species distributional data: a review. *Ecography* 30, 609-628.

Dudko, R. Y., Matalin, A. V., Fedorenko, D. N. (2010). The ground beetle fauna (Coleoptera, Carabidae) of Southeastern Altai. *Entomological Review* 90, 968-988.

Dutilleul, P. (1993). Modifying the t-test for assessing the correlation between two spatial processes. *Biometrics* 49, 305-314.

Fattorini, S. (2002). Biogeography of the tenebrionid beetles (Coleoptera, Tenebrionidae) on the Aegean Islands (Greece). *Journal of Biogeography* 29, 49-67.

Fattorini, S. (2006). Detecting biodiversity hotspots by species-area relationships: a case study of Mediterranean beetles. *Conservation Biology* 20, 1169-1180.

Fattorini, S., Baselga, A., (2012). Species richness and turnover patterns in European tenebrionid beetles. *Insect Conservation and Diversity* 5, 331-345.

Fattorini, S., Ulrich, W. (2012). Drivers of species richness in European Tenebrionidae (Coleoptera). *Acta Oecologica* 43, 22-28.

GADM. (2012). *GADM database of global administrative areas*. URL http://www.gadm.org.

Genuer, R., Poggi, J. M., Tuleau-Malot, C. (2010). Variable selection using random forests. *Pattern Recognition Letters* 31, 2225-2236.

Grubov, V. I. (1982). *Key to the Vascular Plants of Mongolia.* Nauka. Leningrad. [in Russian].

Gschweng, M., Kalko, E. K. V., Berthold, P., Fiedler, W., Fahr., J. (2012). Multi-temporal distribution modelling with satellite tracking data: predicting responses of a long-distance migrant to changing environmental conditions. *Journal of Applied Ecology* 49, 803-813.

Hawkins, B. A., Field, R., Cornell, H. V., Currie, D. J., Guegan, J. F., Kaufman, D. M., Kerr, J. T., Mittelbach, G. G., Oberdorff, T., O'Brien, E. M., Porter, E. E., Turner, J. R. G. (2003). Energy, water, and broad-scale geographic patterns of species richness. *Ecology* 84, 3105-3117.

Hijmans, R., Cameron, S., Parra, J., Jones, P., Jarvis, A. (2005). Very high resolution interpolated climate surfaces for global land areas. *International Journal of Climatology* 25, 1965-1978.

Hijmans, R. J., van Etten, J. (2012). *Raster: Geographic analysis and modeling with raster data.* R package version 2.0-12. URL http://CRAN.R-project.org/package=raster.

Hilbig, W. (1995). *The Vegetation of Mongolia.* SPB Academic Publishing. Amsterdam.

Hof, C., Brändle, M., Dehling, D. M., Munguía, M., Brandl, R., Araújo, M. B., Rahbek, C. (2012). Habitat stability affects dispersal and the ability to track climate change. *Biology Letters.*

Hothorn, T., Hornik, K., Strobl, C., Zeileis, A. (2012). *Package Party: A Laboratory for Recursive Partytioning.* URL http://CRAN.R-project.org/party

Kaszab, Z. (1965). Tenebrionidae Coleoptera Collected by the Polish Expeditions to Mongolia and China in the Years 1962-1963 Anatolica-Pseudacta New Record 70 Known Species. *Fragmenta Faunistica* 11, 417-430.

Kaszab, Z. (1967). Die Tenebrioniden der Westmongolei. *Acta Zoologica Academiae Scientiarum Hungaricae* 13, 279 - 351.

Kaszab, Z. (1977). Results of the Mongolian German biological expeditions since 1962, part 76: complementary data on the fauna of Tenebrionidae of western Mongolia (Coleoptera). *Mitteilungen aus dem Zoologischen Museum in Berlin* 53, 277-280.

Koleff, P., Gaston, K. J., Lennon, J. J. (2003). Measuring beta diversity for presence-absence data. *Journal of Animal Ecology* 72, 367-382.

Konstantinov, A. S., Korotyaev, B. A., Vollkovitsh. (2009). Insect biodiversity in the Palearctic region. In: Foottit, R., Adler, P. (eds.), *Insect Biodiversity: Science and Society*, pp. 107-162. Blackwell Publisher, Chinchester.

Krasnov, B. R., Shenbrot, G. I. (1996). Spatial structure of community of darkling beetles (Coleoptera: Tenebrionidae) in the Negev Highlands, Israel. *Ecography* 19, 139-152.

Krasnov, B. R., Shenbrot, G. I. (1998). Structure of communities of ground-dwelling animals at the junction of two phytogeographic zones. *Journal of Biogeography* 25, 1115-1131.

Kreft, H., Jetz, W. (2010). A framework for delineating biogeographical regions based on species distributions. *Journal of Biogeography* 37, 2029-2053.

Labidi, I., Errouissi, F., Nouira, S. (2012). Spatial and temporal variation in species composition, diversity, and structure of mediterranean dung beetle assemblages (Coleoptera: Scarabaeidae) across a bioclimatic gradient. *Environmental Entomology* 41, 785-801.

Liu, J. L., Li, F. R., Liu, C. A., Liu, Q. J. (2012). Influences of shrub vegetation on distribution and diversity of a ground beetle community in a Gobi desert ecosystem. *Biodiversity and Conservation* 21, 2601-2619.

McIntyre, N. E. (2000). Community structure of *Eleodes* beetles (Coleoptera: Tenebrionidae) in the shortgrass steppe: scale-dependent uses of heterogeneity. *Western North American Naturalist* 60, 1-15.

Medvedev, G. S. (1990). *Keys to the darkling beetles of Mongolia*. USSR Academy of Science. Leningrad. [in Russinan].

Medvedev, G. S., Lobanov, A. L. (1990). A faunistic list of tenebrionids (Coleoptera, Tenebrionidae) of the Mongolian People's Republic with coordinates of the localities. *Nasekomye Mongolii* 11, 139-204.

Moritz, H. (2000). Geodetic reference system 1980. *Journal of Geodesy* 54, 395-405.

Mueller, P., Diamond, J. (2001). Metabolic rate and environmental productivity: well-provisioned animals evolved to run and idle fast. *Proceedings of the National Academy of Sciences* 98, 12550-12554.

Oksanen, J., Kindt, R., Legendre, P., O'Hara, B., Simpson, G. L., Peter, S., Stevens, M. H. H., Wagner, H. (2009). *Vegan: community ecology package*. R package version 1.15-4. (http://cran.r-project.org/).

Paknia, O., Pfeiffer, M., (2011). Steppe versus desert: multi-scale spatial patterns in diversity of ant communities in Iran. *Insect Conservation and Diversity* 4, 297-306.

Paknia, O., Pfeiffer, M. (2012). Productivity alone does not explain species richness of ants – An example from Central Persian deserts. *Journal of Arid Environments* 85, 86-92.

Parmenter, R. R., Parmenter, C. A., Cheney, C. D. (1989). Factors influencing microhabitat partitioning in arid-land darkling beetles (Tenebrionidae): temperature and water conservation. *Journal of Arid Environmenst* 17, 57-67.

Pfeiffer, M., Chimedregzen, L., Ulykpan, K. (2003). Community organization and species richness of ants (Hymenoptera: Formicidae) in Mongolia along an ecological gradient from steppe to Gobi desert. *Journal of Biogeography* 30, 1921-1935.

Pfeiffer, M., Enkhmunkh, B. (2013). Diversity and community pattern of darkling beetles (Coleoptera: Tenebrionidae) in arid Mongolia. *Exploration into the Biological Resources of Mongolia* 12.

Qian, H., Ricklefs, R. E. (2012). Disentangling the effects of geographic distance and environmental dissimilarity on global patterns of species turnover. *Global Ecology and Biogeography* 21, 341-351.

R Development Core Team. (2007). *R: a language and environment for statistical computing.* R Foundation for Statistical Computing, Vienna, Austria. (http://www.R-project.org).

Rangel, T. F., Diniz-Filho, J. A. F., Bini, L. M. (2010). SAM: a comprehensive application for Spatial Analysis in Macroecology. *Ecography* 33, 46-50.

Richter, V. A., Korotyaev, B. A. (2011). In memory of G. S. Medvedev (1931-2009). *Entomological Review* 91, 533-543.

Sackmann, P., Flores, G. E. (2009). Temporal and spatial patterns of tenebrionid beetle diversity in NW Patagonia, Argentina. *Journal of Arid Environments* 73, 1095-1102.

Sagdy, C. T. (1996). Darkling beetle (Coleoptera, Tenebrionidae) ecological complexes of Ubsunur Hollow (Tuva). *Entomological Review* 75, 96-111.

Shi, G. 1993. A comparative study of 39 binary similarity coefficients. *Memoirs of the Association of Australasian Palaeontologists* 15, 329-341.

Soininen, J., Hillebrand, H. (2007). Disentangling distance decay of similarity from richness gradients: response to Baselga (2007). *Ecography* 30, 842-844.

Soininen, J., Lennon, J. J., Hillebrand, H. 2007a. A multivariate analysis of beta diversity across organisms and environments. *Ecology* 88, 2830-2838.

Soininen, J., McDonald, R., Hillebrand, H. (2007b). The distance decay of similarity in ecological communities. *Ecography* 30, 3-12.

Stapp, P. (1997). Microhabitat use and community structure of darkling beetles (Coleoptera, Tenebrionidae) in shortgrass prarie. *American Midland Naturalist* 137, 298-311.

USGS. (2004). Shuttle Radar Topography Mission, 1 Arc Second scene SRTM_u03_ n008e004, Unfilled Unfinished 2.0, Global Land Cover Facility. University of Maryland, College Park, Maryland.

Von Wehrden, H., Wesche, K., Tungalag, R. (2006). Plant communities of the Great Gobi B Strictly Protected Area, Mongolia. *Mongolian Journal of Biological Sciences* 4, 3-17.

Wiens, J. J., Graham, C. H. (2005). Niche Conservatism: Integrating Evolution, Ecology, and Conservation Biology. *Annual Review of Ecology, Evolution, and Systematics* 36, 519-539.

Williams, E. (2012). Aviation Formulary V1.46 williams.best.vwh.net/avform.htm., 24.12. 2012 US.

Williams, J. B., Tieleman, B. I. (2002). Ecological and evolutionary physiology of desert birds: A progress report. *Integrative and Comparative Biology* 42, 68-75.

Wright, D. H. (1983). Species-energy theory: an extension of species-area theory. *Oikos*, 496-506.

WWF. (2012). http://worldwildlife.org/. US.

Zachariasson, K. E. (1996). The water conserving physiological compromise of desert insects. *European Journal of Entomology* 93, 359-367.

In: Steppe Ecosystems
Editors: M. B. Morales Prieto and J. Traba Diaz
ISBN: 978-1-62808-298-2

Chapter 4

DIVERSITY PATTERNS IN THE STEPPE OF ARGENTINEAN SOUTHERN PATAGONIA: ENVIRONMENTAL DRIVERS AND IMPACT OF GRAZING

Pablo Luis Peri,[1,2,*] ***María Vanessa Lencinas,***[3]
Guillermo Martínez Pastur,[3] ***Grant W. Wardell-Johnson***[4]
and Romina Lasagno[2]
[1]Universidad Nacional de la Patagonia Austral (UNPA)-CONICET,
Río Gallegos, Santa Cruz, Argentina
[2]Instituto Nacional de Tecnología Agropecuaria (INTA), Argentina
[3]Centro Austral de Investigaciones Científicas (CADIC) – CONICET, Argentina
[4]Curtin Institute for Biodiversity and Climate, Curtin University, Bentley, WA, Australia

ABSTRACT

The steppe ecosystem, mainly characterised by the presence of tussock, short grasses and shrubs, covers 85% of the total area in Santa Cruz Province and 25% in Tierra del Fuego Island. Most of the land in the Patagonian region has been influenced by domestic livestock grazing for more than 100 years. This has led to a substantial modification of the ecosystem and the original floristic patterns. Erosion and degradation processes have occurred in several areas of Patagonia mainly due to an overestimation of the carrying capacity of these rangelands.

In this chapter we review patterns of plant and insect diversity in relation to environmental drivers and grazing impact in the steppe of Argentinian South Patagonia. In Santa Cruz, results from 141 sites indicated significant interactions between grazing and the abiotic environment (mainly water avalilability) on plant diversity. The

* Corresponding author: Pablo Luis Peri. Universidad Nacional de la Patagonia Austral (UNPA)-CONICET-Instituto Nacional de Tecnología Agropecuaria (INTA). CC 332 (CP 9400), Río Gallegos, Santa Cruz, Argentina. E-mail: pperi@correo.inta.gov.ar.

complexity of these interactions indicated the need for examining patterns of species turnover at different spatial scales.

Analysis of the steppe vegetation patterns from 113 sites along Tierra del Fuego Island demostrated differences related to geographical zones (North, Center, East and South), dominant vegetation types (grasslands, peatlands or shrublands) and disturbance impact (grazing, beavers or burned areas). Because insect diversity of Tierra del Fuego steppes is poorly known, coleopterans were selected as potential indicators of biodiversity using pitfalls traps in the same vegetation survey sites. We found significant changes in ground-active beetle assemblages generated by grazing and livestock activities, both in grassland, peatlands and shrublands. Therefore, this group of insects could be useful indicators of biodiversity conservation and ecosystem management.

Keywords: Floristic patterns; grassland; insect diversity; peatlands; plant diversity; shrublands; species richness; water availability

INTRODUCTION

Argentinian southern Patagonia includes Santa Cruz and Tierra del Fuego Provinces. Santa Cruz Province has an area of 243,943 km^2 and extends from latitudes 46º to 52º30'S. Tierra del Fuego Province includes many islands of varied size and the Argentine Antartic Sector. The main island has an area of 20,180 km^2, and extends between 52° and 56° S. In Santa Cruz, there are three main ecosystems: the Andes Mountains, the steppe and the valleys. Rainfall decreases from 800-1000 mm to 200 mm/year from west to east across the Andes Mountains, that act as an orographic barrier to moist winds coming from the west. In Santa Cruz, the native forest covers a narrow (100 km wide) and long (1000 km) strip of land. In Tierra del Fuego, the Andes Mountains decrease in height, and twist to the west. Therefore rainfall decreases from north to south, and forests occupy zones where the annual rainfall rises from 400 to 800 mm, south of the steppe. The southern beeches, lenga (*Nothofagus pumilio*), ñire (*N. antarctica*) and guindo (*N. betuloides*) are the dominant forest species in Argentinian southern Patagonia, covering 1,269,796 ha (535,889 ha in Santa Cruz and 733,907 ha in Tierra del Fuego Provinces).

The steppe ecosystem, mainly characterised by the presence of tussock (*Festuca*, *Stipa*), short grasses (*Poa*, *Carex*) and shrubs, covers 85% of the total area in Santa Cruz and 25% in Tierra del Fuego. The vegetation physiognomy of the Santa Cruz and Tierra del Fuego steppes is illustrated by Figures 1 and 2, respectively.

The main activity in this ecosystem is extensive sheep production, with stocking rates ranging from 0.13 to 0.75 heads ha^{-1} year^{-1}. The average size of properties is 12,400 ha and the largest station is 179,000 ha. During the last 70 years, a degradation process of the steppe (desertification) has occurred in the central part of Santa Cruz and partialy in the north of Tierra del Fuego, resulting from a combination of overgrazing and drought. Grazing impacts have been particularly profound in high latitude environments such as Patagonia, which has led to substantial ecosystem modification, in particular a significant increase in bare areas and change from the original floristic patterns (Bisigato and Bertiller, 1997). Erosion and degradation processes have occurred in several areas of Patagonia due to an overestimation of the carrying capacity of these rangelands, inadequate distribution of animals in very large and heterogeneous paddocks, and year-long continuous grazing (Golluscio et al., 1998). There are more than 6.5 million hectares affected by desertification (del Valle et al., 1995), where annual pasture production does not exceed 40 kg dry matter.ha^{-1}.

It has also been reported that grazing intensity on these extensively managed grasslands has affected ecosystem C levels. Peri (2011) reported that C stock in grasslands decreased from 130 Mg C. ha^{-1} under low grazing intensity (0.10 ewe ha^{-1} yr^{-1}) to 50 Mg C.ha^{-1} at a heavy stocking rate (0.70 ewe ha^{-1} yr-1) mainly due to an decline in plant cover and C lost from soil (mainly the organic layer in increasingly bare areas) as a consequence of soil erosion by strong winds.

There is some previous information about the distribution and physiognomic floristic description of the Patagonian flora at the regional, province and landscape level (Boelcke et al., 1985; Ares et al., 1990; León et al., 1998). The vegetation of continental Patagonia is in general low in plant cover (10-60%) and dominated by shrubs and perennial grasses (Bertiller and Bisigato, 1998). Precipitation decreases from the western mountains towards the east, and this distinct precipitation gradient substantially influences patterns of vegetation distribution (Jobbágy et al., 1995). Local edaphic and topographic variation in Patagonia is also correlated with local species turnover and structural patterns of vegetation (Soriano, 1983; Bisigato and Bertiller, 2004). Most of the floristic and structural inventories of Patagonian vegetation under different grazing regimes and disturbance intensities have been carried out in the extra-Andean regions such as the steppes of northern Patagonia (Ares et al., 1990; Beeskow et al., 1995; Bisigato and Bertiller, 1997; Aguiar and Sala, 1998). However, the influence of grazing disturbance on the floristic patterns of southern Patagonia remains limited, despite the vast area influenced by grazing. Also, these findings have to be interpreted with caution, as knowledge on interactions between grazing, environmental factors (edaphic and climatic variables) and vegetation type is negligible. An understanding of these interactions is, however, crucial for the design and implementation of viable management options towards the conservation of rangeland biodiversity and function (Ciblis and Coughenour, 2001; Golluscio et al., 1999).

In this chapter we review observed patterns of diversity in the steppe of South Patagonia in relation to environmental drivers and grazing impact. Specifically, we present observational data of vegetation diversity in both Santa Cruz and Tierra del Fuego, highlighting the influence of grazing on plant diversity in Santa Cruz and on plant and insect diversity in Tierra del Fuego.

PLANT DIVERSITY

There are 1,378 vascular plant species recorded from arid and semi-arid Patagonia (Correa, 1971), mainly represented by angiosperms, of which around 30% are endemic and 340 are exotic. Sheep grazing has been shown to reduce vascular plant diversity in several Patagonian ecosystems, both by promoting local extinction of preferred forage plants and by altering the relative abundance of species in the grazed plant communities (Aguiar and Sala, 1998; Bertiller and Bisigato, 1998). Soriano et al. (1995) assembled a list of 76 endangered species in Patagonia, of which a quarter are grasses, under heavy grazing conditions. There are only two natural reserves in arid and semi-arid Argentinian southern Patagonia steppe, although many of the National Parks along the Patagonian Andes include areas of steppe-forest ecotone. These are Bosques Petrificados Natural Monument (61,200 ha) and Monte León National Park (61,700 ha) both located in the province of Santa Cruz.

Environmental and Disturbance Patterns on Steppe Plant Diversity in Santa Cruz Province

Patagonian vegetation is especially susceptible to grazing pressures by domestic livestock as it presumably evolved under light grazing by native animals and is thus typified by limited adaptations to high grazing intensities. Modified vegetation patterns in response to high-intensity grazing are commonly distinguished by a shift from palatable grasses towards unpalatable woody plants (Paruelo and Golluscio, 1993).

The intensity of the replacement in life-form types depends on the abiotic characteristics and constraints of the affected ecosystem. This process and its different stages of life-form replacement has been related to the "state-and-transition" model for non-equilibrium rangeland ecosystems (Paruelo and Golluscio, 1993). Following this model, vegetation systems in Patagonia are not shifting linearly towards an anticipated climax state but, in contrast, alternate between fixed states. Regional analysis linked to local environmental variables and grazing intensity increases our capacity to predict vegetation responses to natural or human-induced environmental changes at a global scale (Körner, 1994). Vegetation patterns in Patagonia and differentiation between plant community types are typically correlated with biotic and abiotic factors, while abiotic habitat factors are in turn strongly affected by levels of grazing intensity (Milchunas et al. 1988). In this context, the main environmental and management factors that affect the plant diversity in the steppe ecosystem in Southern Patagonia are presented.

Study Area and Data Collection

A study of the relationship between floristic pattern and environmental factors under different levels of grazing disturbance commenced in 2007. This research was carried out in 141 sites in Santa Cruz Province, across latitudinal (46º 00'–52º 23' S) and longitudinal (65º 43'–73º 35' W) transects, corresponding to temperature and rainfall gradients, respectively (Figure 3).

The locations of sampling sites were selected to cover different levels of grazing (undisturbed vegetation, low, medium and high stocking rate) across the environmental variation in southern Patagonia. The intensity of grazing in each site was estimated by discriminating the mean sheep stocking rates for the main four ecological areas in Santa Cruz Province, such as (1) Central Plateau, Andean Vegetation, Humid Magellanic Grass Steppe, (2) Mata Negra Matorral Thicket, (3) Shrub steppe of Golfo San Jorge and Mountains and Plateaus, and (4) Dry Magellanic Grass Steppe and Sub-Andean Grassland (Table 1). The estimation of carrying capacity is based on the biomass production of short grasses and forbs that grow in the space among tussocks of each ecological area and the requirements of 530 kg DM.yr^{-1} for 1 Corriedale ewe of 49 kg of live weight which represents a "Patagonian sheep unit equivalent (PSUE)" (Borrelli, 2001). At each sampling location, plant, soil properties and environmental variables were measured in a 20 m × 50 m quadrat (1000 m^2). This plot size enables regional comparisons in diversity-associated factors for the broad vegetation types (e.g. grasslands, shrublands and forests) of Patagonia. All quadrats were permanently marked and assessed at least once during the flowering period (spring-summer) for accurate plant identification. We recorded the percentage cover of each vascular plant species occurring in each plot.

Figure 1. Tussock grassland in the Sub-andean Grassland ecological areas (Santa Cruz Province).

Table 1. Mean sheep stocking rates (± standard deviation) for different ecological areas in Southern Patagonia (Santa Cruz Province) calculated for 1 Corriedale ewe (49 kg of live weight, requiring 530 kg DM. yr^{-1})

	Stocking rate (ewe.ha^{-1}.yr^{-1})		
Ecological areas	Low	Medium	High
Central Plateau area	0.05 ±0.03	0.12 ± 0.02	0.24 ±0.10
Andean Vegetation and Humid Magellanic Grass Steppe areas	0.25 ±0.08	0.40 ±0.05	1.20 ±0.40
Mata Negra Matorral thicket, Shrub steppe Golfo San Jorge and Mountains and Plateaus areas	0.10 ±0.02	0.17 ±0.03	0.52 ±0.22
Dry Magellanic Grass Steppe and Sub-andean Grassland areas	0.16 ±0.05	0.26 ±0.05	0.62 ±0.16

Climatic parameters for each site were derived from the WorldClim data set (Hijmans et al., 2005). WorldClim contains geographic surfaces for 19 different climatic parameters that describe rainfall, temperature and their variability.

Incoming solar radiation (Wh m^{-2}) was calculated using the Solar Radiation tool in ArcGIS version 9.2 (ESRI, California), with topography data downloaded from the NASA Shuttle Radar Topography Mission (SRTM) Digital Elevation Model (DEM) of the globe (www2.jpl.nasa.gov/srtm).

We also calculated two composite climatic variables to use in statistical analyses because they integrate climatic conditions highly relevant to plant growth: WATG that represents mean annual water availability (Wynn et al., 2006) and E-T that is the climatic index derived by Liski et al. (2003).

At each quadrat we collected nine soil samples at 0-30 cm depth using a hand auger. Soil aggregates were extracted from the 10 cm depth samples and finely ground to below 2 µm using a tungsten-carbide mill. The ground aggregates were used to measure the percentage carbon and nitrogen using a LECO auto-analyzer (St. Joseph, US), major cations (Na, Al, P, K, Ca), pH, percentages of clay, silt and sand.

Data Analysis

Patterns derived from the site-by-species matrix were firstly examined using numerical taxonomic analysis approaches (Belbin, 1985) using PATN 3.11 (Belbin, 2006). The initial analyses were performed using the full vegetation data matrix (species x site; FVDM, 141 plots, by 275 taxa) and the full environmental data matrix (FEDM) describing the 141 plots by 33 variables.

Following initial analysis with the FVDM, species occurring only in a single or two quadrats, were excluded from pattern analysis. This procedure was carried out to render greater congruence between results from the three different portrayals of multivariate patterns: cluster analysis, ordination and network analysis.

Dissimilarity between sites based on overall floristic composition was quantified using the Bray-Curtis metric, a method that has performed consistently well in a variety of tests and simulations on different types of data (Faith et al., 1987). Data were clustered using the unweighted pair group method using arithmetic averaging (UPGMA) with beta set at -0.1 (Belbin, 1990). A hierarchical classification of the species was performed using the Two-Step Association Measure (Austin and Belbin, 1982), with Beta set at –0.1. The site classification and species classification were imposed on the data matrix to form a two-way table (see Belbin, 1990).

Thus the site groups can be assessed on the basis of species influence, while species groups can be assessed by their association with sites. The dissimilarity matrix was presented visually through semi-strong hybrid multidimensional scaling ordination (SSH MDS with dissimilarity cut level at 0.9), using PATN 3.11 (Belbin, 2006). The Minimum Spanning Tree (MST) was used in conjunction with UPGMA and SSH MDS to assess congruence between ordination (for trends), cluster analysis (for groups) and networks (nearest neighbor analysis).

Canonical correspondence analysis (CCA) and partial canonical correspondence analysis (pCCA) were employed as multivariate techniques to reveal the influence of environmental variables on the floristic patterns and to partition the information between the groups of variables, respectively. The pCCA analyses were carried out on the full sets of the environmental factors, which were partitioned into climatic, soil, and disturbance sets. Finally, we used the species-by-sample incidence matrix (presence-absence data) to estimate species richness patterns using accumulation curves which serve in assessment of the completeness of samples; the estimated asymptote is used as a species richness estimate

(Colwell et al., 2004). Sample-based rarefaction curves implicitly reflect empirical levels of within-species aggregation of individuals by considering only incidence, thus providing a realistic estimate of species numbers to be found in sets of real-world samples (Gotelli and Colwell, 2001).

Results and Discussion

A total of 275 plant species from 145 genera and 52 families were recorded in the 141 quadrats surveyed across Santa Cruz Province. This represents 12.7% of the flora of Patagonia. Local native plants accounted for 91.3% of the total species (45 endemic species), while alien taxa accounted for 24 species (8.7%).

Results from cluster analysis, nearest neighbor analysis and ordination lacked congruence when species occurring at less than three sites were included in overall analysis. This analysis determined three main community patterns. The first group included 10 wetland sites. These wetlands (locally named *mallines*) are usually found in valleys, which contain soils with high water and carbon contents. They are characterized by spatial and temporal heterogeneity determined from soil moisture and climate variations and are highly regarded by ranchers because of their high productivity for livestock. The wetlands are humid meadows dominated by Juncaceae, Cyperaceae, and Poaceae such as *Juncus* spp., *Carex* spp. and *Festuca pallescens*, found in both the steppe and Andean topography of Patagonia. The second group included 53 grassland (steppe) sites. The vegetation of the steppe is dominated by grasses and sedges (*Bromus*, *Carex*, *Festuca gracillima*, *Hordeum*, *Jarava*, *Poa*, *Rytidosperma virescens*, *Trisetum*) with dwarf shrubs and herbs such as *Nardophyllum*, *Perezia*, *Azorella*, and *Nassauvia* admixed. The vegetation of the grass-shrub steppe is dominated by *Agrostis*, *Festuca*, *Hordeum* and *Trisetum*, although shrubs (*Adesmia*, *Chuquiraga*, *Junellia*, *Mulinum*, *Senecio*) are also frequent. The vegetation of the third group (also 53 sites) corresponded to shrubland or shrub-grass steppe sites dominated mainly by tall shrubs such as *Berberis, Colliguaja intergerrima*, *Chuquiraga*, *Junellia*, *Lepidophyllum cupressiforme*, *Lycium*, *Mulinum* and *Prosopis*, with grass-rich undergrowth including *Bromus*, *Hordeum jarava* and *Poa*.

The CCA of all plots revealed clear separation of the wetland plots along the CCA Axis 2 (Figure 4), with factors such as water content (WATG), magnesium (Mg), organic matter (ORGM) strongly correlated with nitrogen (N) playing a major role in the separation and indicating that these wetlands are rich in organic matter and basic ions.

Analyzing only the steppe vegetation, water availability was the main environmental factor driving the plant diversity and these long-term differences in water availability have created distinct plant communities. This becomes important because alteration of precipitation regimes under climate change may influence species richness, especially in arid and semiarid plant communities where water is a primary limiting resource (Sala et al., 1988). Adler and Levine (2007) reported that long-term increase in precipitation at a site will eventually lead to increases in species richness in water-limited plant communities, while decreases in precipitation will decrease richness. In this context, a climatic-envelope model for 125 woody Florida plant species described by Box et al. (1993), based on comparison with both floristic lists at sites and species range maps, supported the hypothesis that climatic factors may exert important direct and/or indirect control over the natural distribution of plant types, even species and even at local scales.

However, at a global scale, Box (1995) reported that some of the main climate-related constraints on plant/vegetation metabolism and biomass maintenance involved temperature (temperature extremes may damage through limitation of water uptake or by causing excessive water loss) and in less extent water availability (annual climatic moisture balance of precipitation divided by potential evapotranspiration and the average length of the dry season), but being both often interrelated.

Figure 2. Grassland in Tierra del Fuego Province dominated by *Carex curta*, *Poa pratensis* and *Trisetum spicatum*.

Table 2. Information partitioning (% of inertia explained) in pCCA involving full vegetation data matrix (FVDM), analysis of only steppe+wetland plots (SW) and analysis of only steppe plots (S). GRAZ: grazing; soil: groups of variable related to soil characteristics; climate: group of variables describing the climate of the area

	GRAZ	Soil	Climate
FVDM	0.7	16.1	13.7
SW	0.9	19.8	16.9
S	0.1	22.4	19.3

Thus, the main climatic mechanisms limiting plant and vegetation distributions at global scale appear to involve tolerance to maximum and minimum temperatures, requirements for growing season warmth in colder climates; tolerance to desiccation and adaptations to longer-term overall moisture availability and balance (Box, 1995). The factor of grazing (GRAZ) showed a perpendicular (non-correlated) relationship with the factors dominating the clear separation of wetlands.

This suggests a low importance of GRAZ at the regional landscape-level for separation between the wetland and non-wetland communities. This is confirmed through a partial canonical correlation analyses that revealed that grazing pressure (as estimated using a simple ordinal scale) explains only a fraction of the total inertia at landscape levels, for the total vegetation data matrix set (Table 2).

The nature of soils was revealed as the dominating source of information, where climatic characteristics explained more than 15% of the total inertia. However, grazing showed an influence at the lower scales (distinction between woody and non-woody vegetation).

León and Aguiar (1985) reported that in arid and semi-arid grasslands, grass cover decreased exponentially along a grazing gradient and soil erosion and unpalatable shrubs increased towards the heavily grazed situation. Furthermore, the complexity of interactions between environmental variables and grazing disturbance is evident in lagged and frequently irreversible feedback effects and ecosystem sensitivity to rare and extreme events (Westoby et al., 1989). This indicates the need for examining patterns of species turnover at different scales. In our work, CCA indicates significant interactions between grazing and the abiotic environment. Grazing interacts with evaporation stress in particular, and a great variety of climatic variables and the water table during the growing season. Interactions between direct climate change threats (e.g. temperature increases, while rainfall and hydrological increase during extreme weather events) and long standing grazing threats presumably further amplifies these impacts on plant diversity (Parmesan and Yohe, 2003).

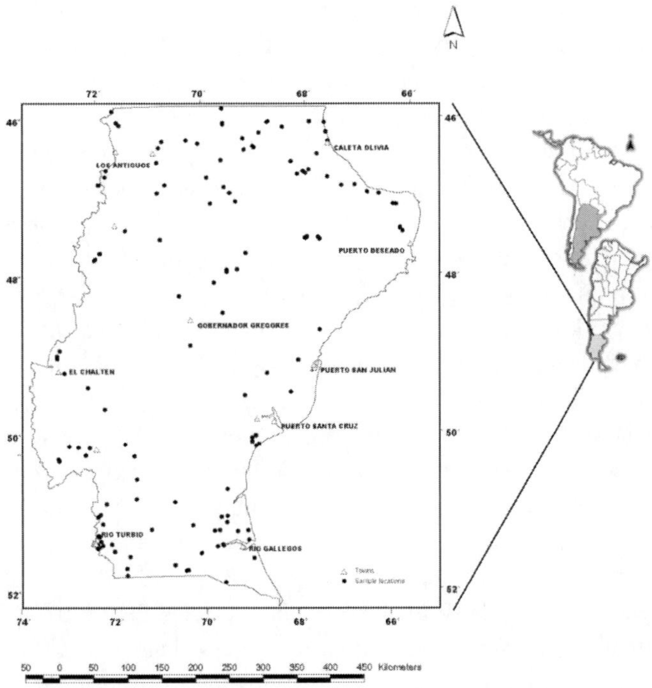

Figure 3. Location of plant biodiversity plots within the Santa Cruz Province, southern Patagonia, Argentina.

The analysis of rarefaction curves (Figure 5) revealed that the number of plots involved in the rarefaction analyses were not sufficient, as all curves have failed to reach saturation (leveling off). The slope of the curves in the grassy steppe vegetation indicated that more intensively grazed vegetation patches were floristically heterogeneous at landscape scales (steeper slope resulting from quicker accumulation of species at given scale). However, the steepest rarefaction curve had the lowest number of plots suggesting that this steepness (increased accumulation of species over small area) may be an artifact of our sampling. Therefore, we can hypothesize that in a small number of well dispersed plots, the number of species not shared will be a function of the distance between these plots (in other words: each plot could have a slightly different regional pool of species).

Environmental and Disturbance Patterns on Steppe Plant Diversity in Tierra Del Fuego Province

Climatic and topographic conditions have an influence over the floristic assemblages (Frangi and Richter, 1992). Also, climate has an influence over soil development, and consequently, over vegetation patterns. According to Moore (1983), the Patagonic steppe of Tierra del Fuego Province is dominated by graminoid grass and shrubs. However, it is possible to find remnant patches of forest communities in the Chilean steppe sector of the Island, including species such as *Drimys winteri*, *Embothrium coccineum* and *Maytenus magellanica*. Northern grasslands of the island are mainly dominated by *Festuca gracillima* Hooker f. asociated with many other grasses and dicotyledonous species. In the more humid sectors, it is common to found communities dominated by *Alopecurus magellanicus*, *Deschampsia antarctica* and *Phleum alpinum*, while in the drier sectors, *Phacelia secunda*, *Valeriana carnosa*, *Rytidosperma virescens*, and *Stipa chrysophylla* are the most abundant taxa.

Humid grasslands or wetlands that ocupy sites along riversides and depressions with clay-soil horizons, are dominated by species such as *Alopecurus magellanicus*, *Deschampsia antarctica*, *D. kingii*, *Hierochloë redolens* and *Hordeum comosum*. Graminoid species such as *Carex*, *Juncus* and other species such as *Acaena*, *Azorella*, *Blechnum*, *Gentianella* and *Gunnera* are also common (Moore, 1983). In southern areas, where rainfall reaches 350 $mm.yr^{-1}$, grasslands gradually change to shrublands of *Chiliotrichum diffusum* (mata negra), an evergreen shrub which also occurs in deciduous and evergreen forests. This species is usually employed as an indicator of ecotone zones between steppe and forests (Moore, 1983). *Sphagnum magellanicum* peatlands is a typical vegetation type in the valleys across the forest and grassland in the southern landscapes in Tierra del Fuego Province (Endlicher and Santana Águila, 1988; Roig, 1998). This characteristic moss community is also accompanied by reeds, grasses and other plants (Roig, 1998), which also provide organic material for the formation of peat.

According to Moore (1983), 417 of the 545 plant species found in Tierra del Fuego are native. Endemic flora is not abundant, representing less than 3% of the native richness, including species of the genus *Atriplex*, *Chiliophyllum*, *Descurainia*, *Festuca*, *Onuris*, *Ourisia*, *Poa*, *Senecio*, *Nassauvia* and *Epilobium*. To date, 128 adventive species were recorded, mainly brought from Europe with cattle (Moore and Goodall, 1974; Goodall, 1981; Moore, 1983). Most of the adventive species were introduced by accident, but many others were introduced to improve the natural grasslands (e.g. *Trifolium repens*) or for human food (*Taraxacum officinale*) (Moore, 1983).

Vegetation diversity of Tierra del Fuego steppes differs among zones. Moreover, several factors influence grassland and shrubland vegetation diversity. Cattle, beavers and fires are the main disturbance factors, generating several degrees of degradation in the main island of Tierra del Fuego.

Study Area and Data Collection

We studied the steppe vegetation patterns from 113 sites along Tierra del Fuego Island (Figure 6), with floristic inventory data from 1999 to 2012, distributed across latitudinal (53° 40'–55° 00' S) and longitudinal (66° 30'–68° 30' W) gradients.

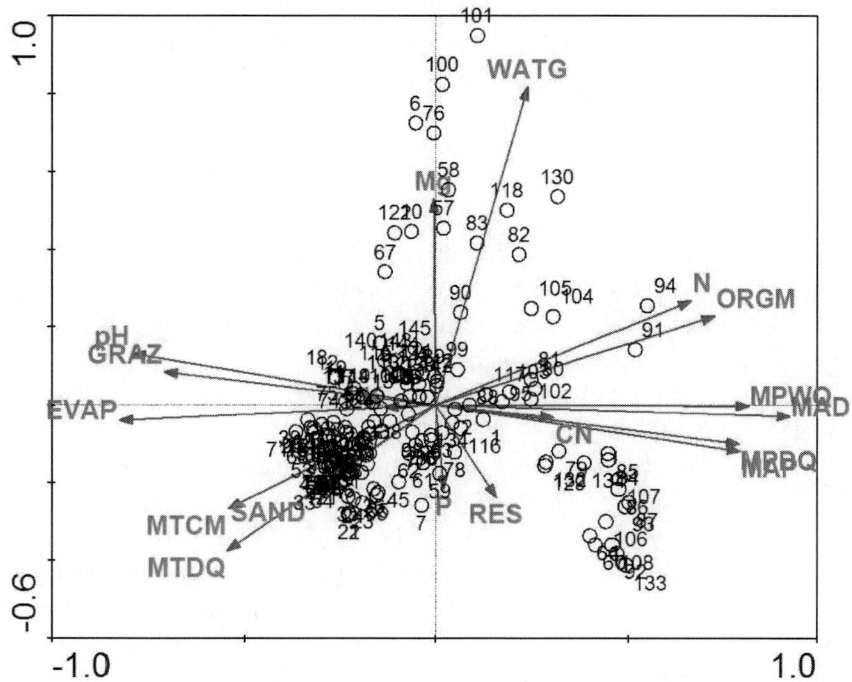

Figure 4. Canonical correspondence analysis (CCA) on the full vegetation data matrix in Santa Cruz Province.

To study differences among zones along Tierra del Fuego, surveys were classified in four zones (North, Center, East and South). To explore differences in diversity patterns, classification of surveys was done according to their dominant vegetation type (grasslands, peatlands, and shrublands).

Finally, the influence of main impact types over the vegetation diversity was evaluated using four categories (without impact or with low impact degree which not modified the main species communities, impacted by grazing, impacted by beavers, and burned areas).

At each sampling site, understory plant species richness and cover (%) were registered by *relevé* methodology (Kent and Coker, 1992). All vascular plants (Dicotyledoneae, Monocotyledoneae and Pteridophyta) were determined to species level, and taxonomy of the species follows the classification proposed by Moore (1983). Hepathic and mosses were considered separately as a single group. Also with vegetation, bare soil cover (soil or litter on the forest floor without vegetation), and debris cover (more than 3 cm diameter branches and dead wood) were estimated.

Data Analysis

Species cover matrices were used for multivariate analyses. A detrended correspondence analysis (DCA) was done (Hill, 1979; Greenacre, 1984; Manly, 1994), without down-weighting of rare species.

It was selected because this is the only ordination technique that simultaneously analyses sampling units and species, allowing the examination of ecological relationships between them in a single-step analysis (Ludwig and Reynolds, 1988), and also due to the unimodal response model of the data set (gradient length in Axis 1=5.65).

Then, a multi-response permutation procedure (MRPP), considering the Euclidean distance, was utilized to analyze group significance. PC-Ord (McCune and Mefford, 1999) software was used for these analyses.

Results and Discussion

A total of 184 vascular plant species were surveyed, where 66% (122 species) were dicotyledons, 32% (59 species) were monocots and 2% (3 species) were ferns. Among these were 24 exotic species (15 dicotyledons and 9 monocotyledons). These observations were complemented with measures of debris, hepatics and mosses, and percentage of bare soil, which were also included in ordination analyses.

In the graphical analysis, significant differences were observed among zones and dominant vegetation types, as well as among the different impact types. In DCA (total inertia= 11.96), the Axis 1 (eigenvalue= 0.6729) was discarded, because it was heavily influenced by only one sample. Therefore, Axes 2 and 3 were selected and presented in Figure 7. These results were confirmed by MRPP (Table 3). According to the zone classification, vascular plant diversity differed among the four categories, making the south zone the most distinct, while the east zone presented more homogeneity than the north and centre zones.

Vascular plant diversity differed also among the three dominant vegetation types (Figure 7 and Table 3), although heterogeneity was greater in grassland than in peatlands and in shrublands. Finally, many differences were observed among impact types over different dominant vegetation types, e.g. beaver impacted grasslands differed from both primary and grazed grasslands, primary grassland differed from both primary peatlands and shrublands, grazed grassland differed from both grazed peatlands and shrublands (Table 3). However, differences were not detected between grasslands without impact and grazed ones, or between pristine and grazed peatlands, or between pristine and burned shrublands. Beside this, differences were not detected between beaver meadows and grazed peatlands (Table 3).

According to these results, vegetation diversity patterns were highly varied among zones, dominant vegetation types and main impacts. Therefore, vegetation diversity response to management in Tierra del Fuego steppes could vary according to environmental and impact drivers. Thus different conservation strategies could be required for the conservation of vascular plant diversity.

Table 3. Multi-Response Permutation Procedures (MRPP) results for vascular plants according to zones (North, Center, East and South), vegetation types (G= grassland, P= peatland, S= shrubland), and impacts (PG =primary or low impacted grassland, GG =grazed grassland, BG =beaver impacted grassland, PP =primary or low impacted peatland, GP =grazed peatland, BP =beaver impacted peatland, PS =primary or low impacted shrubland, GS =grazed shrubland, FS= fired shrubland)

Classification	Group comparisons	A	*p*
Zones		0.0571	<0.0001
	North vs. Center	0.0258	<0.0001
	North vs. East	0.0350	0.0009
	North vs. South	0.0481	<0.0001
	Center vs. East	0.0206	0.0051
	Center vs. South	0.0398	<0.0001
	East vs. South	0.045	<0.0001
Vegetation types		0.1233	<0.0001
	G vs. P	0.1002	<0.0001
	G vs. S	0.0539	<0.0001
	P vs. S	0.1825	<0.0001
Impacts		0.1034	<0.0001
	PG vs. GG	0.0010	0.3072
	PG vs. BG	0.0321	0.0171
	PG vs. PP	0.0916	0.0004
	PG vs. GP	0.0357	0.0211
	PG vs. PS	0.069	0.0002
	PG vs. GS	0.0465	0.0026
	PG vs. FS	0.0995	0.0016
	GG vs. BG	0.0144	0.0032
	GG vs. PP	0.0698	<0.0001
	GG vs. GP	0.0187	0.0009
	GG vs. PS	0.0457	<0.0001
	GG vs. GS	0.0348	<0.0001
	GG vs. FS	0.0371	<0.0001
	BG vs. PP	0.0867	0.0056
	BG vs. GP	0.0392	0.1248
	BG vs. PS	0.123	0.0003
	BG vs. GS	0.1165	0.0004
	BG vs. FS	0.2232	0.0027
	PP vs. GP	-0.0104	0.5775
	PP vs. PS	0.1011	0.0009
	PP vs. GS	0.156	<0.0001
	PP vs. FS	0.1202	0.0033
	GP vs. PS	0.0789	0.0084
	GP vs. GS	0.1312	0.0001
	GP vs. FS	0.1791	0.0057
	PS vs. GS	0.0403	0.0080
	PS vs. FS	0.0383	0.1002
	GS vs. FS	0.0353	0.0465

Insect As Biodiversity Indicators

Arthropods are widely recognized to play a key role in ecosystem processes (e.g. as pollinators, bottom of food chains, maintenance and improvement of soil quality and plague control).

In addition, the loss of insects is rarely appreciated due to their small individual size and usually large populations. Therefore, compared with vascular plants and vertebrates, conservation of insect diversity has been rarely included in ecosystem management planning, since they are considered unpleasant and dangerous for human beings.

However, insects and arachnids are frequently used to evaluate the effects of human activities on biodiversity and environment quality (Kim, 1993; Niemelä, 2001).

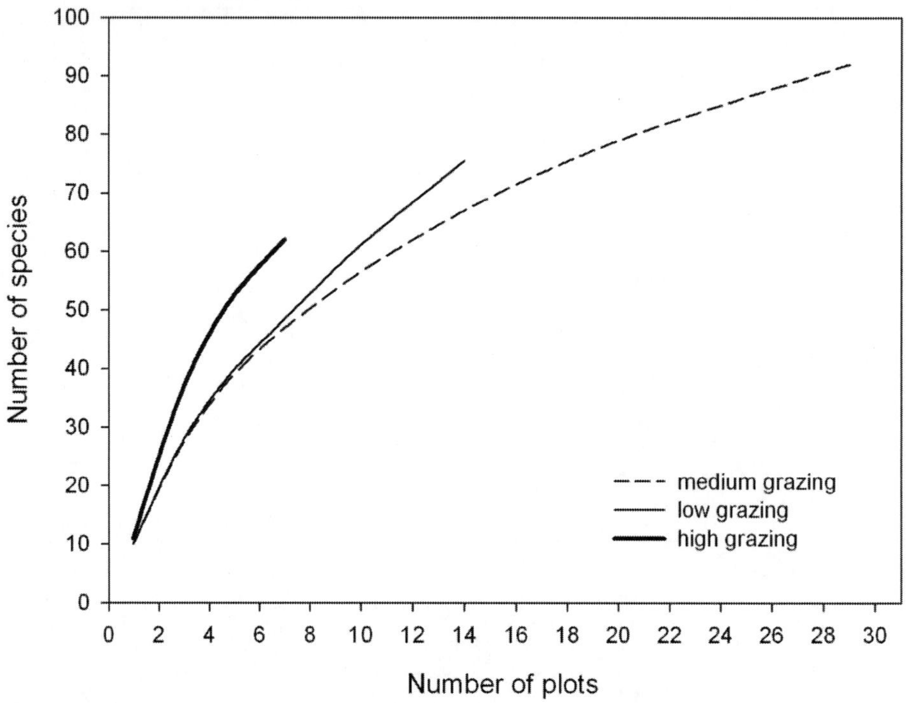

Figure 5. Rarefaction curves in the grassy steppe vegetation of Santa Cruz province for different grazing intensities.

These groups are good indicators at the landscape level (Lewis and Whitfield, 1999) as well, because their abundance, species richness, and occurrence are sensitive to local resource variations (Werner and Raffa, 2000).

The main insect extinction threats or risks in their conservation are habitat loss, exotic species invasion, "chains" of extinction, the effects of pesticides on nontarget species and commercial captures (Romoser and Stoffolano, 1998). Many insect conservationists consider habitat loss due to transformation or habitat fragmentation, the greatest threat to insect species today (Covel, 1976; Pyle, 1976).

Economic human activities as land conversion for intensive agricultural use, deforestation and urbanization significantly affect many vertebrate species, as well as insect communities around the world (Wells et al., 1983). This is frequently described for the tropical forests, but also occurred in the temperate grasslands, where both quantity and quality of habitat were affected. Invasion of introduced species to long-isolated ecosystems (e.g. islands) has resulted in sudden insect species extinctions, since endemic island species evolve under conditions of low pressure from predation, competition and parasitism, being highly susceptible to the arrival of more competitive species. Pest species are sometimes intentionally driven to extinction by the intentional introduction of exotic predators and parasites, a well-known practice called "biological control." Unfortunately, those biological agents for a specific target sometimes exterminate nontarget species as well.

The entomofauna richness of South Patagonia follows the pattern of generally poorer faunal diversity in southern Patagonia (Guzmán et al., 1985-86) compared to northern regions of similar latitudes (Martikainen et al., 2000).

This low diversity is probably related to the short growing season (Roig et al., 2002) and the lower average summer temperatures in subantarctic ecosystems (Ferreyra et al., 1998). Insect diversity of Tierra del Fuego steppes is poorly known and their use as biodiversity indicators must be evaluated. For this purpose, coleopterons were selected as potential indicators.

Study Area and Data Collection

Coleopteran diversity in Tierra del Fuego steppe was sampled with pitfalls traps at the same sites of the 113 vegetation surveys, distributed across latitudinal (53° 40'–55° 00' S) and longitudinal (66° 30'–68° 30' W) gradients. To study differences among zones along Tierra del Fuego, surveys were classified in four zones (North, Center, East, and South). To explore differences in diversity patterns, classification of surveys was done according to their dominant vegetation type (grasslands, peatlands and shrublands). Finally, the influence of main impact types over vegetation diversity was evaluated using four categories (without impact or with low impact degree which did not modify the main communities, impacted by grazing, impacted by beavers, and burned areas).

During January-February, five pitfall traps (Barber, 1931) of 14 cm height and 12 cm diameter were placed in a cross design in each site, separated 5 m from the center. Pitfall or interception traps are mainly oriented to capture ground-walking insects, as Scarabaeidae and Carabidae (Lanfranco, 1977; 1991; Kotze and Samways, 1999; Werner and Raffa, 2000). In each trap, water was used as a retention agent, and commercial detergent was employed to diminish surface tension. Restriction in collection dates is due to seasonality of insect populations in regions with extreme environmental conditions and short growing season (Roig et al., 2002). Traps remained active during a full week, and after this time, the five traps in each site were collected and processed as one. In each sampling, coleopterans were sorted, counted and classified under a binocular dissecting scope (x10-x20) until the lowest possible taxonomic level (Family, Genus or Species) using standard keys (Richards and Davies, 1984; Ross, 1973; Romoser and Stoffolano, 1998). When it was not possible, because Patagonian insect systematics are still incomplete, these were classified as different morphospecies (Oliver and Beattie, 1993). The use of morphospecies instead of formal taxonomic species may be sufficiently close to estimate species richness with average errors below 15% in rapid assessment of biodiversity inventories, monitoring or preliminary ecological studies (Oliver and Beattie, 1993). Likewise, morphospecies have been demonstrated to be a good tool for insect diversity studies in *Nothofagus* forests (Spagarino et al., 2001; Lencinas et al., 2008;

2010). However, it is possible that some insect species of the South Patagonian ecosystems were not studied or captured yet. Therefore real current insect diversity is likely underestimated. The morphospecies were deposited in the permanent reference collection at Centro Austral de Investigaciones Científicas (CADIC-CONICET) at Ushuaia, Argentina. For convenience, the term "species" was used to refer to family, genus, taxonomic species and morphospecies. Richness and abundance was calculated for each sample and species.

Data Analysis

Species abundance matrices with total data for each site were used for multivariate analyses. Detrended correspondence analysis (DCA) was performed (Hill, 1979; Greenacre, 1984; Manly, 1994), without down-weighting of rare species.

Similarly to plants, it was selected to analyze insect diversity because this is the only ordination technique that simultaneously analyses sampling units and species, allowing the examination of ecological interrelationships between them in a single-step analysis (Ludwig and Reynolds, 1988), and also due to the unimodal response model of the data set (gradient length in Axis 1=7.21). Then, a multi-response permutation procedure (MRPP), considering the Euclidean distance, was utilized to analyze group significance. PC-Ord (McCune and Mefford, 1999) software was used for these analyses.

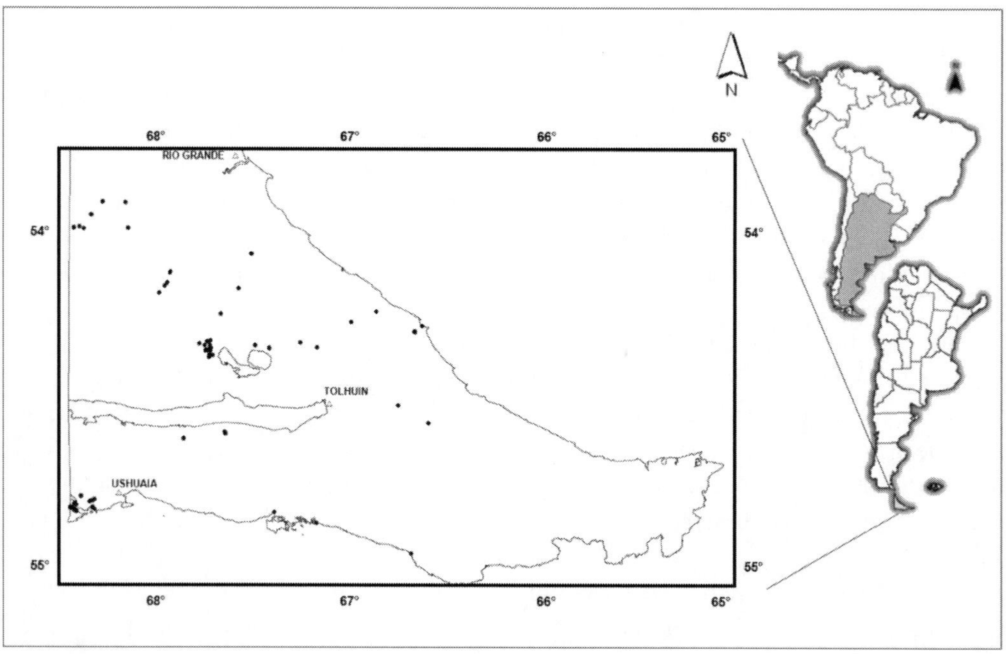

Figure 6. Location of plant biodiversity plots within the Tierra del Fuego Province, Southern Patagonia, Argentina.

Results and Discussion

We captured coleopterans in 106 sites (7 sites did not present coleopterons in the traps), and 85 species were identified along the study. In the Detrended Correspondence Analysis (DCA) and Multi-Response Permutation Procedures (MRPP), 20 species presented only one individual in the whole sampling (commonly called singletons, 24% of the richness), which were excluded in further analysis. Singletons could represent half the richness in other insect inventories (Lewis and Whitfield, 1999; Novotný and Basset, 2000). In the graphical analysis, significant differences were observed among zones and dominant vegetation type, as well as among the different impact types. In DCA (total inertia=12.00), the Axes 1 and 2 were selected and presented in Figure 8. These general results were confirmed by MRPP (Table 4). According to zone classification, ground-active beetle diversity of southern zone significantly differed from the other zones, while North, Centre and East zone did not differed among them (Table 4). About the main vegetation types, statistical differences among grassland, peatlands and shrublands were not detected (Table 4), although general MRPP analysis found significant differences among their ground-active beetle assemblages. However, the peatland and shrubland clusters had greater homogeneity than the grassland cluster (Figure 8).

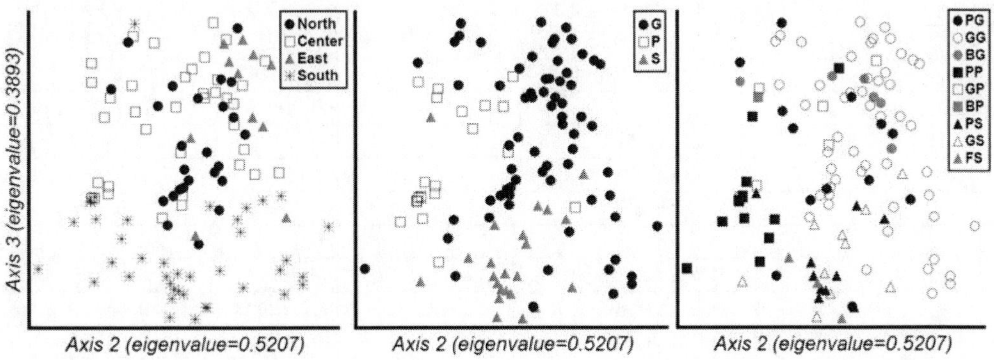

Figure 7. DCA for the vascular plants analysis in Tierra del Fuego, classified by zones (north, centre, east and south), dominant vegetation type (G= grassland, P= peatland, S= shrubland), and impact type (PG= primary or low impacted grassland, GG= grazed grassland, BG= beaver impacted grassland, PP= primary or low impacted peatland, GP= grazed peatland, BP= beaver impacted peatland, PS= primary or low impacted shrubland, GS= grazed shrubland, FS= fired shrubland).

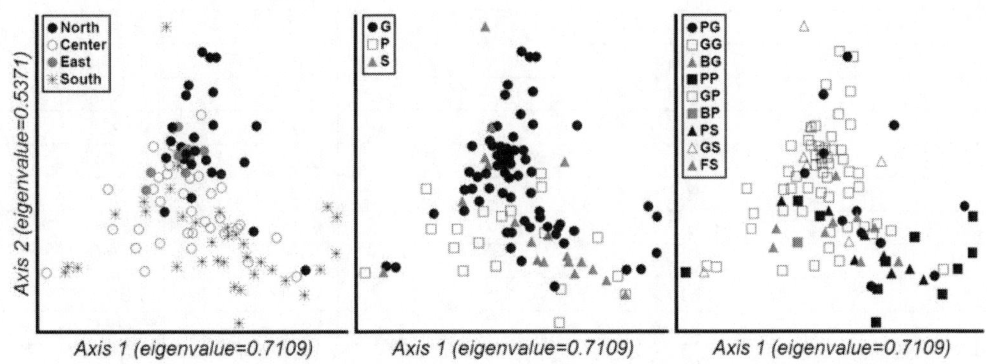

Figure 8. DCA for the epigeon coleopteron analysis in Tierra del Fuego, classified by zones (north, centre, east and south), dominant vegetation type (G= grassland, P= peatland, S= shrubland), and impact type (PG= primary or low impacted grassland, GG= grazed grassland, BG= beaver impacted grassland, PP= primary or low impacted peatland, GP= grazed peatland, BP= beaver impacted peatland, PS= primary or low impacted shrubland, GS= grazed shrubland, FS= fired shrubland).

Table 4. Multi-Response Permutation Procedures (MRPP) results for ground active beetles, according to zones (North, Center, East and South), vegetation types (G= grassland, P= peatland, S= shrubland), and impacts (PG= primary or low impacted grassland, GG =grazed grassland, BG= beaver impacted grassland, PP= primary or low impacted peatland, GP= grazed peatland, BP= beaver impacted peatland, PS= primary or low impacted shrubland, GS= grazed shrubland, FS= fired shrubland)

Classification	Group comparisons	A	*p*
Zones		0.0269	0.0010
	North vs. Center	0.0050	0.1335
	North vs. East	0.0131	0.1600
	North vs. South	0.0189	0.0063
	Center vs. East	0.0133	0.0695
	Center vs. South	0.0191	0.0011
	East vs. South	0.0470	0.0013
Vegetation types		0.0095	0.0395
	G vs. P	0.0051	0.1043
	G vs. S	0.0082	0.0526
	P vs. S	0.0045	0.2099
Impacts		0.0228	0.0230
	PG vs. GG	-0.0032	0.5859
	PG vs. BG	0.0294	0.1541
	PG vs. PP	0.0143	0.1677
	PG vs. GP	0.0052	0.3001
	PG vs. PS	0.0129	0.1882
	PG vs. GS	-0.0149	0.6915
	PG vs. FS	0.0062	0.2520
	GG vs. BG	0.0126	0.0558
	GG vs. PP	0.0160	0.0208
	GG vs. GP	0.0015	0.2457
	GG vs. PS	0.0172	0.0158
	GG vs. GS	-0.0051	1.0000
	GG vs. FS	0.0055	0.0893
	BG vs. PP	0.1200	0.0002
	BG vs. GP	0.0086	0.2850
	BG vs. PS	0.0574	0.0164
	BG vs. GS	0.0230	0.1739
	BG vs. FS	0.0612	0.0671
	PP vs. GP	0.0348	0.0148
	PP vs. PS	0.0404	0.0077
	PP vs. GS	0.0137	0.1668
	PP vs. FS	0.0231	0.0872
	GP vs. PS	0.0116	0.1983
	GP vs. GS	-0.0046	0.5091
	GP vs. FS	0.0177	0.2212
	PS vs. GS	0.0158	0.1924
	PS vs. FS	-0.0142	0.6942
	GS vs. FS	0.0079	0.3228

On the other hand, some impact types significantly influenced insect assemblages (Table 4). The main differences were found between primary peatlands and primary shrublands, as well as among primary peatlands and grazing or beaver impacted grasslands, and among those and grazed peatlands.

However, differences were not found among primary peatlands and primary grasslands, and among primary peatlands and grazed or fired shrublands. Also, differences between grazed shrublands and impacted grasslands (grazed by cattle or impacted by beavers) were found.

From these results, we found significant changes in ground-active beetle assemblages generated by grazing and livestock activities, both in grassland, peatlands and shrublands, which produced more differences than fire or beaver impacts.

Therefore, this group of insects could be used as a useful indicator of biodiversity conservation and ecosystem management.

General Conclusion

In Santa Cruz, water availability was the main environmental factor driving the plant diversity of the steppe vegetation and these long-term differences in water availability have created distinct plant communities. This becomes important because alteration of precipitation regimes under climate change may influence species richness. The factor of grazing explains only a fraction of the total inertia at landscape levels. However, grazing pressure showed an influence at the lower scales (distinction between woody and non-woody vegetation). Results from 141 sites indicated significant interactions between grazing and the abiotic environment (mainly water availability) on plant diversity. The complexity of these interactions indicated the need for examining patterns of species turnover at different spatial scales. Analysis of the steppe vegetation patterns from 113 sites along Tierra del Fuego Island demonstrated differences related to different geographical zones (North, Center, East and South), dominant vegetation types (grasslands, peatlands or shrublands) and disturbance impact (grazing, beavers or burned areas). Thus, vegetation diversity response to management in Tierra del Fuego steppes varied according to environmental and impact drivers. Because of these, management strategies are needed for the conservation of vascular plant diversity.

On the other hand, coleopterans were selected as potential indicators of biodiversity in Tierra del Fuego steppe. We found that grazing and livestock activities in grassland, peatlands and shrublands had a significant impact on ground-active beetle diversity. Therefore, this group of insects could be used as indicators of biodiversity conservation and ecosystem management.

References

Adler, P. B., Levine, J. M. (2007). Contrasting relationships between precipitation and species richness in space and time. *Oikos* 116, 221-232.

Aguiar, M. R., Sala, O. E. (1998). Interactions among grasses, shrubs, and herbivores in Patagonian grass-shrub steppes. *Ecología Austral* 8, 201-210

Ares, J., Beeskow, A. M., Bertiller, M., Rostango, M., Irisarri, M., Anchorena, J., Defossé, G., Merino, C. (1990). Structural and dynamic characteristics of overgrazed lands of northern Patagonia, Argentina. In: A. Breymeyer (ed.), *Managed Grasslands*, pp. 149-175. Elsevier Science Publishers, Amsterdam.

Austin, M. P., Belbin, L. (1982). A new approach to the species classification problem in floristic analysis. *Australian Journal of Ecology* 7, 75-89.

Barber, H. (1931). Traps for cave-inhabiting insects. *Journal Elisha Mitchell Science Society* 46, 259-266.

Bertiller, M. B., Bisigato, A. (1998). Vegetation dynamics under grazing disturbance. The state-and-transition model for the Patagonian steppes. *Ecología Austral* 8, 191-199

Bisigato, A. J., Bertiller, M. B. (1997). Grazing effects on patchy dryland vegetation in northern Patagonia. *Journal of Arid Environments* 36, 639-653.

Bisigato, A., Bertiller, M. (2004). Patch dynamics in the arid southern Argentinean Monte: Microspatial patterns of seedling establishment. *Plant Ecology* 174, 235-246.

Beeskow, A. M., Elissalde, N. O., Rostagno, C. M. (1995). Ecosystem changes associated with grazing intensity on the Punta Ninfas rangelands of Patagonia, Argentina. *J. Range Manage*. 48, 517-522.

Belbin, L. (1985). A background briefing in numerical taxonomy or does your handbag match your shoes? In: *Proceedings of the Digital Equipment Computer Users Society Symposium*, pp. 15-22. Hobart, Australia.

Belbin, L. (1990). The analysis of pattern in bio-survey data. In: Margules, C. R., Austin, M. P. (eds.), *Nature Conservation: Cost Effective Biological Surveys and Data Analysis*, pp. 176–90. CSIRO, Melbourne.

Belbin, L. (2006). *PATN Version 3.11*. Blatant Fabrications, Hobart, Autralia.

Boelcke, O., Moore, D. M., Roig, F. A. (1985). *Transecta botánica de la Patagonia Austral*. CONICET (Argentina), Royal Society (UK) e Instituto de la Patagonia (Chile), Buenos Aires.

Borrelli, P. (2001) Producción animal sobre pastizales naturales. In: Borrelli, P., Oliva, G. (eds.), *Ganadería Sustentable en la Patagonia Austral*, pp. 129-160. INTA, Buenos Aires

Box, E. O., Crumpacker, D. W., Hardin, E. D. (1993). A climatic model for plant species locations in Florida. *Journal of Biogeography* 20, 629--644.

Box, E. O. (1995). Factors determining distribution of tree species and plant functional plants. *Vegetatio* 121, 101-116.

Ciblis, A. F., Coughenour, M. B. (2001) Impact of grazing management on the productivity of cold temperate grasslands of Southern Patagonia - a critical assessment. In: *Procceedings of the XIX International Grassland Congress*, pp. 807-812. Sao Paulo.

Colwell, R. K., Mao, C. X., Chang, J. (2004). Interpolating, extrapolating, and comparing incidence-based species accumulation curves. *Ecology* 85, 2717-2727.

Covell, C. V. Jr. (1976). The Schaus Swallowtail: a threatened subspecies? *Insect World Digest* 3, 21-26.

Correa, M. N. (1971). *Flora Patagónica*. Buenos Aires, Argentina: Colección científica del INTA, Buenos Aires.

Del Valle, H. F., Eiden, G., Mensching, H., Goergen, J. (1995). *Evaluación del estado actual de la desertificación en áreas representativas de la Patagonia. Informe Final Fase I.* Proyecto de Cooperación Técnica entre la República Argentina y la República Federal Alemana. Cooperación técnica argentino-alemana, proyecto INTA-GTZ, Lucha contra la

Desertificación en la Patagonia a través de un sistema de Monitoreo Ecológico. INTA, Buenos Aires

Endlicher, W., Santana Aguila, A. (1988). El clima sur de la Patagonia y sus aspectos ecológicos. Un siglo de mediciones climatológicas en Punta Arenas. *Anales del Instituto de la Patagonia. Serie Ciencias Naturales*. 8, 57-86.

Faith, D., Minchin, P., Belbin, L. (1987). Compositional dissimilarity as a robust measure of ecological distance: a theoretical model and computer simulations. *Vegetatio* 69, 57-68.

Ferreyra, M., Cingolani, A., Ezcurra, C., Bran, D. (1998). High-Andean vegetation and environmental gradients in northwestern Patagonia, Argentina. *Journal of Vegetation Science* 9, 307-316.

Frangi, J., Richter, L. (1992). Los ecosistemas forestales de la Tierra del Fuego. *Vida Silvestre (ICONA, España)* 72, 36-43.

Golluscio, R. A., Deregibus, V., Paruelo, J. M. (1998). Sustainability and range management in the Patagonian steppe. *Ecología Austral* 8, 265-284.

Goodall, R. N. P. (1981). New collections of some rare plants of Tierra del Fuego. *Boletín de la Sociedad Argentina de Botánica* 20, 109-113.

Gotelli, N., Colwell, R. K., (2001). Quantifying biodiversity: Procedures and pitfalls in the measurement and comparison of species richness. *Ecology Letters* 4, 379-391.

Greenacre, M., (1984). *Theory and applications of correspondence analysis.* Academic Press, London.

Guzmán, L., Atalah, C., Venegas, C. (1985-86). Composición específica y estructura de la comunidad de aves de verano en el complejo de la Tundra Magellánica. *Anales del Instituto de la Patagonia, Serie Ciencias Naturales* 16, 75-86.

Hijmans, R. J., Cameron, S. E., Parra, J. L., Jones, P. G., Jarvis, A. (2005). Very high resolution interpolated climate surfaces for global land areas. *International Journal of Climatol*ogy 25, 1965-1978.

Hill, M. O. (1979). *DECORANA. A Fortran program for detrended correspondence analysis and reciprocal averaging*. Publ. Of. Section of Ecology and Systematics. Cornell University, Ithaca, US

Jobbágy, E. G., Paruelo, J. M., León, R. J. C. (1995). Estimación del régimen de precipitación a partir de la distancia a la cordillera en el noroeste de la Patagonia. *Ecología Austral* 5, 47-53.

Kent, M., Coker, P. (1992). *Vegetation description and analysis: a practical approach.* CRC Press-Belhaven Press, London.

Kim, K. (1993). Biodiversity, conservation and inventory: Why insects matter. *Biodiversity and Conservation* 2, 191-214.

Körner, C. (1994). Scaling from species to vegetation: The usefulness of functional groups. In: Schulze, E.-D., Mooney, H. A. (eds.), *Biodiversity and Ecosystem Function*, pp. 117-140. Springer, Heidelberg.

Kotze, D. J., Samways, M. J. (1999). Invertebrate conservation at the interface between the grassland matrix and natural Afromontane forest fragments. *Biodiversity and Conservation* 8, 1339-1363.

Lanfranco, D. (1977). Entomofauna asociada a los bosques de Nothofagus pumilio en la región de Magallanes. 1° parte: Monte Alto (Río Rubens, Ultima Esperanza). *Anales del Instituto de la Patagonia* 8, 319-346.

Lanfranco, D. (1991). Sinopsis de los insectos que atacan bosques de lenga (Nothofagus pumilio (Poepp. et Endl.) Krass.) en Magallanes. *Anales del Instituto de la Patagonia, Serie Ciencias Naturales* 20, 89-93.

Lencinas, M. V., Martínez Pastur, G., Anderson, C. B., Busso, C. (2008). The value of timber quality forests for insect conservation on Tierra del Fuego Island compared to associated non-timber quality stands. *Journal of Insect Conservation* 12, 461-475.

Lencinas, M. V., Martínez Pastur, G., Cellini, J. M., Gallo, E., Busso, C. (2010). Diversidad de lepidópteros en bosques aprovechados: variación en el corto plazo por aplicación de retención variable. *Revista Investigaciones Científicas de la UNERMB* 1, 87-101.

León, R. J. C., Aguiar, M. R. (1985). El deterioro por uso pastoril en estepas herbáceas patagónicas. *Phytocoenología* 13, 181-196.

León, R. J. C., Bran, D., Collantes, M., Paruelo, J. M., Soriano, A. (1998). Grandes unidades de vegetación de la Patagonia extra andina. *Ecología Austral* 8, 125-144.

Lewis, C. N., Whitfield, J. B. (1999). Braconid wasp (Hymenoptera: Braconidae) diversity in forest plots under different silvicultural methods. *Environmental Entomology* 28, 986-997.

Liski, J., Nissinen, A. R. I., Erhard, M., Taskinen, O. (2003). Climatic effects on litter decomposition from arctic tundra to tropical rainforest. *Global Change Biology* 9, 575-584.

Ludwig, J. A., Reynolds, J. F. (1988). *Statistical Ecology: A Primer in Methods and Computing*. John Wiley and Sons Ltd., New York.

Manly, B., (1994). *Multivariate statistical methods. A primer*. Chapman-Hall. London.

Martikainen, P., Siitonen, J., Punttila, P., Kaila, L., Rauh, J. (2000). Species richness of Coleoptera in mature managed and old-growth boreal forests in southern Finland. *Biological Conservation* 94, 199-209.

McCune, B., Mefford, M. J. (1999). *Multivariate analysis of ecological data.* Version 4.0. MjM software. Gleneden Beach (OR), US.

Milchunas, D. G., Sala, O. E., Lauenroth, W. K. (1988). A generalized model of the effects of grazing by large herbivores on grassland community structure. *The American Naturalist* 132, 87-106.

Moore, D. M. (1983). *Flora of Tierra del Fuego*. Anthony Nelson-Missouri Botanical Garden, Saint Louis. Moore, D. M., Goodall, R. P. N. (1974). Further additions to the native vascular flora of Tierra del Fuego. *Botanical Notiser* 127, 38-43.

Niemelä, J. (2001). Carabid beetles (Coleopterae: Carabidae) and habitat fragmentation: A review. *European Journal of Entomology* 98, 127-132.

Novotný, V., Basset, Y. (2000). Rare species in communities of tropical insect herbivores: Pondering the mystery of singletons. *Oikos* 89, 564-572.

Oliver, I., Beattie, A. J. (1993). A possible method for the rapid assessment of biodiversity. *Conservation Biology* 7, 562-568.

Parmesan, C., Yohe, G. (2003). A globally coherent fingerprint of climate change impacts across natural systems. *Nature* 421, 37-42.

Paruelo, J. M., Golluscio, R. A. (1993) Catálogo de estados y transiciones: Estepas graminoso-arbustivas del NW del Chubut. In: Paruelo, J. M., Bertiller, M., Schlichter, T., Coronato, F. (eds.), *Secuencias de deterioro en distintos ambientes patagonicos. Su caracterizacion mediante el modelo de estados y transiciones*, pp. 5-13. Instituto

Nacional de Tecnologia Agropecuria, Deutsche Gesellschaft fuer Technische Zusammenarbeit, Bariloche, Argentina.

Peri, P. L. (2011). Carbon Storage in Cold Temperate Ecosystems in Southern Patagonia, Argentina. In: Islam Atazadeh (ed.), *Biomass and Remote Sensing of Biomass*, pp. 213-226. InTech, Lijeka, Croatia.

Pyle, R. M. (1976). Conservation of Lepidoptera in the United States. *Biological Conservation* 9, 55-75.

Richards, O. W., Davies, R. G. (1984). *Tratado de entomología Imms*. Volumen 2: Clasificación y Biología. Ed. Omega, Barcelona.

Roig, F. (1998). La Vegetación de la Patagonia. In: Correa, M. *Flora Patagónica*, Tomo VIII, vol. I. pp. 48-174. INTA Colección Científica, Buenos Aires

Roig, F. A., Martínez Pastur, G., Moyano, V., Pinedo, L., Lencinas, M. V. (2002). *Cambial activity, phenology and climate in* Nothofagus pumilio *forests of Tierra del Fuego, Argentina*. Sixth International Conference on Dendrochronology. Quebec.

Romoser, W. S., Stoffolano, J. G. (1998). *The science of entomology*. WCB/McGraw-Hill Boston.

Ross, H. (1973). *Introducción a la entomología general y aplicada*. Ed Omega, Barcelona.

Sala, O. E., Parton, W. J., Joyce, L. A., Lauenroth, W. K. (1988). Primary production of the central grassland region of the United States. *Ecology* 69, 40-45.

Soriano, A., (1983). Deserts and semi-deserts of Patagonia. In: N. E. West (ed.), *Temperate Deserts and Semi-deserts*, pp. 423-459. Elsevier, Amsterdam.

Soriano, A., Nogués Loza, M., Burkart, S. (1995). Plant biodiversity in the extra-andean Patagonia: Comparisons with neighboring and related vegetation units. *Actas del Taller Internacional sobre Recursos Fitogenéticos, Desertificación y Uso Sustentable de los Recursos Naturales de la Patagonia*, pp. 36-45. Río Gallegos, Argentina.

Spagarino, C., Martinez Pastur, G., Peri, P. L. (2001). Changes in *Nothofagus pumilio* forest biodiversity during the forest management cycle: 1. Insects. *Biodiversity and Conservation* 10, 2077-2092.

Wells, S. M., Pyle, R. M., Collins, N. M. (eds.). (1983). *The IUCN Invertebrate Red Data Book*. IUCN, Gland and Cambridge.

Werner, S. M., Raffa, K. F. (2000). Effects of forest management practices on the diversity of ground-occurring beetles in mixed norhern hardwood forests of the Great Lakes Region. *Forest Ecology and Management* 139, 135-155.

Westoby, M., Walker, B., Noy-Meir, I. (1989) Opportunistic management for rangelands not at equilibrium. *Journal of Range Management* 42, 266-74.

Wynn, J. G., Bird, M. I., Vellen, L., Grand-Clement, E., Carter, J., Berry, S. L. (2006). Continental-scale measurement of the soil organic carbon pool with climatic, edaphic, and biotic controls. *Global Biogeochemical Cycles* 20, GB1007, doi:10.1029/2005GB 002576.

In: Steppe Ecosystems
Editors: M. B. Morales Prieto and J. Traba Diaz
ISBN: 978-1-62808-298-2

Chapter 5

THE VEGETATION STRUCTURE OF THE MONGOLIAN ALTAI MOUNTAIN STEPPES

Galina N. Ogureeva,[*] Inessa M. Mikljaeva and Maxim V. Bocharnikov
Faculty of Geography, Moscow State University, Moscow, Russia

ABSTRACT

The analysis of vegetation structure of mountain steppes in the Mongolian Altai was carried out within the project "Mongolian Ecosystem Monitoring" with the support of the Joint Russian-Mongolian Complex Biological Expedition of the Russian and the Mongolian Academies of Sciences. The diversity of steppe vegetation is reflected in the structure of the altitudinal zonation of the Mongolian Altai. Central Mongolian-Altai type of altitudinal zonation is characteristic of the highest parts of the Mongolian Altai with their natural differentiation into high-altitude belts. The current coenotic diversity of Mongolian Altai steppes is presented on a medium-scale (1: 1 000 000) map. Different types of vegetation units are displayed on the map. Phytocatenas with a certain proportion and arrangement of plant communities are used as the basic structural units of vegetation on mountain slopes. Combinations of phytocatenas on the slopes of different aspects (expositional combinations) are characteristic of areas with different composition of rocks within an altitudinal vegetation belt. The results of this chapter are expected to help in solving the current problems associated with rapid human-induced transformation of the steppe regions of Mongolia: population growth, changing social and political processes, and introduction of new ways to use land and natural resources.

Keywords: Biodiversity, mountain ecosystems, steppes, structure of mountain vegetation, vegetation mapping

[*] Corresponding author: Galina N. Ogureeva. Faculty of Geography, Moscow State University, Leninskie Gory, 119991, Moscow, Russia. E mail: ogur02@yandex.ru.

INTRODUCTION

Reliable information on the actual state of natural ecosystems and their biodiversity is necessary to solve the problems of mountain ecology. Vegetation cover can be used as a reliable indicator of ecosystems' health at a regional level. Over centuries of human-nature interaction in mountain areas, specific types of resource use by local people have arisen. One example is the use of mountain vegetation as seasonal pastures by nomadic livestock in the Mongolian Altai. Preservation of the natural structure of vegetation cover and its monitoring, effective management of natural resources accounting for the specific features of the mountain systems, conservation and revival of cultural values, people's traditions and customs, along with the general increase of welfare of local populations will help make these territories attractive to people and further development.

The results presented and discussed in this chapter are part of a long-term research of the Daurian-Mongolian sector of the steppe biome carried out thanks to the collaborative efforts of Russian and Mongolian experts of the Joint Russian-Mongolian Complex Biological Expedition of the Russian and Mongolian Academies of Sciences within the program of environmental monitoring "Assessment of the current state of ecosystems of Mongolia", aimed at their preservation and restoration. An important component of steppes monitoring is an on-going vegetation mapping which allows revealing qualitative and quantitative changes of ecosystem components, as well trends in the main natural and anthropogenic processes. The outcomes of these studies are expected to contribute to solving the actual problems related to the rapid transformation of steppe regions of Mongolia under vigorous population growth, to changing social and political processes, and to introducing new approaches to the use of land and natural resources.

MATERIAL AND METHODS

The fieldwork investigation of ecosystems of the Mongolian Altai was carried out in 2010 and 2012 by the mountain group of the Joint Russian-Mongolian Complex Biological Expedition of the Russian and Mongolian Academies of Sciences. While working in the central Mongolian Altai (Figure 1), an inventory of mountain steppe ecosystems in different altitudinal belts in relation the landscape features was made, and the types of vegetation structures within altitudinal belts were determined. It gave the chance to identify regional geobotanical features of mountain steppes and proceed to the mid-scale and overview mapping of mountain vegetation of the Mongolian Altai using remote-sensed imagery and geographic information systems. The obtained regional features of mountain steppe ecosystems were represented on vegetation and ecosystem maps of Mongolia. This chapter synthesizes the floristic analysis and mapping of the information compiled in the assessment of steppe ecosystems of the Mongolian Altai. The Mongolian Ecosystem Map (1:1000000, 1995) served as a basis for the study.

The steppes of Mongolia belong to the geobotanical sub-region of Daurian-Mongolian steppes, located at the eastern part of the Eurasian steppe biome (Lavrenko and Karamysheva, 1993; Yunatov, 1950). The Mongolian-Altai mountain steppes can be considered an independent province with the original types of altitudinal zonation (Yunatov, 1950).

a

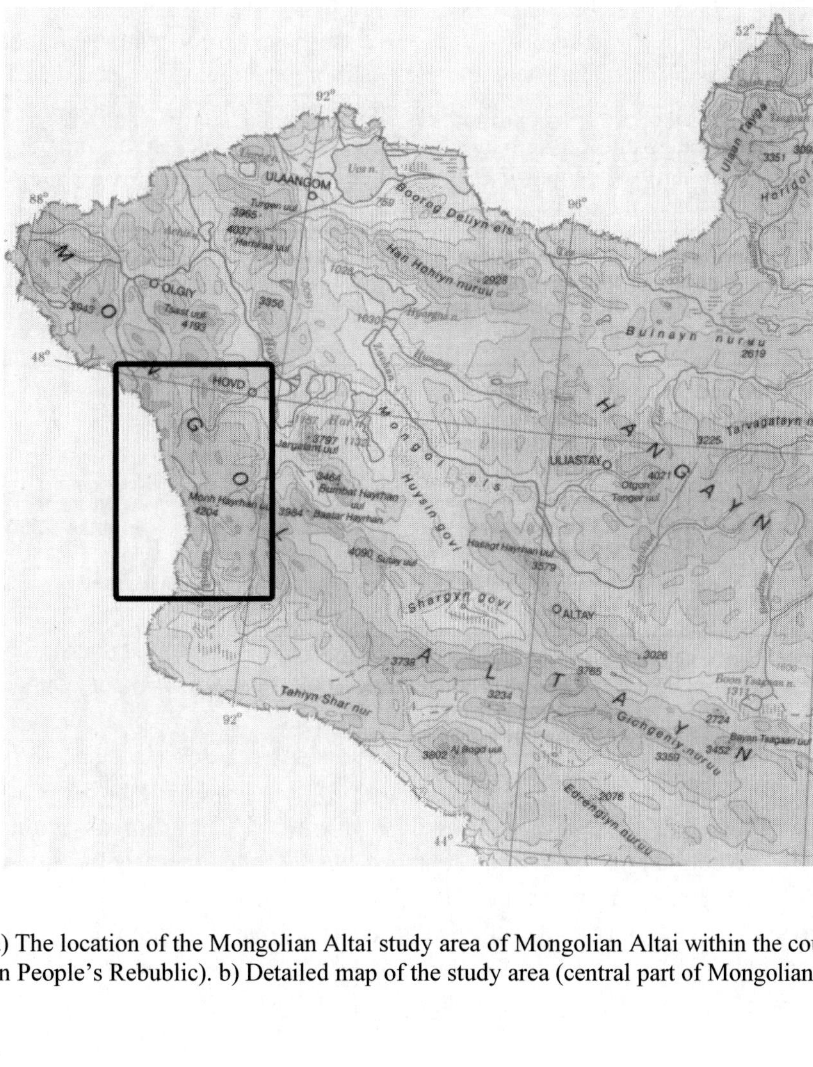

b

Figure 1. a) The location of the Mongolian Altai study area of Mongolian Altai within the country (Mongolian People's Rebublic). b) Detailed map of the study area (central part of Mongolian Altai).

An important formation of granite and slate massifs dominates the central part of the Mongolian Altai, shaping its landscape structure. Climate is continental with moderately dry and dry cool and moderately cold summer, and severe winter (National Atlas, 1990). The average annual precipitation in the higher mountain belts is around 400 mm, and 100–200 mm in the lower belts. The average annual temperature in the higher belts is -6°C and -2°C in the lower belts.

A variety of steppe vegetation is reflected in the structure of altitudinal zonation of the Mongolian Altai, which stretches from the northwest to the southeast of Mongolia for more than 1000 kilometers. The Central Mongolian-Altai type of zonation is characteristic of the highest part of the Mongolian Altai (Munkh-Hayrkhan Mount, 4204 m above sea level). The north of these highlands corresponds to the North-Western Mongolian-Altai altitudinal zone, while the southern part of the ridge corresponds to the the Southern Mongolian-Altai zone (Hilbig, 1995; Ogureeva, 1997; Beket, 2009). Within each zone, mountain steppe ecosystems are classified according to their natural differentiation into altitudinal belts taking into account vertical gradients of temperature and moisture (National Atlas, 1990). Plant communities create a complicated mosaic within each altitudinal level which is described below.

During fieldwork, mountain steppe ecosystems in each altitudinal belt were surveyed. Characteristic types of steppes and their variation depending on rock composition and aspect were defined. For each altitudinal belt, the main structural units of vegetation cover were determined and the characteristics of their components described. The survey was carried out through automobile routes and on key sites, with the aid of remote-sensed imagery. For the central part of the Mongolian Altai 120 standard geobotanical descriptions were made on100 m^2 plots.

The analysis of these descriptions with allocation of constantly present and dominant species of plants made possible the classification of plant communities, the determination of the main types of vegetation structure, and the description of steppe ecosystems. The results of this analysis became the basis for the interpretation of the map of ecosystems of the Mongolian Altai.

RESULTS AND DISCUSSION

Spatial Organization of Steppe Vegetation of the Mongolian Altai

The spatial organization of vegetation cover is represented by its territorial subdivisions (combinations or complexes), which define a stable spatial pattern of plant communities in relation to abiotic conditions and geomorphologic integrity of territorial units (Sochava, 1978). A great variety of ecotopes is typical of mountains because their slopes differ in the steepness, aspect and rock composition. Therefore various types of the vegetation units characteristic of mountain altitudinal belts could be identified. The main structural units for the slopes of an altitudinal belt are geomorphogenic and topo-ecological series or phytocatenas which present a particular composition and arrangement of plant communities along the slope. A phytocatena is, therefore, a sequence of plant communities of a mountain slope integrated through common soil formation and slope processes (an ecological-dynamic series of communities from the upper to the lower parts of a slope).

The serial vegetation of petrophytic communities is usually developed in the top part and the upper parts of the slopes at rubble and stony sites. Transition parts of slopes are covered with steppe communities typical of the altitudinal belt. More mesophytic communities occupy alluvial parts of the slopes and adjacent inter-hill depressions and plains. The most complex phytocatenas are formed on slopes with high degree of ecotopic contrast. Phytocatenas on slopes of different aspect and rock composition differ in vegetation structure and composition. They are characteristic of mountain areas with different rock composition and relief conditions (Ogureeva et al., 2012). Slopes of different aspects generate different different combination of phytocatenas. These expositional combinations are reflected in the map presented in Figure 1. For example, in the strip of desert steppes "Expositional combinations: sere of petrophytic herb-bunchgrass, sedge-grass of northern exposition slopes and bunchgrass desert steppes of the southern exposition slopes on the granite hammocky topography" (N 21, Annex 1) is typical. On pronounced bald reliefs like those of the Altai Mountains, the association of geomorphological units and plant communities constitute a particular type of landscape complex known as a bald mountain complex (Ilyinskaya, 1980; Gromtsev, 2000). For the first time, aggregates as macro-combinations of these bald mountain complexes are identified in the altitudinal belts of the Mongolian Altai, described and depicted in the map of mountain steppes of Mongolia. The series of flood-plain communities dependent on the manifestation and activity of channel processes are characteristic of river valleys. The identification of floodplain series is important for the development of haylands and their management.

Map of Ecosystems of the Mongolian Altai

The actual coenotic diversity of steppes of the Mongolian Altai is shown on a mid-scale ecosystem map (Figure 2) which reflects the relationships between of the actual vegetation cover and the mountain altitudinal gradient. Various types of structural vegetation units within each altitudinal belt are also represented. The fragment of the map included in Figure 2 provides an example corresponding to the highest part of the central Mongolian Altai, and gives an idea of the diversity of existing vegetation types (25 units for all altitudinal belts). Table 1 summarizes the main features of each altitudinal belt in terms of dominant vegetation and extent.

Coenotic Diversity of Mountain Steppes of the Mongolian Altai

The Central Mongolian-Altai vegetation zonation is characterized by a wide distribution of steppes on the slopes of all aspects, while isolated fragments of steppe deserts occur at the bottom of intermountain depressions. The *nival belt* is the area of present-day glaciation. Within the *subnival* and *alpine-tundra belts* groups of cryopetrophytes are widespread at the upper limit of plant distribution. Cryopetrophyte-herb communities and those of mountain meadows and mountain tundra occur in places with more favorable conditions for snow accumulation. A wide distribution of cushion plant formation with *Stellaria pulvinata* is characteristic of the belt; both grasses and sedges are present in the serial communities. The series are represented by two stages: a) petrophytic – on rock outcrops, rock debris and among stone fields; b) well-developed cushion plant formations on the flattened parts of the slopes, on the debris of widespread granites or rubble substratum of bedrocks.

Table 1. Altitudinal vegetation belts present in the Munkh-Hayrkhan Mountain (central Mongolian Altai) and their respective altitude range and area occupied. The percentages within parentheses indicate the corresponding proportion of the total area considered

Belt	Altitude range (m a.s.l.)	Area occupied (km^2)
Nival	3800–4200	110.17 (0.6%)
Vegetation of high-mountain cryoxerophytic communities of subnival and alpine-tundra	3200–4000	3945.41 (22.9%)
Vegetation of high-mountain cryophytic steppes *upper zone*	2700–3200	1542.59 (8.9%)
Vegetation of high-mountain cryophytic *lower zone*	2600–2800	1191.00 (6.9%)
Vegetation of steppe, upper zone of dry steppes	2100-2700	8089.10 (46.9%)
Vegetation of steppe, lower zone of desert steppes	1700-2100	895.46 (5.2%)
Vegetation of desert belt	1300-1900	1479.15 (8.6%)
Total area		17252.88 (100%)

Cryophyte-herb – sedge (*Carex*) - *Kobresia* cushion plant formations are typical of wetter habitats, while cushion plant formations with grasses occupy better drained surfaces.

The belt of *high-mountain cryophytic steppes* is characterized by cryopetrophyte-herb – *Kobresia* – sedge communities and cushion plant formations, in combination with grass – herb, wormwood (*Aretemisia*) – herb – sod grass cryophytic steppes. There are two altitudinal zones within the belt.

In the upper zone (2700-3200 m, 8.9 %) communities of sedge – *Kobresia* meadows and cushion plant formations with grasses and sedges predominate. Cryophyte-herb - grass cushion plant formations mixed with cryophytic steppes on the southern slopes of granite-built denudational hummocks.

In the lower zone (2600-2800 m, 6.9%) *Oxytropis tragacanthoides* series grow in combination with grass cushion plant formations on southern slopes and sedge – *Kobresia* meadows on northern ones. On concave parts of northern slopes of the river valleys small fragments of larch (*Larix sibirica*) woodlands with *Kobresia myosuroides* – sedge cover occur in combination with dwarf birch (*Betula nana*) thickets.

Vegetation of the *mountain steppe* belt covers more than a half of the mountain area (8984.56 km^2 (55.2%). It encompasses two altitudinal zones of the belt, i.e. dry steppes and desert steppes, which include specific sets of communities and their combinations. The vegetation cover is represented by dry sod grass (sheep's fescue *Festuca lenensis* – meadow-grass, feather-grass *Stipa* – wheat grass *Agropyrum*), sedge – sod grass and herb – sod grass steppes. The most complex combinations are formed on the slopes of hummocks due to their high degree of ecotope contrast. The hummocks are often formed in the river valleys within granite and schist massifs, which are widespread in this part of the Mongolian Altai.

The strip o*f dry steppes* (2100-2700 m) is dominated by firm-bunch grass steppes: wheat grass (*Agropyron cristatum*) – feather-grass (*Stipa krylovii*) and wormwood (*Artemisia frigida*) – sheep's fescue (*Festuca lenensis*) steppes.

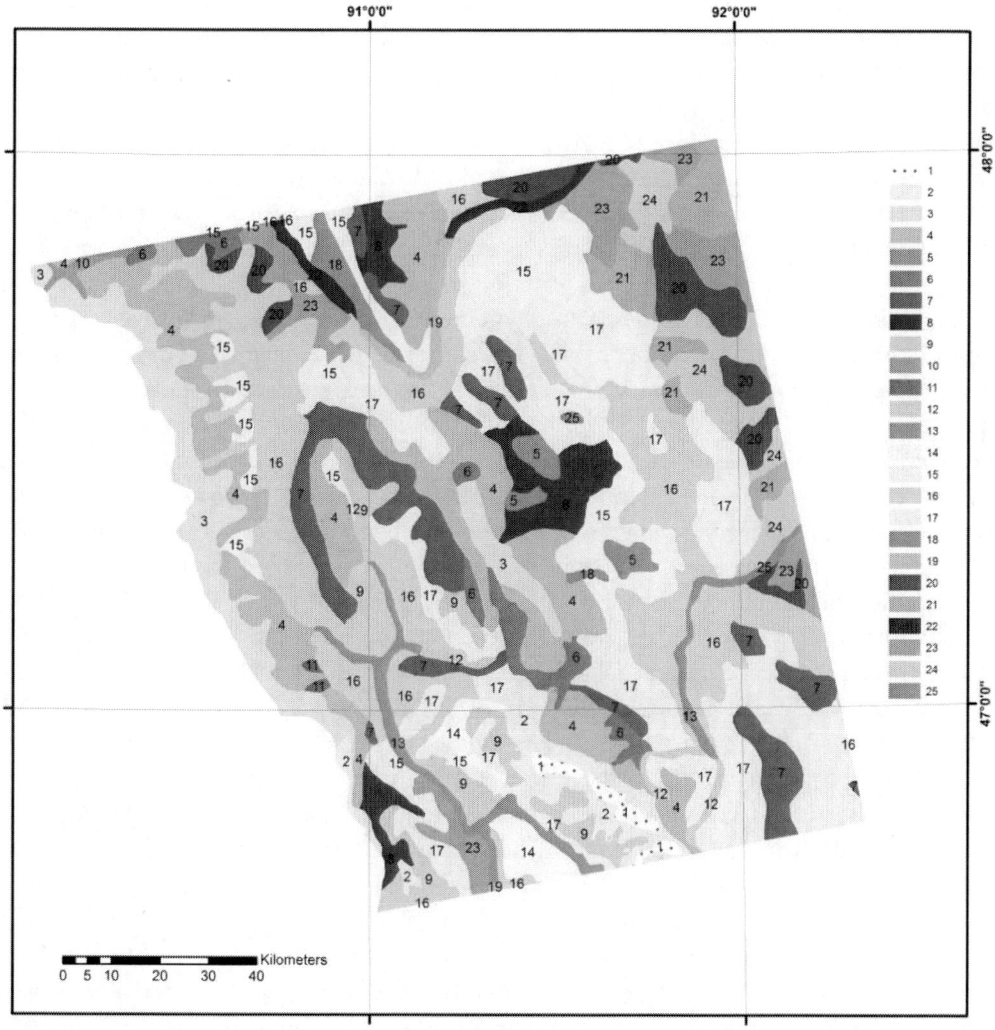

Figure 2. Map of existing ecosystems of the Mongolian Altai (Scale 1:1 000 000). The legend describing each of the numbered vegetation units is provided in Annex 1.

Two invariants of bald mountain complexes for granite massifs and one for schist hummocks were identified. Expositional combinations of phytocatenas on granite massifs are characterized by the presence of pea shrub (*Caragana*) and wheat grass steppes. Pea shrub steppes reach the depressions of the intermountain plains where they cover considerable areas. The combination of wheat grass and feather-grass – wheat grass steppes with tragacanth serial communities is characteristic of the second invariant of bald mountain complexes for granite. The following phytocatena is characteristic of granite massifs in the Bulgan River valley: a) petrophytic steppes (*Oxytropis tragacanthoides, Thymus gobicus, Orostachys spinosa*) in the upper eluvial part of slopes; – b) wheat grass – feather-grass (*Stipa krylovii, Festuca lenensis, Agropyron cristatum*) steppes along the transit parts of slopes; – c) wormwood – blood root – grass – sedge (*Artemisia frigida, A. santolinifolia, Potentilla bifurca, Carex duriuscula*) and blood root – sheep's fescue (*Potentilla acaulis, P. bifurca, Festuca lenensis*) steppes in the alluvial part of slopes.

Sheep's fescue and sheep's fescue – feather-grass (*Festuca lenensis*, *Stipa krylovii*) steppes predominate within schist massifs. Various communities of sedge – herb – grass meadows with halophilous species are grow in river valleys in combination with pea shrub communities.

The zone of *desert steppes* (1700-2100 m) on the western macro-slope of the Mongolian Altai occurs mainly on schist massifs with the typical bald mountain complex of petrophytic-herb – sheep's fescue – wheat grass (*Festuca lenensis*, *Agropyron cristatum*) steppes in combination with wheat grass – needlegrass (*Stipa glareosa*) and wormwood – dwarf sedge – wheat grass – needlegrass desert steppes. The following phytocatena is characteristic of schist massifs: a) in the top part of slopes petrophytic-herb – wheat grass (*Agropyron cristatum*) steppes; – b) in the transit parts desert sedge – sod grass – needlegrass (*Carex duriuscula, Stipa glareosa, Agropyron desertorum, Festuca lenensis*) steppes; – c) in the alluvial part of slopes desert sedge – wheat grass – wormwood (*Carex duriuscula, Agropyron desertorum, Kochia prostrata, Krascheninnikovia ceratoides, Artemisia santolinifolia*) steppes. Desert steppes are widespread in the intermountain plains, occasionally in combination with pea shrub communities (*Caragana bungea, C. pygmaea*). River valleys are covered with grass – sedge and halophyte herbs – grass meadows which are intensively used for haymaking.

Vegetation of the *desert belt* (1300-1900 m) is typical of more southern parts of the Mongolian Altai neighboring the deserts. The desert belt occupies 1479,15 sq. km / 8,6% and includes the zone of steppe deserts. Desert species occur in all plant communities of the belt. The combination of desert steppes (*Stipa glareosa, Agropyron cristatum, A. desertorum*) in the transit parts of slopes and steppe deserts (*Anabasis brevifolia, Kochia prostrata, Stipa glareosa, Convolvulus ammanii*) in the lower parts of slopes and on the deluvial aprons is characteristic for the bald mountain complex of schist massifs. Psammophyte desert communities prevail in the second type of bald mountain complexes where schist massifs gradually plunge under a layer of wind-blown sand. The following phytocatena is characteristic of this type of bald mountain complexes: a) petrophyte-herb series of communities in the top parts of slopes; – b) wormwood – meadow-grass – wheat grass desert steppes on the transit sites of slopes; – c) sedge – Chinese lyme grass (*Leymus chinensis*) steppes in the lower parts of slopes in combination with psammophyte needlegrass – nanophyton (*Nanophyton mongolicum, Stipa glareosa, Cleistogenes soongarica, Reaumuria soongarica, Tribulus terrestris, Allium mongolicum, Krascheninnikovia ceratoides*) steppe deserts. Various types of halophyte meadows occur in the river valleys.

Conclusion

The structural units of each altitudinal belt reflect the diversity of its environments. Structural units of vegetation cover which providing more information about the spatial distribution of particular plant communities were used for the mapping of mountain steppe ecosystems of Mongolian Altai. The structure of vegetation cover demonstrates the actual ecological potential of altitudinal vegetation belts of the Mongolian Altai. The identification and mapping of vegetation units is an essential tool for the assessment of human-induced disturbance on these ecosystems. The extreme climate conditions imposed by continentality and high altitude make mountain steppes particularly vulnerable to grazing intensification.

In this context, the results presented in this chapter can be of great use to accurately assess the current state of the Mongolian Altai ecosystems.

APPENDIX

Description of the vegetation units depicted in Figure 2. Latin names are given by I.A. Gubanov, (1996).

High mountain cryoxerophytic communities (subnival, alpine and alpine-tundra belts) (3200 - 4000 m.a.s.l.)

1 Open cryopetrophytic communities (*Waldheimia tridactylites, Senecio pricei, Saxifraga oppositifolia*) of the glacial-nival and subnival habitats.

2 Sparse cryopetrophytic communities (*Rhodiola quadrifida, Lagotis integrifolia, Bistorta vivipara, Saussurea involucrata, Oxygraphis glacialis, Dracocephalum imberbe*) of the planation surfaces with rock fields.

3 Lichen, lichen-dryad (*Dryas punctata, Salix nummularia*) tundra and herb-grass (*Festuca brachyphylla, Alopecurus turczaninovii, Anthoxanthum odoratum, Deschampsia koelerioides, Cerastium cerastoides, Lagotis integrifolia, Smelowskia alba, Oxytropis chionophylla*) meadows.

4 Cryophytic forb-sagebrush-cushion plant communities (*Stellaria pulvinata, Arenaria capillaris Androsace chamaejasme, Saussurea leucophylla, Oxytropis oligantha, Lagotis integrifolia, Carex rupestris*) with fragments of sedge-kobresia meadows of the tope slopes, planation surfaces and moraines (high mountain cushion plant communities – 3200-3500 m).

5 Cryophytic forb communities with *Carex*, *Kobresia*, *Kobresia*-grass-sedge (*Carex rupestis, C. sempervierens, Kobresia myosuroides, K. smirnovii, Koeleria altaica, Saxifraga hirculus, Bistorta vivipara, Thalictrum alpinum, Leontopodium ochroleucum*) tundra and grasslands of the slopes and the planation surfaces.

6 Grass-sedge-kobresia, kobresia-sedge wet tundra (*Carex rupestris, C. melanantha, Kobresia myosuroides, K. smirnovii, Festuca kryloviana, Koeleria altaica, Lagotis integrifolia, Melandrium apetalum, Saxifraga hirculus, Thalictrum alpinum, Gentiana algida, Leontopodium campestre, Erigeron eriocalyx*) of the planation surfaces.

High mountain cryophytic steppe vegetation (2600-3200 m.a.s.l.)
The upper strip (2700-3200 m.a.s.l.)

7 Expositional combinations: Cryophytic forb-grass cushion plant communities (*Festuca kurtschumica*, *Poa altaica*, *Helictotrichon asiaticum*, *Trisetum spicatum*, *Stellaria pulvinata*, *Saussurea leucophylla*, *Androsace incana*, *Oxytropis oligantha*, *Smelovskia alba*, *Amblynotus rupestris*) of the northern exposition slopes and cryophytic steppes (*Festuca lenensis*, *Poa attenuata*, *Koeleria altaica*, *Artemisia frigida*, *Potentilla nivea*, *Chamaerhodos altaica*) of the southern exposition slopes on the granite hummocky topography.

8 Expositional combinations: Cryophytic forb-Kobresia-sedge (*Carex rupestris*, *C. sempervirens*, *Kobresia myosuroides*, *K. smirnovii*, *Lagotis integrifolia*, *Melandrium apetalum*, *Saxifraga macrocalyx*) with cushion plant communities (*Stellaria pulvinata*, *Amblyonotus rupestris*, *Oxytropis oligantha*, *Sibbaldianthe adpressa*, *Arenaria meyeri*) of the northern exposition slopes and cryophytic forb-grass steppes (*Poa altaica*, *P. attenuata*, *Koeleria altaica*, *Festuca brachyphylla*, *F. kurtschumica*) of the southern exposition slopes on the granite hummocky topography.

The lower strip (2600-2800 m.a.s.l.)

9 Expositional combinations: cryophytic bunchgrass (*Agropyron cristatum*, *Festuca kyrtschumica*, *F. lenensis*)-forb-bunchgrass (*Agropyron cristatum*, *Chamaerhodos altaica*, *Allium altaicum*, *Arctogeron gramineum*), sagebrush-bunchgrass (*Artemisia obtusiloba*, *A. frigida*) steppes of the northern exposition slopes and forb-bunchgrass (*Poa attenuata*, *Festuca kyrtschumica, F. lenensis*, *Carex duriuscula*, *Potentilla aphanes*) steppes of the southern exposition slopes with fragments of cushion plant communities (*Stellaria pulvinata*) on the granite hummocky topography.

10 Subalpine bush (*Betula rotundifolia*, *Lonicera hispida*, *Spiraea alpine*, *Lagotis integrifolia*, *Luzula sibirica*, *Saxifraga hirculus*), willow plant communities (*Salix hastata*, *S. berberifolia*, *S. glauca*) in combination with larch open woodlands.

11 Larch (*Larix sibirica*) woodlands with grass-Kobresia, Kobresia-sedge (*Kobresia myosuroides*, *Carex rupestris*, *Ptilagrostis mongholica*, *Festuca sibirica*, *Alopecurus pratensis*, *Poa pratensis*, *Bistorta vivipara*, *Hedysarum alpinum*, *Aconitim volubile*) plant communities and larch open woodlands with dwarf birch (*Betula rotundifolia*) on the concave part of northern exposition slopes.

The vegetation of the river valleys of the belt

12 Sedge-Kobresia (*Kobresia myosuroides*, *K. smirnovii*, *Carex repestris*, *C. sempervirens*), grass-sedge (*Poa pratensis*, *Calamagrostis macilenta*, *Bistorta vivipara*, *Saxifraga hirculis*, *Oxyria dygina*), dwarf birch (*Betula rotundifolia*) plant communities on the frozen soils along the streams.

13 Grass-sedge (*Carex duriuscula*, *Poa pratensis*), sometimes in combination with shrub communities (*Salix glauca*, *Dasiphora fruticosa*) by the flood plains.

The vegetation of the steppe belt
The strip of dry steppes (2100-2700 m)

14 Wormwood-grass (*Koeleria altaica*, *K. crisata*, *Festuca lenensis*, *Artemisia frigida*, *A. rhutifolia*, *Potentilla sericea*), sedge-bunchgrass (*Carex pediformis*, *C. duriuscula*, *Helictotrichum altaicus*, *Aster altaicus*, *Gentiana decumbens*, *Erithichium rupestris*) steppes with larch open woodlands of northern exposition slopes.

15 Phytocatena: petrophytic herb-bunchgrass (*Agropyron cristatum*, *Stipa krylovii*, *Cleistogenes squarrosa*, *Gypsophila desertorum*, *Ptilotrishum canescens*, *Lagochilus diacanthophyllus*) steppes on the top eslopes –herb-bunchgrass (*Stipa krylovii*, *Koeleria cristata*, *Sibbaldianthe adpressa*, *Amblinotus rupestris*, *Bupleurum bicaule*,

Dontostemon integrifolius) steppes of the transition parts of the slopes – sedge-sagebrush –bunchgrass steppes (*Carex duriuscula, Artemisia frigida, Agropyron cristatum*) on the lower parts of slopes and deluvial fans with shrub (*Caragana pygmaea, C. bungei*) of southern exposition slopes on the granite.

16 Expositional combinations: tragacanth petrophytic (*Oxytropis tragacanthoides, Dracocephalum peregrinum, Festuca lenensis, Orostachys spinosa*) steppes of the topes slopes – bunchgrass, wormwood-bunchgrass, herb-bunchgrass steppes (*Stipa krylovii, S. glareosa, Agropyron cristatum, A. desertorum, Krascheninnikovia ceratoides, Allium mongolicum, Carex duriuscula, Artemisia frigida, A. santolinifolia*) steppes of deluvial fans of southern exposition slopes and bunchgrass (*Festuca lenensis, Stipa krylovii, Koeleria cristata, Aster alpinus, Potentilla acaulis, P. bifurca*) of the northern exposition slopes on the granite hummocky topography.

17 Phytocatena: sere of petrophytic herb (*Sibbaldianthe adpresssa, Helictotrichon schellianum, Festuca lenensis, Agropyron cristatum, Koeleria cristata*) of the tope slopes – bunchgrass steppes (*Agropyron cristatum, Festuca lenensis, Koeleria cristata*) of the transition part of the slopes – sagebrush-bunchgrass-sedge (*Artemisia frigida, Potentilla bifurca, Carex duriuscula*) steppes of the lower part of slopes of southern exposition slopes on the schistose complex.

18 Bunchgrass steppes (*Festuca lenensis, Poa attenuata, Stipa krylovii, Koeleria cristata*) with shrub (*Caragana bungei, C. pygmaea*) on intermontane plains.

Vegetation of river valleys of the dry steppes

19 Grass-sedge, forb-grass meadows (*Ptilagrostis mongholica, Leymus chinensis, Hordeum brevisubulatum, Agrostis mongholica, Carex duriuscula, Rheum compactum, Stellaria dichotoma*) in the flood plains.

The strip of desert steppes (1700-2100 m.a.s.l.)

20 Expositional combinations: sere of petrophytic herb-bunchgrass (*Artemisia santolinifolia, A. rutifolia, Aster alpinus, Orostachys spinosa*), small cushion plant communities (*Artemisia pycnorhyza, Oxytropis trichophysa, Arenaria cappilaris, Astragalus galactites, Hedysarum pumilum, Rhinactinidia eremophyla*) of the top eslopes – herb-bunchgrass (*Stipa krylovii, S. glareosa, Agropyron cristatum, Krascheninnikovia ceratoides, Stellaria graminifolia, Dontostemon integrifolius*) of the southern exposition slopes and wormwood-sedge-bunchgrass (*Festuca lenensis, Agropyron cristatum, Artemisia frigida, A. santolinifolia, Carex duriuscula, Panzerina lanata, Kochia prostrata, Bupleurum bicaule*) desert steppes of northern exposition slopes sometimes with shrub communities (*Caragana bungei, C. pygmaea*) on the schistose complex.

21 Expositional combinations: sere of petrophytic herb-bunchgrass (*Stipa krylovii, Agropyron cristatum, Poa attenuata, Cleistogenes squarrosa, Orostachis spinosa, Dontostemon perennis*), sedge-grass (*Carex duriuscula, Leymus chinensis*) of northern exposition slopes and bunchgrass (*Stipa glareosa, S. krylovii, Agropyron cristatum, Kochia prostrateprostrata, Artemisia frigida*) desert steppes of the southern exposition slopes on the granite hummocky topography.

The vegetation of the river valleys

22 Sedge-galophytichalophytic-grass (*Hordeum brevisubulatum*, *Puccinellia tenuiflora*, *Calamagrostis macilenta*, *Carex duriuscula*,) meadows and plant communities with *Achnatherum splendens*, *Iris lacteal* and *Caragana Bungei*.

Vegetation of the desert belt (1300-1900 m)

23 Phytocatena: petrophytic forb-bunchgrass (*Stipa glareosa*, *Agropyron cristatum*, *A. desertorum*, *Elytrigia geniculata*, *Kochia prostrata*, *Heteropappus hispidus*, *Dontostemon perennis*) desert steppes of the tope slopes – bunchgrass (*Stipa glareosa*, *Artemisia rutifolia*, *Krascheninnikovia ceratoides*, *Kochia prostrata*) desert steppes of the transit parts of slopes – bunchgrass with desert plants (*Anabasis brevifolia*, *Kochia prostrata*, *Stipa glareosa*, *Convolvulus ammanii*)- semi-desert of the lower parts of slopes and diluvial fans on the schistose complex.
24 Psammophytic desert steppes (*Nanophyton mongolicum*, *Stipa glareosa*, *Cleistogenes soongorica*, *Convolvulus gortschakovii*, *C. ammanii*, *Reaumuria songarica*, *Tribulus terrestris*, *Allium mongolicum*, *Krascheninnikovia ceratoides*, *Artemisia scoparia*, *Zygophyllum pterocarpum*, *Astragalus monophyllus*).

The vegetation of the river valleys

25 Halophytic meadows (*Hordeum brevisubulatum*, *Agrostis mongolica*, *Phragmites australis*, *Achnatherum splendens*).

ACKNOWLEDGMENTS

The authors express their sincere gratitude to the leaders of the Joint Russian-Mongolian Complex Biological Expedition of the Russian Academy of Sciences and of the Mongolian Academy of Sciences (MAS), Professor Doctor Gunin (Doctor in Sciences, Professor), and Ch. Dugarzhavu (Member of the Mongolian Academy of Sciences MAS) for the opportunity to lead this work.

REFERENCES

Beket, U. (2009). *The Vegetation of the Mongolian Altai*. Problems of sustainable land use and nature conservation. Bad Godesberg, Bonn, Germany.

Ecosystems of Mongolia. (1995). Map scale 1:1 000 000. Russian Academy of Science, Moscow.

Gromtsev, A. N. (2000). *Landscape ecology of boreal forests: Theoretical and applied aspects*. Forest Institute of the Karelian Research Centre of the Russian Academy of Science Petrozavodsk, Russia. [in Russian].

Gubanov, I. A. (1996). *Conspectus of Flora of Outer Mongolia (Vascular Plants).* Valang press, Moscow. [in Russian].

Hilbig, W. (1995). *The vegetation of Mongolia.* SPB Academic Publishing, Amsterdam.

Ilyinskaya, S. A. (1980). Landscape complexes of forest types. *Lesovedeniye* 4, 20-29. [in Russian].

Lavrenko, E. M., Karamysheva, Z. V. (1993). Steppes of the former Soviet Union and Mongolia. *Ecosystems of the World* 8, 3-59.

Namzalov, B. B. (1994). *Steppes of the Southern Siberia.* Novosibirsk-Ulan-Ude. [in Russian].

National Atlas, *Mongolian People's Republic.* (1990). Ulan-Bator-Moscow. [in Russian]

Ogureeva, G. N. (1994). Structural and Dynamic Categories in the Vegetation of Mountainous Areas. *Bulletin of the Moscow Society of Nature Scientists* 99, 76-85. [in Russian].

Ogureeva, G. N. (1997). Structure and dynamic of a vegetation cover of high mountain ecosystems of the Mongolian Altai. *Arid Ecosystems* 7, 119-133.

Ogureeva, G. N., Miklyaeva, I. M., Bocharnikov, M. V. (2012). Present-day state and the trends of transformation of the mountain ecosystems in Mongolia, *Moscow University* 5, 28-34. [in Russian].

Sochava, V. B. (1978). *Introduction to the Teaching of Geosciences.* Nauka, Novosibirsk. [in Russian].

Yunatov, A. A. (1950). *Main features of the plant cover of the Mongolian People's Republic.* Nauka, Moscow-Leningrad. [in Russian].

In: Steppe Ecosystems
Editors: M. B. Morales Prieto and J. Traba Diaz
ISBN: 978-1-62808-298-2

Chapter 6

RESILIENCE OR VULNERABILITY? VEGETATION PATTERNS OF A CENTRAL TIBETAN PASTORAL ECOTONE

Georg Miehe,[1,*] Sabine Miehe,[1] Kerstin Bach,[1] Karsten Wesche,[2] Elke Seeber,[3] Lena Behrendes,[1] Knut Kaiser,[4] Christoph Reudenbach,[1] Jasmin Nölling,[1] Jan Hanspach,[5] Mark Herrmann,[6] Ma Yaoming[7] and Volker Mosbrugger[6]

[1]Faculty of Geography, University of Marburg, Marburg, Germany
[2]Senckenberg Museum of Natural History Görlitz, Goerlitz, Germany, Germany
[3]Faculty of Biology – Geobotanik und Botanischer Garten, University of Halle-Wittenberg, Halle (Saale), Germany
[4]Acatech - German Academy of Science and Engineering, Berlin, Germany
[5]Helmholtz Zentrum für Umweltforschung UFZ, Leipzig, Germany
[6]Forschungsinstitut und Naturmuseum der Senckenbergischen Naturforschenden Gesellschaft, Frankfurt, Germany
[7]Institute of Tibet Plateau Research, Chinese Academy of Sciences, P.R. China

ABSTRACT

The question of the resilience or vulnerability of Tibetan highland pastures is not only of regional importance but also of global relevance for the Earth's energy budget and atmospheric circulation. A climate and grazing driven environmental change of the Tibetan highland albedo could induce feedback effects within an ecotone of approximately 200 km between Cyperaceae-dominated grazing-lawns and alpine steppes stretching over 2000 km between the Qilian Shan (38°N) and the Himalaya (28°N). Understanding the vegetation structure of this region would enable accurate modeling approaches for global change scenarios.

---

* Corresponding author: Georg Miehe. Faculty of Geography, University of Marburg, Deutschhausstraße 10, D-35032 Marburg, Germany. E-mail: miehe@geo.uni-marburg.de.

We analyzed the Central Tibetan pasture vegetation in the wider Nam Co area using phytosociological vegetation surveys and remote sensing techniques (ASTER, SRTM). The survey described in this chapter recorded a total of 105 vegetation relevés, with a total of 233 species. The species-based approach using DCA and phytosociological classification distinguishes six plant communities including juniper woodlands and juniper open dwarf shrublands, alpine steppe communities, *Kobresia pygmaea* communities and transitional mosaics with fragments of *Kobresia* turf and alpine steppe. The most common community was the transitional mosaic of *Kobresia pygmaea* mats and alpine steppe. Except for the phanerophytic communities and the grazing lawns of *Kobresia pygmaea*, plant communities are characterized by low overall plant cover with two thirds of species showing values lower than 1%. An analysis of plant functional types shows that most species are adapted to intense grazing regimes. The prevalence of dwarf growth forms, rhizomes and repellent characteristics shows that the pastures are grazing resilient. By contrast, the fragmented turf cover is vulnerable and endangered by intense grazing, digging by small mammals, turf removal and periglacial weathering.

Keywords: Alpine steppe, grazing, *Kobresia pygmaea*, plant functional types, Qinghai-Tibet Plateau, remote sensing

INTRODUCTION

The central Tibetan highlands are a sensitive ecotone between the closed, golf course-like pastures of a tiny endemic sedge, *Kobresia pygmaea,* in the south-eastern plateau and the semi-arid alpine steppe in the north-western plateau. This ecotone occupies a diagonal area of more than 2000 km long and around 200 km wide. It is located between the Qilian Shan in the north-east and the Inner Valleys of the Himalayas in the south-west (Figure 1). The radiation feedback effects of the Tibetan highlands have a crucial impact on the monsoon regime. Any potential changes in the Tibetan surface properties may have consequences for the global climate. Thus, the vulnerability and resilience of the Tibetan pastures are not only of regional but also of global importance. Both, resilience and vulnerability are related approaches in life sciences and the social sciences. Here we understand resilience and vulnerability as complementary concepts (see review of Miller et al., 2010) referring to resilience as the (plant) communities' resistance to disturbance, whereas vulnerability is understood as the condition encompassing characteristics of exposure susceptibility. Specifically, a shift from a closed plant cover to a degraded surface with bare soil would have detrimental effects on evapotranspiration, albedo and surface roughness. It is in this sense understood as vulnerable, whereas its regenerative capacity would be understood as resilience.

The knowledge of the impact of humans on the Tibetan highlands is limited and controversial. Climate-driven shifts of Holocene vegetation have been reconstructed based on lake sediment cores (Co Ngion, 31°24'N/91°28'E, Shen et al., 2008; Nam Co, Zhu et al., 2008; Herrmann et al., 2009).

However, like nearly all climate-sensitive ecotones of the world, this area has also been shaped by humans and their livestock. Nevertheless, some scientists still regard the vegetation of the Tibetan highlands as natural (e.g. Ni, 2000; Song et al., 2004; Yu et al., 2001; Herzschuh et al., 2006).

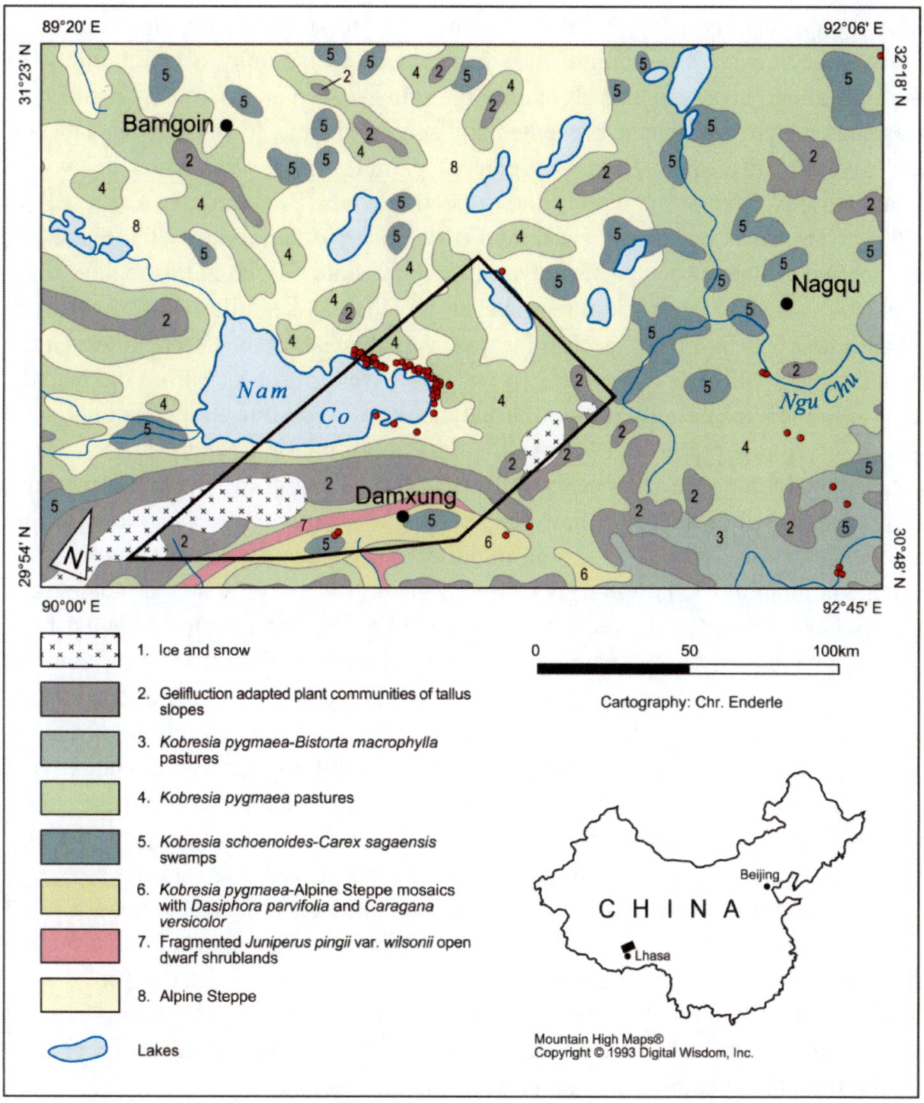

Figure 1. The vegetation of the study area (according to Zhang 1988, changed) with locations of vegetation records (red dots). The black frame indicates the position of the land cover map in Figure 916.

It is remarkable that many modeling approaches of past or future scenarios ignore the possibility of any human impact. Such attempts could generate scenarios that are no better than the starting assumption (cp. Song et al., 2005).

In fact, there is plenty of experimental and palaeoecological evidence that the biomes of at least the humid south-eastern half of the highlands are a human-made plagioclimax that replaced natural tall grassland, forests, and woodlands (Miehe et al., 2008a, b, 2009).

The earliest archaeological records of humans on the Tibetan plateau correspond to undated, probably epipalaeolithic artifacts.

The archaeological exploration of Tibet (Aldenderfer, 2007; Brantingham et al., 2007; Bellezza, 2008) has revealed some evidence that ecosystems were shaped by foragers and herders since the Late Pleistocene.

These groups were certainly capable of using fire, and evidence for fires may serve as a proxy of human impact. Fires might also have occurred naturally after lightning events. However, lightning occurs exclusively during the rainy season and is usually associated with torrential rains which would have prevented any spread of fires. Moreover, the Tibetan flora does not have any fire-adapted plants. Charcoal particles have been found in two palaeo-ecological analyses, even at the base of the respective cores (Damxung: 13 ka cal BP, Miehe et al., 2009; Nam Co: 8 ka cal BP, Herrmann et al., 2009; Adamczyk, 2010). These findings indicate that humans have been present at least since the Late Glacial in this region.

It is unknown when livestock grazing began in Tibet. It is similarly unknown when one of the most important livestock species, the yak, was domesticated (Olsen, 1990; Guo et al., 2006; Rhode et al., 2007), and when sheep and goats were introduced from the Middle East (Flad et al., 2007; Miehe et al., 2009). The palynological record is so far the only source of information to provide proxies wherewith the onset of animal husbandry in the area can be reconstructed. Palynological data (Kaiser et al., 2008; Miehe et al., 2009; Herrmann et al., 2009) both for the valley south of the Nyenquentangla Shan and the Nam Co basin provide the first evidence of grazing weeds during the mid-Holocene climatic optimum. These estimations are ambiguous, however, because all grazing weeds of Tibet are endemic to the highland, and because wild mammals could potentially have had an impact similar to that of livestock. A comparison of vegetation patterns and climate between the more humid south-eastern and the drier north-western Tibetan plateau may shed light on the effects of land use in this region. Theory predicts that in equilibrium systems, like in the south-eastern highlands, livestock grazing should be detrimental to ecosystem stability. Here, precipitation is higher and fodder availability for livestock husbandry much more reliable. Under these conditions, relatively large herds can be maintained but high livestock numbers can lead to ecosystem degradation. By contrast, the non-equilibrium conditions in the drier north-west are highly variable (coefficients of variation in the annual precipitation are greater than 30%). Here, herds are likely to experience non-equilibrium dynamics, with high mortality in drought years but relatively low mortality on average. As a consequence, degradation in these plant communities should be low (Ellis and Swift, 1988; Vetter, 2005). Moreover, vegetation in this region could remain stable even under permanent, intense grazing, if key species possess functional traits adapted to high grazing pressure (Diaz et al., 2001).

To what extent current livestock grazing pressure might exceed the natural grazing impact of wild herbivores is not known. It is possible that grazing induced a long-lasting selection process, in which unpalatable and dwarf plants were favoured over palatable forage plants. Surprisingly, the grazing impact of large wild herbivores in Tibet is rarely mentioned in the literature. This effect can hardly be studied as wildlife has been displaced to the marginal parts of the alpine steppe, in the north-western highlands (Schaller, 1998). Today, the most important remaining wild mammal across Tibet is the endemic plateau pika (*Ochotona curzoniae*), a territorial, winter-active lagomorph of 150 to 200 g (Smith and Foggin, 1999; Pech et al., 2007). Pikas inhabit *Kobresia* pastures in large numbers, especially in the most heavily degraded pastures around settlements ("black soil", Ma et al., 1999). Their role in the destruction of *Kobresia* turfs has not yet been investigated but it is striking that their main distribution area coincides with the distribution of *Kobresia* pastures.

The overall aim of this work is to investigate current and past vegetation patterns in the Tibetan highlands that could serve as reference for future modeling of environmental changes at a regional and global scale.

In this chapter, we use vegetation records and remote sensing techniques to classify and analyze the plant communities of Nam Co, in Tibet. We also conducted interviews with local herders in order to describe the past and present rangeland management system in this area. Our hypotheses are:

Both the grazer-driven equilibrium system of the humid *Kobresia pygmaea* pastures and the rain-driven non-equilibrium system of the arid alpine steppes are both resilient to degradation due to a high proportion of grazing-tolerant plants.

By contrast, the felty turf-cover of the *Kobresia pygmaea* pastures– alpine steppe ecotone is fragmented due to superimposed abiotic, biotic and cultural impacts, which is an indication of its vulnerability.

METHODS

Study Area

The study area is located in the vicinity of Tibet's second largest lake, the Nam Co, in China (1920 km², Zhu et al., 2008; Figure 1). This region occupies the watershed areas among the upper catchment of the northern Kyi Chu river tributaries (Lhasa River, Yarlung Zhangbo/ Brahmaputra drainage), the northern tributaries of the Salween (Nu Jiang), and the interior drainage systems of the central Tibetan highlands, the "Changtang" (i.e. northern plain), behind the mountain range of the Nyenquentangla Shan. Climate in this region is a transition between the humid conditions favoured by the drainage system in the south and the dry conditions generated in the endorheic basins in the north. In the south, river gravel beds are located at 4200 m, the valleys of the upper Salween tributaries at 4500 m, and the Nam Co basin at 4700 m altitude. The area belongs to the Damxung county, Lhasa prefecture, and the Nagqu county, Nagqu prefecture.

Climate and Weather

Precipitation is highest from May to September during the Asian summer monsoon, which delivers humid air-masses from the south-east. In winter, climate is controlled by the dry and cold westerlies (Böhner, 1996). Mean annual temperature in the study area is slightly below 0 °C. Minimum and maximum monthly temperatures at the Bamgoin climate station range from -11.8 °C in January to +9.4° C in July, and at the Nagqu station from -10.2° C in January to +9° C in July (Figure 2). Annual precipitation is low and variable (around 300 mm, Figure 3). The difference in precipitation between the two stations reflects the influence of the monsoon system. The mean precipitation is lower and more variable at Bamgoin, in the north-west (mean: 320, SD: ±133 mm) than in Nagqu, in the south-east (450 ± 112 mm), especially in the growing season (JJA; 206 ± 109 mm vs. 267 ± 87 mm).

Recent studies suggest increasing precipitation totals (Morrill, 2004; Xu et al., 2007); but totals may be less important than the seasonal distribution. Winter precipitation has clearly increased over the last 40 years; i.e. snowfall has significantly increased while summer precipitation has remained more or less stable (Figure 3). The higher frequency of snowfall events has an indirect impact on both the vegetation and rangeland quality, because snowfall may cause loss of livestock and reduced grazing pressure.

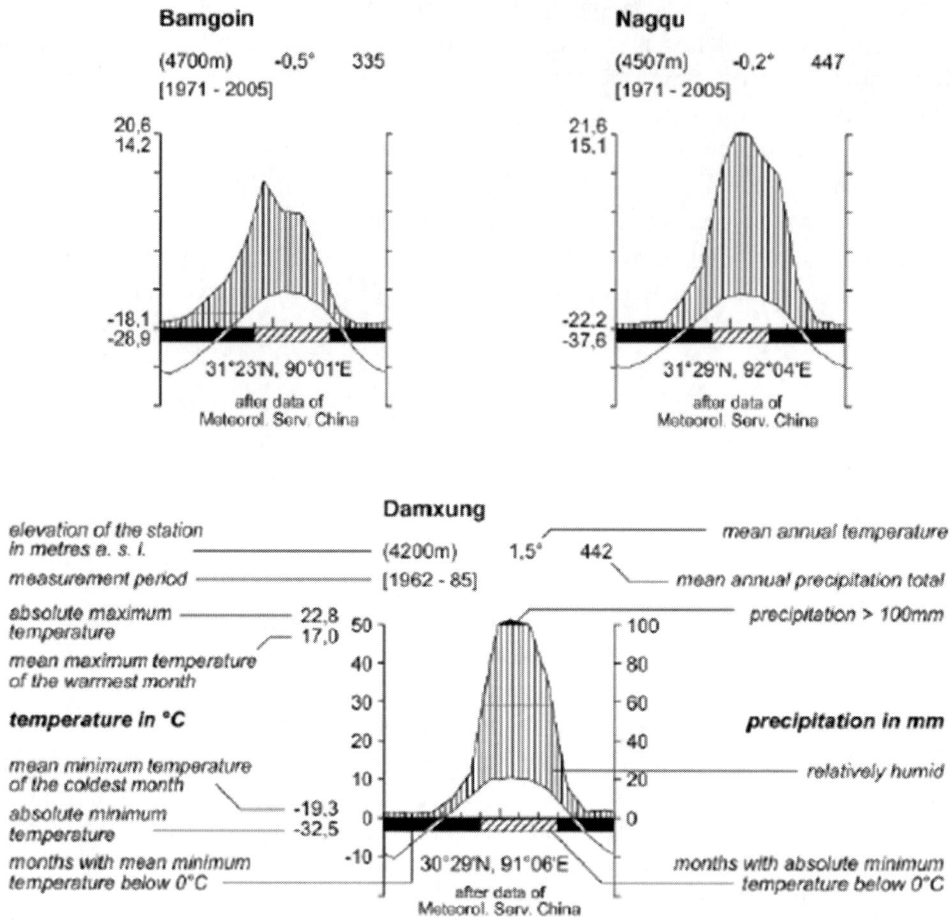

Figure 2. Climatic diagrams from the study area. Data source: Meteorological Services of China (processed by C. Reudenbach, design by C. Enderle).

Geology, Landforms and Soils

The study area is divided by the Nyenquentangla Shan. Its granites and slates are shaped by glacial erosion and periglacial processes. Lower slopes are covered by large side moraines of different age and by talus accumulations. The rolling hills north-east and north of the Nyenquentangla consist of Cretaceous limestone and Jurassic clay rich sandstone. The loess cover on the ridges has been widely eroded and has accumulated as colluvium at their bases. The intramontane basins are characterized by loess-covered pediments and gravel terraces. Valley floors and depressions have stagnant water with Cyperaceae turfs growing in rhizo-hydromors.

Slope deposits, developed from schist, are the dominant substrate on slope ridges (see transect in Figure 4.) Foothills are mainly covered by moraine sediments. Lacustrine pebbles, gravels and sands occur up to nearly 30 m above the present lake level. The highest shoreline has formed a marked cliff into the moraines of mid-Pleistocene age. Numerous lake terraces and shorelines indicate that the lake shrank during the Holocene (Zhu et al., 2008). Palaeosoils in and below the widespread, thin loess cover are evidence of increasing aridity during the mid- and late Holocene.

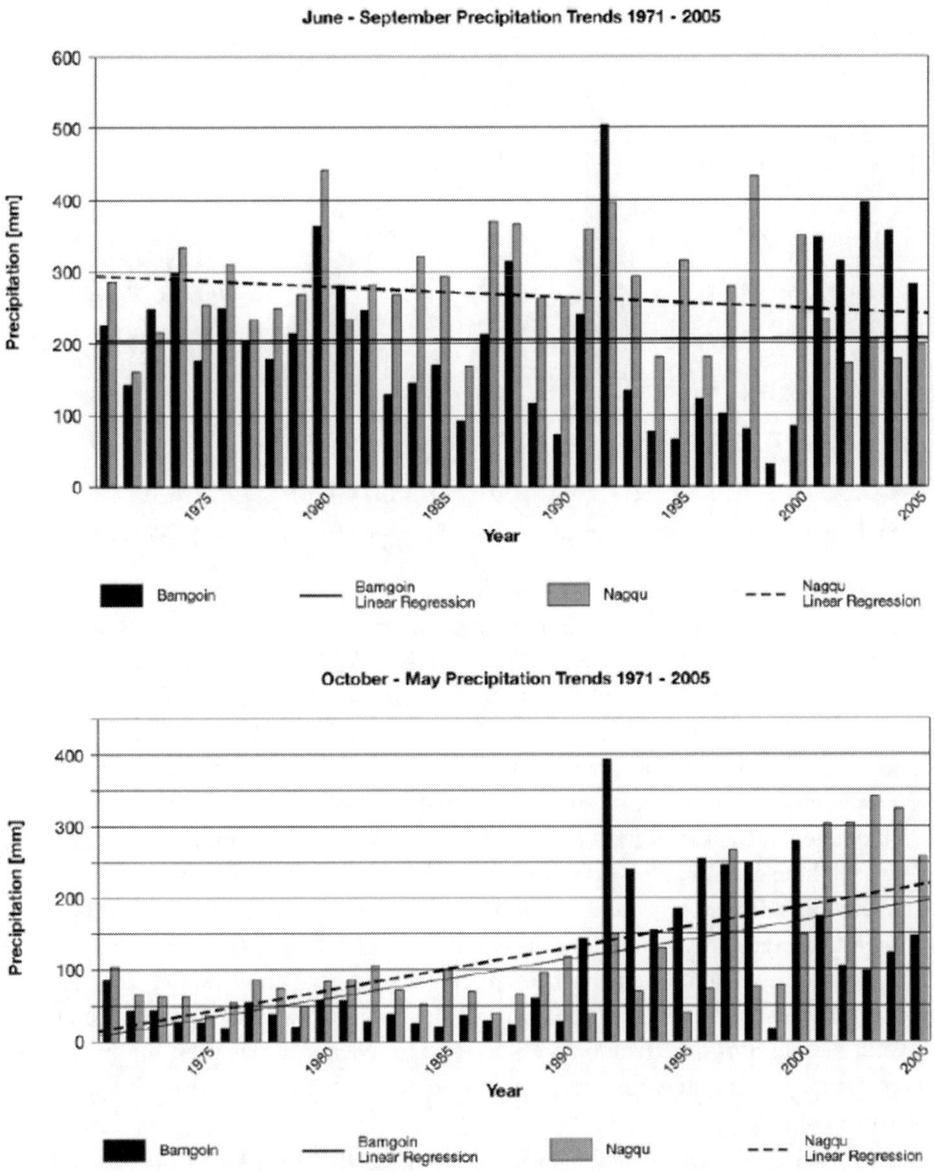

Figure 3. Seasonal precipitation trends for Bamgoin and Nagqu for the period 1971-2005, for the growing season of June - Sept. (a) and Oct. - May (b). Data are based on CWS 2008 (own calculation). The dashed lines are trends predicted by linear regression analysis.

A stagnic cambisol occurs near the ridge. Its stagnic properties might be related to the underlying permafrost. The slopes below the ridges are dominated by weakly developed cambisols that are at least partly covered by a layer of *Kobresia pygmaea* roots and rhizomes (Rhizomull, “Afe”). Moist to wet alluvial fans have gleysols. Thick lacustrine deposits are overlain by loess covers (up to 50 cm thickness) with well developed cambisols. As a rule, the deeper the lacustrine sediments, the thinner the loess cover.

Finally, a sequence of beach ridges next to the lake shore is characterised by arenosols and gleysols.

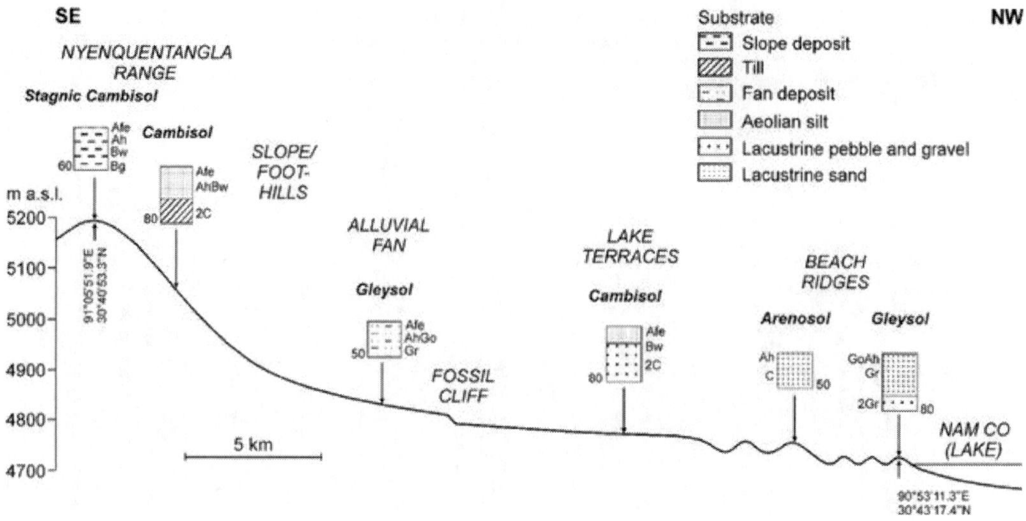

Figure 4. Soil transect from the Nyenquentangla Pass to the Nam Co (after Kaiser 2004, changed).

Past and Present Rangeland Management

Little is known about past rangeland management in the study area. Presumably, a wide range of diverse rangeland management systems have existed. Literature surveys and our own fieldwork surveys suggest that land use systems were variable and locally adapted. In the recent past, they have undergone a major transformation under the Chinese administration.

Herders developed a mobile grazing system in response to the low and highly variable summer precipitation, periodic high snowfalls and different vegetation patterns. Herders used to live in black tents to assure a high degree of daily and seasonal mobility. To exploit further resources, each household had an extra tent (gur) which enabled herders to split the herd and establish a further starting point for daily movements. Since the 1960s, the national rangeland policy aimed to increase production (Goldstein et al., 1990). This policy lead to a destruction of traditional trading systems, sedentarization, changes in herd composition, and a reduction of herd size (Sheehy et al., 2006). Today, most herders have a permanent house at least in winter. With the onset of privatization and large scale fencing of pastures, mobility has been further restricted. In some regions, keeping livestock has been given up entirely (Sheehy et al., 2006; own data).

In the eastern catchment of the Nam Co (Figure 5), livestock is still the only source of income. Herds are mainly made up of sheep, yak (17-30 % of the herds), and a few goats and horses. People live in traditional black tents for most of the year. It is only during the winter months that they move into their houses, near the *Kobresia schoenoides* swamps north of the lake. The herders live here from October to either February or March and await snowmelt in the spring and summer pastures. Once spring has arrived, people and livestock move to the eastern edge of the lake where they stay for one or two months. In early summer, they relocate to the southern shore of the lake. By mid-July, when all snow in the northern valleys of the Nyenquentangla Shan has melted, herds move northwards and reach altitudes of 5100 m. In September, herds are driven to the eastern shore, before they return to their winter places. Every year and every season, each household returns to the same place. Approximately 150 to 200 households follow this lifestyle (herders' direct communications and data based on counted summer tents and winter houses).

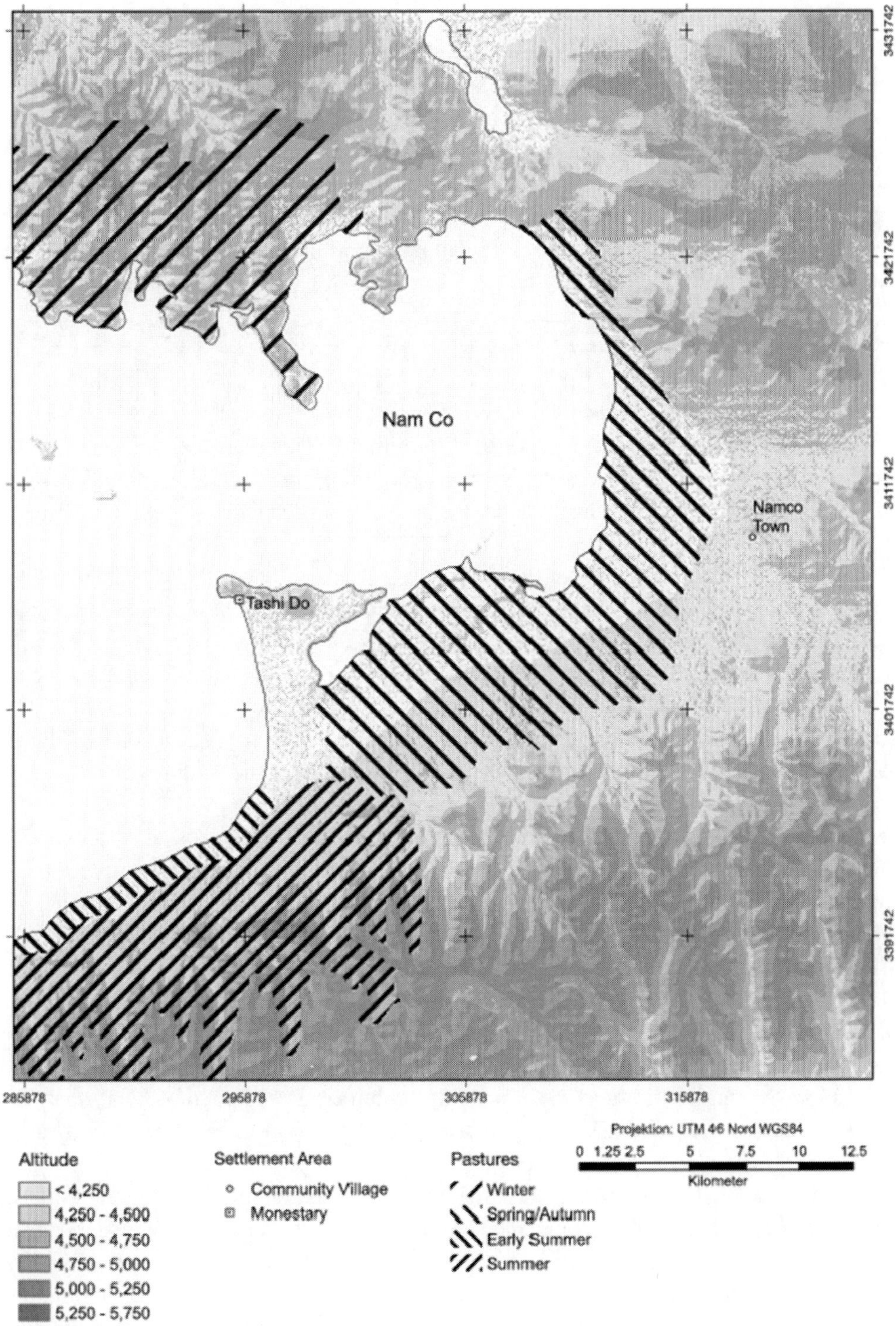

Figure 5. Seasonally changing livestock grazing pastures in the north-eastern Nam Co basin (after Behrendes 2008, changed).

According to local herders, this cyclic movement was established because of a) the high risk of snowfall in the valleys of the northern slope of the Nyenquentangla Shan and also the fairly high risk of snow east of the lake, b) the existence of *Kobresia schoenoides* swamps in the almost snow-free valleys north of the lake, c) the predominantly open vegetation and the absence of *Kobresia schoenoides* swamps in most of the eastern region, and d) the existence of widespread *Kobresia pygmaea* pastures along glacier-fed streams south of the lake.

As the vegetation shows no apparent gradients in grazing weed abundance, it is likely that all available pastures are exploited. It is to be feared that sedentarization will lead to smaller grazing radii leading to pasture degradation around the settlements and undergrazing on remote slopes.

Plant Community Characterization

The first vegetation survey of the area was undertaken by the „Comprehensive Science Investigation Team of the Chinese Academy of Sciences" between 1973 and 1976 (Zhang, 1988, with a vegetation map at a scale of 1:3 million). The plant cover was classified according to dominant species. Vegetation maps for China on a scale of 1:1 million were edited by Hou (2001). Grazing impact assessments based on floristic vegetation records are still not available.

The records for this chapter extend over an area of 150 km from SW to NE and 200 km from E to W. In this direction, summer precipitation decreases from 267 to 204 mm (Figure 2), and altitude increases from 4350 m S of Damxung to 4750 m North of Nam Co, and to 5100 m in the Northeast watershed. The southernmost parts belong to the montane belt as shown by the northernmost outposts of the juniper forests of southern Tibet (Miehe et al., 2008d). The most north-eastern vegetation records, at 5100 m, are part of the alpine *Kobresia pygmaea* ecosystem, in the humid south-eastern highlands. The north-western records represent alpine steppe.

Vegetation surveys were conducted between 1995 and 2009. Plot size was 10 x 10 m. We recorded all vascular plant species in the plots and estimated their absolute cover percentages. We transformed abundance percentages to presence/ absence data during vegetation classification. Original cover percentages were used only as additional information to the classification of unranked communities. In total, 105 records are listed in Tables Appendix 1 and in Appendix 12. We chose plot sites in a deliberate manner, using those that were homogenous both in their species set in the abiotic habitat structure (e.g. slope inclination, exposure, soils). Sites were geo-referenced using a Garmin 12 XL GPS. Plant identification follows "Flora Xizangica" (Wu, 1983-87) or the volumes of "Flora of China" that have appeared so far (Wu and Raven, 1994 ff.). In a few cases, we had to combine closely related taxa (e.g. *Carex ivanoviae/ C. montis-everestii*, *Stipa roborowskyi/ S. purpurea*).

We applied a phytosociological classification to vegetation types following Mueller-Dombois and Ellenberg (1974). Arrangements of plot and species groups were supported by fidelity values using the phi coefficient of association (Φ) as implemented in JUICE 6.5, a software to sort phytosociological tables (Tichý, 2002). In order to analyze how communities were separated along environmental gradients, we performed a detrended correspondence analysis (DCA, Hill and Gauch, 1980), in the software PC-ORD (McCune and Mefford, 1999). Data were standardized by logarithmic transformation prior to the analysis (rare taxa were not downweighted). DCA was conducted with default program parameters (rescaled axes, rescaling threshold (0), and 26 segments).

It is almost impossible to differentiate altitudinal belts based on vegetation patterns in the Tibetan highlands due to the high human impact. This is especially true for the alpine belt (e.g. Herzschuh, 2006). We defined all taxa growing at altitudes higher than 5100 m as alpine species.

In order to assess to what degree communities were resilient or vulnerable to grazing, we assigned species to five different plant functional types (see Figure 9): plants (A) forming flat cushions, (B) growing with buds closely attached to the ground, or (C) with rosettes flattened to the ground, (D) being poisonous or unpalatable for herbivores during at least part of the year, and (F) taxa vulnerable to grazing, including most graminoids. E refers to the cover of all resilient plant functional types (sum of A to D). Vegetation cover of these types has been calculated for each plot as the sum of all single cover values in arid sites and as overlapping but independent cover values in the humid sites.

Satellite-Based Land Cover Mapping Assessment

Gaining information on the vegetation cover is one of the most pressing aims in studies of arid and semi-arid environments. Satellite-based assessments can provide valuable information but, unfortunately, these methods still need to be tested in many regions. The range of previously applied satellite-based methods varies from simple, unsupervised clustering algorithms to more sophisticated, supervised approaches like the widely used maximum likelihood estimation of neural networks (John et al., 2008; Lu, 2006; Mas and Flores, 2008; Govender et al., 2007; von Wehrden et al., 2006). Neural approaches have an advantage because they can handle the central problems of remote sensing in arid environments; according to Mas and Flores (2008), Artificial Neural Networks (ANN) can integrate the nonlinear relationships between the dependent and predictor variables in a learning process, and tend to generalize noisy and imprecise data. They incorporate *a priori* knowledge and constraints during the analysis, and handle data types other than remote sensing, as well as ancillary data from other sources.

We restricted the satellite-based land cover mapping to the central part of the study area where adequate satellite data were available (Figure 1, 916). Vegetation classification was based on ASTER (Advanced Spaceborne Thermal Emission and Reflection Radiometer) images and a Digital Terrain Model (DTM) from the Shuttle Radar Topography Mission (SRTM) dating from February 2000 (Rabus et al., 2003). These data were well calibrated and processed from reliable resources (http://lpdaac.usgs.gov). The used DTM version 4 (V4) from the Consortium for Spatial Information of the Consultative Group on International Agricultural Research (CGIAR-CSI), with a 3 arcsec (90m) spatial resolution (Jarvis et al., 2008). ASTER data were obtained from the Land Processes Distributed Active Archive Center (LP DAAC, USGS/EROS, Sioux Falls, SD, US). ASTER imagery has been available since 2000 from the NASA Terra satellite. In the current study, ASTER level 1B data (orthorectified) scene was acquired on November 10th, 2004 (UTM 46N, WGS-84, granules SC: AST_L1A.003: 2026149457, SC:AST_L1A.003:2026149457) and converted to radiances. The ASTER sensor acquires data in three band subsystems: Visible and Near Infrared (VNIR) bands 1, 2, 3 (resolution 15 m), Short Wave Infrared (SWIR) bands 4-9 (resolution 30 m), and Thermal Infrared (TIR), bands 10-1 (resolution 90 m). Additionally, a processed ASTER Digital Elevation Model (DEM) is available that is based on the VNIR bands 3b and 3n. It provides a significantly better resolution (15-30 meters) than the SRTM tiles but fails in areas of cloud cover (Toutin, 2008).

For the classification process, we derived from SRTM DTM V4 three basic terrain variables (elevation, slope, and aspect) that correlate significantly to distinct vegetation types and abundances, with an approximate resolution of 90 m. Moreover, we used the DEM derived from ASTER in the classification process to improve the cloud border identification.

Due to the high correlation between band subsystems, a Principal Component Analysis (PCA) was applied to remove redundant information. The first two VNIR components, band 1 (wavelength: 520-600 nm) and band 2 (630-690 nm) explained 99% of the image variability (Eklundh and Singh, 1993).

As the mosaic of *Kobresia schoenoides* mats was composed of open water, soil and vegetation, we calculated the Perpendicular Vegetation Index (PVI), an advanced soil-corrected Normalized Differenced Vegetation Index (NDVI). Specifically, we used the modified PVI of Bannari et al., (1995) to account for the water and soil background within the communities, The PVI was calculated using the ASTER bands 2 (WL 630 – 690 nm) and 3N (WL 760 – 860 nm) with the following equation:

$$PVI = \frac{(NIR - a)\times(Red + b)}{\sqrt{1 + a^2}}$$

where *Red* is the reflectance of ASTER band 2, Near Infrared (*NIR)* is the reflectance of ASTER band 3N, a represents the intercept of the soil line (a= -6.35), and b indicates the slope of the soil line (b= 0.89). Given that the image was taken in late autumn, thermal information can be expected to play a significant role in the classification process. During this time, *Kobresia pygmaea* is often brownish with low chlorophyll content.
Nevertheless, the turf blocks of *Kobresia pygmaea* can be expected to emit thermal data different from that of bare soil and alpine steppe. In the PCA, band 4 explained more than 99% of the thermal information in the SWIR subtype group of the sensor, and was thus chosen for vegetation classification.

Neural Networks have to pass through a learning process to obtain memory of patterns and iteratively identify the relationship between cause and effect (Hayking, 1999). There are no rules for the configuration and design of ANNs, and optimized results can be best achieved by structured trial and error (Chen and McNairn, 2006). Therefore, the performance and accuracy of ANNs have been evaluated during each learning step.

For training and testing, a Multi-layer Perceptron (MLP) procedure was choosen which provides an arbitrarily specified number of up to 100 pixels per class. The order of significance of each data layer was identified by a Canonical Component Analysis (Zhao and MacLean, 2000). This approach was preferred because the order of bands feeding in the MLP has a major influence on the quality of the training process and hence on the classification (Mas and Flores, 2008).

In a final step, we compared these results with independent ground truth data and we constructed an error matrix calculation for data validation. With respect to the mentioned shortcomings of georeferencing, record plots were treated as areas instead of points. Furthermore, we calculated Cramer's V index for data validation, which is an effective measure of similarity (Rees, 2008).

RESULTS

The Plant Communities

The analysis identified six plant communities (Appendix 1) that included a total of 229 species. Community 2, the dwarf shrublands of *Juniperus pingii* var. *wilsonii*, and community 6, the closed *Kobresia pygmaea* pastures, differ from the other communities because they present a single dominant species (Community 2: *Juniperus pingii* var. *wilsonii*, 36% mean cover; Community 6: *Kobresia pygmaea*, 87% mean cover). The remaining plant communities cover more than 90% of the study area and are characterized by a high proportion of bare soil and the absence of a closed plant cover (59% of the 177 species present a 1% cover or less).

The DCA indicates more than one complete species turnover between the most dissimilar sites (length of gradient is 6.2, Figure 6). Axis 1 shows a correlation with slope inclination (r= 0.54) and the cover of unpalatable plant species (r= 0.78). Axis 2 has a shorter gradient (2.9) and is correlated with elevation (r= 0.44).

Community 1: Juniperus Tibetica tibetica Woodlands

Community 1 represents open and degraded *Juniperus tibetica* woodlands at the northern treeline ecotone. All eight records are located in the broad valley south of the Nyenquentangla Shan, at 4250 - 4550 m altitude, on exposed slopes with open rocky soil. At present, only a few turf patches with *Kobresia pygmaea* can be found on the south-facing slope, where *Juniperus* reaches its highest cover values (up to 70%).

Species richness is variable (5 - 31 species/ plot). The community structure has three layers, a tree, a shrub, and an herb layer. The gnarled trees hardly attain 3 m in height and exhibit a high proportion of dead wood. We detected only a few seedlings, possibly due to harsh climate and strong grazing and trampling.

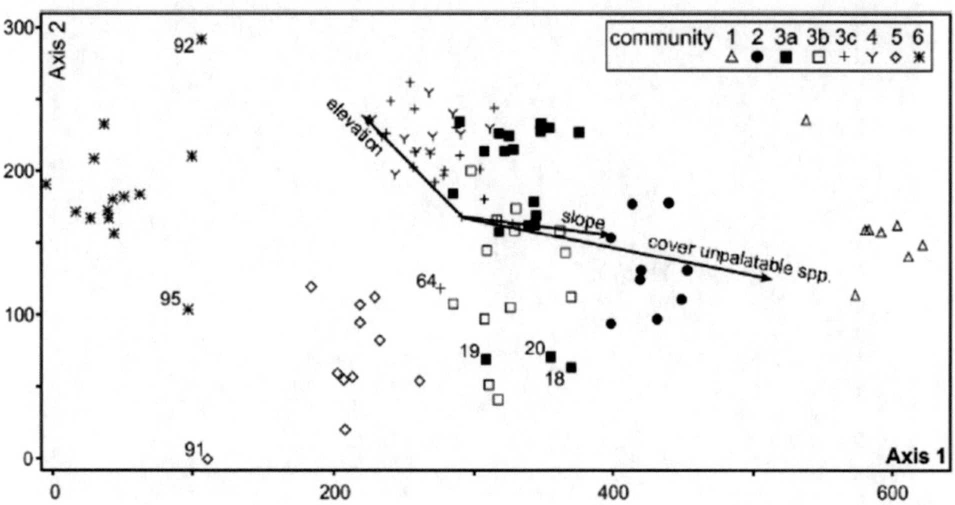

Figure 6. Detrended Correspondence Analysis (total inertia = 9.59, eigenvalue/ length of gradient axis 1 = 0.69/ 6.2, axis 2 = 0.28/ 2.9). Symbols indicate communities. Numbered records refer to the running numbers of the sites in the appendix.

Many of these sites are sacred for Tibetan Buddhists (especially the hill on which vegetation plots 1 to 7 were recorded, see Figure 7), which may have contributed to its conservation. According to the local community, these juniper forest remnants were once more widespread in the valley south of the mountain chain (Miehe et al., 2008d). Molecular evidence suggests this site is an Ice Age refugium because trees represent a unique haplotype (Opgenoorth et al., 2010). The pollen record supports these findings because *Juniperus* pollen was present deep in the studied core (13,000 years BP, Miehe et al., 2009). The pollen record indicates that grazing has prevailed in the last 8800 years and has increased in the last 2000 years (Miehe et al., 2009).

The shrub layer consists of *Juniperus pingii* var. *wilsonii* (3-70%) and a few dwarf shrubs, like *Caragana versicolor*, *Rosa sericea* and *Ephedra gerardiana*. These shrubs are the easternmost occurrence of semiarid thorny dwarf scrub that extends further westwards along the Yarlung Zhangbo into Ladakh (Hartmann, 1987). Even though *Caragana versicolor* reaches 4900 m in western Tibet, the shrub is absent around Nam Co.

The species composition of the herb layer is highly heterogeneous, with a mixture of species that are typically found in alpine steppe (*Arenaria bryophylla*, *Carex montis-everestii*), *Kobresia pygmaea* pastures (*K. pygmaea*, *Potentilla saundersiana*), montane pastures partly with C4 grasses (*Tripogon liouae*, *Orinus thoroldii*, *Pennisetum flaccidum*), and montane wastelands (*Dolomiaea calophylla*, *Elsholtzia ciliata*, *Persicaria glacialis*, *Chenopodium nepalense*). Similar communities occur around Lhasa (3650 m). Poisonous and unpalatable plants are the most dominant functional type (mean cover of 57%, Figure 89). The mean cover of all grazing-resilient plants is 62%. These plants are never used as fodder, not even during late winter, when forage is scarce (*Stellera chamaejasme*, *Astragalus monbeigii, Iris potaninii, Delphinium* spp.). The abundance of grazing-resilient plants reflects pasture degradation and indirectly, strong human impact.

Figure 7. The northernmost woodlands of *Juniperus tibetica* (Community 1), with *Juniperus pingii* var. *wilsonii* and *Caragana versicolor* dwarf scrub on a south-facing slope. South-west of Damxung, 4250 m. September, 2009.

In the DCA, all sites of this community overlap with high disturbance and steep slope inclination. All sites are located on exposed slopes near settlements, thus receiving the highest impact of grazing and trampling.

Community 2: Juniperus Pingii pingii Varvar. Wilsonii wilsonii Open Dwarf Shrublands

Community 2 represents a dwarf juniper scrub (Figure 8) which is typical for the treeline ecotone of semiarid southern Tibet (Zhang, 1988). It is the northernmost vegetation type dominated by phanerophytes and the only source of wood in the region. Its character species, *Juniperus pingii* var. *wilsonii*, is a shrubby variant with a controversial taxonomical status (Farjon, 2005). It forms dense patches of up to 10 m diameter and 1.3 m height, and grows on rocky slopes with southern aspects. Individual shrubs may attain ages up to several hundreds of years (Liang et al., 2011). Growth is obviously rainfall limited as shown by dendrochronological analyses (Liang et al., 2011). Given that there are plenty sites around Nam Co with the same environmental properties, the vegetation of Community 2 may have once been much more widespread. On the southern slope of the Nyenquentangla, individual shrubs may form extensive, closed thickets. By contrast, around Nam Co and especially in the hills of its north-eastern catchment, only few isolated patches of this community can be found. Given that this vegetation type can be regularly found in the region, it is most probably natural. The vegetation composition reflects decreased humidity and stands in contrast to the southern slope of the Nyenquentangla Shan where dwarf shrubs such as *Lonicera rupicola, L. hispida, Ribes glaciale*, and *Rosa sericea* occur.

Figure 8. *Juniperus pingii* var. *wilsonii* open dwarf scrub on a southern exposed slope (Community 2). The juniper stands reach 6 m in diameter and 1.3 m tall. The surrounding pastures have a high cover of unpalatable plants. North-east of Nam Co., 4850 m. Sep. 2005.

Trees of *Juniperus tibetica* have so far not been discovered around Nam Co and it remains doubtful if trees were ever present here during the Holocene (cp. Herrmann et al., 2009). Community 2 has a mosaic structure with dense patches of shrubby junipers surrounded by open soil and isolated herbs and graminoids. The plant cover around the juniper patches ranges from 20 to 33%. On average, 39 species occur in each plot. Species of the *Kobresia* pastures are nearly absent, whereas alpine steppe species are commoner (*Stipa purpurea, Carex montis-everestii, Incarvillea younghusbandii*). These mosaics share a greater number of species with the juniper woodlands of southern Tibet (*Phlomis younghusbandii, Thalictrum rutifolium, Dracocephalum tanguticum, Elymus schrenkianus, Pedicularis alaschanica, Cryptothladia kokonorica*). The community is characterized by the abundance (71%) of poisonous and unpalatable plants which is five times greater than in the herbaceous communities 3 - 6 (Figure 9). Cushion plants are nearly absent from Community 2 (less than 1%). The mosaic structure and the intermediate species composition (see DCA, Figure 6) indicate a transitional status of this community that mediates between open woodland pastures and alpine steppe pastures. At present, stands are only used as winter pastures, when livestock feeds on the dry aboveground plants. Thus, the current level of degradation is relatively low, and the present species composition may reflect more intense grazing by livestock or wild mammals in the recent past.

Community 3: Alpine Steppe

Community 3 was recorded in 51 plots and is located in the north-eastern Nam Co catchment, at 4660 - 4890 m altitude (42% of all records were recorded at 4720 - 4750 m). Sites are located on moderately inclined, south-facing slopes, or in pediments and lake terraces. Loess is the common substrate on these sites. The average vegetation cover is 31%. Most plant species (65%) have a total cover of 1% or less. The community attains the highest graminoid cover (30%), Cyperaceae reach only a 9% cover. The altitude and the presence of cushion plants (Figure 10) in this community match the description of the 'High cold steppe' by Wang (1988) and justify its classification as alpine steppe. Only few species are taller than 10 cm (*Stellera chamaejasme, Cryptothladia kokonorica, Dracocephalum tanguticum, Iris potaninii, Thermopsis lanceolata, Stipa koelzii, Pennisetum flaccidum*). These taxa are mainly poisonous plants or grazing weeds reaching their altitudinal limits here. The majority of dicots are shorter than 5 cm.

The 51 plots belonging to this community can be further divided into three sub-communities. Sub-community 3a (running no. 18 to 37, see Appendix 1) is characterized by *Stellera chamaejasme*. It is mostly found in the hills and pediments of the northern side of Nam Co; 40% of the records are on south-facing slopes. Plants avoided by livestock and disturbance indicator species (*Dracocephalum tanguticum, Euphorbia stracheyi, Astragalus tribulifolius, Youngia simulatrix, Cryptothladia kokonorica, Sedum perpusillum, Lasiocaryum munroi, Rhodiola smithii*) reach high frequency values. It is possible that these stands replaced the *Juniperus pingii* var. *wilsonii* thickets from the south-facing slopes in the north-eastern Nam Co catchment that had been cleared for firewood.

Sub-community 3b is characterized by *Kengylia thoroldiana* (running no. 38 to 50). This community type was mainly recorded at the south-eastern sandy plains of Nam Co. The most common species are *Kengylia thoroldiana, Kobresia robusta, Dontostemon glandulosus*, and *Dracocephalum heterophyllum*. They are character species of eastern alpine steppes on sand dunes and are common in the sand fields around Madoi in the north-eastern highlands.

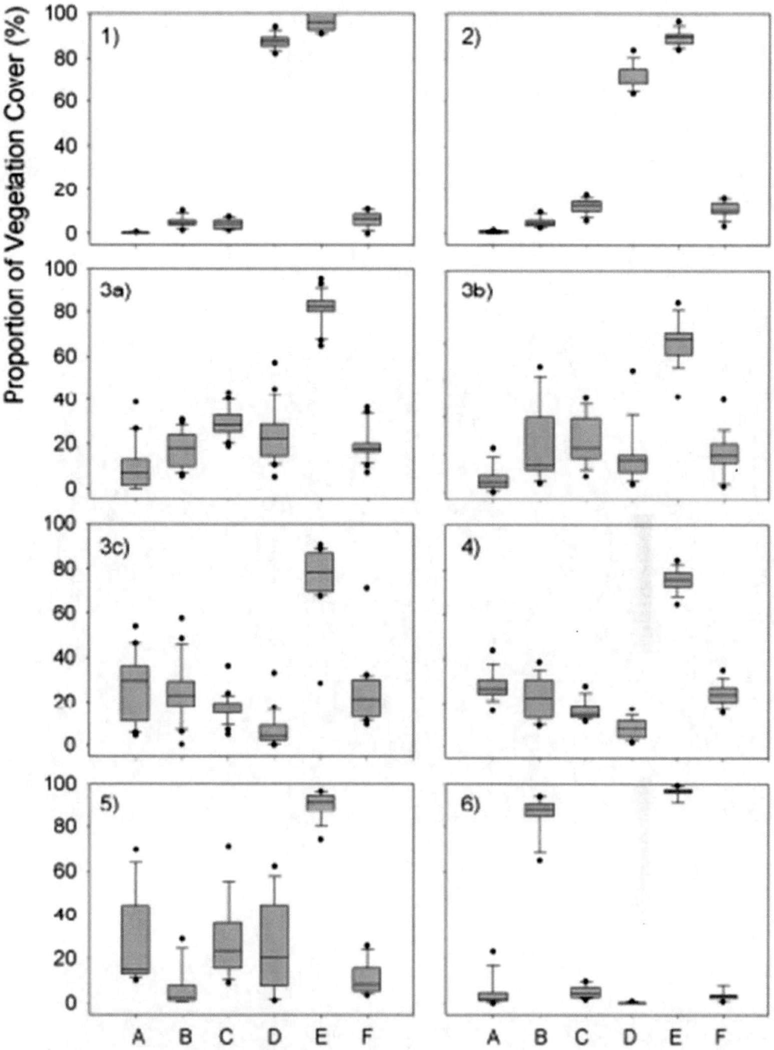

Figure 9. Mean cover values for the plant functional types in the communities. A-D and F represent plant functional types as described in the text. E refers to the cover of all resilient plant functional types (sum of A-D).

Sub-community 3c (running no. 51 to 68) is mainly distributed on loess covered hills and sandy plains north of Nam Co and reaches a mean richness of 25 species/ plot and mean vegetation cover of 36%. Graminoids, such as *Festuca valesiaca, Koeleria litvinowii*, and *Poa albertii* are frequent, with a mean cover of 17% (46% of the total plant cover). Cushion plants (*Astragalus arnoldii, Sibaldiante adpressa*) have mean cover values of 10%. It is distinguished from sub-community 3b by four species (*Artemisa stricta, Astragalus arnoldii, Potentilla exigua, Koeleria litvinowii*).

Community 4: Alpine Steppe – Kobresia Pygmaea pygmaea Mosaic

This community is the most common vegetation type in the 2000 km ecotone. On a larger scale, it can be regarded as a transitional vegetation type with strong affinities to alpine steppes of Community 3 (Figure 6) and more limited affinities to closed *Kobresia* pastures in the south-east (Community 6, Figure 6). The mean plant cover is 42%. Its intermediate nature is also reflected in the vegetation structure that is characterized by a mosaic of isolated turfs and *Kobresia pygmaea* patches, surrounded by open soil with scattered graminoids, rosettes and cushions. The turf sods resemble stepping stones and are scattered across the landscape (Figure 11). They are all 5 - 10 cm high and likely the remnants of a once closed turf cover (see description of Community 6 below).

Figure 10. Alpine steppe in an open plain. The total plant cover is 25%. Six graminoid species cover 20%. 31°23'N/94°40'E, 4680 m. August 1993.

Figure 11. *Kobresia pygmaea* turf that has been fragmented into stepping-stone-like turf sods (1). Pikas dug holes underneath the sods (arrow). The open soil among the turf sod is colonized by alpine steppe species (2). Qinghai (34°37'N/98°00'E), 4240 m. August 2009.

Humans have certainly had a negative impact in the recent past, by cutting sods for construction purposes (Figure 12). Yaks tear up the turf cover with their horns to wallow in the open soil. The closed *Kobresia* turfs, also called the "rhizomull", are mostly resilient to trampling, a globally unique feature for an ecosystem that is subject to grazing impact and desertification (Kaiser et al., 2007). Abiotic processes, however, seem to have a negative effect. In the highest pastures of the region, gelifluction can break sods up during freeze-thaw events (Miehe, 1988).

Figure 12. *Kobresia pygmaea* sods have been cut to construct a protecting wall against precipitation in the vicinity of a newly built winter village. Nam Co, 4750 m. September 2007.

The remaining turf sods mostly have steep sides with dry roots, and lack any higher plants, algae, mosses, or lichens. This structure largely resembles the turf exfoliation cliffs caused by wind and needle-ice (Troll, 1973). The sides of these turf cakes sods are often undermined by burrowing pikas which probably take advantage of the stable sods and their insulating effects during winter. On steep slopes, the sides of such sods are even used by sheep for rubbing.

A widespread pattern throughout the entire altitudinal range of the *Kobresia* turfs are polygonal cracks caused by heavy frost (Zhang, 1988), or desiccation (Miehe et al., 2008b). They are widened by the joint impact of gelifluction, deflation, and digging by pikas (Figure 13). The relatively high richness (mean 38 species) and the mosaic structure in this community reflects its transitional position between the two major biomes, the relatively mesic *Kobresia pygmaea* grasslands and the alpine steppe. This notion is supported by the fact that plant functional types of both the alpine steppe and *Kobresia pygmaea* pastures are prevalent. Two of the most widespread grazing resilient species, *Androsace tapete* and *Leontopodium pusillum* attain highest frequency and cover degrees in this community.

The spatial species distribution follows a regular pattern. Whereas typical *Kobresia* mat species are confined to the turf sods, alpine steppe species grow among the sod fragments. However, the open patches around the *Kobresia pygmaea* stands are colonized by alpine steppe species. Both findings suggest that alpine steppe species could slowly replace the *Kobresia pygmaea* mats.

The plot data thus show the same aridification trend as the large-scale proxies derived from the palynological sediments in Tibetan lakes (e.g. Academia Sinica, 1990; Avouac et al., 1996; Zhu et al., 2008) which estimated a general trend towards lower monsoonal rainfall since 5600 years calibrated before present (Shen et al., 2008).

Figure 13. *Kobresia pygmaea* turf, split by polygonal cracks. Pikas have undermined the turf cliffs (knife). *Kobresia* fragments are carved by soil erosion (arrow). South of Nagqu, 4540 m. September 2009.

This explanation might be too simple, though. Increased levels of humidity in parallel with a cooling trend may enhance needle ice frequency and promote turf exfoliation. The same is true for a cooling trend: During the period 1951-80, a drop in temperature of 0.6-1.2 K (Böhner, 1996) resulted in intensified gelifluction as shown by screes overriding turfs or intensified frost breaking up the turf (Miehe, 1988). Probably, a long-term late-Holocene trend of decreasing humidity is currently being superimposed to short term thermal oscillations. Wind erosion might also have played a role in the retreat of exposed turf cliffs. A long-lasting deflation effect, however, is not evident because deflation pavements are missing in the study area. One possible conclusion is that the process of turf retreat is relatively recent and/ or deflation is not as strong as in the southern Tibetan Foehn-exposed rain shadow of the High Himalayas (Miehe, 1988). In summary, Community 4 seems to represent an intermediate succession stage, from the closed alpine *Kobresia* turfs typical of a humid biome to the alpine steppe characteristic of a semiarid one.

Among the 132 species found in Communities 3 and 4, 45% form cushions (11%, e.g. *Androsace tapete*) or rosettes that are closely appressed to the surface (34%, e.g. *Taraxacum sherriffii*). Poisonous plants or grazing weeds avoided by livestock due to repellent characters (e.g. *Incarvillea younghusbandii*, *Dracocephalum heterophyllum*) make up 21 % of the species with a 15% cover. Graminoid species represent 18%, with an average cover of 13% (37% plant cover, cp. Figure 14 a, b). Their average growth height ranges between 3 to 5 cm (e.g. *Carex montis-everestii*) and 10 to 15 cm (e.g. *Stipa purpurea*, *Carex moorcroftii*). The latter are the main forage resource. Some 12% of the species with a cover percentage of 22% are resilient against overgrazing because their buds are closely attached to the surface and thus well protected (e.g. *Stipa purpurea*) or because they have rhizomes that are hidden in the soil (e.g. *Carex moorcroftii*). Only bunch grasses like *Festuca* or *Poa* spp. (2% of the species and 9% of the cover percentage) are prone to grazing. Thus, most alpine steppe species in the study area are more or less tolerant to grazing. The resilience against grazing and the absence of grazing gradients has been similarly detected in the mountain desert steppes of the Gobi Altai, in southern Mongolia (Stumpp et al., 2002), which share physiognomy and related flora.

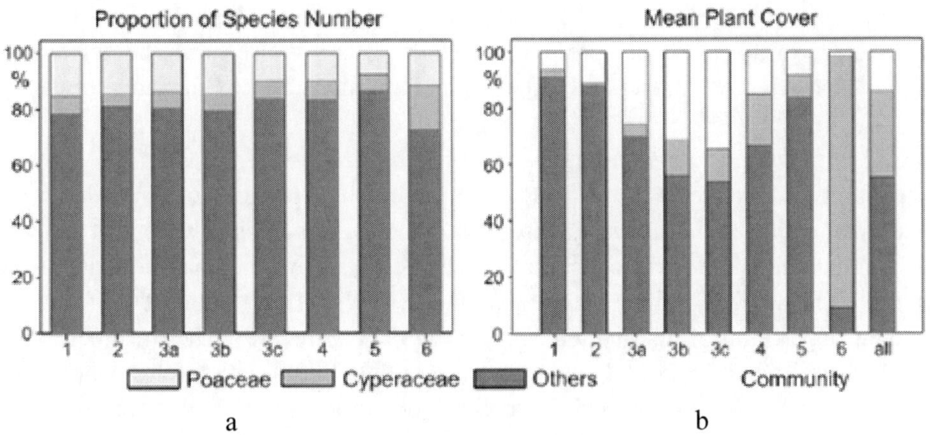

Figure 14. Proportion of species richness (a) and vegetation cover (b) by Poaceae, Cyperaceae and other families in each community.

Community 5: Lancea Tibeticatibetica-Wastelands

This community (11 plots) is a common vegetation type in Tibet's alpine pastures. It represents the final degradation stages following the destruction of a protecting turf cover. This vegetation type is also called "black soil" (Ma et al., 1999) in the north-eastern highlands due to the high humus content of the exposed loess. The community tends to be more common on exposed slopes, where inclination, shallow soils and presumably higher grazing pressure may increase damage and erosion. According to the DCA, this community is a transition between alpine steppes and *Kobresia pygmaea* pastures. The majority of sites are located at the left side of axis 1 with higher variability along axis 2 (Figure 6). They share plant species and environmental conditions with Community 3b and with some plots of Community 2 and 3a (running no. 18-20). Community 5 differentiates from Community 2 and 3a along an elevation gradient.

Dwarf shrubs (*Dasiphora parvifolia*) and *Artemisia* sub-shrubs (*A. tridactyla*, *A. santolinifolia*) occur in this community, the latter often heavily grazed. Some common species have rhizomes, like *Thalictrum rutifolium* and *Artemisia tridactyla*, which allows plants to cope with the unstable substrate.

A high proportion of the species of Community 5 (43%) are annual rosette plants, especially members of the Brassicaceae, Boraginaceae, Polygonaceae and Chenopodiaceae. This group reaches a high frequency in this vegetation type (26% of the plots). Some typical Tibetan grazing indicator plants are also present, including *Potentilla bifurca*, *Lancea tibetica,* and annual *Artemisia* species. (*A. hedinii*, *A. stricta*). Where pikas are less common, small mats of *Leontopodium pusillum* spread and develop an "edelweiss semi-desert" (Holzner and Kriechbaum, 2000). Nearly all species in this community are endemic and some are even monospecific genera (e.g. *Microgynoecium*, Wu et al., 1981). More than 66% of the species of community 5 are found above 5100 m. They occur on regularly disturbed and open soils. Such sites are generated by large herbivores, especially the Tibetan wild ass and the wild yak, which wallow in open soil, and by smaller digging mammals like marmots and pika. Thus, these niches must have been available for long periods of time.

Vegetation structure is patchy in this community. Total plant cover is not very high (mean: 62%, range: 27 - 83%) and mean species richness is 29 species/ plot. Forage plants, if present, are heavily grazed by livestock and pikas. Pika density is highest in this community with an average of 34 pika holes per plot. Whether pikas actually cause the degradation or just follow the destructive impact of livestock is an important question for future research.

Community 6: Kobresia Pygmaea pygmaea Grazing Lawns

Community 6 (14 plots) is the intact, golf course-like pastures of *Kobresia pygmaea* which are commonly found in the south-eastern Tibetan highlands (Figure 15), the world's largest humid alpine biome. The community is characterized by a number of distinct features. It represents the world's highest plant community (Miehe, 1989) with 70% of the species having their upper limit of distribution above 5100 m (Figure 719). The rhizomes and roots build a characteristic soil type ("rhizomull", Kaiser et al., 2008) with a felty A-horizon. Soil blocks are so tough that they are used for construction purposes (cp. Figure 12). All species in this community are highly adapted to grazing. The species set differs from the other communities, as shown by the DCA (Figure 6).

This community type was recorded mainly in the north-eastern study area, in valley bottoms with loess and gravel substrates, at 4480 - 5100 m.

Figure 15. Detail of closed *Kobresia pygmaea* grazing mats (Community 6). 30°48'N/92°35'E, 4770 m. July 2004.

The community is made up almost entirely of a single plant species, *Kobresia pygmaea*, which covers more than90% in nine of 14 samples. Species richness is low (mean: 17 species/ plot). Plants are small, with their above-ground biomass near the soil surface.

Land Cover Assessment

The absolute classified area covers approximately 5656 km^2 (Table 1). In addition to gravel, snow, ice, clouds and water, six different vegetation classes were identified (Figure 16), based on 71 training areas. These classes were deliberately chosen so that the floristically classified vegetation types were represented.

Table 1. Relative and absolute cover percentages of the land cover classes

Type	Area/ km^2	Percentage
Kobresia pygmaea pastures	295.75	5.0%
Kobresia pygmaea turf/ alpine steppe mosaic	1612.12	27.1%
Dense alpine steppe	431.95	7.3%
Sparse alpine steppe	61.34	1.0%
Alpine scree plant communities	153.84	2.6%
Kobresia schoenoides swamps	650.19	10.9%
Gravel	420.19	7.1%
Snow/ ice	1462.17	24.6%
Clouds	229.64	3.9%
Water	635.33	10.7%

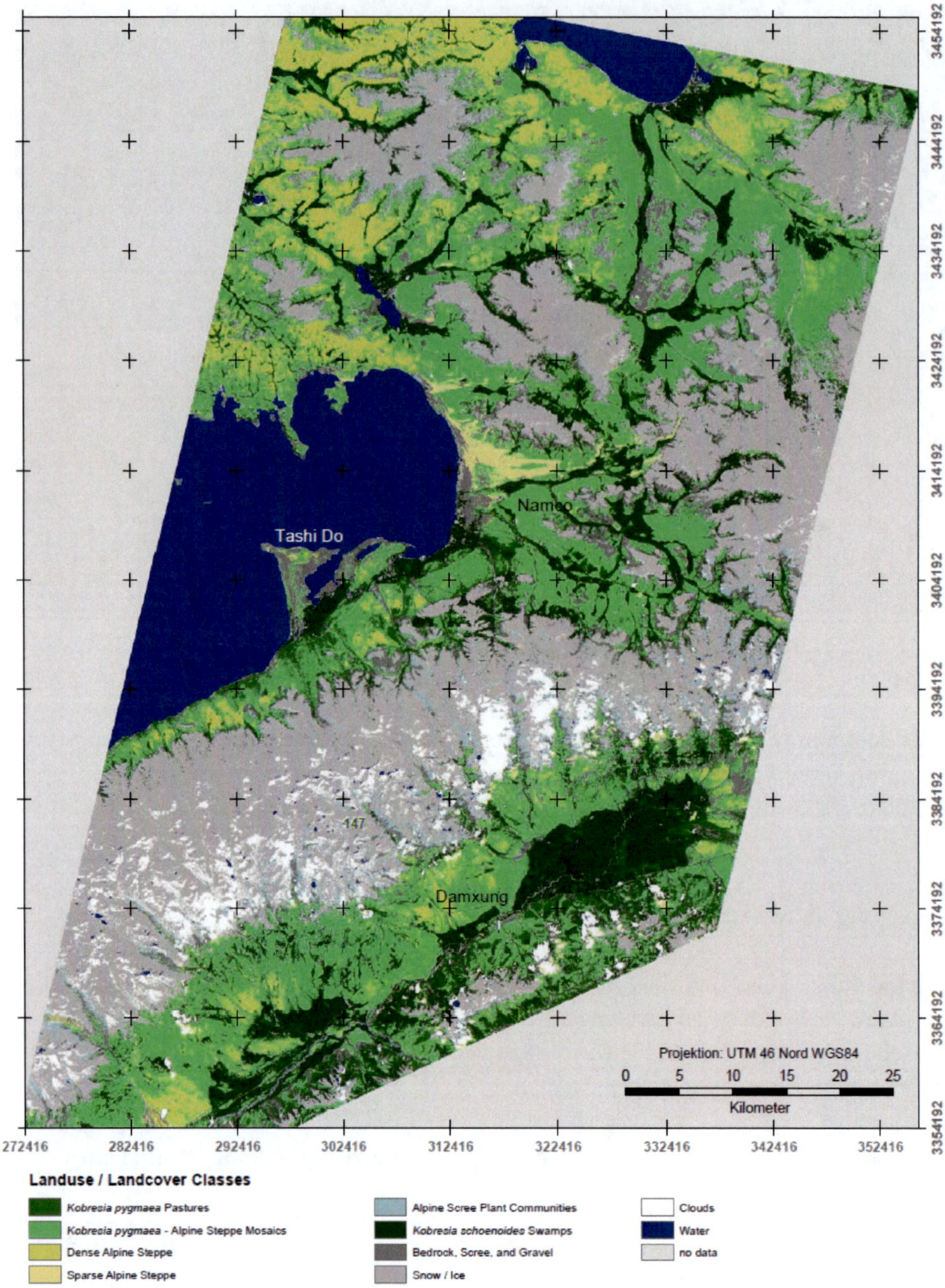

Figure 16. Land use/ land cover classes derived from supervised classification of satellite imagery.

Several runs of the Mulit Layer Perceptron procedure were performed, yielding classification accuracies of 69% - 79%. Finally, the VNIR bands 1 and 2, the scaled PVI (PVI*100), a PCA image of the ASTER DEM and the SRTM DEM and SWIR band 4 were used in the given order. The obtained accuracy of the training process was 76.2% and showed no tendency for overtraining.

Cramer's V index shows a correlation of 0.46 (p= 0.02) which shows that the classification was accurate. Furthermore, it provides a reliable estimation of the correspondence between plant community composition and vegetation formations.

However, whereas the floristic classification of plant communities focuses on species composition, the satellite based classification has to deal with both the response in vegetation cover and the abiotic surface signals. Thus, some of the floristically derived vegetation types had to be fused while others could be differentiated further. For example, Community 3 could not be distinguished from Community 4 using the spectral images from November. The *Kobresia pygmaea* turf/ alpine steppe mosaic and the *Kobresia schoenoides* swamps covered the largest areas of all identified vegetation types.

DISCUSSION

Is There Resilience or Vulnerability in an Equilibrium/ Non-Equilibrium Ecotone?

The study area stretches from a climatically stable equilibrium system in the east (Community 6, cp. Figure 1) to a highly variable non-equilibrium system in the west (Community 3).

Major parts of the area belong to an ecotone where species of both biomes intermix. The *Kobresia* pastures in the east have a closed plant cover, which stands in contrast to the naturally open alpine steppe in the west. Thus, the percentage of open substrate should be used as an indicator for rangeland health only in the east. The land cover map (Figure 16, see also Table 1) shows a high proportion of vegetation mosaics which can possibly be attributed to Community 4.

The vulnerability or resilience of a pasture is reflected in the prevalent plant functional types (Duckworth et al., 2000). In Communities 1 and 2, 19% of all species are graminoids which have buds closely attached to the ground or hidden in the soil. This group makes up 10% of the cover in this vegetation type (Figure 14a,b). Rosette species are especially common in Communities 3-6 (Community 3: 37%; covering 25%, Community 4: 33%; 17%, Community 5: 43%; 28%, Community 6: 45%; 5%). Taller plants, which are constantly avoided by livestock, like *Stellera chamaejasme*, *Cryptothladia* spp., and *Meconopsis horridula*, make up 6% of all species in the data set, having a mean cover of 1% across all records. These species reach their highest proportions in Communities 1 and 2 (24% and 86% of the cover, respectively). Poisonous grazing weeds attain very high cover percentages in some plots.

Today, most *Kobresia* pastures in the study area lack closed turf cover but it is possible that they once formed closed vegetation (see conceptual model in Figure 17). However, this hypothesis has yet to be proven. Compared to the alpine steppe, the *Kobresia* turfs in the eastern equilibrium system are likely to be more vulnerable than the non-equilibrium system of the west.

The answer to the question of whether a plant community is vulnerable or resilient to grazing relies on the chosen plant functional types. The present distribution, however, can be interpreted in different ways, dependening on the question of the state of naturalness.

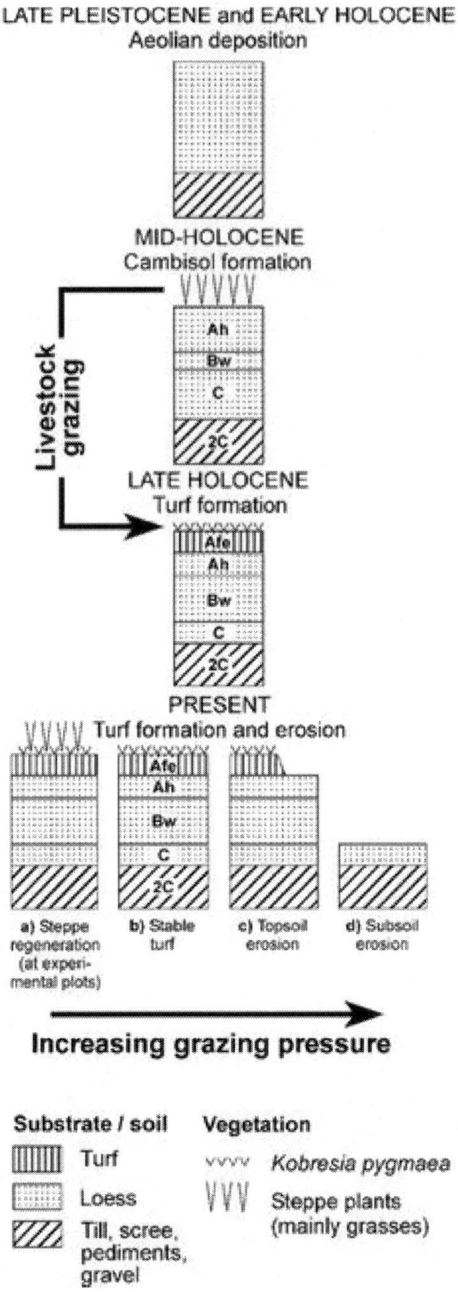

Figure 17. Conceptual model showing possible vegetation and soil dynamics in the mosaic of *Kobresia pygmaea* mats and alpine steppe in the Nagqu area (ca. 4500 m). Adapted from Kaiser et al., 2008.

The current physiognomy of the studied communities corresponds to what is generally defined as "alpine" plant communities (Körner, 1999). Consequently, it is not surprising that most authors regard the current state of Tibetan highlands as natural. However, such an assumption might not be correct for the *Kobresia* communities. Grazing exclosure experiments indicate that this community might have formed due to strong grazing effects, at least for the equilibrium systems of the *Kobresia* pastures (Miehe et al., 2008b, c).

Most of the current *Kobresia* pastures of the Tibetan highlands are at altitudes potentially suitable for tree growth, and most probably represent a synanthropic plagioclimax (Miehe et al., 2008a, c). Grazed plots resemble "alpine" short grasslands, whereas fenced plots differ by having tall grasses that overgrow the small sedge mats, rosette and cushion plants. Data from grazing exclosure experiments of truly alpine altitudes are still not available for the Tibetan highlands, because exclosure experiment plots in the southern (Reting 30°18'N/91°31'E) and north-eastern highlands (Xinghai 35°30'N/99°49'E) are located around 400 m below the upper tree line (Miehe et al., 2008b, c). However, the recently commenced privatization of rangeland with the fencing of winter pastures at alpine altitudes can be seen as a large scale fencing experiment. Even at 4340 m, the area outside the fence used as all-year round common grazing ground was covered by "natural alpine" *Kobresia pygmaea* vegetation. By contrast, the area inside the fence, where grazing is now excluded, was dominated by tall grassland that had overgrown the small "alpine" *Kobresia* mats (Figure 18).

Most of the species in this data set are endemic to the Tibetan highlands or at least to High Asia (Wu, 1983-87; Liu, 1996-99) which indicates that species must have their natural habitats here. This alpine endemism must have developed during the uplift and isolation of the highlands into alpine altitudes since the Tertiary (Rowley and Currie, 2008; Spicer et al., 2003; Tapponnier et al., 2001), as well as during the climatic changes in the Quaternary. Plants have likely adapted to the cold and partly dry conditions. The larger herbivores of the highlands are also endemic, and it is possible that the endemic plants have co-evolved with these herbivores.

It is known that plant species from different natural communities can be found in synanthropic communities (e.g. like in Central European meadows, Ellenberg, 1996).

Figure 18. Fenced winter grazing reserve in *Kobresia pygmaea* pastures. Outside the fence, "alpine" *Kobresia pygmaea* mats are maintained by grazing, whereas inside the fence, they have increasingly been overgrown by grasses. Qinghai (35°45'N/95°46'E), 4340 m. Sep. 2002.

As the *Kobresia*-pastures of the study area have a high number of endemic species and are in part a grazing-climax, the question arises of where these species may have their natural habitats. According to own records and regional accounts (Wu, 1983-87; Liu, 1996-99), 53% of the species listed in the supplement reach their upper distribution limit at 5100 m altitude or higher (Figure 19).

In Community 6, such "high alpine" species account for 70% of the regional species pool and attain a mean cover of 90%. They dominate the upper limit of the alpine belt and are typical species of the world's highest plant communities at 5960 m, on the northern escarpment of Mt. Everest (Miehe, 1989).

There, they form dense cushions made up mostly of *Kobresia pygmaea, Festuca tibetica, Saussurea graminea* var. *ortholepis, Arenaria bryophylla, Cortiella caespitosa*, and *Saussurea leontodontoides*, surrounded by boulders and scree of a periglacial environment. Approximately 35% of these species are rosette plants that closely hug the ground.

Free range grazing is the traditional rangeland management in the Tibetan highlands and it is still being practiced today. Animals, especially goats and sheep, selectively graze palatable plants that are available during the different seasons.

Taller and palatable forage plants are skimmed off except for their basal buds. Grazing favors shorter plants, like *Kobresia pygmaea*, over taller grasses that have more than three quarters of their above-surface biomass within the grazing reach of livestock.

It is likely that dwarf high alpine species and those from disturbed sites (e.g., pika burrows) have spread in response to high grazing pressure and reassembled to form the present pastures. The *Kobresia* pastures can thus be regarded as an assemblage of secondary vegetation composed of grazing tolerant plants (Grime, 1979; Diaz et al., 1992).

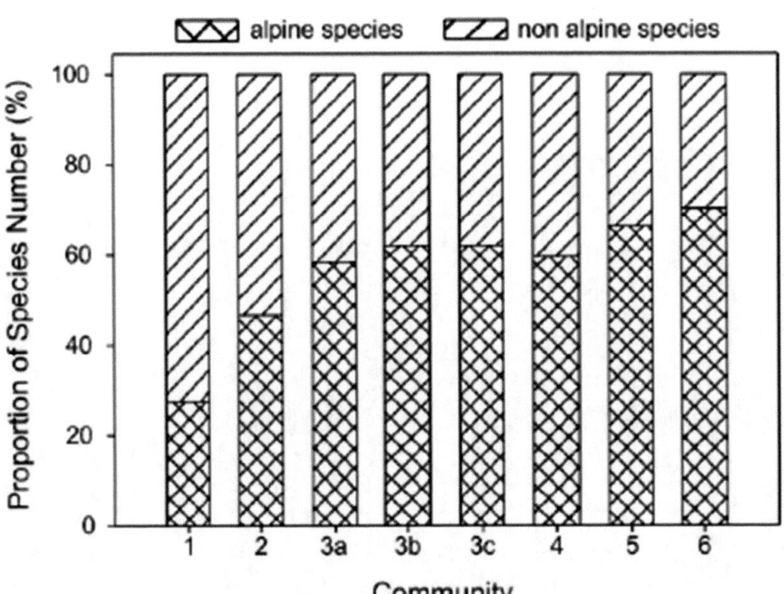

Figure 19. Proportion of alpine species (growing higher than 5100 m) within the communities (community numbers according to the table in the supplement).

The prime example is *Kobresia pygmaea* itself as an endemic and grazing adapted constituent of the pastures that is extremely resilient both against trampling and against grazing.

In conclusion, the eastern part of the study area is dominated by synanthropic *Kobresia pygmaea* rangelands, in which grazing adapted plant functional types prevail. This vegetation type is comparable to the *Cynodon dactylon* grazing lawns of the East African highlands (McNaughton, 1984).

Although the proportion of grazing adapted species is high, the fragmented turf cover indicates that habitat destruction is beyond the resilience of this system. Up to now, no evidence of turf cover recovery could be detected in the area of investigation. Possible detrimental effects are imposed by climate (gelifluction and needle ice, deflation, desiccation), animals (pika, destructive impacts of livestock), and humans (turf extraction). How much turf cover was destroyed has not been quantified yet, but probably extends throughout the entire biome in this region.

The natural extension of this vegetation type can only be reconstructed with the help of palaeo-ecological studies and long-term experiments.

In contrast to the *Kobresia* pastures, the alpine steppe vegetation is more open, having an average cover of 39% in the study plots. It is not known to what extent this vegetation type is natural. Data from grazing exclosure plots are not yet available for these communities. We also do not know to which extent the present livestock numbers exceed the former number of large wild herbivores as a natural disturbance factor. Climatic data imply the presence of a non-equilibrium system which is largely rain-driven.

We can assume that episodic heavy snowfall in winter is another factor to control livestock numbers. Most alpine steppe species in the study area are High Asian endemics. In Communities 3 and 4, 45% are cushion or rosette plants, while 20% of the species are poisonous or grazing-resistant weeds.

Graminoids are the main forage resource and make up 18% of the species, with an average cover of 13%. Their average growth height is 3 - 15 cm. Most graminoids (12% of all species in the community, 22% cover) are resilient to overgrazing.

Only bunch grasses are prone to grazing. Thus, alpine steppe species in the study area are more or less tolerant to grazing. We did not detect any clear grazing gradients and thus assume that the alpine steppe is a Central Asian non-equilibrium climax.

The high proportion of grazing-adapted plant functional types in *Kobresia* mats and alpine steppe suggests that both biomes are relatively resilient against degradation. Human impact may be monitored using the cover of unpalatable species as a measure of pasture degradation.

In contrast to the grazing resilience of the pastoral resources, the widespread *Kobresia pygmaea* turf is more vulnerable and subject to fragmentation. Future studies need to investigate whether this degradation process is natural or whether it was induced by overgrazing in the last decades.

Appendix 1

Vegetation table of relevés at lake NamCo comprising 105 records and 299 species

Running Number	1	2	3	4	5	6	7	8	9	10	11	12	13	14	15	16	17	18	19	20	21	22	23	24	25	26	27	28	29	30	31	32	33	34	35	36	37
Altitude (m.s.l.)	4285	4252	4286	4280	4288	4293	4350	4550	4700	4770	4780	4800	4730	4760	4740	4710	4740	4710	4790	4730	4720	4730	4720	4760	4720	4810	4720	4720	4750	4870	4750	4850	4750	4830	4740	4680	4680
Exposition	SW	SSW	SSW	S	S	SSW	S	S	SE	SW	SE	SE	S	SE	S	S	S	SE	W	SE	NW	SE	SW	SW	NW	S	E	N	S	S	N	NE	0	S	S	0	0
Inclination (°)	0	39	42	0	40	38	10	30	30	20	17	10	30	27	16	27	38	14	15	17	10	8	10	20	16	7	3	20	11	4	23	3	0	3	10	0	0
Vegetation Cover (%)	70	65	50	50	70	70	85	50	85	65	70	65	55	60	55	50	50	35	30	43	40	45	35	45	35	30	38	40	40	42	35	30	25	30	25	15	5
Species Number	25	21	25	22	22	19	31	5	38	36	42	43	48	31	31	39	42	27	22	29	34	28	26	28	35	38	35	40	28	39	36	24	18	24	34	28	14
Community No.	1								2									3a																			
Juniperus tibetica	27	25	25	23	30	37	+	.	.	.	.	.	.	.	.	.	.	.	.	.	.	.	.	.	.	.	.	.	.	.	.	.	.	.	.	.	.
Caragana versicolor	6	5	+	7	2	4	5	.	.	.	.	.	.	.	.	.	.	.	.	.	.	.	.	.	.	.	.	.	.	.	.	.	.	.	.	.	.
Dolomiaea calophylla	2	+	2	1	8	3	+	.	.	.	.	.	.	.	.	.	.	.	.	.	.	.	.	.	.	.	.	.	.	.	.	.	.	.	.	.	.
Rhodiola sacra	+	2	1	+	+	1	+	.	.	.	.	.	.	.	.	.	.	.	.	.	.	.	.	.	.	.	.	.	.	.	.	.	.	.	.	.	.
Sedum roborowskii	+	+	.	+	+	+	+	.	.	.	.	.	.	.	.	.	.	.	.	.	.	.	.	.	.	.	.	.	.	.	.	.	.	.	.	.	.
Piptatherum gracile	1	2	+	1	+	+	.	.	.	.	.	.	.	.	.	.	.	.	.	.	.	.	.	.	.	.	.	.	.	.	.	.	.	.	.	.	.
Poa hylobates	1	+	+	2	1	+	1	.	.	.	.	.	.	.	.	.	.	.	.	.	.	.	.	.	.	.	.	.	.	.	.	.	.	.	.	.	.
Stellaria dianthifolia	+	+	+	+	+	+	+	.	.	.	.	.	.	.	.	.	.	.	.	.	.	.	.	.	.	.	.	.	.	.	.	.	.	.	.	.	.
Pleurospermum foetens	2	1	+	+	1	2	.	.	.	.	.	.	.	.	.	.	.	.	.	.	.	.	.	.	.	.	.	.	.	.	.	.	.	.	.	.	.
Kobresia macrantha	.	1	.	.	1	.	4	.	+	.	.	.	.	.	.	.	+	.	.	.	.	.	.	.	.	.	.	.	.	.	.	.	.	.	.	.	.
Pennisetum flaccidum	1	1	1	1	+	1	5	.	1	.	.	.	.	.	.	.	.	1	.	.	.	.	.	.	.	.	.	.	.	.	.	.	.	.	.	.	.
Ephedra gerardiana	1	.	3	+	1	+	.	.	.	.	.	.	.	.	.	.	.	.	.	.	.	.	.	.	.	.	.	.	.	.	.	.	.	.	1	.	.
Urtica hyperborea	.	+	.	+	.	.	3	.	.	.	+	.	.	2	+	.	+	.	.	.	.	.	.	.	.	.	.	.	.	.	.	.	.	.	.	.	.
Allium fasciculatum	+	+	+	+	1	.	.	.	r	.	.	.	+	.	.	.	+	.	.	.	.	.	.	.	.	.	.	.	.	.	.	.	.	.	+	.	.
Astragalus tribulifolius	.	.	2	.	+	1	2	4	+	.	.	.	+	.	.	+	+	.	.	.	.	5	+	+	.	.	+	.	.	.	.	.	.	1	.	.	.
Juniperus pingii v. wilsonii	17	10	8	.	13	3	70	30	60	40	40	40	30	35	35	25	17	.	.	.	.	.	.	.	.	.	.	.	.	.	.	.	.	.	.	.	.
Phlomis younghusbandii	2	+	+	+	1	1	+	.	1	1	1	2	2	.	1	+	.	4	.	2	.	.	.	4	.	.	2	3	.	.	1	r	.	.	.	.	.
Dracocephalum tanguticum	4	3	+	1	1	2	1	.	+	+	r	r	+	1	.	+	3	.	r	+	+	.	+	+	1	+	r	1	2	1	+	.	.	.	+	.	.
Stellera chamaejasme	2	+	5	+	5	.	2	.	+	+	+	.	1	.	1	2	+	.	.	1	4	4	3	3	1	1	4	4	+	+	+	.	3	2	.	2	+
Artemisia moorcroftiana	.	.	.	.	.	.	.	.	+	+	.	.	+	6	+	+	5	.	.	.	.	.	+	.	.	.	.	.	.	.	.	.	.	.	.	.	.
Poa albertii ssp. poophagorum	.	.	.	.	.	.	.	.	1	.	+	.	+	.	+	+	+	.	.	.	.	.	.	.	.	.	.	.	.	.	.	.	.	.	.	.	.
Poa araratica s.l.	.	.	.	.	.	.	.	.	1	+	3	1	1	+	2	2	2	.	.	.	.	.	.	.	.	.	.	.	.	.	.	.	.	.	.	.	.
Poa asperifolia	.	.	.	.	.	.	.	.	1	4	+	.	.	.	.	.	1	.	.	.	.	.	.	.	.	.	.	.	.	.	.	.	.	.	.	.	.
Stephanachne cf. pappophorea	.	.	.	.	.	.	.	.	1	.	+	.	.	.	+	1	.	.	.	.	.	.	.	.	.	.	.	.	.	.	.	.	.	.	.	.	.
Gentianella azurea	.	.	.	.	.	.	.	.	.	+	+	+	.	.	.	r	.	.	.	.	.	.	.	.	.	.	.	.	.	.	.	.	.	.	.	.	.
Rhodiola tangutica	.	.	.	.	.	.	.	.	+	1	r	+	+	.	1	+	.	.	.	.	.	.	.	.	.	.	.	.	.	.	.	.	.	.	.	.	.
Allium przewalskianum	.	.	.	.	.	.	.	.	r	+	.	.	+	.	.	.	.	.	.	.	.	.	.	.	.	.	.	.	.	.	.	.	.	.	.	.	.
Stipa penicillata v. penicillata	.	.	.	.	.	.	.	.	+	.	+	.	+	+	+	.	.	+	.	.	.	.	.	.	.	.	.	.	.	.	.	.	.	.	.	.	.
Microula tangutica	.	.	.	.	.	.	.	.	+	1	1	+	+	+	+	+	.	+	.	+	.	+	.	.	+	.	.	.	.	.	.	.	.	.	.	.	.
Salsola monoptera	.	.	.	.	.	.	.	.	.	.	.	+	r	+	.	.	.	r	.	+	.	.	.	.	.	.	.	.	.	.	.	.	.	.	.	.	.
Dontostemon glandulosus	.	.	.	.	.	.	.	.	r	.	.	.	+	+	+	.	1	.	.	1	.	.	.	.	.	.	.	.	+	.	.	.	.	.	.	+	.
Corydalis hookeri	.	.	.	.	.	.	.	.	r	+	+	+	+	+	+	r	+	+	+	+	.	.	.	.	.	.	.	.	+	.	.	+	.	.	.	.	.
Trikeraia oreophila	.	.	.	.	.	.	.	.	.	+	.	+	+	.	.	.	+	.	.	+	.	+	.	+	.	.	.	.	.	.	.	.	.	.	.	.	.
Anaphalis nubigena	.	.	.	.	.	.	.	.	+	+	+	.	.	.	.	+	.	+	+	+	+	.	2	2	.	2	.	.	.	1	4	.	.	.	.	.	.
Silene caespitella	.	.	.	.	.	.	.	.	+	+	.	+	r	.	.	+	+	.	.	.	+	.	.	.	+	+	.	.	.	.	+	r	.	.	.	.	.
Pleurospermum aromaticum	.	.	.	.	.	.	r	.	+	+	+	+	+	+	+	+	.	.	.	.	.	.	.	.	.	.	.	.	.	.	.	+	.	.	.	.	.
Ajuga lupulina	.	.	.	.	.	.	.	.	r	1	+	.	+	.	1	.	.	+	6	4	.	.	.	.	.	.	.	.	+	.	.	.	.	.	.	.	.
Thalictrum rutifolium	.	.	.	.	.	.	.	.	+	+	r	+	+	+	+	+	+	+	+	+	.	.	.	.	.	.	.	.	+	.	.	.	.	.	.	.	.
Artemisia tridactyla	.	.	.	.	.	.	.	.	+	1	+	1	1	+	1	+	+	+	4	1	.	.	.	.	.	+	+	.	+	+	+	+	.	.	.	.	.
Pedicularis cheilanthifolia	.	.	.	.	.	.	.	.	.	+	r	.	r	.	.	+	+	.	.	.	1	+	.	+	+	+	+	.	.	+	.	+	.	.	+	.	.
Euphorbia stracheyi	.	.	.	.	.	.	.	.	+	+	3	+	+	2	1	+	r	10	+	10	1	.	4	.	.	1	+	+	+	.	+	.	.	.	+	3	.
Lasiocaryum munroi	.	.	.	.	.	.	.	.	+	+	+	+	+	.	+	.	.	+	.	+	+	+	+	+	r	+	r	.	.	.	.	+	.	+	.	+	.
Carex stenophylla	.	.	.	.	.	.	.	.	.	.	.	+	.	.	.	+	.	.	.	.	+	+	+	+	2	+	1	+	+	+	+	1	+	+	+	+	+
Rhodiola smithii	.	.	.	.	.	.	.	.	.	.	.	+	.	.	.	r	.	.	.	.	.	+	.	+	r	+	+	.	.	+	.	r	.	+	.	.	.
Oxytropis stracheyana	.	.	.	.	.	.	.	.	.	.	.	.	.	.	+	.	r	.	.	.	.	.	.	.	r	1	.	.	2	+	.	.	2	.	+	+	+
Youngia simulatrix	.	.	.	.	.	.	.	.	.	.	.	+	.	.	.	.	+	.	.	+	+	+	+	+	2	+	r	3	3	+	+	1	.	.	.	+	+
Saussurea leiocarpa	.	.	.	.	.	.	.	.	.	+	.	+	.	.	.	+	.	.	.	.	3	7	5	7	3	2	+	1	6	4	.	.	.	2	+	+	+
Festuca cf. valesiaca	.	.	.	.	.	.	.	.	.	.	.	.	.	.	.	.	.	.	.	.	6	4	6	4	4	+	2	4	6	+	3	3	6	5	1	+	.
Lomatogonium brachyantherum	.	.	.	.	.	.	.	.	.	.	.	.	.	.	.	.	.	.	.	.	+	1	.	+	+	.	+	+	+	+	+	+	+	1	1	+	.
Kengyilia thoroldiana v. thoroldiana	.	.	.	.	.	.	.	.	.	.	.	.	.	.	.	.	.	.	.	.	.	.	.	.	.	.	.	+	.	.	+	.	+	.	+	+	+
Kobresia robusta	.	.	.	.	.	.	.	.	.	.	.	.	.	.	.	.	.	.	.	.	.	.	.	.	.	.	.	.	.	.	.	.	+	+	+	+	.
Deyeuxia zangxiensis	.	.	.	.	.	.	.	.	.	.	.	.	.	.	.	.	.	.	.	.	.	.	.	.	.	.	.	+	.	.	.	.	.	.	+	+	+
Artemisia stricta	.	.	.	.	.	.	.	.	.	.	.	.	.	.	.	.	.	.	.	.	+	.	.	.	.	.	.	+	.	.	.	.	.	.	.	.	.
Astragalus arnoldii	.	.	.	.	.	.	.	.	.	.	.	.	.	.	.	.	.	.	.	.	.	.	.	.	.	.	.	.	.	+	.	.	.	.	.	.	.
Potentilla exigua	.	.	.	.	.	.	.	.	.	.	.	+	.	.	.	.	.	.	.	.	.	.	.	.	.	.	.	.	.	.	.	.	.	.	.	.	.
Koeleria litvinowii ssp. argentea	.	.	.	.	.	.	.	.	.	.	.	.	.	.	.	.	.	.	.	.	.	.	.	.	.	.	.	.	.	.	1	.	.	.	.	.	.
Sibbaldianthe adpressa	.	.	.	.	.	.	.	.	.	.	.	+	.	.	.	.	.	.	.	.	+	.	+	.	.	.	r	.	.	.	.	.	+	.	.	.	.
Taraxacum sherriffii	.	+	.	.	.	.	.	.	.	.	.	.	.	.	.	.	.	.	.	.	+	.	.	.	+	.	.	+	+	.	r	r	.	.	.	.	.
Allium cyaneum	.	.	.	.	.	.	.	.	.	.	.	.	.	+	.	.	+	.	.	.	.	+	.	.	+	.	.	.	.	+	+	.	.	.	+	r	.
Comastoma pedunculatum	.	.	.	.	.	.	.	.	.	.	+	.	.	.	.	.	.	.	.	+	.	.	.	+	.	.	.	.	.	.	.	.	+	.	.	.	.
Delphinium caeruleum	.	.	.	.	.	.	.	.	.	.	.	r	.	.	.	.	.	.	.	.	.	.	.	.	+	.	.	.	+	+	r	.	.	.	.	.	.
Draba zangbeiensis	.	.	.	.	.	.	.	.	.	.	.	+	.	.	.	.	.	.	.	.	.	.	.	.	.	.	.	.	.	.	.	.	.	.	+	.	.
Aster yunnanensis	.	.	.	.	.	.	.	.	.	.	.	+	.	.	.	.	+	.	.	.	.	.	.	.	.	+	.	.	.	.	.	.	.	.	.	.	.
Kobresia capillifolia	.	.	.	.	.	.	.	.	.	.	.	.	.	.	.	+	.	.	.	.	.	.	.	.	.	+	.	.	.	.	.	.	.	.	.	.	.
Eritrichium tangkulaense	.	.	.	.	.	.	.	.	.	.	.	+	+	.	.	.	.	.	.	.	.	.	.	.	+	.	.	.	.	.	.	.	.	.	.	.	.
Meconopsis horridula	.	.	.	.	.	.	.	.	.	.	.	.	.	.	.	.	.	.	.	.	.	.	.	.	.	.	.	.	.	.	.	.	.	.	.	.	.
Dasiphora parvifolia	.	.	.	.	.	.	.	.	.	.	5	.	.	.	.	.	.	.	.	.	.	.	.	.	.	.	.	.	.	.	.	.	.	.	.	.	.
Kobresia pygmaea	.	.	.	.	.	.	+	.	.	.	.	.	.	.	.	.	.	.	.	.	.	.	.	.	.	.	.	.	.	.	.	.	.	.	.	.	.
Potentilla saundersiana	.	.	.	.	.	.	.	.	.	.	.	.	.	.	.	.	.	.	.	.	.	.	.	.	.	.	.	.	.	+	.	.	.	.	.	.	.

Running Number	38	39	40	41	42	43	44	45	46	47	48	49	50	51	52	53	54	55	56	57	58	59	60	61	62	63	64	65	66	67	68
Altitude (m.s.l.)	4760	4690	4720	4690	4800	4820	4730	4800	4720	4690	4690	4740	4660	4730	4710	4730	4710	4790	4715	4730	4700	4725	4740	4740	4810	4820	4710	4890	4850	4840	4860
Exposition	E	SE	0	0	SE	NE	0	NE	N	SW	0	S	0	NW	0	0	0	NW	0	0	0	0	0	0	W	N	0	S	N	N	NW
Inclination (°)	10	2	0	0	10	10	0	20	5	40	0	11	0	14	0	0	0	15	0	0	0	0	0	0	5	1	0	4	2	1	5
Vegetation Cover (%)	37	15	25	25	20	38	15	35	5	15	5	30	35	45	45	25	35	35	25	25	15	35	56	15	45	60	25	30	45	35	45
Species Number	31	25	16	16	23	32	19	29	18	24	23	15	20	31	27	22	29	31	27	22	29	17	18	19	24	20	21	25	34	33	26
Community No.	3b													3c																	
Juniperus tibetica	.	.	.	.	.	.	.	.	.	.	.	.	.	.	.	.	.	.	.	.	.	.	.	.	.	.	.	.	.	.	.
Caragana versicolor	.	.	.	.	.	.	.	.	.	.	.	.	.	.	.	.	.	.	.	.	.	.	.	.	.	.	.	.	.	.	.
Dolomiaea calophylla	.	.	.	.	.	.	.	.	.	.	.	.	.	.	.	.	.	.	.	.	.	.	.	.	.	.	.	.	.	.	.
Rhodiola sacra	.	.	.	.	.	.	.	.	.	.	.	.	.	.	.	.	.	.	.	.	.	.	.	.	.	.	.	.	.	.	.
Sedum roborowskii	.	.	.	.	.	.	.	.	.	.	.	.	.	.	.	.	.	.	.	.	.	.	.	.	.	.	.	.	.	.	.
Piptatherum gracile	.	.	.	.	.	.	.	.	.	.	.	.	.	.	.	.	.	.	.	.	.	.	.	.	.	.	.	.	.	.	.
Poa hylobates	.	.	.	.	.	.	.	.	.	.	.	.	.	.	.	.	.	.	.	.	.	.	.	.	.	.	.	.	.	.	.
Stellaria dianthifolia	.	.	.	.	.	.	.	.	.	.	.	.	.	.	.	.	.	.	.	.	.	.	.	.	.	.	.	.	.	.	.
Pleurospermum foetens	.	.	.	.	.	.	.	.	.	.	.	.	.	.	.	.	.	.	.	.	.	.	.	.	.	.	.	.	.	.	.
Kobresia macrantha	.	.	.	.	.	+	.	.	.	.	.	.	.	.	.	.	.	.	.	.	.	.	.	.	.	.	.	.	.	.	.
Pennisetum flaccidum	.	.	.	.	.	.	.	.	.	.	.	2	.	.	.	.	.	.	.	.	.	.	.	.	.	.	.	.	.	.	.
Ephedra gerardiana	.	.	.	.	.	.	.	.	.	+	.	.	.	.	.	.	+	.	r	.	+	.	.	.	.	.	.	.	.	.	.
Urtica hyperborea	.	.	.	.	.	.	.	.	.	.	.	.	.	.	.	.	.	.	.	.	.	.	.	.	.	.	.	.	.	.	.
Allium fasciculatum	.	.	.	.	.	.	.	.	.	.	.	.	.	.	.	.	.	.	.	.	.	.	.	.	.	.	.	.	.	.	.
Astragalus tribulifolius	.	.	.	.	.	2	.	.	.	.	.	.	.	.	.	.	.	.	.	.	.	.	.	.	.	.	.	.	.	.	.
Juniperus pingii v. wilsonii	.	.	.	.	.	.	.	.	.	.	.	.	.	.	.	.	.	.	.	.	.	.	.	.	.	.	.	.	.	.	.
Phlomis younghusbandii	.	.	.	.	.	.	.	.	.	.	.	.	.	.	.	.	.	.	.	.	.	.	.	.	.	.	.	.	.	.	.
Dracocephalum tanguticum	.	.	.	.	.	.	.	+	.	.	.	.	.	.	.	.	.	.	.	.	.	.	.	.	.	.	.	.	.	.	.
Stellera chamaejasme	+	1	+	.	.	.	.	.	.	.	.	.	.	.	1	.	.	.	.	.	.	.	.	.	.	.	.	.	.	.	.
Artemisia moorcroftiana	.	.	.	.	.	.	.	.	.	.	.	.	.	.	.	.	.	.	.	.	.	.	.	.	.	.	.	.	.	.	.
Poa albertii ssp. poophagorum	.	.	.	.	.	.	.	.	.	.	.	.	.	.	.	.	.	.	.	.	.	.	.	.	.	.	.	.	.	.	.
Poa araratica s.l.	.	.	.	.	.	.	.	.	.	.	.	.	.	.	.	.	.	.	.	.	.	.	.	.	.	.	.	.	.	.	.
Poa asperifolia	.	.	.	.	.	.	.	.	.	.	.	.	.	.	.	.	.	.	.	.	.	.	.	.	.	.	.	.	.	.	.
Stephanachne cf. pappophorea	.	.	.	.	.	.	.	.	.	.	.	.	.	.	.	.	.	.	.	.	.	.	.	.	.	.	.	.	.	.	.
Gentianella azurea	.	.	.	.	.	.	.	.	.	.	.	.	.	.	.	.	.	.	.	.	.	.	.	.	.	.	.	.	.	.	.
Rhodiola tangutica	.	.	.	.	.	.	.	.	.	.	.	.	.	.	.	.	.	.	.	.	.	.	.	.	.	.	.	.	.	.	.
Allium przewalskianum	.	+	.	.	.	.	.	.	.	.	.	.	.	.	.	.	.	.	.	.	.	.	.	.	.	.	.	.	.	.	.
Stipa penicillata v. penicillata	.	.	.	.	.	.	.	+	.	+	.	.	.	.	.	.	.	.	.	.	.	.	.	.	.	.	.	.	.	.	.
Microula tangutica	+	.	.	.	+	+	.	.	.	.	.	.	r	.	.	.	.	.	.	.	.	.	.	.	.	.	.	.	.	.	.
Salsola monoptera	.	+	.	+	+	.	.	.	.	.	r	.	.	.	.	.	.	.	.	.	.	.	.	.	.	.	.	.	.	.	.
Dontostemon glandulosus	r	+	+	.	r	.	+	.	.	+	.	.	r	+	.	.	.	.	.	.	.	.	.	.	.	.	.	.	.	.	.
Corydalis hookeri	.	.	.	.	.	.	.	+	.	.	.	.	.	.	.	.	.	.	.	.	.	.	.	.	.	.	.	.	.	.	.
Trikeraia oreophila	.	.	.	.	.	.	.	.	.	.	.	.	.	.	.	.	.	.	.	.	.	.	.	.	.	.	.	.	.	.	.
Anaphalis nubigena	.	.	.	.	.	.	.	.	.	.	.	.	.	.	.	.	.	.	.	.	.	.	.	.	.	.	.	.	.	.	.
Silene caespitella	.	.	.	.	.	.	.	.	.	.	.	.	.	.	.	.	.	.	.	.	.	.	.	.	+	+	.	.	.	.	.
Pleurospermum aromaticum	.	.	.	.	.	.	.	.	.	.	.	.	.	.	.	.	.	.	.	.	.	.	.	.	.	.	.	.	.	.	.
Ajuga lupulina	.	.	.	.	.	.	.	.	.	.	.	.	.	+	.	.	.	.	.	.	.	.	.	.	.	.	.	.	.	.	.
Thalictrum rutifolium	r	.	.	.	.	.	.	+	.	+	r	.	.	+	.	.	.	.	.	.	.	.	.	.	.	.	.	.	.	.	.
Artemisia tridactyla	.	+	.	.	.	.	.	.	.	+	+	.	.	.	.	.	.	.	.	.	.	.	.	.	+	.	.	.	.	.	.
Pedicularis cheilanthifolia	+	.	.	.	.	.	.	+	.	.	r	.	.	.	.	.	.	.	.	.	.	.	.	.	.	.	.	.	.	+	.
Euphorbia stracheyi	.	1	+	.	.	.	.	.	r	+	.	17	.	.	.	.	.	.	.	.	.	.	.	.	.	.	+	.	.	.	.
Lasiocaryum munroi	+	+	+	.	.	+	.	.	.	.	+	.	.	.	.	.	+	+	+	+	+	.	.	.	.	.	+	.	+	.	.
Carex stenophylla	+	.	1	.	.	+	+	.	r	.	.	.	+	+	+	+	+	+	+	+	+	3	.	+	+	+	.	+	+	+	+
Rhodiola smithii	.	.	.	.	.	+	.	.	.	.	r	.	.	.	.	.	r	.	+	.	r	.	.	+	.	.	.	.	.	.	.
Oxytropis stracheyana	+	+	.	.	.	.	2	.	+	+	r	1	.	.	.	r	.	.	.	+	.	.	.	1	.	.	r	.	.	+	.
Youngia simulatrix	1	.	.	.	1	.	.	+	.	+	.	2	.	+	3	.	.	.	.	.	.	.	.	.	+	.	.	+	.	+	2
Saussurea leiocarpa	.	1	.	.	.	.	.	.	.	.	+	.	.	.	.	.	.	.	.	.	.	.	.	.	.	.	.	.	r	+	+
Festuca cf. valesiaca	+	+	.	.	.	.	1	+	+	+	.	.	.	.	2	2	3	3	3	3	2	2	8	3	+	5	.	+	3	4	4
Lomatogonium brachyantherum	1	+	.	r	.	1	.	.	.	.	.	+	.	+	1	+	1	+	.	.	+	.	.	+	.	+	.	+	+	+	.
Kengyilia thoroldiana v. thoroldiana	+	+	.	+	+	+	+	+	+	+	.	+	.	+	+	+	.	+	+	+	+	.	.	+	.	.	.	.	.	.	.
Kobresia robusta	+	+	.	.	+	+	+	+	+	.	.	+	.	.	.	+	.	+	.	+	.	+	.	.	.	.	.	.	.	.	.
Deyeuxia zangxiensis	.	.	1	+	.	2	1	.	.	+	.	.	+	2	2	.	.	.	.	1	.	+	.	.	.	.	.	.	.	.	.
Artemisia stricta	.	.	+	.	.	+	.	.	.	.	.	.	.	.	.	.	+	+	.	+	r	.	+	+	+	+	+	+	+	.	.
Astragalus arnoldii	.	.	.	.	.	.	.	.	.	.	.	.	.	+	+	4	2	7	4	2	1	1	.	+	.	.	+	+	+	.	.
Potentilla exigua	.	.	.	.	.	.	.	.	.	.	.	.	5	.	1	r	+	+	+	.	r	5	6	.	3	+	.	+	1	+	.
Koeleria litvinowii ssp. argentea	.	.	4	.	.	.	.	.	.	.	.	.	.	+	1	+	+	2	+	.	+	.	3	.	12	+	18	+	1	.	.
Sibbaldianthe adpressa	.	.	.	.	.	.	2	.	.	.	.	.	.	.	.	+	+	+	+	+	1	+	1	.	.	+	.	.	.	.	.
Taraxacum sherriffii	.	.	.	.	.	.	.	.	.	.	r	.	.	+	+	.	.	.	.	.	.	.	.	.	.	.	.	+	+	+	+
Allium cyaneum	.	.	.	.	.	.	.	+	.	+	.	.	.	.	.	.	.	.	.	.	.	.	.	.	.	.	.	+	.	+	+
Comastoma pedunculatum	.	.	.	.	.	.	.	.	.	.	.	.	.	.	.	.	1	.	.	.	.	.	.	.	.	.	.	.	+	+	.
Delphinium caeruleum	.	.	.	.	.	.	.	.	.	.	.	.	.	.	.	.	.	.	.	.	.	.	.	.	.	.	.	.	.	+	+
Draba zangbeiensis	.	.	.	.	.	.	.	.	.	.	.	.	.	.	.	.	.	.	+	.	.	.	.	.	.	.	.	.	.	.	+
Aster yunnanensis	.	.	.	.	.	.	.	.	.	.	.	.	.	.	.	.	.	.	.	.	.	.	.	.	.	.	.	.	.	.	.
Kobresia capillifolia	.	.	.	.	.	.	.	.	.	.	.	.	.	.	.	.	.	.	.	.	.	.	.	.	.	.	.	.	.	.	.
Eritrichium tangkulaense	.	.	.	.	.	.	.	.	.	.	.	.	.	.	.	.	.	.	.	.	.	.	.	.	.	.	.	.	.	.	.
Meconopsis horridula	.	.	.	.	.	.	.	.	.	.	.	.	.	.	.	.	.	.	.	.	.	.	.	.	.	.	.	.	.	.	.
Dasiphora parvifolia	.	.	.	.	.	.	.	.	.	3	.	.	.	.	.	.	.	.	.	+	.	.	.	.	.	.	.	.	.	.	.
Kobresia pygmaea	.	.	.	.	.	.	.	.	.	.	.	.	.	.	.	.	.	.	.	.	.	.	.	.	.	.	.	.	.	.	.
Potentilla saundersiana	.	.	.	.	.	.	.	r	.	.	.	.	.	.	.	.	.	+	.	.	.	.	+	.	.	.	.	.	+	+	.

Running Number	69	70	71	72	73	74	75	76	77	78	79	80	81	82	83	84	85	86	87	88	89	90	91	92	93	94	95	96	97	98	99	100	101	102	103	104	105	
Altitude (m.s.l.)	4870	4770	4810	4860	4830	4840	4800	4820	4850	4850	4910	4860	4535	4480	4470	4460	4460	4470	4470	4470	4480	4480	4750	4810	5100	4770	4480	4700	4620	4750	4750	4650	4760	4810	4770	4750	4930	
Exposition	SW	0	N	NW	0	0	NE	NE	NE	S	NW	NW	E	E	ESE	E	E	ENE	ENE	ENE	NE	NE	0	0	0	S	NE	0	0	0	SE	S	S	SW	S	S	S	
Inclination (°)	15	0	17	17	0	0	20	5	11	10	25	15	4	8	12	2	2	13	13	13	12	12	0	0	0	2	2	0	0	0	3	5	6	2	1	0	5	
Vegetation Cover (%)	35	40	40	35	40	35	45	45	55	45	35	58	55	52	61	54	78	63	63	66	27	83	85	84	90	95	92	90	98	100	97	96	98	85	95	98	90	
Species Number	38	33	40	37	42	35	39	39	46	38	35	38	23	22	32	26	29	31	33	30	33	33	23	22	24	16	18	9	18	18	17	20	22	18	15	15	11	
Community No.	4												5											6														
Juniperus tibetica	.	.	.	.	.	.	.	.	.	.	.	.	.	.	.	.	.	.	.	.	.	.	.	.	.	.	.	.	.	.	.	.	.	.	.	.	.	7
Caragana versicolor	.	.	.	.	.	.	.	.	.	.	.	.	.	.	.	.	.	.	.	.	.	.	.	.	.	.	.	.	.	.	.	.	.	.	.	.	.	7
Dolomiaea calophylla	.	.	.	.	.	.	.	.	.	.	.	.	.	.	.	.	.	.	.	.	.	.	.	.	.	.	.	.	.	.	.	.	.	.	.	.	.	7
Rhodiola sacra	.	.	.	.	.	.	.	.	.	.	.	.	.	.	.	.	.	.	.	.	.	.	.	.	.	.	.	.	.	.	.	.	.	.	.	.	.	7
Sedum roborowskii	.	.	.	.	.	.	.	.	.	.	.	.	.	.	.	.	.	.	.	.	.	.	.	.	.	.	.	.	.	.	.	.	.	.	.	.	.	6
Piptatherum gracile	.	.	.	.	.	.	.	.	.	.	.	.	.	.	.	.	.	.	.	.	.	.	.	.	.	.	.	.	.	.	.	.	.	.	.	.	.	6
Poa hylobates	.	.	.	.	.	.	.	.	.	.	.	.	.	.	.	.	.	.	.	.	.	.	.	.	.	.	.	.	.	.	.	.	.	.	.	.	.	7
Stellaria dianthifolia	.	.	.	.	.	.	.	.	.	.	.	.	.	.	.	.	.	.	.	.	.	.	.	.	.	.	.	.	.	.	.	.	.	.	.	.	.	7
Pleurospermum foetens	.	.	.	.	.	.	.	.	.	.	.	.	.	.	.	.	.	.	.	.	.	.	.	.	.	.	.	.	.	.	.	.	.	.	.	.	.	6
Kobresia macrantha	.	.	.	.	.	.	.	.	.	.	.	.	.	.	.	.	.	.	.	.	.	.	.	.	.	.	.	.	.	.	.	.	.	.	.	.	.	6
Pennisetum flaccidum	.	.	.	.	.	.	.	.	.	.	.	.	.	.	.	.	.	.	.	.	.	.	.	.	.	.	.	.	.	.	.	.	.	.	.	.	.	10
Ephedra gerardiana	r	.	.	.	.	.	+	.	.	.	.	.	.	.	.	.	.	.	.	.	.	.	.	.	.	.	.	.	.	.	.	.	.	.	.	.	.	12
Urtica hyperborea	.	.	.	+	.	.	.	.	.	+	.	.	.	.	.	.	.	.	.	.	.	.	.	.	.	.	.	.	.	.	.	.	.	.	.	.	.	9
Allium fasciculatum	.	.	.	.	+	.	.	.	.	.	.	.	.	.	.	.	.	.	.	.	.	.	.	.	.	.	.	.	.	.	.	.	.	.	.	.	.	10
Astragalus tribulifolius	.	.	.	.	.	.	.	.	.	.	.	.	.	.	.	.	.	.	.	.	.	.	.	.	.	.	.	.	.	.	.	.	.	.	.	.	.	14
Juniperus pingii v. wilsonii	.	.	.	.	.	.	.	.	.	.	.	.	.	.	.	.	.	.	.	.	.	.	.	.	.	.	.	.	.	.	.	.	.	.	.	.	.	15
Phlomis younghusbandii	.	.	.	.	+	.	r	.	.	+	.	.	.	.	.	.	.	.	.	.	.	.	.	.	.	.	.	.	.	.	.	.	.	.	.	.	.	23
Dracocephalum tanguticum	r	+	+	.	+	+	+	.	+	.	.	.	.	.	.	.	.	.	.	.	.	.	.	.	.	.	.	.	.	.	.	.	.	.	.	.	.	34
Stellera chamaejasme	+	r	+	.	3	1	2	+	+	.	.	.	.	.	.	.	.	.	.	.	.	.	.	.	.	.	.	.	.	.	.	.	.	.	.	.	.	39
Artemisia moorcroftiana	.	.	.	.	.	.	.	.	.	.	.	.	.	+	.	.	.	.	.	.	.	.	.	.	.	.	.	.	.	.	.	.	.	.	.	.	.	9
Poa albertii ssp. poophagorum	.	.	.	.	.	.	.	.	.	.	.	.	.	.	.	.	.	.	.	.	.	.	.	.	.	.	.	.	.	.	.	.	.	.	.	.	.	6
Poa araratica s.l.	.	.	.	.	.	.	.	.	.	.	.	.	.	.	.	.	.	.	.	.	.	.	.	.	.	.	.	.	.	.	.	.	.	.	.	.	.	9
Poa asperifolia	.	.	.	.	.	.	.	.	.	.	.	.	.	.	.	.	.	.	.	.	.	.	.	.	.	.	.	.	.	.	.	.	.	.	.	.	.	4
Stephanachne cf. pappophorea	.	.	.	.	.	.	.	.	.	.	.	.	.	.	.	.	.	.	.	.	.	.	.	.	.	.	.	.	.	.	.	.	.	.	.	.	.	4
Gentianella azurea	.	.	.	.	.	.	.	.	.	.	.	.	.	.	.	.	.	.	.	.	.	.	.	.	.	.	.	.	.	.	.	.	.	.	.	.	.	4
Rhodiola tangutica	.	.	.	.	.	.	.	+	.	.	.	.	.	.	.	.	.	.	.	.	.	.	.	.	.	.	.	.	.	.	.	.	.	.	.	.	.	8
Allium przewalskianum	.	.	.	.	.	.	.	.	.	.	.	.	.	.	.	.	.	.	.	.	.	.	.	.	.	.	.	.	.	.	.	.	.	.	.	.	.	4
Stipa penicillata v. penicillata	.	.	.	.	.	.	.	.	.	.	.	.	.	.	.	.	.	.	.	.	.	.	.	.	.	.	.	.	.	.	.	.	.	.	.	.	.	8
Microula tangutica	.	.	.	.	.	.	.	.	.	.	.	.	.	.	.	.	.	.	.	.	.	.	.	.	.	.	.	.	.	.	.	.	.	.	.	.	.	15
Salsola monoptera	.	.	.	.	.	.	.	.	.	.	.	.	.	.	.	.	.	.	.	.	.	.	.	.	.	.	.	.	.	.	.	.	.	.	.	.	.	9
Dontostemon glandulosus	.	.	.	.	.	.	.	.	.	.	.	.	.	.	.	.	.	.	.	.	.	.	.	.	.	.	.	.	.	.	.	.	.	.	.	.	.	15
Corydalis hookeri	.	.	.	.	.	r	.	.	.	+	.	.	.	.	.	.	.	.	.	.	.	.	.	.	.	.	.	.	.	.	.	.	.	.	.	.	.	16
Trikeraia oreophila	.	.	.	.	+	+	.	.	.	+	.	.	.	.	.	.	.	.	.	.	.	.	.	.	.	.	.	.	.	.	.	.	.	.	.	.	.	10
Anaphalis nubigena	4	.	.	+	+	3	1	2	1	+	.	.	.	.	.	.	.	.	.	.	.	.	.	.	.	.	.	.	.	.	.	.	.	.	.	.	.	20
Silene caespitella	r	r	+	.	+	+	+	+	r	.	r	r	.	.	.	.	.	.	.	.	.	.	.	.	.	.	.	.	.	.	.	.	.	.	.	.	.	22
Pleurospermum aromaticum	.	.	.	.	+	+	+	+	r	.	.	.	.	.	.	.	.	.	.	.	.	.	.	.	.	.	.	.	.	.	.	.	.	.	.	.	.	14
Ajuga lupulina	.	.	.	.	.	.	.	.	+	.	.	.	.	.	.	+	.	10	8	4	.	+	.	.	.	.	.	.	.	.	.	.	.	.	.	.	.	15
Thalictrum rutifolium	.	.	.	.	.	+	.	.	.	.	.	.	.	+	.	+	.	+	+	+	r	.	.	.	.	.	.	.	.	.	.	.	.	.	.	.	.	24
Artemisia tridactyla	.	1	+	.	+	+	+	+	+	.	.	.	18	.	.	+	.	18	25	30	1	1	.	.	.	.	.	.	.	.	.	.	.	.	.	.	.	34
Pedicularis cheilanthifolia	+	+	+	.	+	+	+	+	+	.	.	+	.	.	.	.	.	.	.	.	.	.	.	.	.	.	.	.	.	.	.	.	.	.	.	.	.	26
Euphorbia stracheyi	r	.	.	.	.	.	.	.	.	.	.	.	.	.	.	.	.	.	.	.	.	.	.	.	.	.	.	.	.	.	.	.	.	.	.	.	.	27
Lasiocaryum munroi	.	.	.	.	.	.	+	.	.	.	.	.	.	.	+	.	1	1	1	1	.	1	.	.	.	.	.	.	.	.	.	.	.	.	.	.	.	35
Carex stenophylla	.	+	+	+	+	+	+	+	+	+	.	+	.	.	.	.	.	.	.	.	.	.	.	.	.	.	.	.	.	.	.	.	.	.	.	.	.	49
Rhodiola smithii	.	.	.	.	r	.	.	.	+	.	.	.	.	.	.	.	.	.	.	.	.	.	.	.	.	.	.	.	.	.	.	.	.	.	.	.	.	17
Oxytropis stracheyana	+	+	+	.	.	.	.	.	.	1	.	.	.	.	.	.	.	.	.	.	.	.	.	.	.	.	.	.	.	.	.	.	.	.	.	.	.	25
Youngia simulatrix	+	+	1	+	+	+	.	.	r	+	+	2	.	.	.	.	.	.	.	.	.	.	.	.	.	.	+	.	.	.	.	.	.	.	.	.	.	37
Saussurea leiocarpa	2	3	2	.	1	+	1	1	1	2	.	.	.	.	.	.	.	.	.	.	.	.	.	.	.	.	.	.	.	.	.	.	.	.	.	.	.	30
Festuca cf. valesiaca	.	10	9	5	.	.	+	+	.	1	3	5	.	.	.	.	.	.	.	.	.	.	.	.	.	.	.	.	.	.	.	.	.	.	.	.	.	44
Lomatogonium brachyantherum	+	.	+	.	+	+	.	+	+	+	.	.	.	.	.	.	.	.	.	.	.	.	.	.	.	.	.	.	.	.	.	.	.	.	.	.	.	35
Kengyilia thoroldiana v. thoroldiana	.	.	.	+	r	.	+	.	.	+	.	.	.	.	.	.	.	.	.	.	.	.	.	.	.	.	.	.	.	.	.	.	.	.	.	.	.	27
Kobresia robusta	.	.	.	.	.	.	.	.	.	.	.	.	.	.	.	.	.	.	.	.	.	.	.	.	.	.	.	.	.	.	.	+	.	.	.	.	.	16
Deyeuxia zangxiensis	.	.	.	.	.	.	.	.	.	.	.	.	.	.	.	.	.	.	.	.	.	.	.	.	.	.	.	.	.	.	.	.	.	.	.	.	.	13
Artemisia stricta	.	.	.	.	.	.	.	.	.	.	.	.	.	.	+	.	.	.	.	.	.	+	.	.	.	.	.	.	.	.	.	.	.	.	.	.	.	16
Astragalus arnoldii	.	.	.	6	5	6	.	.	.	+	3	+	.	.	.	.	.	.	.	.	.	.	.	.	.	.	.	.	.	.	.	.	.	.	.	.	.	19
Potentilla exigua	.	.	+	+	.	.	.	.	.	r	+	+	.	+	2	.	+	.	+	+	+	+	.	.	.	+	+	.	.	.	.	.	.	.	.	.	.	28
Koeleria litvinowii ssp. argentea	.	.	.	1	.	.	+	+	+	+	+	1	.	.	1	1	3	5	4	2	2	+	.	.	.	.	.	.	.	.	.	.	.	.	.	.	.	29
Sibbaldianthe adpressa	.	.	+	.	.	.	.	.	.	+	.	.	.	.	.	.	.	+	.	r	.	.	+	.	.	+	.	.	.	+	.	.	+	.	+	+	.	24
Taraxacum sherriffii	.	+	+	.	r	+	.	.	r	.	+	+	.	.	.	.	.	.	.	.	.	.	.	+	.	r	.	.	.	.	.	.	.	+	+	.	.	24
Allium cyaneum	r	+	+	+	+	+	+	+	+	+	+	.	.	.	.	.	.	.	.	.	.	.	.	.	.	.	.	.	.	.	.	.	.	.	.	.	.	23
Comastoma pedunculatum	r	1	+	1	.	.	+	+	+	.	+	+	.	.	.	.	.	.	.	.	.	.	.	.	+	.	.	.	.	.	.	.	.	.	.	.	.	16
Delphinium caeruleum	r	+	+	.	.	.	+	+	.	.	r	+	.	.	.	.	.	.	.	.	.	.	.	.	.	.	.	.	.	.	.	.	.	.	.	.	.	13
Draba zangbeiensis	.	.	.	+	.	.	.	+	.	+	+	+	.	.	.	.	.	.	.	.	.	.	+	.	.	.	.	.	.	.	.	.	.	.	.	.	.	10
Aster yunnanensis	r	.	.	+	+	.	.	+	+	.	+	.	.	.	.	.	.	.	.	.	.	.	.	.	.	.	.	.	.	.	.	.	.	.	.	.	.	9
Kobresia capillifolia	+	.	+	.	+	.	2	+	+	.	+	1	.	.	.	.	.	.	.	.	.	.	.	.	.	.	.	.	+	.	.	.	.	.	.	.	.	10
Eritrichium tangkulaense	.	.	.	.	+	+	+	+	+	+	.	+	.	.	.	.	.	.	.	.	.	.	.	.	.	.	.	.	.	.	.	.	.	.	.	.	.	10
Meconopsis horridula	.	.	.	r	r	+	+	+	+	+	.	+	.	.	.	.	.	.	.	.	.	.	3	.	.	.	.	.	.	.	.	.	.	.	.	.	.	9
Dasiphora parvifolia	.	.	+	1	1	2	.	+	1	1	.	.	1	.	.	.	.	.	.	.	.	.	10	.	.	.	+	2	3	.	.	.	.	.	.	.	.	14
Kobresia pygmaea	.	+	+	+	+	5	6	10	18	10	.	4	1	+	.	3	5	+	.	.	1	1	6	70	75	90	90	90	96	90	95	85	93	80	90	98	80	31
Potentilla saundersiana	1	+	+	+	+	+	2	+	2	+	3	3	1	.	1	.	+	.	.	+	+	+	.	.	3	1	.	.	4	2	+	8	1	3	3	+	3	33

Species	Total
Veronica ciliata	20
Artemisia santolinifolia	12
Axyris prostrata	16
Galium exile	11
Persicaria glacialis	13
Heracleum millefolium	9
Draba ellipsoidea	7
Saussurea andryaloides	8
Stellaria decumbens s.l.	7
Lagotis brachystachya	7
Microgynoecium tibeticum	7
Anaphalis xylorhiza	5
Lancea tibetica	12
Astragalus strictus	10
Lamiophlomis rotata	10
Primula walshii	9
Cortiella caespitosa	8
Kobresia vidua	5
Saussurea graminea v. ortholepis	6
Gentiana spp. indet.	4
Pedicularis muscoides	4
Oxytropis pauciflora	4
Festuca cf. forrestii	3
Thalictrum alpinum	15
Stipa rohmooiana	3
Pedicularis roylei	3
Aster flaccidus	12
Kobresia royleana	12
Oreosolen wattii	30
Leontopodium pusillum	60
Androsace tapete	51
Saussurea leontodontoides	41
Potentilla bifurca s.l.	52
Incarvillea younghusbandii	56
Stipa roborovskyi/ purpurea	73
Poa albertii ssp. albertii	71
Carex montis-everestii	60
Chenopodium nepalense	31
Iris potaninii v. potaninii	30
Astragalus monbeigii	35
Elymus schrenkianus	35
Cryptothladia kokonorica	44
Gentiana hyalina	47
Heteropappus semiprostratus/ gouldii	49
Sedum perpusillum	43
Pedicularis alaschanica	40
Callianthemum pimpinelloides	39
Dracocephalum heterophyllum	31
Hypecoum leptocarpum	16
Stracheya tibetica	21
Ajania tenuifolia	18
Arenaria bryophylla/ kansuensis	24
Viola kunawurensis	30
Oxytropis chiliophylla	21
Astragalus confertus	25
Gentiana pseudoaquatica	19
Carex moorcroftii	23
Pleurospermum hedinii	10
Corydalis mucronifera	8
Anthoxanthum cf. glabrum	6
Polygonum sibiricum	4
Thermopsis lanceolata	4
Artemisia xigazeensis	4
Anemone imbricata	4
Oxytropis deflexa	6
Stipa koelzii	8
Draba ladyginii	9
Kobresia humilis	4
Androsace erecta	1
Silene subcretacea	2
Galium aparine	1
Arenaria festucoides	2
Campanula aristata	1
Elsholtzia ciliata	1
Parietaria micrantha	1
Allium spicatum	1
Fagopyrum tataricum	2
Tripogon licuae	1
Orinus thoroldii	1
Onosma hookeri	1
Berberis stiebritziana	1
Delphinium kamaonense	3
Platygyria waltonii	2
Stipa subsessiliflora	2
Microula diffusa	1
Ranunculus indivisus	1
Poa versicolor s.l.	1
Anaphalis subumbellata	1
Lepidium capitatum	1
Hedinia tibetica	2
Sisymbrium heteromallum	1
Draba stenocarpa	1
Ranunculus membranaceus	2
Kobresia stolonifera	1
Gentianella pygmaea	3
Ranunculus nephelogenes	2
Trikeraia hookeri	3
Dilophia salsa	1

APPENDIX 1 (CONTINUED)

Astragalus sp. 5				r					1
Corispermum falcatum				+ r					2
Oxytropis tatarica				r +					2
Lonicera rupicola				2					1
Viola bulbosa				+					1
Lomatogonium carinthiacum					r +				2
Potentilla anserina				+	+ +				3
Gentiana globosa					r				1
Artemisia demissa					+	+			2
Astragalus heydei					1				1
Astragalus sp. 2					4				1
Artemisia stracheyi				+	5				2
Artemisia minor		1				r + +			4
Saxifraga flagellaris				r		+			2
Rhodiola bupleuroides						+			1
Puccinellia iliensis						+			1
Carex supina						+			1
Gentianopsis barbata v. barbata						r			1
Astragalus sp. 3						+			1
Bistorta macrophylla			+			+			2
Saussurea lhunzhubensis							4		1
Actinocarya acaulis							1	+	2
Gentiana pseudosquarrosa							+ +		2
Plumbagella micrantha		r		+			+ +		4
Artemisia hedinii							+ + +		3
Kobresia pusilla							+ + +	6	4
Potentilla sinonivea								4	1
Draba altaica								+	1
Ranunculus brotherusii								+ +	2
Oxytropis humifusa								2 1 +	3
Pedicularis nana								+ +	2
Ptilagrostis concinna								1	1
Oxytropis reniformis								2	1
Rheum pumilum								+	1
Lagotis brevituba								+	1
Saxifraga sinomontana								+	1
Trisetum spicatum								+ +	2
Microcaryum pygmaeum								+ +	2
Gentiana sp. vel aff. sino-ornata								+	1
Microula tibetica							1 + 8	1	4
Gentiana cf. lawrencei v. farreri								2	1
Carex melanantha								+ +	2
Gentiana micantiformis								+	1
Carex przewalskii								+ +	2
Gentiana prostrata v. karelinii								2 +	2
Gentiana veitchiorum								+	1
Gueldenstaedtia himalaica								+	1
Draba lasiophylla								+	1
Stipa regeliana								+	1
Festuca tibetica								+	1
Kobresia tibetica								+	1
Hippolytia kennedyi								+	1
Poa attenuata								+ +	2
Astragalus sp. indet.								+	1
Kobresia schoenoides								+	1
Veronica cephaloides								+	1
Przewalskia tangutica				r			2		2
Poa szechuensis v. rossbergiana							+ + + + + 4		6
Stellaria umbellata							1		1
Aconitum pendulum							6		1
Boraginaceae sp. indet. (04-060-07)							+		1
Ptilagrostis dichotoma							+		1

Shown are cover degrees in percent for the species in each record plot ('+' = cover < 1%; 'r' = rare incidence of the species; '.' = no occurance of the species in this record).

APPENDIX 2

Categorical synoptic table for Communities 1 - 6. All species with frequency > 3% (I=0.1-19%; II=20-39%; III=40-59%; IV=60-79%; V=80-100%) are included

Community No.	1	2	3a	3b	3c	4	5	6	
No. of relevés	8	9	20	13	18	12	11	14	freq.
Character species of Community 1									
Juniperus tibetica	V	.	.	.	.	.	.	.	6.7
Caragana versicolor	V	.	.	.	.	.	.	.	6.7
Dolomiaea calophylla	V	.	.	.	.	.	.	.	6.7
Rhodiola sacra	V	.	.	.	.	.	.	.	6.7
Sedum roborowskii	IV	.	.	.	.	.	.	.	5.7
Piptatherum gracile	IV	.	.	.	.	.	.	.	5.7
Poa hylobates	V	.	.	.	.	.	.	.	6.7
Stellaria dianthifolia	V	.	.	.	.	.	.	.	6.7
Pleurospermum foetens	IV	.	.	.	.	.	.	.	5.7
Character species of Community 1 and 2									
Kobresia macrantha	II	II	.	I	.	.	.	.	5.7
Pennisetum flaccidum	V	I	I	I	.	.	.	.	9.5
Ephedra gerardiana	IV	.	I	I	I	I	I	.	12.4
Urtica hyperborea	II	III	.	.	.	I	.	.	8.6
Allium fasciculatum	IV	II	I	.	.	I	.	.	9.5
Astragalus tribulifolius	IV	III	II	I	.	.	.	.	14.3
Juniperus pingii var. wilsonii	V	V	.	.	.	.	.	.	15.2
Phlomis younghusbandii	V	IV	II	.	.	II	.	.	22.9
Dracocephalum tanguticum	V	V	IV	I	.	III	.	.	34.3
Stellera chamaejasme	IV	IV	IV	II	I	IV	.	.	39
Character species of Community 2									
Artemisia moorcroftiana	.	IV	I	.	.	.	I	.	8.6
Poa albertii ssp. poophagorum	.	IV	.	.	.	.	.	.	5.7
Poa araratica s.l.	.	V	.	.	.	.	.	.	8.6
Poa asperifolia	.	III	.	.	.	.	.	.	3.8
Stephanachne cf. pappophorea	.	III	.	.	.	.	.	.	3.8
Gentianella azurea	.	III	.	.	.	.	.	.	3.8
Rhodiola tangutica	.	IV	.	.	.	I	.	.	7.6
Allium przewalskianum	.	II	.	I	.	.	.	.	3.8
Other species in Communities 2 – 4									
Stipa penicillata var. penicillata	.	III	I	I	.	.	.	.	7.6
Microula tangutica	.	V	I	II	.	.	.	.	15.2
Salsola monoptera	.	II	I	II	.	.	.	.	8.6
Dontostemon glandulosus	.	III	I	III	I	.	.	.	15.2
Corydalis hookeri	.	V	II	I	.	I	.	.	16.2
Trikeraia oreophila	.	III	I	.	.	II	.	.	9.5
Anaphalis nubigena	.	III	III	.	.	IV	.	.	20
Silene caespitella	.	IV	II	.	I	V	.	.	21.9

Appendix 2 (Continued)

Community No.	1	2	3a	3b	3c	4	5	6	
No. of relevés	8	9	20	13	18	12	11	14	freq.
Pleurospermum aromaticum	I	V	I	.	.	III	.	.	14.3
Ajuga lupulina	.	III	I	.	I	I	III	.	15.2
Thalictrum rutifolium	.	V	I	II	I	I	III	.	23.8
Artemisia tridactyla	.	V	III	II	I	III	IV	.	34.3
Pedicularis cheilanthifolia	.	III	III	II	I	IV	.	.	25.7
Euphorbia stracheyi	.	V	III	II	I	I	.	.	26.7
Lasiocaryum munroi	.	IV	III	II	II	I	III	.	35.2
Carex stenophylla	.	II	V	III	V	V	.	.	48.6
Rhodiola smithii	.	II	II	I	II	I	.	.	17.1
Oxytropis stracheyana	.	II	II	III	II	II	.	.	24.8
Youngia simulatrix	.	II	IV	II	II	V	.	I	37.1
Saussurea leiocarpa	.	II	IV	I	I	IV	.	.	29.5
Character species of Community 3 and 4									
Festuca cf. valesiaca	.	.	IV	III	V	IV	.	.	43.8
Lomatogonium brachyantherum	.	.	IV	II	IV	III	.	.	35.2
Kengyilia thoroldiana var. thoroldiana	.	.	II	IV	III	II	.	.	26.7
Character species of Community 3									
Kobresia robusta	.	.	I	IV	II	.	.	I	16.2
Deyeuxia zangxiensis	.	.	I	III	II	.	.	.	13.3
Artemisia stricta	.	.	I	I	IV	.	I	.	16.2
Character species of Community 3 and 4									
Astragalus arnoldii	.	.	I	.	IV	III	.	.	19
Potentilla exigua	.	I	.	I	IV	III	IV	I	27.6
Koeleria litvinowii subsp. *argentea*	.	.	I	I	IV	III	IV	.	28.6
Sibbaldianthe adpressa	.	I	I	I	III	I	II	II	23.8
Taraxacum sherriffii	I	.	II	I	II	III	.	II	23.8
Allium cyaneum	.	II	II	I	I	V	.	.	22.9
Character species of Community 4									
Comastoma pedunculatum	.	I	I	.	I	IV	.	I	16.2
Delphinium caeruleum	.	I	I	.	I	III	.	.	13.3
Draba zangbeiensis	.	I	I	.	I	III	I	.	9.5
Aster yunnanensis	.	II	I	.	.	III	.	.	8.6
Kobresia capillifolia	.	I	I	.	.	IV	.	I	10.5
Eritrichium tangkulaense	.	II	I	.	.	III	.	.	9.5
Character species of Community 4 and 5									
Meconopsis horridula	.	.	.	.	.	IV	I	.	8.6
Dasiphora parvifolia	.	I	.	I	I	III	I	II	14.3
Kobresia pygmaea	I	.	.	.	.	V	IV	V	31.4
Potentilla saundersiana	.	.	I	I	II	V	III	IV	33.3
Veronica ciliata	.	.	I	.	II	I	V	II	20
Artemisia santolinifolia	.	II	I	.	I	II	II	.	12.4

Community No.	1	2	3a	3b	3c	4	5	6	
No. of relevés	8	9	20	13	18	12	11	14	freq.
Character species of Community 5									
Axyris prostrata	.	I	.	II	II	.	IV	I	16.2
Galium exile	IV	.	.	I	.	.	III	.	11.4
Persicaria glacialis	I	I	I	.	.	.	V	I	13.3
Heracleum millefolium	.	.	I	I	I	.	III	.	8.6
Draba ellipsoidea	I	.	.	.	.	.	III	.	6.7
Saussurea andryaloides	.	.	.	.	.	.	IV	.	7.6
Stellaria decumbens s.l.	.	.	.	.	.	.	III	I	6.7
Lagotis brachystachya	.	.	.	.	.	.	III	I	6.7
Microgynoecium tibeticum	.	.	.	.	.	.	IV	.	6.7
Anaphalis xylorhiza	.	.	.	.	.	.	II	I	4.8
Character species of Community 5 and 6									
Lancea tibetica	.	.	.	.	.	.	V	II	12.4
Astragalus strictus	I	.	.	.	.	.	II	III	10.5
Lamiophlomis rotata	.	.	.	.	.	.	III	II	9.5
Character species of Community 6									
Primula walshii	.	.	.	.	.	.	.	IV	8.6
Cortiella caespitosa	.	.	.	.	.	.	.	III	7.6
Kobresia vidua	.	.	.	.	.	.	.	II	4.8
Saussurea graminea var. ortholepis	.	.	.	.	.	I	.	II	5.7
Gentiana spp. indet.	.	.	.	.	.	.	.	II	3.8
Pedicularis muscoides	.	.	.	.	.	.	.	II	3.8
Oxytropis pauciflora	.	.	.	.	.	.	.	II	3.8
Festuca cf. forrestii	.	.	.	.	.	.	.	II	2.9
Thalictrum alpinum	.	I	.	.	I	II	I	III	15.2
Stipa rohmooiana	.	.	.	.	.	.	.	II	2.9
Pedicularis roylei	.	.	.	.	.	.	.	II	2.9
Aster flaccidus	.	.	.	.	.	I	I	IV	12.4
Kobresia royleana	.	.	.	.	.	.	I	IV	12.4
Other species									
Oreosolen wattii	.	III	II	I	II	II	I	III	30.5
Leontopodium pusillum	.	.	IV	II	V	V	V	II	60
Androsace tapete	.	IV	II	I	III	V	III	IV	51.4
Saussurea leontodontoides	.	II	III	IV	II	II	I	IV	41
Potentilla bifurca s.l.	II	IV	III	III	III	IV	V	II	52.4
Incarvillea younghusbandii	.	IV	V	IV	III	IV	IV	.	56.2
Stipa roborovskyi/ purpurea	I	IV	V	V	V	V	III	I	73.3
Poa albertii subsp. *albertii*	.	IV	V	III	V	V	V	I	71.4
Carex montis-everestii	II	III	IV	III	V	V	II	I	60
Chenopodium nepalense	IV	III	II	III	II	.	III	I	31.4
Iris potaninii var. *potaninii*	III	.	II	I	III	II	IV	.	29.5
Astragalus monbeigii	V	II	II	III	IV	I	.	I	35.2
Elymus schrenkianus	III	V	II	I	II	V	.	.	35.2
Cryptothladia kokonorica	.	V	IV	IV	II	IV	.	.	43.8

APPENDIX 2 (CONTINUED)

Community No.	1	2	3a	3b	3c	4	5	6	
No. of relevés	8	9	20	13	18	12	11	14	freq.
Gentiana hyalina	.	III	IV	II	V	V	.	I	46.7
Heteropappus semiprostratus/ gouldii	II	IV	IV	III	III	I	V	.	48.6
Sedum perpusillum	.	IV	IV	II	III	III	II	.	42.9
Pedicularis alaschanica	.	V	III	IV	II	III	.	.	40
Callianthemum pimpinelloides	.	II	IV	III	III	III	IV	.	39
Dracocephalum heterophyllum	.	II	II	V	I	II	IV	.	31.4
Hypecoum leptocarpum	.	III	I	II	I	.	II	.	16.2
Stracheya tibetica	.	II	II	II	I	II	II	.	21
Ajania tenuifolia	.	.	II	I	I	I	IV	.	18.1
Arenaria bryophylla/ kansuensis	.	.	I	I	II	III	IV	I	23.8
Viola kunawurensis	.	II	III	II	III	I	.	I	30.5
Oxytropis chiliophylla	.	.	II	I	II	IV	.	.	21
Astragalus confertus	.	.	II	I	III	III	I	.	24.8
Gentiana pseudoaquatica	.	II	II	.	II	I	.	II	19
Carex moorcroftii	.	.	I	II	IV	III	.	I	22.9
Pleurospermum hedinii	.	.	I	I	I	II	.	.	9.5
Corydalis mucronifera	.	I	I	II	.	.	.	.	7.6
Anthoxanthum cf. glabrum	.	.	I	II	.	.	.	.	5.7
Polygonum sibiricum	.	.	I	I	I	.	.	.	3.8
Thermopsis lanceolata	.	.	I	.	I	.	.	.	3.8
Artemisia xigazeensis	.	I	I	.	I	I	.	.	3.8
Anemone imbricata	.	.	I	.	.	I	.	I	3.8
Oxytropis deflexa	.	.	.	.	I	II	.	.	5.7
Stipa koelzii	.	I	I	I	I	I	.	.	7.6
Draba ladyginii	.	II	.	I	I	I	I	.	8.6
Kobresia humilis	.	.	.	.	I	.	.	II	3.8

ACKNOWLEDGMENTS

Fieldwork and evaluation was made possible through a grant of the German Research Council (DFG).

Plant identifications were kindly provided by R. Søreng, NMNH, Washington DC (*Poa*), M. Olonova (*Poa*), Tomsk; Zhang Shuren, Beijing (*Kobresia* and *Carex)*; I. Al-Shebaz, St. Louis, Missouri (Brassiaceae); E. v. Raab-Straube, Berlin (*Saussurea*); E. Aitken, RBG Edinburgh (Gentianaceae); Wang Hong, KIB, Kunming (*Pedicularis)*; the late J. Sójak, Praha (*Potentilla)*; M. Watson, RBG Edinburgh (Apiaceae); B.W. Dickoré, München (*Leontopodium,* Boraginaceae); A. Farjon, RBG Kew (Cupressaceae); and E. Hörandl, Wien (*Ranunculus*). C. Enderle kindly provided the maps and figures, G. Schulz did her best to translate the text.

REFERENCES

Adamczyk, S. (2010). Vegetationsgeschichtliche Untersuchung zur Geschichte des menschlichen Eingriffs in den Haushalt der Natur Ost-, Zentral-Tibets und West-Sichuans, sowie zum Problem der eiszeitlichen Waldrefugien. Ph.D. Hohenheim University, Stuttgart.

Aldenderfer, M. (2007). Modelling the Neolithic on the Tibetan Plateau. *Developments in Quaternary Sciences* 9, 151-165.

Academia Sinica – Institute of Geography (1990). *Atlas of Tibet Plateau.* Beijing. [in Chinese].

Avouac, J. P., Dobremez, J. F., Bourjot, L. (1996). Palaeoclimatic interpretation of a topographic profile across middle Holocene regressive shorelines of Langmu Co (Western Tibet). *Palaeoclimatology, Palaeogeography, Palaeoecology* 120, 93-104.

Bannari, A., Morin, D., Huette, A. R., Bonn, F. (1995). A review of vegetation indices. *Remote Sensing Reviews* 13, 95-120.

Bellezza, J. V. (2008). Zhang Zhung. *Foundation of civilization in Tibet.* Österreichische Akademie der Wissenschaften, Philosophisch-Historische, Wien.

Böhner, J. (1996). Säkulare Klimaschwankungen und rezente Klimatrends Zentral- und Hochasiens. Göttinger Geographische Abhandlungen 101, Göttingen

Brantingham, P. J., Gao, X., Olsen, J. W., Ma, H. Z., Rhode, D. E., Zhang, H. Y., Madsen, D. B. (2007). A short chronology for the peopling of the Tibetan Plateau. *Developments in Quaternary Sciences* 9, 129-150.

Diaz, S., Acosta, A., Cabido, M. (1992). Morphological analysis of herbaceous communities under different grazing regimes. *Journal of Vegetation Science* 3, 689-696.

Diaz, S., Noy-Meir, I., Cabido, M. (2001). Can grazing response of herbaceous plants be predicted from simple vegetative traits? *Journal of Applied Ecology* 38, 497-508.

Duckworth, J. C., Kent, M., Ramsay, P. M. (2000). Plant functional types: an alternative to taxonomic plant community description in biogeography? *Progress in Physical Geography* 24, 515-542.

Eklundh, L., Singh, A. (1993). A comparative analysis of standardised and unstandardised principal components analysis in remote sensing. *International Journal of Remote Sensing* 14, 1359-1370.

Ellenberg, H. (1996). *Vegetation Mitteleuropas mit den Alpen.* Ulmer. Stuttgart.

Ellis, J. E., Swift, D. M. (1988). Stability of African pastoral ecosystems: Alternate paradigms and implications for development. *Journal of Range Management* 41, 450-459.

Farjon, A. (2005). *A monograph of Cupressaceae and Sciadopitys.* Royal Botanic Gardens, Kew.

Flad, R. K., Yuan, J., Li, S. C. (2007). Zooarchaeological evidence for animal domestication in northwest China. *Developments in Quaternary Sciences* 9, 167-204.

Foggin, J. M., Smith, A. T. (1999). The plateau pika (Ochotona curzoniae) is a keystone species for biodiversity on the Tibetan plateau. *Animal Conservation* 2, 235-240.

Goldstein, M. C., Beall, C. M., Cincotta, R. P. (1990). Traditional nomadic pastoralism and ecological conservation on Tibet's northern Plateau. *National Geographic Research* 6, 139-156.

Govender, M., Chetty, K., Bulcock, H. (2007). A review of hyperspectral remote sensing and its application in vegetation and water resource studies. *Water S.A.* 33, 145-152.

Grime, J. P. (1979). Evidence for the existence of three primary strategies in plants and its relevance to ecological and evolutionary theory. *The American Naturalist* 111, 1169-1194.

Guang Zhao, MacLean, A. L. (2000). A comparison of canonical discriminant analysis and principal component analysis for spectral transformation. *Photogrammetric engineering and remote sensing* 66, 841-847.

Guo, S., Savolainen, P., Su, J., Zhang, Q., Qui, D., Zhou, J., Zhong, Y., Zhao, X., Liu, J. (2006). Origin of mitochondrial DNA diversity of domestic yaks. *BMC Evolutionary Biology* 6, 73.

Hartmann, H. (1987). Pflanzengesellschaften trockener Standorte aus der subalpinen und alpinen Stufe im südlichen und östlichen Ladakh. *Candollea* 42, 277-326.

Hayking, S. (1999). *Neural Networks: a comprehensive foundation.* Prentice Hall, New Jersey.

Herrmann, M., Lu, X. M., Berking, J., Schütt, B., Mosbrugger, V. (2010). Reconstructing Holocene vegetation and climate history of Nam Co area (Tibet) using pollen and other palynomorphs. *Quaternary International* 218, 45-57.

Herzschuh, U., Winter, K., Wünnemann, B., Li, S. (2006). A general cooling trend on the central Tibetan Plateau throughout the Holocene recorded by the lake Zigetang pollen spectra. *Quaternary International* 154/155, 113-121.

Hill, M. O., Gauch, H. G. (1980). Detrended correspondence analysis: an improved ordination technique. *Vegetatio* 42, 47-58.

Holzner, W., Kriechbaum, M. (2000). Pastures in south and central Tibet (China). I. Methods for a rapid assessment of pasture conditions. *Austrian Journal of Agricultural Research* 51, 259-266.

Jarvis, A., Reuter, H. I., Nelson, A., Guevara, E. (2008). *Hole-filled seamless SRTM data V4*, International Centre for Tropical Agriculture (CIAT). http://srtm.csi.cgiar.org (28.7.2010)

John, R., Chen, J., Lu, N., Guo, K., Liang, C., Wei, Y. (2008). Predicting plant diversity based on remote sensing products in the semi-arid region of Inner Mongolia. *Remote Sensing of Environment* 112, 2018-2032.

Kaiser, K., Miehe, G., Barthelmes, A., Ehrmann, O., Scharf, A., Schult, M., Schlütz, F., Adamczyk, S., Frenzel, B. (2008). Turf-bearing topsoils on the central Tibetan Plateau, China: Pedology, botany, geochronology. *Catena* 73, 300-311.

Kaiser, K., Schoch, W. H., Miehe, G. (2007). Holocene paleosols and colluvial sediments in Northeast Tibet (Qinghai Province, China): properties, dating and paleoenvironmental implications. *Catena* 69, 91-102.

Körner, C. (1999). Alpine Plant Life. *Functional Plant Ecology of High Mountain Ecosystems.* Springer, Berlin.

Liang, E., Lu, X. M., Ren, P., Li, X. X., Zhu, L. P., Eckstein, D. (2011). Annual increments of juniper dwarf shrubs above the tree line on the central Tibetan Plateau: a useful climatic proxy. *Annals of Botany* 109, 721-728.

Liu, S. W. (ed., 1996-99). *Flora Qinghaiica*, 4 vols. Qinghai People's Publishing House. Xining.

Lu, D. (2006). The potential and challenge of remote sensing based biomass estimation. *International Journal of Remote Sensing* 27, 1297-1328.

Ma, Y. S., Lang, B. N., Wang, Q. J. (1999). Reviews and prospect of the study on "Black Soil" deterioration grassland. *Prataculture Science* 16, 5-8.

Mas, J. F., Flores, J. J. (2008). The application of artificial neural networks to the analysis of remotely sensed data. *International Journal of Remote Sensing* 29, 617-663.

McCune, B., Mefford, M. J. (1999). *PC-ORD. Multivariate Analysis of Ecological Data,* Version 4. MjM Software Design, Gleneden Beach (OR) US.

McNaughton, S. J. (1984). Grazing lawns: animals in herds, plant form, and coevolution. *The American Naturalist* 124, 863-886.

Miehe, G. (1988). Geoecological reconnaissance in the alpine belt of Southern Tibet. *GeoJournal* 17, 635-648.

Miehe, G. (1989). Vegetation patterns on Mt. Everest as influenced by monsoon and foehn. *Vegetatio* 79, 21-32.

Miehe, G., Miehe, S., Kaiser, K., Liu, J. Q., Zhao, X. Q. (2008a). Status and dynamics of the *Kobresia* pygmaea ecosystem on the Tibetan plateau. *Ambio* 37, 272-279.

Miehe, G., Kaiser, K., Sonam, Co., Zhao, X. Q., Liu, J. Q. (2008b). Geo-ecological transect studies in northeast Tibet (Qinghai, China) reveal human-made mid-Holocene environmental changes in the upper yellow river catchment changing forest to grassland. *Erdkunde* 62, 187-199.

Miehe, G., Miehe, S., Kaiser, K., Wesche, K., Zhao, X., Liu, J. (2008c). "Alpine meadows" of the Tibet Plateau are a synantrophic pseudoclimax. *VIII International Rangeland Congress* (29 June-5 July 2008). Hohot, China

Miehe, G., Miehe, S., Will, M., Opgenoorth, L., La Duo, Tsering D., Liu, J. Q. (2008d). An Inventory of forest relics in the pastures of Southern Tibet (A.R.Xizang, China). *Plant Ecology* 194, 157-177.

Miehe, G., Miehe, S., Kaiser, K., Reudenbach, C., Behrendes, L., La Duo, Schlütz, F. (2009). How old is pastoralism in Tibet? An ecological approach to the making of a Tibetan landscape. *Palaeogeography, Palaeoclimatology, Palaeoecology* 276, 130-147.

Miller, F., Osbahr, H., Boyd, E., Thomalla, F., Bharwani, S., Ziervogel, G., Walker, B., Birkmann, J., van der Leeuw, S., Rockström, J., Hinkel, J., Downing, T., Folke, C., Nelson, D. (2010). Resilience and Vulnerability: Complementary or Conflicting Concepts? *Ecology and Society* 15: 11, online.

Morrill, C. (2004). The influence of Asian summer monsoon variability on the water balance of a Tibetan lake. *Journal of Paleolimnology* 32, 273-286.

Mucina, L., Rutherford, M. C. (2006). *The Vegetation of South Africa, Lesotho and Swaziland.* Nantional Botanical Institute, Pretoria.

Mueller-Dombois, D., Ellenberg, H. (1974). *Aims and methods of vegetation ecology*. Wiley and sons, New York.

Ni, J. (2000). A simulation of biomes on the Tibetan Plateau and their response to global climate change. *Mountain Research and Development* 20, 80-89.

Olsen, S. J. (1990). Fossil ancestry of the yak, its cultural significance and domestication in Tibet. *Proceedings of the Academy of Natural Sciences of Philadelphia* 142, 73-100.

Opgenoorth, L., Vendramin, G. G., Mao, K. S., Miehe, G., Miehe, S., Liepelt, S., Liu, J. Q., Ziegenhagen, B. (2010). Tree endurance on the Tibetan Plateau marks the world's highest known tree line of the Last Glacial Maximum. *New Phytologist* 185, 332-342.

Pech, R. P., Jiebu, A. D., Zhang, Y. M., Liu, H. (2007). Population dynamics and responses to management of plateau pikas Ochotona curzoniae. *Journal of Applied Ecology* 44, 615-624.

Rabus, B., Eineder, M., Roth, A., Bamler, R. (2003). The shuttle radar topography mission R. – a new class of digital elevation models acquired by spaceborne radar, Photogramm. *International Journal of Remote Sensing* 57, 241-262.

Rees, W. G. (2008). Comparing the spatial content of thematic maps. *International Journal of Remote Sensing* 29, 3833-3844.

Rhode, D., Madsen, D. B., Brantingham, P. J., Tsultrim, D. (2007). Yaks, yak dung and prehistoric human habitation of the Tibetan plateau. *Developments in Quaternary Sciences* 9, 205-226.

Rowley, D. B., Currie, B. S. (2006). Paleo-altimetry of the late Eocene to Miocene Lunpola basin, Central Tibet. *Nature* 439, 677-681.

Schaller, G. B. (1998). *Wildlife of the Tibetan Steppe*. University of Chicago Pres, Chicago.

Sheehy, D. P., Miller, D., Johnson, D. A., (2006). Transformation of traditional pastoral livestock systems on the Tibetan steppe. *Sécheresse* 17, 142-151.

Shen, C. M., Liu, K. B., Morrill, C., Overpeck, J. T., Peng, J. L., Tang, L. Y. (2008). Ecotone shift and major droughts during the mid-late Holocene in the central Tibetan plateau. *Ecology*, 89, 1079-1088.

Shi, Y. F., Yu, G., Liu, X. D., Li, B. G., Yao, T. D. (2001). Reconstruction of the 30-40 ka BP enhanced Indian monsoon climate based on geological records from the Tibetan plateau. *Palaeogeography, Palaeoclimatology, Palaeoecology* 169, 69-83.

Smith, A. T., Foggin, J. M. (1999). The plateau pika (Ochotona curzoniae) is a keystone species for biodiversity on the Tibetan plateau. *Animal Conservation* 2, 235-240.

Song, M., Zhou, C., Hua, Q. (2004). Distributions of dominant tree species on the Tibetan Plateau under current and future climate scenarios. *Mountain Research and Development* 24, 166-173.

Song, M., Zhou, C., Hua, Q. (2005). Simulated distribution of vegetation in response to climate change on the Tibetan Plateau. *Journal of Vegetation Science* 16, 341-350.

Spicer, R. A., Harris, N. B. W., Widdowson, M., Herman, A. B., Guo, S., Valdes, P. J., Wolfe, J. A., Kelley, S. P. (2003). Constant elevation of Southern Tibet over the past 15 million years. *Nature* 421, 622-624.

Stumpp, M., Wesche, K., Retzer, V., Miehe, G. (2005). Impact of grazing livestock and distance from water points on soil fertility in southern Mongolia. *Mountain Research and Development* 25, 244-251.

Tapponnier, P., Xu Z. Q., Roger, F., Meyer, B., Arnaud, N., Wittlinger, G., Yang, J. S. (2001). Oblique Stepwise Rise and Growth of the Tibet Plateau. *Science* 294, 1671-1677.

Tichy, L. (2002). JUICE, software for vegetation classification. *Journal of Vegetation Science* 13, 451-453.

Toutin, T. (2008). ASTER DEMs for geomatic and geoscientific applications: a review. *International Journal of Remote Sensing* 29, 1855-1875.

Troll, C. (1973). Rasenabschälung (Turf Exfoliation) als periglaziales Phänomen der subpolaren Zonen und der Hochgebirge. *Zeitschrift für Geomorphologie N.F.*, Suppl. 17, 1-32.

Vetter, S. (2005). Rangelands at equilibrium and non-equilibrium: recent developments in the debate. *Journal of Arid Environments* 62, 321-341.

Wang, J. T. (1988). The steppes and deserts of Xizang Plateau (Tibet). *Vegetatio* 75, 135-142.

Wehrden v., H., Wesche, K., Miehe, G., Reudenbach, C. (2006). Vegetation mapping in Central Asian dry ecosystems using Landsat ETM+. A case study on the Gobi Gurvan Saikhan National Park. *Erdkunde* 60, 261-272.

Wu, C. Y. (1983-86). *Flora Xizangica*. Science Press. Beijing.

Wu, C. Y. and Raven, P. H. (1994 ff). *Flora of China*. Sciences Press. Beijing.

Wu, Z. Y., Tang, Y. C., Li, X. W., Wu, S. G., Li, H. (1981). Dissertations upon the origin, development and regionalization of Xizang Flora through the floristic analysis. In: Liu, D. S. (ed.), *Geological and Ecological Studies of Qinghai-Xizang Plateau*, Vol. II, pp. 1219-1244. Science Press, Beijing.

Xu, Z. X., Gong, T. L., Li, J. Y. (2007). Decadal trend of climate in the Tibetan Plateau – regional temperature and precipitation. *Hydrological Processes* 22, 3056-3065.

Yu, G., Chen, X., Ni, J. (2000). Palaeovegetation of China: a pollen data-based synthesis for the mid-Holocene and Last Glacial Maximum. *Journal of Biogeography* 27, 635-664.

Zhang, J. W. (1988). Vegetation of Tibet. Science Press, Beijing.

Zhao, G., MacLean, A. L. (2000). A comparison of canonical discriminant analysis and principal component analysis for spectral transformation. *Photogrammetric engineering and remote sensing* 66, 841-847.

Zhu, L. P., Wu, Y. H., Wang, J. B., Liu, X., Ju, J. T., Xie, M. P., Li, M. H., Mäusbacher, R., Schwalb, A., Daut, G. (2008). Environmental changes since 8.4 ka reflected in the lacustrine core sediments from Nam Co, central Tibetan Plateau, China. *The Holocene* 18, 831-839.

In: Steppe Ecosystems
Editors: M. B. Morales Prieto and J. Traba Diaz
ISBN: 978-1-62808-298-2

Chapter 7

ESTIMATION OF CARBON DIOXIDE EXCHANGE OF CASCADE GEOCHEMICALLY CONJUGATED STEPPE ECOSYSTEMS IN SALINITY CONDITION

Aleksei Naumov* and Elena Smolentseva
Institute of Soil Science and Agrochemistry
Siberian Branch of Russian Academy of Sciences, Novosibirsk, Russia

ABSTRACT

Current world climatic change is likely to have a strong influence upon soil geochemical processes in sub-arid and arid territories and on their water-saline functioning. In southern West Siberia, due to insufficient atmospheric precipitation, saline soils, salty and bitterly salty lakes are frequent phenomena. The forecasted increase in climate aridity and the accumulation of salts in landscapes are considered risk factors capable of causing qualitative shifts and transformations of the steppe ecosystem. The intensity of soil CO_2 emissions is regarded as an integral parameter, characterizing the functional condition on steppe catenas.

This study was conducted in 2009-2011 on an 1100 metres steppe catena, located in a salt lake depression. For comparison, similar measurements were carried out in a field of wheat on south chernozem soil (Calcic Chernozem). The results show that the rate of carbon dioxide emission on geochemical conjugated catenas normally falls from the top of the slope to accumulative areas down the slope. The rate of CO_2 emission was between 110 and 750 mg CO_2 /m^2·hr.

The above-ground plant matter deposition made 30-50 % of the integral flux in the steppe and meadow-steppe communities. On scarce vegetation sites carbon dioxide emissions were 20-50 % lower. In the salt lake coast band between saltwort patches, the intensity of soil respiration was at its lowest. It is possible that biological and physical-chemical processes could generate CO_2 there.

Keywords: Arid zone, catena, evaporation, soil respiration, steppe, Western Siberia

---

* Institute of Soil Science and Agrochemistry. Siberian Branch of Russian Academy of Sciences, Novosibirsk, Russia. Email: a.naum@ngs.ru.

INTRODUCTION

Steppes are usually considered to be a type of vegetation in the temperate zone, where perennial winter-hardy xerophilous gramineous plants are dominant (Walter, 1968). In Western Siberia, the steppe zone occupies the southern part of the plain and is characterised by a continental climate with high summer temperatures and little precipitation. A significant portion of steppe territory is plowed up, as a consequence of which the steppe has become a disappearing landscape.

Natural steppe vegetation is currently found only in small patches on low ridges and flat slopes or depressions along temporary streams that shrink in summer. These steppe remains present a relatively poor floristic diversity.

In the current climate and land-use change context, the prediction of structural and functional changes in landscapes, ecosystems, and the entire biosphere should be prioritised. The principles of landscape structure identification have been well developed by many authors (Dale, 1970; Farina, 2006; Neef, 1967; Forman and Godron, 1986; Naveh and Liberman, 1994; Sochava, 1978; Solntsev, 1981; Turner et al., 1991). At the same time, an ecosystem approach makes it possible to study the mechanisms of interaction between the dynamic components of the system that change in time and the relatively constant ones. The classification of landscapes with respect to the conditions of the migration of chemical elements, contributed to the development of a landscape-geochemical concept of catena, according to which a series of elementary landscapes that alternate with each other from the uppermost point of local depressions to the lowest point and are connected by lateral migration flows, form a catena (Glazovskaya, 1988; Gerrard, 1992; Thwaites, 2005). These catenas constitute the simplest cascade landscape-geochemical systems.

Among the ecological problems faced by steppe ecosystems, territory aridisation and soil salinisation have been the focus of much scientific research. However, little attention is paid to the comprehensive assessment of the functional state of conjugated elements in steppe landscape-geochemical systems. In this context, information on integral parameters that characterise an aggregate manifestation of many reactions and processes inside the systems may be of great use.

The processes of carbon mass exchange between soil, vegetation and atmosphere in a sequence of the ecosystems conjugated in a steppe catena are responsible for the functional stability of the carbon biogeochemical cycle. In this chapter, the emission flows of carbon dioxide and total evaporation are considered as integral parameters to characterise the functional state and the function of individual components of a cascade landscape-biogeochemical steppe catena.

METHODS

Study Area

The studied area (53.52 – 54.0° N and 77.05 – 77.20° E) is located in the Northern Kulunda geo-morphological province, in the steppe biome of Western Siberia. The climate of the region is classified as arid with a precipitation to evaporation ratio of 0.69. Summer droughts are frequent. Mean air temperatures are -20º C (January) and 19º C (July). The annual mean temperature is -0.2°C. Annual precipitation averages 310 mm, about one-third of which falls in winter and snow cover persists for about 5 months of the year.

Fieldwork was performed on a 1100 metres long landscape-geochemical catena in 2009-2011. The catena represented alternating vegetation communities and soils of different degrees of salinity (Figure 1). Data collection was carried out in mid-summer (the end of June-beginning of July), when the vegetation growth and biological activity of soil biota were highest.

There are several estimates of soil respiration for steppe ecosystems. A curve with a maximum in mid-summer is usually typical of the seasonal dynamics of steppe soil respiration (Coleman et al., 1976; de Jong et al., 1974; Kucera, and Kirkham, 1971; Naumov, 2009). A comparative analysis of actual values of emission of carbon dioxide (R) and total evaporation (E), which are integral parameters obtained in different years and characterise the most intense period of steppe ecosystem functioning, was performed in order to monitor the ecological state of the soil of an arid steppe territory under conditions of salinisation. Based on this, the period of field observations were selected. All the measurements on the catena were carried out during 1-2 days, in order to avoid the influence of different weather conditions.

The classification system of the World Reference Base (WRB) for Soil Resources (IUSS Working Group, 2006) was used as diagnostic basis. The soil cover in this steppe zone is very diverse. There are Chernozems on elongated hills and flat plains (as zonal type), and Solonchaks and Solonetzs in local depressions among salt lakes. The parent material of the soil is mostly composed by Quaternary deposits: aeolian (calcareous loess), re-washed aeolian (calcareous loess loams) and lacustrine-alluvial sediments of different granulometric composition.

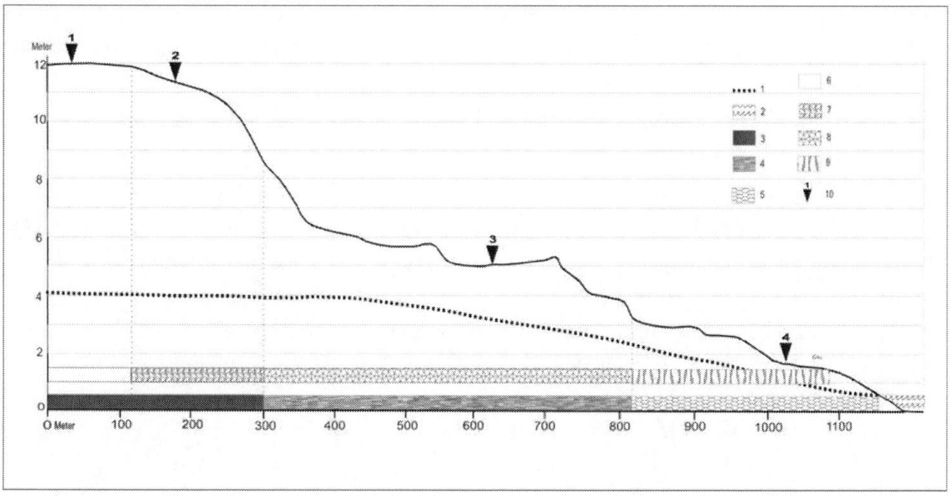

Figure 1. Vertical profile of of the study landscape-geochemical system or catena. 1 – groundwater level, 2 – salt lake; soils: 3 – Calcic Chernozem (Molliglossic), 4 – Calcic Solonetz (Glossalbic), 5 – Gleyic Solonchak (Chloridic); vegetation: 6 – arable, wheat field; 7 – herb-bunchgrass steppe, 8 – saline herb meadow, 9 – halophytes communities, 10 – measurement points.

Field Measurements

The vertical profile of the study catena site is presented in Figure 1. Measurement points were distributed, from top to bottom, in the following vegetation types:

- Point 1: a wheat field on top of the catena
- Point 2: herb-bunch steppe dominated by *Stipa capillata* L. and *Festuca valesiaca* Gaudin.
- Point 3: halophyte meadow on a flat area at mid-slope
- Point 4: sparse glasswort (*Salicornia europaea* L.) formation alt the lake shore (see Korolyuk et al., 2008 for a description of vegetation of salt lake depression in the Kulunda steppe).

To measure the rate of soil CO_2 exchange, a static chamber method was used. For the measurement time, the exposition chamber, made of organic glass with a volume of 60.85 l, was placed on a stainless steel collar sized 37 x 37cm that was previously embedded 10 cm deep in soil. Two chambers were placed in each vegetation formation. The system was hydrosealed and to avoid overheating of the chamber due to absorption of solar radiation, it was protected by a lightproof cover with a layer of reflecting aluminum foil. The air was circulated in the chamber by a built-in fan and the system was powered by a boost battery. The exposition time was 15 minutes. The cover was removed for a short time to measure the rate of apparent photosynthesis in daylight.

The emission rate was calculated as the change in the carbon dioxide content inside the chamber using a linear regression model. The concentration of carbon dioxide, pressure, temperature, and relative humidity of air inside the chamber was recorded automatically by a Testo-435 multifunctional analyser. The determination coefficients R^2 of linear regressions were in the range 0.94 - 0.99, which confirmed the reliability of the results. The temperature of the soil was measured by a T1 pin-type digital thermometer produced by the same company.

The rate of total evaporation was estimated as the change in relative air humidity inside the exposition chamber. To approximate the experimental time dependence of humidity, a fourth and fifth degree polynomial function was used. The rate of the process was calculated as the value of the first derivative of this function related to the beginning of the experiment. For comparison, similar measurements were performed in the wheat field (point 1).

The soil diagnostic horizons and material were characterised using particle size, pH, organic carbon, carbonate content, and exchangeable Ca, Mg, Na and K basic ions, as recommended by Van Reeuwijk (2006). Carbonate, bicarbonate, chloride, sulphate anions and cations of Ca, Mg, Na, K were also analyzed in saturate extracts of soil salt type determination.

RESULTS AND DISCUSSION

Tables 1 and 2 show the general features of soils in the study steppe catena. A zonal type of soil, namely sandy loam Chernozem, is represented by arable plots and virgin steppe (points 1 and 2). The upper mollic horizon in these soils has a thickness of 35 cm with content of organic carbon at 0.8 and 1.5 %, respectively. The Bk horizon contains secondary carbonates in the form of a soft layer on the structured surface. Calcium dominates among the exchange cations (70-80 %). Its distribution over the profile has accumulative behavior with a maximum in the Ah horizon. The content of exchangeable magnesium is approximately equal in all horizons. The content of exchangeable sodium is low, only 2-3 %.

Solonetz occurs on the slope in point 3. The diagnostic natric horizon (Btn) of Solonetz is very dense, well-structured with a clay type subsurface stratum. This horizon has a rough column or prismatic structure (Figure 2). The structural elements are characterised by a rounded and often whitish upper part. Pedogenic accumulation of calcium carbonate occurs in the lower part of the Btn horizon, its content being greater than in the deep subsoil and overlying horizons. The subsurface layers have a strongly alkaline reaction, which indicates the presence of sodium carbonate (Na_2CO_3). Carbonate ion was not found in other soils of the catena. A specific feature of this horizon is the high content of exchangeable sodium (60 %) and magnesium (31 %). The total content of water soluble salts is low.

Table 1. Main physicochemical characteristic of the soils sampled along the studied catena. Granulometry values of soil types are provided. Organic carbon is indicated as Corg

Horizon	Depth (cm)	pH (H_2O)	% Corg	% $CaCO_3$	% Sand (2000-63μm)	% Silt (63-2μm)	% Clay (<2μm)	% Water
Point 1. Calcic Chernozem (Molliglossic), arable land								
Ah	0-23	6.2	0.79	0.0	72.5	11.0	16.5	6.8
AhB	30-40	7.9	0.48	0.0	71.7	12.9	15.4	5.5
Bk	60-70	8.2	0.35	5.8	71.5	7.8	20.7	7.3
Ck	90-100	8.3	0.12	4.6	70.3	8.5	21.2	8.0
Point 2. Calcic Chernozem (Molliglossic), virgin soil								
Ah	0-10	6.7	2.30	0.0	70.8	17.1	12.1	5.3
Ah	10-20	6.8	0.62	0.0	65.8	13.2	21.0	3.5
Bk	40-50	7.3	0.30	5.3	70.9	10.7	18.4	7.9
Ck	90-100	8.4	0.11	4.5	66	16.6	17.4	8.8
Point 3. Calcic Solonetz (Glossalbic)								
A	0-5	8.7	0.72	2.5	78.7	10.7	10.6	1.3
Bn	6-16	9.2	0.54	1.5	56.6	12.8	30.6	9.7
Bn	20-30	9.7	0.23	3.2	43.2	16.9	39.9	8.9
Bnk	38-48	9.5	0.14	12.7	47.5	15.5	37.0	10.0
Ck	80-90	9.7	0.10	9.0	57.6	13.2	29.2	9.3
Point 4. Gleyic Solonchak (Chloridic)								
Az	0-17	9.1	0.33	2.9	91.9	4.0	4.1	4.6
Az	17-26	9.0	0.32	5.0	72.8	8.9	18.3	15.5
Brz	30-40	9.2	0.27	9.6	57.4	23.2	19.4	14.7
Crz	50-60	9.5	0.12	8.4	63.4	19.1	17.5	14.1

Table 2. Ionic composition of the saturate soil extracts from the studied catena. Ion concentration is provided in cmol$_c$ kg^{-1}

Depth (cm)	Ca^{2+}	Mg^{2+}	Na^{+}	HCO_3^-	Cl^-	SO_4^{2-}
Point 1. Calcic Chernozem (Molliglossic), arable land						
0-23	0.20	0.20	0.86	0.24	0.42	0.60
30-40	0.40	0.40	0.60	0.32	0.64	0.44
60-70	0.40	0.60	0.73	0.32	0.85	0.56
90-100	0.30	0.40	0.94	0.40	0.85	0.39
Point 2. Calcic Chernozem (Molliglossic), virgin soil						
0-10	0.12	0.08	0.14	0.16	0.12	0.06
10-20	0.12	0.11	0.15	0.12	0.20	0.06
40-50	0.36	0.12	0.20	0.48	0.18	0.02
90-100	0.38	0.20	0.28	0.46	0.32	0.08
Point 3. Calcic Solonetz (Glossalbic)						
0-5	0.34	0.46	0.82	0.92	0.58	0.12
6-16	0.21	0.25	5.34	2.24	2.38	1.18
20-30	0.29	0.16	9.20	1.88	4.22	3.35
38-48	0.25	0.13	7.60	1.76	3.54	2.48
80-90	0.13	0.07	5.45	1.76	2.36	1.29
Point 4. Gleyic Solonchak (Chloridic)						
0-17	0.19	4.77	51.32	0.32	45.22	10.74
17-26	0.40	2.46	37.41	0.24	34.36	5.67
30-40	0.23	1.68	23.15	0.44	20.50	4.12
50-60	0.32	2.02	21.36	0.32	21.20	2.18

Figure 2. Coarse columnar element of soil structure in the natric horizon of Solonetz (point 3).

Gleyic Solonchak (point 4) occupies the lowest position on the catena near the lake. This soil is characterised by a saline crust formed on the surface and a high content of soluble salts in all horizons of the soil profile. The main source of salts is the saline lake and underground flowing water. Salts are accumulated due to intense summer evaporation of soil moisture. Sodium cation and a chlorine anion dominate in saturate soil extracts (Table 2). The chlorine-sodium type of soil salinisation corresponds to the salt composition of the water in the lake.

A high concentration of salts in soils is toxic for plants and micro-organisms. Halophytes are rarely found on solonchak, although glasswort is abundant here. The content of organic carbon in this soil is very low (0.33 %). All horizons of Gleyic Solonchak have an alkali reaction. The accumulation of salts in the inland lake basins of the Kulunda steppe is caused by an excess of evaporation over precipitation, low winter temperatures, and additional feeding by the salty underground flowing waters (Lebedeva et al., 2008).

The content of salt in Bagan Lake is 300 g/l. According to ion composition, its water is classified in the chlorine-saline type. Sodium and magnesium chlorines are dominant in the salt composition (197 g/l and 54 g/l, respectively). These results are consistent with previous works in the study area (Lebedeva et al., 2008).

The sequence of plant communities in the catena from the upper position (point 1) to the lower one (point 4) is determined mainly by the influence of soluble salts that accumulate in the upper soil horizons (Table 2). The reserve of phytomass living above ground and the quantity of steppe plant litter decreases along the catena and the vegetation cover of solonetz becomes thinner. The halophyte plants form different sized patches further down in the catena under the influence of high amounts of soil salts (Figure 3). Bare soil occupies up to 60-70%.

Figure 4 shows seasonal variation of the air temperature and precipitation in the years under study. The seasonal variation in the air temperature was very similar in the different years considered, and was consistent with the average known long-term tendency. The highest values of the average monthly temperature were 19.0 °C in June 2010 and 20.9 °C in June 2011. The average temperature in June 2009 was only 15.4 °C. The amounts of precipitates for the warm period of the year were 299 mm in 2009, 122 mm in 2010 and 182 mm in 2011. The respective annual temperatures were 1.2, 0.7 and 1.9 °C.

Figure 3. General view of halophytes community around the salty Bagan Lake.

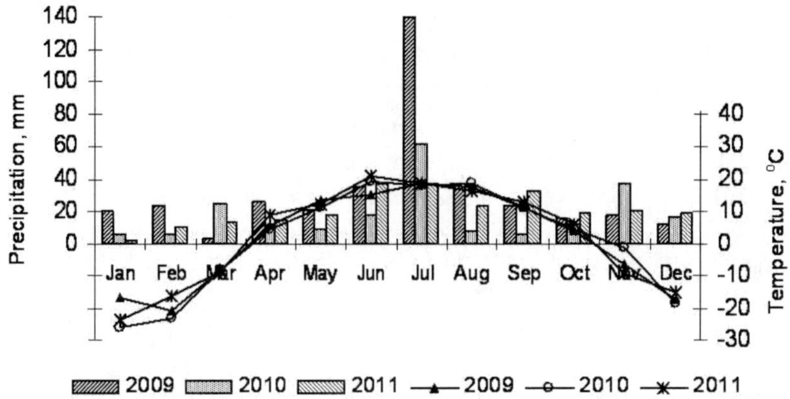

Figure 4. Climadiagram of the study area (Karasuk meteostation data, 60 km in distance, free on WEB-site http://rp5.ru/archive).

Considering the sum of active temperatures (t> 10 °C), the warmest year was 2011 (2630 °C) and the coldest one was 2009 (2260 °C). Therefore, according to the hydrothermal conditions that control soil biological processes, the most favourable vegetation growth season occurred in 2011.

The temperature conditions during the observation periods were favorable for the soil biota activity. At a depth of 10 cm, the temperature of the soil varied from 15 to 22°C in the different ecosystems. The highest values were recorded in the soils located in transition of the catena. The soils of eluvial and accumulative positions were cooler. Table 3 shows the soil temperatures, respiration and evaporation values recorded at the different ecosystems sampled over the three years of study. The highest value of evaporation in 2009 (4.8 mm/day) was recorded in the wheat field.

Table 3. CO_2 emission (R, mg/m^2/h), evaporation (□, mm/day) and soil temperature (t, °C) at 10 cm depth in the study steppe catena. □ (point 1): wheat crop on Calcic Chernozem; □$_0$ (point 2): wormwood-gramineous steppe on Calcic Chernozem; □$_1$ (point 2): feather-grass forb steppe (Calcic Chernozem); □$_2$ (point 3): salted gramineous - mixed herb meadow (Calcic Solonetz), □$_3$ (point 3): thinned *Limonium-Puccinellia* community on a crust Solonetz; □$_4$ (point 4): coastal strip of a salt lake with saltwort (Gleyic Solonchak). The aboveground part of the plants and the plant litter were removed during the R and E measurement in 2010 and 2011

Year	Index	Ecosystem					
		A	T_0	T_1	T_2	T_3	T_4
	t	18.4	18.3	21.9	n.m.	22	n.m.
2009	R	720±196	379±60	361±25	n.m.	253±121	n.m.
	E	4.8±0.1	1.5±0.2	1.3±0.5	n.m.	1.4±0.8	n.m.
	t	16.6	19.1	20.5	25.2	25.7	20.9
2010	R	189±78	526±55	596±215	612 ± 94	399±22	n.m.
	E	1.7±1.1	2.1±0.6	2.6±0.3	2.6 ± 0.3	2.0	2.6±0.7
	t	16.9	19.9	21.0	21.4	18.6	14.8
2011	R	577±20	565±242	489±67	527 ± 39	234±32	112±2
	E	0.9	2.1±0.2	2.2±0.6	0.9 ± 0.1	0.7±0.4	0.8±0.1

In 2010 the evaporation from the soil surface on the catena was 2.0-2.6 mm/day with relatively small variability in the index. The moisture expiration in the arable field was somewhat lower (1.7 mm/day), but showed great dispersion in the index values.

In addition to the main factors, such as temperature and soil moisture, the CO_2 emission from the soil to the atmosphere is influenced by vegetation (Raich and Tufekcioglu, 2000). Vegetation cover favours a soil microclimate and, together with the physical properties of soil strata, influences the activity of soil microorganisms through organic matter that enters the soil with plant residues (Naumov, 2009). The contribution of the plant root system respiration to the total soil-atmosphere CO_2 flux can be considerable. In different balance models, their contribution is conventionally accepted to equal 15-30 % (Raich and Schlesinger, 1992; Stolbovoi, 2003). In steppe phytocoenoses, the share of respiration of underground organs in the total balance of carbon is quite relevant because of the great reserves of living roots. In the arid steppe ecosystems, this emission amounted to 50-60% of the gross photosynthesis value during the maximum plant stand in the summer (Naumov, 1988).

Results showed that CO_2 emission in the landscape-geochemical catena decrease from the eluvial position towards the accumulative one (Table 3). The range of emission was 110 to 750 mg $CO_2/m^2/hr$. The contribution of above-ground plant biomass to the total flux amounted to 30-50 % in the steppe and herb meadow communities. The measured flows differed by 20-50% in the sections with thinned vegetation. In the coastal strip among the saltwort, the emission of carbon dioxide is lower than in the other ecosystems present in the catena. However, the low content of living roots and organic carbon in the soil do not point to a dominant role of biological soil constituents in the emission of CO_2. It is expected that carbon dioxide released into the atmosphere in this part of the landscape biogeochemical system can originate from both biological and physicochemical processes. The study of this issue is of great importance for understanding the “working mechanism” of local sources of carbon dioxide in a steppes zone. The measurements of visible photosynthesis in 2009 showed that in daylight (partly cloudiness, light intensity 56-95 klux) and with an air temperature of 25-27 °C, the steppe phytocoenosis (T_0) accumulated 340-490 mg of $CO_2/m^2/hr$, and the wheat crop 800-1220 mg $CO_2/m^2/hr$. The contribution of soil respiration to the total carbon balance was 40 % for both ecosystems.

The dispersion of emission flows of carbon dioxide and soil water evaporation in the arable field (southern chernozem Calcic Chernozem) was caused mainly by sowing in rows. The emission of CO_2 in the exposition chambers placed in the space between the rows was lower than in the rows. This difference was expressed by values that differed by 1.5 - 2 times.

The annual carbon flow to the atmosphere due to the soil respiration was calculated using the maximum emission rate in the steppe ecosystem (T_0) is 0.33 kg C $/m^2/yr$. However, in view of the character of the soil cover of the cascade landscape-geochemical steppe catena, the average annual carbon emission into the atmosphere may be less than 0.2 kg C $/m^2/yr$ in the studied system.

CONCLUSION

The functioning of the studied cascade landscape-geochemical system is characterised by high variability of indices in different years, as well as by the mobility of the boundaries of the identified ecosystems. Current global climate change may exert a substantial influence on the soil-geochemical processes and the water-salt regime of sub-arid and arid territories. Saline soils, solonetz and solonchak soils, salt and bitter lakes are prevalent under a shortage of atmospheric precipitation in the southern areas of Western Siberia. Aridisation of climate and accumulation of salts in soils should be considered risk factors that can cause qualitative shifts or even transformation of landscape-geochemical conjugated systems of different hierarchical levels.

ACKNOWLEDGMENTS

This research was supported by a grant from the Presidium of Russian Academy of Sciences (Project 4.14).

REFERENCES

Coleman D.C., Andrews R., Ellis, J.E., Singh, J.S. (1976). Energy flow and partitioning in selected man-managed and natural ecosystems. *Agro-Ecosystems* 3, 45-54.

Dale, M.B. (1970). Systems analysis and ecology. *Ecology* 51(1), 2-16.

De Jong, E., Schappert, H.J.V., MacDonald, K.B. (1974). Carbon dioxide evolution from virgin and cultivated soil as affected by management practices and climate. *Canadian Journal of Soil Science* 54, 299-307.

Farina, A. (2006). *Principles and methods in landscape ecology: towards a science of the landscape*. Springer. Dordrecht, the Netherlands.

Forman, R.T.T., Godron, M. (1986). *Landscape ecology*. Wiley, New York.

Gerrard, J. (1992). *Soil geomorphology: An integration of pedology and geomorphology*. Chapman and Hall, London.

Glazovskaya, M.A. (1988). *Geochemistry of natural and technogenic landscapes of the USSR*. Vishaya skola, Moscow. [in Russian].

IUSS Working Group WRB. (2006). *World reference base for soil resources 2006*. 2nd edition. World Soil Resources Reports No. 103. FAO, Rome.

Korolyuk, A.Y., Smolentzev, B.A., Lashchinsky, N.N. (2008). Soil and vegetation cover of the circumlake plain of Lake Kulunda. *Vegetal Life of Asian Russia* 2, 80-88.

Kucera, C., Kirkham, D. (1971). Soil respiration studies in tallgrass prairie in Missouri. *Ecology* 52, 912-915.

Lebedeva, M.P., Lopukhina, O.V., Kalinina, N.V. (2008). Specificity of the chemical and mineralogical composition of salts in Solonchak playas and lakes of the Kulunda Steppe. *Eurasian Soil Science* 41(4), 416-428.

Naumov, A.V. (1988). *Respiratory gas-exchange and productivity of steppe phytocoenoses*. Nauka, Novosibirsk. [in Russian].

Naumov, A.V. (2009). *Soil Respiration: constituents, ecological functions, geographic patterns*. SB RAS Publishing House, Novosibirsk. [in Russian].

Naveh, Z., Lieberman, A.S. (1994). *Landscape ecology: theory and application*. Springer, New York.

Neef, E. (1967). *Die theoretischen Grundlagen der Landschaftslehre*. Gotha/Leipzig, 152 S.
Raich, J.W., Schlesinger, W.H. (1992). The global carbon dioxide flux in soil respiration and its relation to vegetation and climate. *Tellus B* 44, 81-99.
Raich, J.W., Tufekcioglu, A. (2000). Vegetation and soil respiration: correlations and controls. *Biogeochemistry* 48, 71-90.
Sochava, V.B. (1978). *Introduction to geosystem studies*. Nauka. Novosibirsk. [in Russian].
Soil Survey Laboratory Methods Manual (2004). *Soil Survey Investigations Report* No. 42, Version 4.0, Lincoln, USA.
Solntsev, V.N. (1981). *System organization of landscape*. Mysl, Moscow. [in Russian].
Stolbovoi V. (2003). Soil respiration and its role in Russia's terrestrial C flux balance for Kyoto baseline year. *Tellus B* 55, 258-269.
Thwaites, R.N. (2005). Soil geomorphology In: R. Lal (ed.), *Encyclopedia of Soil Science*, Second edition. CRC Press, Boca Raton (FL), USA.
Turner, M.G., Gardner, R.H., O'Neil, R.V., (2001). *Landscape Ecology in Theory and Practice: Pattern and Process*. Springer, New York.
Van Reeuwijk, L.P. (2006). *Procedures for Soil Analysis*. 7th Ed. Technical Report 9. Wageningen.
Walter, H. (1968). *Die Vegetation der Erde in öko-physiologischer Betrachtung. Bd. 2. Die gemässigten und arktischen Zonen*. VEB Gustav Fischer Verlag, Jena.

In: Steppe Ecosystems
Editors: M. B. Morales Prieto and J. Traba Diaz
ISBN: 978-1-62808-298-2

Chapter 8

THE LATE QUATERNARY PALEOECOLOGY AND ENVIRONMENTAL HISTORY OF HORTOBÁGY, A UNIQUE MOSAIC ALKALINE STEPPE FROM THE HEART OF THE CARPATHIAN BASIN

***Pál Sümegi*[1,2,*], *Gábor Szilágyi*[3], *Sándor Gulyás*[1], *Gusztáv Jakab*[4] *and Attila Molnár*[3]**

[1]University of Szeged, Department of Geology and Paleontology, Szeged, Hungary
[2]Institute of Archeology, Hungarian Academy of Science, Budapest, Hungary
[3]Headquaters of the Hortobágy National Park, Debrecen, Hungary
[4]University of Szent István, Tessedik Campus, *Szarvas,* Hungary

ABSTRACT

The first national park in Hungary was established in 1973 in the area of Hortobágy with the aim of protecting dry and humid alkaline steppe areas and concomitant fauna. In 1999 the Hortobágy National Park was included in the world heritage list of the UNESCO.

The outstanding importance of the park comes from dominantly non-arboreal, steppe vegetation harboring a unique avifauna and highly variable alkaline and chernozem soils displaying a complex mosaic-like spatial patterning. In addition, pastoralism in Hortobágy dates back several hundred or even thousands of years. Several hypotheses were postulated mainly by botanists and pedologists, which blamed human activities related to river regulations starting during the second half of the 19th century as being responsible for alkalinization and the emergence of the modern ecosystem. Conversely, paleoecological and paleobiological studies starting roughly 30 years ago unambiguously pointed to a natural origin of the alkaline steppe dating back to the end of the last ice age. Human activities with the arrival of the first farmers and agropastoralism hardly affected the natural evolution of the area for millennia. Drainage of marshlands and inundated areas initiating during the 19th century and followed by the introduction of rapidly

* University of Szeged, Dept of Geology and Paleontology. H-6722 Szeged Egyetem u.2-6. Email: sumegi@geo.u-szeged.hu.

intensifying farming, however, had significant impact on the landscape. The present chapter would like to highlight the past natural evolution of this unique alkaline steppe ecosystem with special attention to impacts related to former and present-day human activities, including conservation measures

Keywords: Quaternary paleoecology, Pannonian alkaline steppe, environmental mosaics, Hortobágy National Park, Central Europe, Hungary

INTRODUCTION

The origin and spatio-temporal evolution of lowland alkaline areas within the Carpathian Basin, including the alkaline grasslands of Hortobágy has been a subject of continuous debate among various groups of Hungarian natural scientists for almost a century. According to the view of the most influential group of botanists in scientific forums, alkalinization is the outcome of the late 19th century flood control and drainage measures in the area of the Great Hungarian Plain.

An alternative view, minoritary for several decades, states that the referred measures only revived this process dating back much earlier in natural landscape evolution.

Unfortunately, no consensus was reached until the 1990s. Partly, because multi-disciplinary and comprehensive investigations making a consistent application of methodologies and meeting international standards (i.e. pollen and plant macrofossil analysis on undisturbed samples with reliable absolute chronologies) that could have provided univocal evidence in the issue of alkaline landscape formation were scarce. Nevertheless, results of the few scattered studies applying such methodologies in the area of Hortobágy were evaluated with a strong bias from the prevailing scientific views. The purpose of the present work was thus to clarify such debate by highlighting the history of alkaline landscapes in the Carpathian Basin with a special focus on the alkaline steppes of Hortobágy using available multi-disciplinary hard data.

In order to fully understand the origin of alkaline landscapes, one must be acquainted with the causes of the process. The concept of alkalinization differs among the various groups of scientists. Geoscientists, soil scientists and experts in agriculture generally refer to the accumulation of primarily sodium salts in the near surface (sodium-carbonates and bicarbonates: $Na_2CO_3.10H_2O$) leading to the emergence of saline and or alkali soils (Treitz, 1898, 1917, 1925; Scherf, 1935; Arany, 1956; Sigmond, 1906, 1934; Szabolcs, 1961; Várallyay, 1967, 1993, 1999). Botanists, on the other hand, think about the presence of saline or salt-loving and salt-resistant alkaline plants when they talk about the process of salinization (Kerner, 1863; Borbás, 1886; Rapaics, 1916).

However, both processes are tightly interconnected. Not surprisingly, works dealing with the question generally look for the origin of salts, mainly sodium salts in soils and salt lakes, salt marshes, as well as salt steppes.

The importance of sodium excess and sodium accumulation in the process of alkalinization in the case of the alkaline soils of the Great Hungarian Plain was first highlighted by Sámuel Teschedik in the early 19th century (Teschedik, 1804). Therefore, as a first step, it is necessary to examine where the material yielding the salts of alkali elements (mainly sodium) comes from.

Geological Basis of Alkalinization and Potential Sources of Alkaline Salts

Sodium accumulation in alkaline areas has been initially attributed to plant matter decay (Muraközy, 1902; Sigmond, 1923). This assumption is rather controversial since plants incorporate sodium from some other source. Thus the alkaline and halophyl plants inhabiting the salt lakes, salt steppes and salt marshes cannot be considered as primary sources of sodium. Sodium is present in various forms in these habitats: dissolved in groundwater or attached to the surface of soil colloids. It may also appear in solid form in deposits precipitated from saline soils, saline lakes or marshes (Bernátsky, 1905; Tóth et al. 2001; Szendrei andTóth, 2006). Therefore, alkaline elements, primarily sodium as a main driver of the emergence of alkaline soils can derive from various sources. In this context, it is important to highlight what Hungarian scientists thought of potential primary sodium sources in the evolution of alkaline areas. József Szabó, a Hungarian geologist working during the second half of the 19th century, was the first to state that alkaline salts derive from the weathering of sodium present in minerals from the volcanic rocks surrounding the basin (Szabó, 1864). He interpreted the presence of the so-called "white muddy soil" found in the alkaline meadows of Heves County, in the valley of the Laskó creek, as the weathering byproduct of rhyolite tuff. Another view by Jenő Kvassay connected sodium sources to the Tertiary salt deposits of surrounding mountains, whose weathered material reached the center of the basin via fluvial transport (Kvassay, 1876). Conversely, Galgóczy (1876-1877) pointed out the local Tertiary marine salt deposits buried by younger Quaternary fluvial sequences as potential alkaline and sodium sources for the alkalinization of the Great Hungarian Plain. Researchers working on the topic later on used some of these initial hypotheses or a combination of them in their works regarding sources of sodium.

Nevertheless, as early as 1930, a consensus was reached stating that alkaline salts derived from the weathering of near-surface rocks and minerals of the Great Hungarian Plain after having been transported into their present location via fluvial and aeolian means (Inkey, 1895, Sigmondi, 1906; Arany, 1934, Sümeghy, 1937; Kreybig, 1944; Endrédi, 1941; Székyné Fux and Szepesi, 1959). Results of extensive geochemical studies and salt source modeling dated to the second half of the 20th century corroborated the assumptions regarding the origin of alkaline elements for alkali salts (Sümegi, 1989, 1997; Szöőr et al. 1991; Sümegi et al. 2000; Tóth et al. 2001). Erosion of the surrounding mountain belt, primarily of carbonated and siliceous rocks provided material for the Quaternary sequences of the Great Hungarian Plain (Molnár, 1960, 1964, 1965, 1967, 1973; Sümegi, 1989, 1997, 2004, 2005; Timár et al. 2005). However, the presence of bedrock material providing sources of alkaline elements can not grant salinization or alkalinization alone. An interaction of the groundwater with the bedrock is necessary for the appearance of Ca^{2+}, Mg^{2+}, Na^{+}, $HCO3^{-}$ and Cl^{-} ions (Várallyay, 1967; Scherf, 1935; Székyné Fux and Szepesi, 1959; Sümegi, 1989, 1997; Szöőr et al. 1991, Fórizs, 2003). The nature of this interaction is determined by the hydrogeology and geomorphology of a given area.

Hydrogeological and Geomorphological Basis of Alkalinization

According to the findings of available geochemical studies and models, the initial step of this alkalinization process is the dissolution of finely dispersed calcite found in calcareous sandy and loessy deposits of the Great Hungarian Plain. Carbonate dissolution favors the solution of silicates via turning the pH higher (Cox, 1991; Wright, 2003). As a result, sodium-bearing silicates, mainly those of plagioclase feldspars fall apart into their constituents releasing free sodium in groundwater (Székyné Fux and Szepesi, 1959; Sümegi, 1989, 1997, 2004; Sümegi et al. 2000; Sümegi and Szilágyi, 2010; Tóth et al. 2001).

Extensive isotope studies on aquifers of alkaline landscapes have univocally proved that potential salt sources are dominantly connected to the shallow, local upwellings of the groundwater (Deák, 1974, 1975; Sümegi, 1997; Tóth et al. 2001; Fórizs, 2003), in contrast to former theories which linked the process to upwellings from deeper deposits or gas exhalation along faults and fractures (Treitz, 1924; Scherf, 1947, 1949; Stegena and Szebényi, 1949; Tóth, 1963; Tóth and Almási, 2001, Mádlné Szőnyi et al. 2005). Thus the higher dissolved salt content and high alkalinity of the alkaline landscapes of Hortobágy derive from the chemical weathering and dissolution of the Quaternary bedrock (Székyné Fux and Szepesi, 1959; Sümegi, 1989; Tóth et al. 2001) rather than from the older, buried salt deposits found at greater depths in the basin.

Based on available data, local upwelling or discharge zones of the groundwater (Fórizs, 2003; Fórizs et al. 2006; Tóth et al. 2001) connected to interdune depressions (Mucsi, 1963), abandoned, infilled riverbeds (Molnár and Szónoky, 1976), or the margin of the floodplains (Kovács et al. 2006; Rakonczai, 2006; Rakonczai and Kovács, 2006) promote alkalinization, generating the hydrogeological and geomorphological prerequisites of the process. The height of the groundwater table is fundamentally influenced by the local morphology (Mados, 1943; Szabolcs, 1957).

Thus the local morphological endowments (Strömpl, 1931) as well as the perched depth of the groundwater table (Mados, 1943), i.e. relative elevational differences, are important components of the system yielding alkalinization (Strömpl, 1931; Mados, 1943; Szabolcs, 1957). These elevational differences are quite pronounced along the margins of the floodplain as well as those of abandoned, infilled riverbeds (Kovács et al. 2006; Rakonczai, 2006; Rakonczai and Kovács, 2006). Therefore, it is not surprising that the highest degree of alkalinization is observed along the margin of these natural depressions (Stefanovits, 1956). These areas favor the rapid emergence of a groundwater table critical for the initiation of alkalinization (Kerényi, 2003). On the other hand, precipitation of alkaline salts at the surface and in the near-surface deposits is hampered by a permanent upwelling of groundwater, where continuous flushing removes dissolved salts and elements. However, evaporation of groundwater with high dissolved salt content, together with rainfall in natural discharge areas with no surface runoff, can bring about alkalinization. Such processes are characteristic in a part of the alkaline landscapes of Hortobágy.

Upwelling of dissolved salts is driven by local discharge areas found in front of local recharge areas. It is also influenced by the development and nature of the capillary fringe found above the groundwater table, which is controlled by the cyclical fluctuations of the groundwater table, insolation, surface vegetation cover, morphology and microclimatic endowments (Sümegi, 1997). Capillary drive is an important component of alkalinization, as it enables the upwelling of salt-rich groundwater against gravity as well. This ensures a constant or temporarily link between the sodium-rich groundwater and the near-surface aquisystems present in alkaline soils, salt lakes and salt marshes. In areas, where the groundwater table is deeper than two meters, capillary lift cannot bring up waters rich in dissolved salts and sodium to the surface, hampering alkalinization (Mados, 1943).

The Climatic Basis of Alkalinization

The nature of the capillary rise and the emergence of a capillary fringe are largely influenced by the climatic endowments of the Carpathian Basin. Dynamic fluctuations in the groundwater table were characterized by three maxima connected to precipitation peaks occurring during the spring snowmelt, the early summer and early fall periods preceding the 19th century flood control measures. Among the various components influencing the climate, high continentality with a frequency of 30-40 % at the centennial scale plays the most important role in creating capillary action responsible for alkalinization (Lóczy, 1910; Bacsó, 1961; Karuczka, 1999; Pálfai, 1989, 1993, 1994; Pálfai et al. 2003; Réthly, 1933; Szelepcsényi et al., 2009a, b). As a result of high continentality, the ocurrence of intense droughts in periods between major floods mainly confined to the second half of the summer favors the emergence of a capillary fringe in the uppermost one or two meters of soil.

Thus the highly variable climate of the Carpathian Basin, characterized by highly alternating wet and dry periods driven and controlled by the intensity of continentality is a major asset to alkalinization. On one hand, it contributes to the development of the capillary fringe fostering the upward movement of sodium-rich waters in soils. On the other hand, it also fosters the emergence of oversaturated waters in alkaline salts via enhanced evaporation in potential discharge areas.

Hypotheses and Conventional Themes Regarding the Evolution of Alkaline Landscapes of the Carpathian Basin

Based on the information presented in the previous sections, alkalinization can be regarded as an outcome of the unique interplay of geology, geomorphology, hydrogeology and climate in the Carpathian Basin (Sümegi et al., 2000). Nevertheless, the present paper would like to focus on the evolution of alkaline grasslands and forest steppes of Hortobágy, which is not an easy task. A part of the problem comes from the fact that proposed hypotheses were not clearly supported by evidence and have been taken as *fait accompli* by many leading Hungarian botanists.

A major pitfall faced by studies discussing the historical aspect of alkalinization comes from the fact that vegetation, soils and hydrology of the Great Hungarian Plain experienced fundamental transformations as a result of the 19th century flood control measures taken before the original conditions could have been mapped and understood in detail (Dapsy, 1869; Boros and Bíró, 1999). Therefore the majority of propositions were aimed at deciding whether or not alkalinization was human-induced, and thus the byproduct of the flood control measures (Molnár, 1996, 1997, 2003, 2007, 2008-2009, 2009; Molnár and Bíró, 1997). Scientific observations preceding the period of the river regulations (Balogh, 1840; Towson, 1797; Kitaibel, 1796 in Molnár 2007; Teschedik, 1804; Zipser, 1817) univocally attested the presence of alkaline soils and plants inhabiting these habitats (Molnár and Borhidi, 2003; Molnár, 2007; Sümegi et al. 2000; Timkó, 1934) in the area of the Great Hungarian Plain.

An incorrect translation of written historical ecological records from Latin might have contributed to a misinterpretation regarding the origin of alkaline landscapes (Molnár, 2008-2009, 2009). Based on this source, the presence of extensive woodlands was presumed for the period preceding the flood control measures. Thus extensive deforestation and dessication was blamed to initiate alkalinization. However, Molnár (2009) first pointed out that no information was given regarding the presence of woodlands in Hortobágy in the referred charters. Environmental historical and paleoecological research may yield more reliable information of the history of alkalinization than historical ecological records dated to the 18-19th centuries.

Paleoecological Data on the Origin and History of Alkalinization of Hortobágy and the Carpathian Basin

As it has been stated in the previous section, one of the major problems in understanding the historical aspect of alkalinization in the Carpathian Basin comes from the fact that researchers working on the topic did not correctly identify the sources of alkalinization. Furthermore, the general principles drawn from observations in relation with the modern alkaline flora and fauna were used to tackle potential sites of alkalinization in the past. This would be no major problem in itself. However, theoretical considerations grounded on these principles completely ruled out the possibility of alkalinization during the ice ages, stating that environmental conditions were not favorable in the Great Hungarian Plain for the process itself (Miháltz, 1947).

As a result nobody would actually assume the presence and or look for ice age alkaline deposits or signs referring to alkalinization preceding historical times in the Great Hungarian Plain until 1988. In 1988 a multi-proxy paleoecological study of an undisturbed core sequence from the eastern margin of Hortobágy managed to correctly identifiy an alkaline paleosol horizon dated between 30-40 kys under ice age loess deposits (Sümegi, 1989, 1997, 2001, 2003; Szöőr et al. 1991, 1992).

The referred paleosol was described as steppe alkaline soil even in the field (Sümegi, 1989). Later analyses revealed the presence of minerals characteristic of alkaline soils such as gypsum, polialuminates, amorphous silica gel, corroborating postulations made in the field (Sümegi, 1989, 1997; Szöőr et al. 1991, 1992).

All these data univocally attested that conditions favoring alkalinization could have emerged during the last glacial cycle dated to MIS 3 as part of an intensive short interstadial warming (Dansgaard – Oeschger cycle; Sümegi, 1989). Cores taken parallely with these investigations on Copper Age burial mounds have also revealed the presence of Early Holocene fixed, buried chernozem and alkaline soils in the referred area (Sümegi, 1988; Sümegi and Szilágyi, 2011).

Findings of these initial investigations have unambiguously proven that the question of alkalinization in the Great Hungarian Plain is far from settled. In order to correctly elucidate the full story multiproxy studies using sedimentological, geochemical, palynological, paleozoological, and plant macrofossil investigations are needed within a reliable independent chronological framework. The data thereby obtained enable us to test and refute or modify former ideas and hypotheses put forth in connection with the origin and evolution of alkaline landscapes in Hungary. In order to achieve these goals, multidisciplinary investigations were implemented on several undisturbed continuous core sequences taken as part of numerous independent campaigns from 1999 onwards in the area of Hortobágy with the aim of highlighting the vegetation history of the area (Sümegi, 1997; Sümegi et al. 2000, 2005, 2006).

MATERIAL AND METHODS

Alkaline habitats cover an area of ca. 10,000 km^2 in the Carpathian Basin. Extensive alkaline landscapes are found in several parts of the Great Hungarian Plain including the Danube-Tisza Interfluve and the Körös-Maros Interfluve. The area of Hortobágy is the largest coherent occurrence of an alkali habitat type in Europe covering an area of ca. 2300 km^2. The Puszta in Hortobagy is of important natural and historical-cultural value, which has been recognised as a UNESCO World Heritage site since 1999. In 1973 it was declared the first national park in Hungary (Figure 1).

Both the landscape and vegetation of alkaline areas in Hortobágy show considerable variation with the presence of Achillea salt steppes (*Achilleo setaceae–Festucetum pseudovinae*), Artemisia salt steppes (*Artemisio santonici-Festucetum pseudovinae*), dense and tall *Pucinellia* swards *(Puccinellietum limosae)*, the so-called "blind-szik" (*Camphorosmetum annuae*), and salt marshes (European slough grass meadows - *Agrostio - Beckmannietum eruciformis*, foxtail meadows- *Agrostio - Alopecuretum pratensis*; Borhidi, 2007).

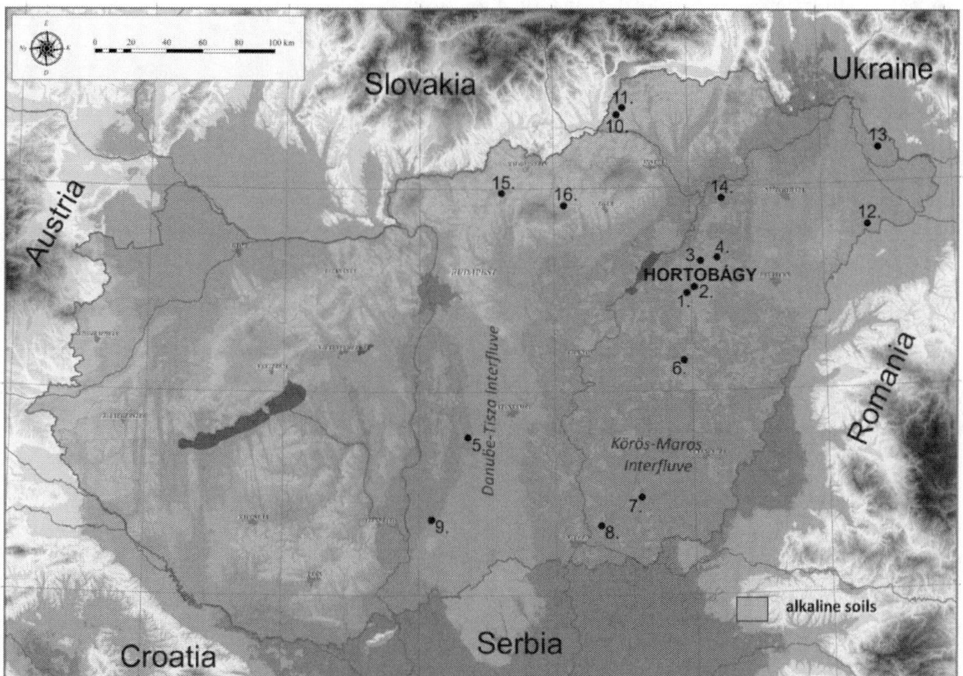

Figure 1. The geographical distribution of alkaline regions in Hungary with locations of the Hortobágy study area and boreholes referred in this chapter. 1 = Nagyiván, Kunkápolnás site, 2 = Zám, Halasfenék (based on minimum 500 pollen grains), 3 = Máta, Pap-ere, 4 = Hortobágy, Fecskerét, 5= Lake Kolon (Sümegi et al. 2011), 6 = Ecsegfalva, Lake Kiri (Willis, 2007), 7 = Kardoskút, Lake Fehér (Sümegi et al. 1999), 8. Maroslele – Pana (Sümegi et al. 2012), 9.Hajós (Jakab et al. 2004), 10./Kelemér, Kis-Mohos (Willis et al. 1997), 11. Kelemér, Nagy-Mohos (Magyari et al. 1999), 12./Bátorliget, protected marshland (Willis et al. 1995), 13./Csaroda, Nyíres Lake (Sümegi, 1999), 14./Tiszadob, Sarlóhát (Magyari et al. 2010), 15./Nagybárkány, Nádas Lake (Sümegi et al. 2010), 16./Sirok, Nyírjes Lake (Gardner, 1999).

In order to reconstruct the evolution of these alkali habitats, catchment basins found between salt steppes, representing former paleochannels and depressions as well as man-made earth mounds were systematically sampled via undisturbed coring (Figure 2). For the assignment of sampling sites geographical maps prepared in the 18th century, the maps of the so-called 1st Austrian Military Survey have been used (Figure 3). These maps represent the close-to natural conditions of the landscape preceding the largest anthropogenic disturbance related to the 1847-1867 flood control measures. The sampling sites correspond to abandoned and infilled paleochannels fringing the open vegetation alkaline grasslands (Figure 4). In order to eliminate potential pollen contamination, corings were implemented during the winter. The retrieved samples were subjected to sedimentological, geochemical, palynological, malacological and plant macrofossil analyses along with ^{14}C chronological investigations (Sümegi, 1997; Sümegi et al. 2000, 2005, 2006). Based on the established independent chronology, paleoenvironmental changes could be traced back to 50,000 years.

Undisturbed sedimentary sequences from 9 different sites (Figure 2) including the 10 m long core of the Nagyiván-Kunkápolnás marshland, marked by No. 1 on Figure 1 and No.2 on Figure 2, were sampled using a 10 cm diameter special double walled core-head. The main lithostratigraphic features of the sedimentary sequence were determined and analyzed. For the description of the cores, the internationally accepted system and symbols of Troels-Smith developed for unconsolidated sediments was adopted (Troels-Smith 1955).

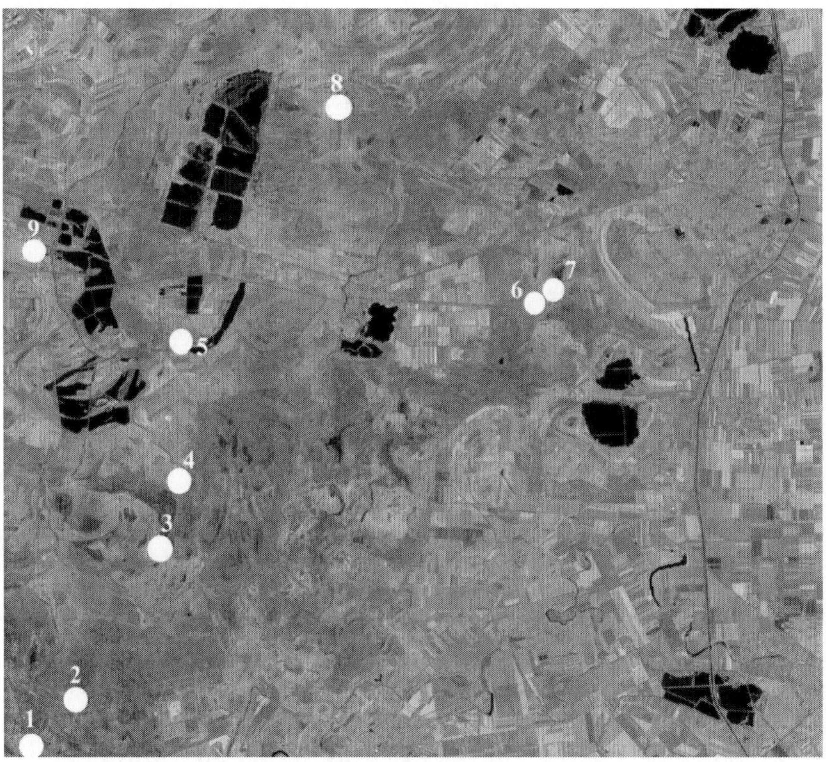

Figure 2. Aerial view of Hortobágy with location of the studied cores. Sampling points: 1./Nagyiván – Ecsehalom (mound), 2./Nagyiván – Kunkápolnás (paleochannel), 3./Zám, Halasfenék (paleochannel), 4./Faluvég, Faluvég-halom (mound), 5.Hortobágy, Kun György Lake (paleochannel), 6./Hortobágy, Szálkahalom (mound), 7./Hortobágy, Fecskerét (paleochannel), 8./Máta, Pap-ere (paleochannel), 9./Hortobágy, Csipő-halom (mound).

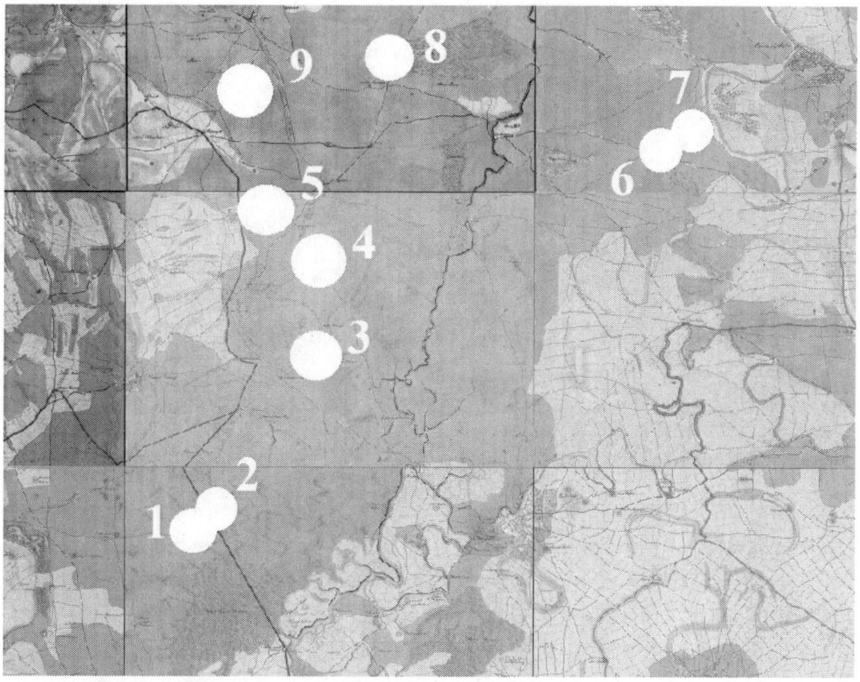

Figure 3. The area of Hortobágy on the First Austrian Military Survey Map (1782) with location of the studied boreholes referred to in Figure2.

Figure 4. The coring process on the frozen surface of the paleochannel in the Kunkápolnás marsh at Nagyiván.

Samples were prepared for elemental composition analyses using the sequential extraction method of Dániel et al. (1996) and Dániel (2004). Concentrations of selected major and trace elements were determined via flame and graphite furnace atomic absorption spectroscopy. Heavy mineral analysis of the sandy deposits in the bedrock level of the core was carried out in the Hungarian Geological Institute.

Radiocarbon dates of the sequences were obtained by both bulk and AMS (Accelerator Mass Spectrometry) analyses. Four bulk samples were analysed for radiocarbon ages at the Nuclear Research Centre of the Hungarian Academy of Sciences, Debrecen, Hungary. Additional samples of plant macrofossils were analysed using the AMS technique at the Pozńan Radiocarbon Lab, Poland. The raw dates were calibrated using the Oxcal v.3.9 software package (Bronk and Ramsey 2000), using the athmospheric data of Stuiver et al. (1998). The original dates (^{14}C) are indicated as uncal BP, while the calibrated dates are indicated as cal BC or cal BP.

The retrieved cores were also subsampled at 2-cm/4-cm intervals for pollen analysis. A volumetric sampler was used to obtain 1 cm^3 samples, which were then processed for pollen (Berglund and Ralska-Jasiewiczowa, 1986). Some pollen samples were examined using the Zólyomi–Erdtman $ZnCl_2$ procedure, the most generally applied method in Hungary (Zólyomi, 1952), as this procedure offers better results than other methods in the case of oxbow lake sediments (Magyari, 2002). A known quantity of exotic pollen was added to each sample in order to determine the concentration of identified pollen grains (Stockmarr, 1971). A minimum count of 300 grains per sample (excluding exotics) was made in order to ensure a statistically treatable sample size (Maher, 1972). Charcoal abundances were determined using the point count method (Clark, 1982). Tablets with a known *Lycopodium* spore content (supplied by Lund University, Sweden) were added to each sample to enable calculation of pollen concentrations and accumulation rates. The pollen types were identified and modified according to Moore et al. (1991), Beug (2004), Punt (1980), and Kozáková and Pokorny (2007), supplemented by examination of photographs in Reille (1992,1995, 1998) and of reference material held in the Hungarian Geological Institute, Budapest.

Percentages of terrestrial pollen taxa, excluding Cyperaceae, were calculated using the sum of all those taxa. Percentages of Cyperaceae, aquatics and pteridophyte spores were calculated relative to the main sum plus the relevant sum for each taxon or taxon group. Calculations, numerical analyses and pollen diagrams were performed using the software package Psimpoll 4.26 (Bennett, 2005). Local pollen assemblage zones (LPAZs) were defined using optimal splitting of information content (Birks and Gordon, 1985), zonation being performed using the 20 terrestrial pollen taxa that reached at least 5% in at least one sample.

For the description of macrofossils, a modified version of the QLCMA (semi-quantitative quadrat and leaf-count macrofossil analysis) technique by Barber et al. (2004) has been adopted (Jakab et al. 2004).

Results of pollen, macrobotanical and malacological analyses were used to make inferences regarding the past vegetation around the former river beds analysed, and in the wider Hortobágy region. Such inferences, however, must also consider the area from which pollen reaches the catchment site.

According to modelling and empirical studies for medium sized lacustrine basins with a diameter of ca. 100-150 m (Sugita, 1994; Soepboer et al., 2007), the correlation between pollen abundances and vegetation composition does not improve for vegetation found at distances larger than 400–600 m. The regionally ‘uniform’ background pollen component representing the vegetation between 600 m and several tens of kilometres from the lake, accounts for ca. 45% of the total pollen spectrum (Soepboer et al., 2007). Pollen data from the paleochannels of Hortobágy thus provide us an integrated palaeovegetation record for the landscape around the wetland considered and the surrounding region, with pollen from predominant extra-local and regional sources (Jacobson and Bradshaw, 1981; Prentice, 1985).

Nevertheless, it must also be acknowledged that catchment basins found on floodplains experiencing recurrent floods receive large quantities of so-called "alien" pollen in the watershed originating from distant areas, largely distorting the final pollen spectrum (Fall, 1987; Hall, 1989).

Consequently, these paleochannels are far from ideal pollen traps. The degree of "pollen contamination" is largely dependent on the depth of the floodwaters, their extent and the vegetation of the flooded area, which may bias the reconstruction of local and regional vegetations. In order to control and limit potential bias as much as possible, analysis of plant macrofossils yielding information on the vegetation perished and preserved *in situ* can be and was used in our work. This way elements of the once locally grown flora could be told apart from potential regional and extraregional elements.

Palaeovegetation can be reconstructed from pollen data using various approaches. Given our objectives, assessing the extent to which forest steppe or steppe, as opposed to closed forest, occupied in the surrounding landscape and region is the key question (Magyari et al., 2010). In our work the so-called biomization approach (Prentice et al., 1996) was adopted complemented by an indicator taxa approach to make inferences on the potential local presence of steppe (Magyari et al., 2010; Magyari, 2011).

Based on the biomization approach (Prentice et al., 1996) the steppe-indicator pollen taxa predominantly consist of herbaceous taxa typical of steppe grasslands. Although we have used their occurrence as additional evidence of the local presence of open stands, such inferences must be made with caution. Many steppe indicator taxa (herbs) are insect-or self-pollinated species and produce relatively small quantities of pollen (e.g. *Allium, Astragalus, Euphoria, Verbascum*). They are consequently under-represented in pollen spectra. Other steppe indicators are wind-pollinated and produce abundant pollen (e.g. Artemisia, Gramineae, Chenopidaceae), thus being over-represented.

According to the works of Beug (2004), Kozáková and Pokorný (2007), and Magyari et al. (2010) the following steppe indicator pollen taxa were identified in the core sequence of the analysed palaeochannels: *Ajuga, Allium,* Compositae (including *Artemisia*, Aster-type species and members of the subfamily Cichorioideae), Caryophyllaceae (including undetermined and Dianthus-type species), Chenopodiaceae (*Atriplex*, *Kochia*), *Euphorbia*, Gramineae, *Helianthemum, Inula*, Matricaria type species (including *Achillea, Anthemis, Matricaria*), *Plantago lanceolata, Plantago major/P. media, Thalictrum*, *Astragalus*, *Trifolium pratense* type species, *Trifolium repens* type species and *Verbascum*.

The final aim of our work was to give a reliable reconstruction of vegetation evolution of Hortobágy based on the investigation of local catchment basins (Sümegi et al., 2000, 2005, 2006, 2012).

There have been recent attempts to extend the pollen results of oxbow lakes located on the distant floodplain of the Tisza river (ca. 60 kms) to the area of Hortobágy as well (Magyari, 2011). These long-distance inferences are rather ambiguous partly due to the mentioned taphonomical problems on the one hand. On the other hand, geologically speaking the modern floodplain of the Tisza is much younger (15-18 kys) and characterized by a morphological and geological evolution utterly different from that of Hortobágy (Timár et al., 2005).

RESULTS AND DISCUSSION

On Alkaline Soil and Vegetation Evolution for the Terminal Part of the Ice Age and the Holocene

The 10 m long core sequence taken near Kunkápolnás (Figure 2) yields us information about the paleoenvironmental and paleovegetational changes of the study area during the past 50 kys (Figure 5). According to the results obtained, the studied paleochannel was characterized by relatively uniform and slow accumulation of clay silty material (As_2Ag_2) reflecting natural flood cycles for the past 45 kys. Signs indicating alterations in the sedimentary facies with an increase in the organic and clay content is restricted to the upper part corresponding to the past 5 kys. All this indicates the prevalence of relatively uniform sedimentological processes for the major part of the evolution of the channel, which is advantageous from a paleoecological point of view as the observed fluctuations and differences in the pollen and macrobotanical spectrum must reflect changes independent from varying geological processes (varying erosional base, selective pollen accumulation and preservation). The recorded increase in the clay and organic content in the upper part of the sequence may indicate anthropogenic disturbances in the environment of the catchment basin attributable to the appearance of food producing cultures (Willis et al., 1998), as this level seems to be coeval with the appearance of representatives of the Pit Grave Culture in the area (Sümegi and Szilágyi, 2011). Representatives of this culture are characterized by extensive animal husbandry and the construction of earth burial mounds (kurgans). One of these burial mounds is located roughly 500 meters southwest of the studied core of Kunkápolnás. But scattered mounds related to this culture were identified within a radius of ca. 1 km of other studied profiles in the area of Hortobágy as well.

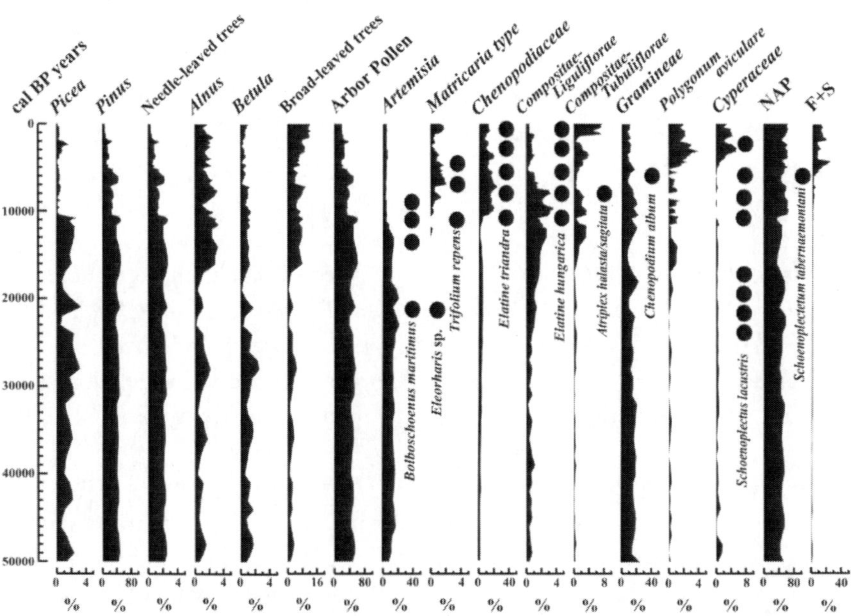

Figure 5. Pollen diagram of the Kunkápolnás core with macrobotanical remains depicted (selected taxa, minimum 500 pollen grains).

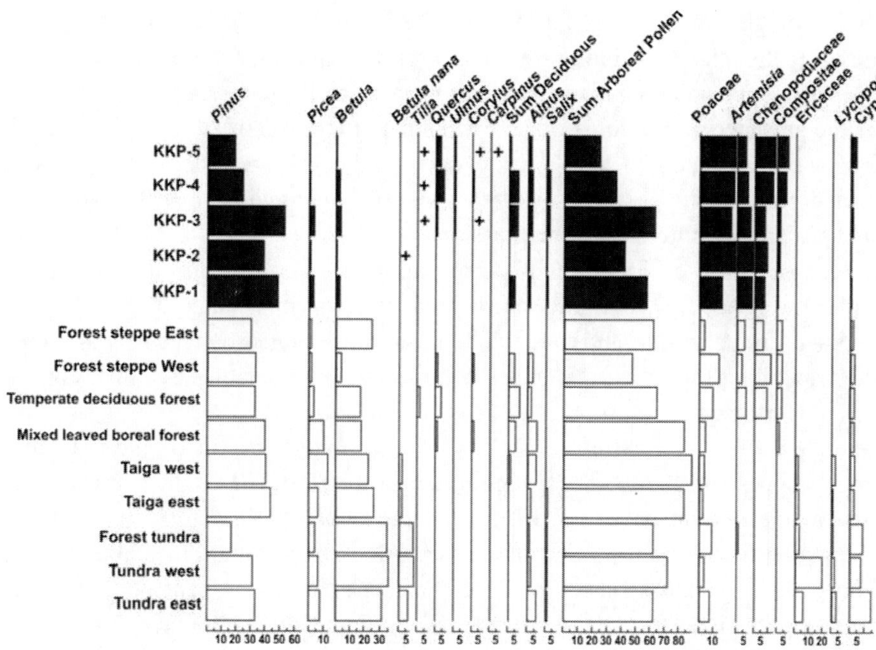

Figure 6. Zone-average pollen frequencies of selected pollen types from Kunkápolnás marsh plotted alongside mean values of major pollen types in various vegetation zones of the former Soviet Union. Surface pollen spectra are redrawn from Peterson (1983). KKP-1 = pollen spectra from Dansgaard – Oschger (Greenland Interstadial) events in the sequence of the Kunkápolnás marsh. KKP-2 = pollen spectra from Heinrich (Greenland Stadial) events in the sequence of Kunkápolnás marsh. KKP-3 = pollen spectra from late-glacial age in the sequence of Kunkápolnás marsh. KKP-4 = pollen spectra from early postglacial time (Early Holocene Age) in the sequence of Kunkápolnás marsh. KKP-5 = pollen spectra from late postglacial time (Late Holocene Age) in the sequence of Kunkápolnás marsh.

Going back to the original Kunkápolnás core sequence, the section dated between 50,000 and 12,000 calendar years is characterized by wavelike fluctuations in the dominances of arboreal and non-arboreal pollen grains probably corresponding to cyclical coolings and warm-ups recorded in other ice core, ocean core and terrestrial records (Buiron et al. 2012, Bond et al. 1992, Genty et al. 2003, Grimm et al. 1993, 2006, Grootes et al.1993, Svensson et al. 2008, Timmermann et al. 2010, Yiou et al. 1995). The periods representing cold maxima (Heinrich or Greenland Stadial, GS, events, Björck et al., 1997; Walker et al. 1999; Svensson et al. 2008) are all characterized by an increase of Artemisia and goosefoot (Chenopodiaceae), as well as grass (Gramineae) pollen grains. Conversely, the warm-up periods (Dansgaard – Oeschger or Greenland Interstadial, GIS, events) (Björck et al., 1997; Walker et al., 1999; Svensson et al., 2008) are marked by minor increases in coniferous pollen ratio (*Abies, Pinus, Picea*), as well as that of certain deciduous taxa, primarily alder and willow (*Alnus, Salix*) accompanied by decreases of Non Arboreal Pollen (NAP, Figure 6.).

Although there are marked differences in the pollen composition of the referred horizons corresponding to cyclical coolings and warmings, they seem to indicate the emergence of mixed vegetation, whose modern analogues can be found in the eastern and western parts of the Eurasian forest-steppe belt (Figure 6).

According to Magyari et al. (2010), there is no simple relationship between the proportion of pollen of woody taxa in records from medium-sized lakes and the proportion of woodland cover on the surrounding landscape. Yet both Allen (2000) and Magyari et al. (2010) come up with a classification system based on the ratio of Arboreal to Non Arboreal Pollen (AP/NAP), which could be adopted in our work as well. According to this system, one can presume the presence of closed woodland when the proportion of AP is higher than 70 %. If this value is between 50 and 70 % it indicates the presence of forest-steppe type vegetation. Between 40 and 50 % a steppe with scattered spots of woodlands can be assumed. Below 40% one can speak about pure steppes or grasslands (Tarasov et al. 1998, 2000). The ratio of AP was between 40-50 % in the horizons corresponding to coolings in our studied profile, indicating the emergence of grassland dominated forest-steppes and or grasslands with scattered woodland spots in our study area. The same ratio ranged between 50 and 60 % in the levels corresponding to the warmings referring to the emergence of classical continental forest-steppes in Hortobágy.

However, there are some differences between the pollen spectra of modern classical Eurasian forest steppes and the ice age pollen spectrum of Hortobágy. These differences are well recorded in the proportion of *Artemisia* and goosefoot (*Chenopodium*) pollen grains exceeding 5 % as well as the proportion of grasses (Graminae) with values above 10 % for the studied horizons (Figure 6.). Based on these characteristics, the pollen spectrum seems to show a strong correspondence with the steppe zones found in the southern margin of the Eurasian forest-steppe belt, namely those in northern Mongolia and the foothill region of the Altai Mountains (Tarasov et al., 1998, 2000, see also Chapters 3 and 5 in the present volume). In addition similarities with the pollen spectra of the dry intramontane basins of the Altai and Saian Mountains could be presumed (Chytrý et al., 2008).

At the same time based on the marked appearance of coniferous pollen grains (dominantly *Pinus, Picea*) along with other AP during the warming periods, a good correspondence is found with the composition of the modern Eurasian forest-steppe belt. Therefore, during the warmer periods of the last glaciation a more pronounced appearance of woody taxa, primarily pine, but also narrow-leafed deciduous elements (alder, willow, birch) may be proposed, as it can be currently found in part of the modern Eurasian forest-steppe zone.

Taxa belonging to the genus *Pinus* are collectively depicted on the diagrams, due to uncertainties in differentiating between Swiss stone pine (*Pinus cembra*) and mugo pine (*Pinus mugo*) produced by poor pollen preservation. Yet pollen grains of both taxa are present in the studied material besides those of Scots pine (*Pinus sylvestris*). Also besides the prevalence of Norway spruce (*Picea abies*), grains of Serbian spruce (*Picea omorica*) have been detetected in the studied material. Based on this and other data, this taxon presently limited to the Dinarides was a character element of the vegetation of the Great Hungarian Plain between MIS4 and MIS 2 (ca. 70-25 kya, Járainé-Komlódi, 1970; Sümegi et al. 1999; Magyari, 2002). Scattered occurrence of cold-resistant larch (*Larix*) was also recorded in the levels corresponding to cooling periods (Greenland Stadial 5-2 events).

Besides the dominant elements marking the presence of a special forest-steppe vegetation, pollen grains characteristic of tundra, woody tundra, alpine tundra and boreal high tussocks, as well as Arctic marshland meadows subordinately (*Betula nana*, *Pinus mugo*, *Thalictrum*, *Sanguisorba*, *Angelica*, *Campanula*, *Filipendula*, *Armeria maritima, Selaginella selaginoides, Pinguicula, Hippophae, Ephedra)* were also identified in the profile.

Among the NAP, sunlight-tolerant steppe elements like *Helianthemum, Taraxacum* type species also turned up. Besides the referred NAP, the dominant elements were grasses,

Artemisia and goosefoot implying the presence of a mixed cold continental steppe and tundra as well as boreal taiga vegetation elements in the area of Hortobágy during the last glaciation.

Based on the pattern obtained from the pollen spectrum of the studied profiles, it is possible to infer that the refered mosaic-like complexity of vegetation was preserved throughout the major coolings and warmings of the entire glacial period. It was only the size of individual mosaics that changed over the period alternating milder, drier and cooler, wetter phases.

Certain extreme cold-loving elements (*Larix, Pinus cembra, Betula nana, Pinus mugo, Selaginella selaginoides, Hippophae)* might have disappeared to reappear as the cimate turned colder again during the major cold maxima.

A very similar system of species-rich environmental mosaics characterized the vegetation of the major part of the Great Hungarian Plains during the last glacial cycle (Magyari, 2002, 2011; Magyari et al. 2010; Sümegi et al. 1999, 2011, 2012; Willis, 1997, 2007). Nevertheless, significant differences were identified in the pollen profiles of Hortobágy when compared to the rest of the studied profiles of the Great Hungarian Plain. Namely, certain taxa including various grass types, *Artemisia* and goosefoots turned up in larger proportions here throughout the entire ice age (Figure 6.). This marked difference was recorded in all studied profiles of Hortobágy (Figure 7.).

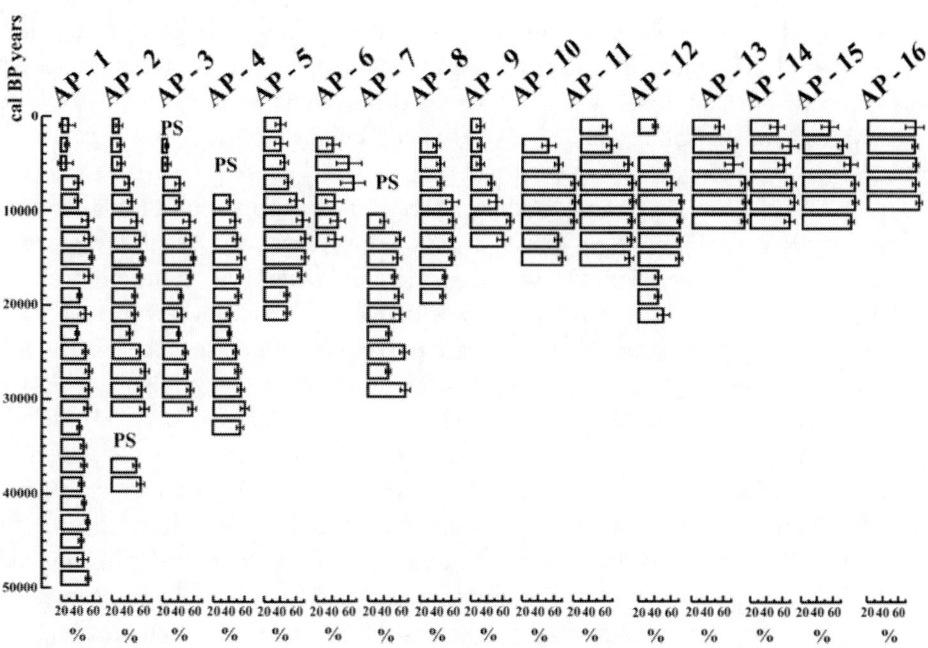

Figure 7. Abundance of pollen of woody taxa in Holocene records from Hortobágy and the Carpathian Basin. The range of values, as well as an indication of the most frequent value, is plotted for each site for 2 millennial intervals from 50,000 to 0 cal BP years. AP = Arboreal Pollen 1 = Nagyiván, Kunkápolnás site, 2 = Zám, Halasfenék (based on minimum 500 pollen grains), 3 = Máta, Pap-ere, 4 = Hortobágy, Fecskerét, 5= Lake Kolon (Sümegi et al. 2011), 6 = Ecsegfalva, Lake Kiri (Kiritó) (Willis, 2007), 7 = Kardoskút, Fehér Lake (Sümegi et al. 1999), 8. Maroslele – Pana (Sümegi et al. 2012), 9.Hajós (Jakab et al. 2004), 10./Kelemér, Kis-Mohos (Willis et al. 1997), 12./Bátorliget, protected marshland (Willis et al. 1995), 13./Csaroda, Nyíres Lake (Sümegi, 1999), 14./Tiszadob, Sarlóhát (Magyari et al. 2010), 15./Nagybárkány, Nádas Lake (Sümegi et al. 2010), 16./Sirok, Nyírjes Lake (Gardner, 1999).

Based on this information, the expansion of the mosaics of *Artemisia* grasslands and more humid continental steppes must have been more important in the area of Hortobágy and the surrounding areas of Hajdúhát and Hajdúság than in other parts of the Great Hungarian Plain. This glacial period grassland communities have probably hosted spots of alkaline habitats. Their correct identification using pollen data (Magyari, 2011), however, is quite difficult and problematic. Reliable information may come from the analysis of plant macrofossils, which represent traces of the local vegetation.

Remains of the halophilous saltmarsh bulrush (*Bolboschoenus maritimus*), together with other modern salt marsh character elements (*Eleocharis, Schoenoplectus lacustris, Juncus*) were retrieved from the horizon corresponding to the Last Glacial Maximum (LGM) between 20-24 calendar kys of the studied profile. This type of vegetation inhabits the lower lying parts of the modern Hortobágy, experiencing constant inundation during most part of the growth season, and it is characterized by extreme alkalinity forming the so-called salt marshes. All these records unambiguously indicate the emergence of salt marsh components in a vegetation mosaic of continental steppes, boreal floodplain vegetation and birch-pine taiga patches during a cold phase of the terminal part of the last glaciation. While pollen data only gave us a hint on the possible presence of alkaline habitats during the LGM, macrofossils provided univocal evidence (Figure 5).

Marked changes can be observed both in the pollen spectrum and plant macrofossil composition of the horizon dated to the early postglacial period ca. 13-10,000 calendar years. These changes clearly indicate the emergence of salt steppes, that is, salt marshes within a mosaic of grasslands and marshes in the referred period. According to the available records these early postglacial alkaline habitats have been continuously present for the past ca. 13 kys (Figure 5).

During 13 kys there has been a gradual increase in the proportion of coniferous pollen grains, mainly *Pinus* and *Picea*, up to 10.4 kys before present. As a result the proportion of AP reached a value between 55 and 63 %, characteristic of classical Eurasian forest steppes, but still below the limit of 70% marking the transition to woodlands (Magyari et al., 2010).

The proportion of *Artemisia* and Graminae remained high among the NAP during the Early Holocene, when a gradual increase of other Compositae, goosefoots and, *Polygonum aviculare* taxa also took place (Figure 5). Parallel with these changes, *Matricaria* type pollen grains (including *Achillea, Anthemis, Matricaria*) also turned up in this part of the Kunkápolnás profile. These periodic fluctuations seem to show a good correspondence with the climatic fluctuations characteristic of the early postglacial period. During the transitionary phase of the early postglacial (Pleistocene/Holocene transition) a marked transformation can be identified in the composition of the forest steppe vegetation of Hortobágy, characterized by a rapid increase in richness values but preserving the dominance of non-arboreal elements inhabiting dry steppes as well alkaline steppes. Macrobotanical remains of the halophilous saltmarsh bulrush (*Bolboschoenus maritimus*) turned up here again in the profile parallel with the previously mentioned changes (Figure 5). Furthermore, several character species of the modern saltmarshes (*Elatine triandra*, *Elatine hungarica*, *Schoenoplectus lacustris, Juncus*) are likewise present in the early post-glacial part of the Kunkápolnás. All these data univocally indicate the emergence of alkaline habitats with the closure of the ice age in Hortobágy.

Besides the more humid, alkaline habitats, drier loessy and alkaline grasslands were equally present as marked by the increases of *Matricaria* type and other Compositae pollen grains as well as the appearance of a marker taxon, *Trifolium repens*, characteristic of these drier grasslands in the macrofossil record. As it has been stated earlier, favorable climatic changes during the early postglacial period, mainly gradual increases of the temperature must have fostered these vegetational transformations (Willis et al., 1995, 1997) reflected in the increase of AP. Nevertheless, the appearance of new taxa among the NAP, and a higher percentage of *Artemisia* type pollen grains may indicate the lengthening of arid periods during the growth season.

The iterative droughts may be attributed to the gradual increases in continentality of the summer periods. Nevertheless, a major hydrological transformation related to the birth of the modern Tisza river valley at the margin of the study area before the early post-glacial period might have also contributed to these changes (Sümegi et al. 2000, Timár et al. 2005). Before the configuration of the modern Tisza valley, located at a lower elevation than Hortobágy, watercourses originating from the adjacent watershed area of the Carpathian Mountains could freely transport waters to Hortobágy. Now that these watercourses were isolated from Hortobágy and were forced to charge into the Tisza, watersupply was strictly dependant on the seasonal flooding of the Tisza.

Seasonal floods in the watershed of the Tisza River are controlled by the amount of rainfall reaching the area, as well as the timing of the spring snowmelt. So the first part of the growth season, from early spring to early summer, is characterized by frequent floods and more extensive inundation in the adjacent areas of the floodplain. The second half is a relatively floodfree drier period lasting from early summer to early fall. All in all besides a general increase in the temperature, the appearance of these cyclical floods and droughts might have had a significant impact on the evolution of the early postglacial vegetation of Hortobágy.

During the early Holocene, vegetation evolution seemed to follow the path starting in the early post-glacial, characterized by a significant reduction of pines parallel with the expansion of deciduous arboreal elements (*Quercus, Tilia, Ulmus*). As the sum of AP was below 50 % we may postulate the emergence of a grassland dominated forest steppe in Hortobágy during this period. The observed shift in the pollen spectrum was so unusual that results were cross-checked by several independent palynologists who eventually arrived at the same conclusion (Zsófia Medzihradszky, Elvira Bodor, pers. comm.). Namely, there was no significant increase in the proportion of AP during the early postglacial and Early Holocene in the area of Hortobágy, based on the patterns observed in Kunkápolnás borehole profile.

Although there is a slight decrease in the proportion of *Artemisia* pollen grains in our studied borehole for the Early Holocene compared to the Pleistocene values, the referred taxon was present in higher abundances in Hortobágy than at other studied pollen profiles of the Carpathian Basin (Gardner, 1999; Jakab et al., 2004; Magyari et al., 1999, 2010; Sümegi, 1999; Sümegi et al., 1999, 2010, 2011, 2012; Willis, 2007; Willis et al., 1995, 1997). Parallelly there was an increase in the ratio of Chenopodiaceae, *Matricaria* type taxa and other Compositae and Gramineae, in the referred horizon. This pollen composition indicates the presence of wet and dry meadows, alkaline steppes, and seasonally dessicating salt marshes in the vicinity of the studied Kunkápolnás borehole during the Early Holocene. These interpretations are further corroborated by results of plant macrofossil analyses indicating the presence of saltmarsh elements (*Bolboschoenus maritimus, Elatine triandra, Elatine hungarica, Schoenoplectus lacustris, Juncus*), as well as character species of drier loess and alkaline steppes such as *Trifolium repens* and *Atriplex hastata*.

The amount of arboreal vegetation must have been minimal, with a dominance of so-called edaphic grassland patches within the mosaic of salt marshes, wet and dry meadows and salt steppes (Figure 6). The inferred Early Holocene vegetation of Hortobágy was in sharp contrast with the reconstructed coeval vegetation of the foothill and mid-mountain regions, the valleys of watercourses enjoying better water supply, or the forested sand dunes of the Great Hungarian Plain (Willis et al. 1995, 1997, Magyari et al. 2011) (Figure7). Similar conditions were traced only regionally in such areas of the Great Hungarian Plain, where the unique geology and geomorphology of the landscape favored the more extensive emergence of loessy grasslands and alkaline habitats (Jakab et al. 2004, Sümegi et al. 2011, 2012, Willis, 2007).

The vegetation picture for the past 8000 years in the vicinity of the studied borehole remained similar to the Early Holocene, composed of adjacent mosaics of alkaline marhses, dry and wet meadows, woodland patches, salt steppes, and grass-dominated wetlands. Plant macrofossils characteristic of alkaline marshes (softstem bulrush *Schoenoplectus tabernaemontani*), as well as drier grasslands all turned up in the uppermost part of the studied core sequence dated to the past 8000 years.

The only difference in the Holocene is recorded in an increase of weed pollen grains marking human impact attributed to increased pressure on the landscape by grazing, crop farming, or treading. The rapid increase in weed pollen grains representing such taxa as *Polygonum aviculare, Plantago lanceolata, Plantago major/media, Rumex, or Verbascum* was observed in the horizon coeval with the Pit Grave Culture (Figure 5). The intensification of human influences on the landscape is recorded in the accelerated paludification of the studied paleochannel, turning the original oxbow lake system into a marshland ca. 5000 years ago. These wetland conditions have been preserved since then.

CONCLUSION

The present chapter was aimed at describing the historical evolution of loessy, alkali grassland-wetland complexes found in the area of Hortobágy in NE Hungary. These Pannonian vegetation complexes are of outstanding natural and historical-cultural value and have been recognised as a UNESCO World Heritage site since 1999. It must be noted though that protection was granted under the category of cultural landscape. In 1973 the first national park in Hungary was established here. Initial thoughts regarding the evolution of this unique flora were postulated on the basis of observations made in the modern landscape. These were later complemented by paleoecological (mainly pollen) data extrapolated from other distant sites to the referred study area, which again questions the reliability of the interpretations. In contrast, another school starting ca. 25 years ago focused on the on-site sampling and multi-proxy paleoecological analysis of Quaternary deposits to find an answer to the problem (Sümegi, 1989, 2005; Sümegi and Szilágyi, 2011; Sümegi et al. 2000). The paleobotanical investigation of the Kunkápolnás paleochannel borehole was a central part of the campaign, whose results dating back to the past 50 kys were discussed in the present chapter. According to the findings of these extensive studies, the grassland dominated character of Hortobágy has not been fundamentally transformed by human communities for the past 5000 years. Thus, the evolution of the modern alkaline grassland-wetland, mosaic-like, habitat complexes dates back several millennia. During the last glacial cycle between MIS4 and MIS1 (ca. 70 kys), the area of Hortobágy formed an integral part of the vegetation mosaic belt composed of tundra, continental grassland, taiga and alkaline elements, which stretched from the Atlantic in the west all the way to Siberia and Eastern China to the east. It is not suprinsing that ice age vegetation was composed of a continental grassland-dominated forest steppe in the area of Hortobágy, which also included patches of taiga and tundra elements. This complex system of adjacent vegetation mosaics seemed to have been preserved throughout the entire period, regardless of climatic fluctuations recorded in the form of coolings (Heinrich or GS evenst), and warmings (Dansgaard –Oschger or GIS events). It was only the size of the individual mosaics that changed over time, alternating milder, drier and cooler, or wetter climatic periods. Certain extreme cold-loving elements might have disappeared to later reappear as the climate turned colder again during the major cold maxima.

A very similar system of species-rich environmental mosaics characterized the vegetation of the major part of the Great Hungarian Plain during the last glacial cycle. Nevertheless, based on the higher percentage of non-arboreal elements and *Artemisia*, the referred forest steppe vegetation mosaics of Hortobágy seem to show good correspondence with the modern forest steppes of Central Asia and Southern Siberia, as well as the intramontane dry basins of these regions.

Alkaline marshes and alkaline plants turned up in the area of Hortobágy even during the last glacial maximum. Yet, the continuous presence of these alkaline habitats could have been attested only from the early postglacial period. According to our findings, the geological and geomorphological changes connected to the emergence of the modern course of the Tisza valley north of Hortobágy resulted in a transformation of the hydrology and seasonally fluctuating water supply to the studied landscape. Along with a gradual global increase of temperatures, it might have triggered the emergence of alkaline marshes, steppes, dry loess steppes and mixed woodlands in Hortobágy during the Pleistocene/Holocene transition. The Holocene in characterized by an advent of deciduous tree taxa in general. Yet, the unique endowments of the landscape favored the preservation of forest steppe-like bioms characterized by an NAP value above 50%. Arboreal elements only periodically turned up along the minor watercourses harboring gallery forests.

So the biome structure of the ice age boreal forest steppe was fundamentally preserved. It was only the specific composing elements that changed due to the arrival of submediterreneam, pontian, subcontinental arboreal elements turning the flora into a temperate oak forest steppe during the Holocene. The development of this unique vegetation mosaic, extremely rich in species, was controlled not only by the climate system of the Carpathian Basin, but also by regional and local edaphic factors (orography, bedrock geology, hydrology, soils). The mainly open forest-steppe landscape with scattered spots of alkaline habitats dating to the Pleistocene/Holocene transition was preserved in the younger periods of the Holocene as well. Human-influences, marked by the first appearance of the representatives of the Late Copper Age and Early Bronze Age Pit Grave Culture generally resulted in an accelerated erosion of orographic highs and soils leading to the rapid infilling of the catchment areas of depressions, marshes and former paleochannels. But the productive cultures engaged in extensive animal farming could not alter the basic vegetation and landscape structure, as no changes were introduced in the unique hydrology of the area.

Dramatic changes occurred only roughly 160 years ago, when the expansion of crop farming and flood control, as well as drainage measures, fundamentally altered the hydrology of the area. This process, which was a major threat to natural vegetation, including the mosaics of alkaline habitats dating back ca 12 kys, was enhanced as a result of forced collectivization in the agriculture during the communist era and the introduction of alien species in crop farming (rice or *Taraxacum glaucanthum*). Fortunately, the foundation of the national park 40 years ago gave a halt to these negative processes along with the numerous measures aimed to restore and preserve the natural vegetation mosaics as part of multiple correlative campaigns. These and other measures in the future will hopefully enable the restoration of this unique vegetation stucture composed of mosaic-like microhabitats, along with the harbored fauna, for the future generations.

REFERENCES

Allen, J.R.M., Watts, W.A.,Huntley, B. (2000). Weichselian palynostratigraphy, palaeovegetation and palaeoenvironment: the record from Lago Grande di Monticchio, southern Italy. *Quaternary International* 73-74, 91–110.

Arany, S. (1934). A hortobágyi szikes talajok. (Alkaline soils of the Horotbágy. In: Sajó, E., Trummer, A. (eds.), *A magyar szikesek. (Hungarian alkali landscapes),* pp. 98-108. Patria Nyomda, Budapest.

Arany, S. (1956). *A szikes talaj és javítása. (Melioration of alkali soils).* Mezőgazdasági Kiadó, Budapest.

Bacsó, N. (1959). *Magyarország éghajlata. (The climate of Hungary).* Akadémiai Kiadó, Budapest.

Balogh, J. (1840). *A magyarországi szikes vidékek természettudományi tekintetben. (Hungarian alkaline sites from a natural scientific perspective).* Magyar Tudományos Társaság, Buda.

Barber, K. E., Chambers, F. M., Maddy, D., Brew, J. (1994). A sensitive high resolution record of the Holocene climatic change from a raised bog in northern England. *The Holocene* 4, 198-205.

Bennett, K.D. (2005). *Psimpoll manual.* Available at: http://www.kv.geo.uu.se/psimpoll.html (accessed 12 February 2008).

Bernátsky, J. (1905). A magyar Alföld szíklakó növényzetéről. (Saline vegetation of the Great Hungarian Plain) *Annales Musei Nationalis Hungarici* 3, 121-174.

Berglund, B.E., Ralska-Jasiewiczowa, M. (1986). Pollen analysis and pollen diagrams. In: Berglund, B. E. (eds.). *Handbook of Holocene Palaeoecology and Palaeohydrology.* J. Wiley and Sons Ltd., Chichester-Toronto.

Beug, H.J. (2004). *Leitfaden der Pollenbestimmung für Mitteleuropa und angrenzende Gebiete. (Textbook of pollen identification for Central Europe and adjacent regions).* Pfeil, München.

Björck, S., Walker, M.J.C., Cwynar, L.C., Johnsen, S., Knudsen, K.L., Lowe, J.J., Wohlfarth, B., and INTIMATE Members. (1998). An event stratigraphy for the Last Termination in the North Atlantic region based on the Greenland ice-core record: A proposal by the INTIMATE group. *Journal of Quaternary Science* 13, 282–292.

Birks, H.J.B., Gordon, A.D. (1985). *Numerical methods in Quaternary pollen analysis.* Academic Press, London.

Borbás, V. (1886). *A magyar homokpuszták növényvilága, meg a homokkötés. (Flora of Hungarian sandy steppes).* Personal publication, Budapest.

Bond G.C., Heinrich, H., Broecker, W., Labeyrie, L., McManus, J., Andrews, J., Huon, S., Jantschik, R., Clasen, S., Simet, C., Tedesco, K., Klas, M., Bonani, G., Ivy, S. (1992). Evidence for massive discharges of icebergs into the North Atlantic ocean during the last glacial. *Nature* 360, 245-249.

Borhidi, A. (2007). *Magyarország növénytársulásai.* (*Plant communities of Hungary*) Akadémiai Kiadó, Budapest.

Boros, E., Bíró, Cs. (1999). A Duna-Tisza-közi szikes tavak ökológiai állapotváltozásai a XVIII-XX. századokban. (Ecological changes in natron lakes of the Danube-Tisza Interfluve between the 18th and 20th centuries) *Acta Biologica Debrecina Oecologica Hungarica* 9, 81–105.

Bronk Ramsey, C. (2001). Development of the Radiocarbon Program OxCal, *Radiocarbon* 43, 355-363.

Buiron, D., Stenni, B., Chappellaz, J., Landais, A., Baumgartner, M., Bonazza, M., Capronc, E., Frezzotti, M., Kageyama, M., Lemieux-Dudon, B, Masson-Delmotte, B., Parrenin, F., Schilt, A., Selmo, E., Severi, M., Swingedouwc, D., Udisti R. (2012). Regional imprints of millennial variability during the MIS 3 period around Antarctica. *Quaternary Science Reviews* 48, 99-112.

Cox, P.A., (1991). *The Elements – Their origin, abundance and distribution.* Oxford University Press, Oxford.

Clark, R.L. (1982). Point count estimation of charcoal in pollen preparations and thin sections of sediments. *Pollen et Spores* 24, 523-535.

Chytrý, M., Danihelka, J., Kubešová, S., Lustyk, P., Ermakov, N., Hájek, M., Hájková, P., Kočí, M., Otýpková, Z., Roleček, J., Řezníčková, M., Šmarda, P., Valachovič, M., Popov, D., Pišút, I. (2008). Diversity of forest vegetation across a strong gradient of climatic continentality: Western Sayan Mountains, southern Siberia. *Plant Ecology* 196, 61-83.

Dapsy, L. (1869). A Tiszaszabályozás befolyása a magyar talajra. (The effects of river control measures of the Tisza River on Hungarian soils) *Természettudományi Közlöny* 1, 97-108.

Dániel, P., (2004). Methods of the five-step extraction-digestion method. In: Sümegi, P., Gulyás, S. (eds.), *The geohistory of Bátorliget Marshland*, pp. 53-56. Archaeolingua Press, Budapest.

Deák, J. (1974). Izotóptechnikai módszerek alkalmazása a Tiszántúlon, a talaj-, a mélységi- és a felszíni víz kapcsolatának vizsgálatára. *Beszámoló a VITUKI 1972. évi munkájáról* 3, 317-334.

Deák, J. (1975). Use of environmental isotopes to investigate the connection between surface and subsurface waters in the Nagykunság region, Hungary. *Isotope Techniques in Groundwater Hydrology*, IAEA, Vienna, 157-167.

Deák, J., Fórizs, I., Dese, É., Hertelendi, E. (1992). The origin of groundwater and dissolved ammonium in SE-Hungary by environmental isotope data. *Tracers in Hydrology* 215, 117-124.

Endrédy, E. (1941). A szikes talajok képződésnek kérdéséről. (On the origin of alkali soils) *Öntözésügyi Közlemények* 3, 207 - 217.

Great Hungarian Plain, I. (2003). Isotopes as natural tracers in the water cycle: Examples from the Carpathian Basin. Studia Universitatis Babeş-Bolyai, *Physica Special Issue* 48, 69-77.

Fórizs, I., Tóth, T., Kuti, L. (2006). A szikesedés izotóp-hidrogeokémiai vizsgálata a hortobágyi Nyírő-laposon. (Isotopegeochemical studies on alkalinization on the Nyírő Flats of Hortobágy). In: Szendrei, G. (ed), *Magyarország környezetgeokémiai állapota. MTA Földrajztudományi Osztály*, pp. 83-90. Környezetgeokémiai Albizottság. Budapest.

Földvári, A. (1956). "Hidroaerolit" kőzetek a magyarországi negyedkor lerakódásaiban. ("Hydroaerolite" Rocks in the Quaternary Deposits of Hungary) *Földtani Közlöny* 86, 357-360.

Földvári, A. (1958). "Hydroaerolite" Rocks in the Quaternary Deposits of Hungary. *Acta Geologica Hungarica* 5, 287-292.

Galgóczy, K. (1876-1877). *Pest, Pilis és Solt törvényesen egyesült vármegye monográfiája.* I -III. Országos Monográfia Társaság, Budapest.

Gallet, S., Jahn, B., Van Vliet Lanoë, B., Dia, A, Rosselo, E. (1998). Loess geochemistry and its implications for particle origin and composition of the upper continental crust. *Earth and Planetary Science Letters* 156, 157-172.

Gardner, A.R. (1999). The impact of Neolithic agriculture on the environments of south-east Europe. *Ph.D. Thesis,* University of Cambridge.Cambridge

Genty, D., Blamart, D., Ouahdi, R., Gilmour, M., Baker, A., Jouzel, J., Van Exter, S. (2003). Precise dating of Dansgaard-Oeschger climate oscillations in Western Europe from stalagmite data. *Nature* 421, 833-837.

Grimm, E.C., Jacobson, G.L., Watts, W.A, Hansen, B.C.S., Maasch, K.A. (1993). A 50,000-Year Record of Climate Oscillations from Florida and Its Temporal Correlation with the Heinrich Events. *Science* 261, 198-200.

Grimm, E.C., Watts, W.A., Jacobson, G.L., Hansen, B.C.S., Almquist, H.R., Dieffenbacher-Krall, A.C. (2006). Evidence for warm wet Heinrich events in Florida. *Quaternary Science Reviews* 25, 2197–2211.

Grootes, P.M., Stuiver, M., White, J.W.C., Johnsen, S.J., Jouzel, J. (1993). Comparison of oxygen isotope records from the GISP2 and GRIP Greenland ice cores. *Nature* 366, 552-554.

Inkey, B. (1895). Jelentés az 1895. évben Csongrád és Csanád megyékben végzett földtani felvételekről. (Report on geological surveys implemented in Csongrád and Csanád counties in 1895) *Magyar Királyi Földtani Intézet Jelentése 1895-ről*, 100-108.

Jacobson, G.L., Bradshaw, R.H.W. (1981). The selection of sites for palaeovegetational studies. *Quaternary Research* 16, 80–96.

Jakab, G., Sümegi, P. (2004). A lágyszárú növények tőzegben található maradványainak határozója mikroszkópikus bélyegek alapján. *Kitaibelia* 9, 93-129.

Jakab, G., Sümegi, P., Magyari, E. (2004). New Quantative Method for the Paleobotanical Description of Late Quaternary Organic Sediments (Mire-Development Pathway and Paleoclimatic Records from Southern Hungary*). Acta Geologica Hungarica* 47, 373-409.

Járainé-Komlódi, M. (1970). Studies on the vegetational history of *Picea omorica* Panc. on the Great Hungarian Plain. *Annales Universitatis Scientiarum Budapestinensis* de *Rolando Eötvös Nominatae, Sectio Biologica* 12, 143 – 156.

Karuczka, A. (1999). Időjárási viszonyok hatása a szikes talaj sómérlegére. (Climate influencing the salt budget of soils). *Agrokémia és Talajtan* 48, 459-467.

Kerényi, A. (2003). *Környezettan - Természet és társadalom - globális szempontból. (Environmental science for natural and social scientists- a global perspective).* Mezőgazda Kiadó, Budapest.

Kerner, A. (1863). A Duna-menti országok növényvilága. A magyar Alföld és a Bihar-hegység. In: Oroszi S. (ed.), 2004: *Erdészettörténeti Közlemények, 62.* pp. 63-65. Fordította: Madas, L., Budapest.

Kovács, F., Szatmári, J., Rakonczai, J. (2006). Assessment of the special soil degradation (bench erosion) with GIS methods from the Great Hungarian Plain. In: Suárez, J., Márkus, B. (eds.) *Shaping the future of Geographic Information Science in Europe*, pp. 29-35. Proceedings of 9th AGILE Conference on Geographic Information Science.

Kozáková, R., Pokorný, P. (2007). Dynamics of the biotopes at the edge of a medieval town: pollen analysis of Vltava river sediments in Prague, Czech Republic. *Preslia* 79, 259-281.

Kreybig, L. (1944). *Magyar tájak tájismereti és terméléstechnikai leírása, 1.* Magyar Állami Földtani Intézet kiadványa, Budapest.

Kvassay, J. (1876). Über die Natron und Székboden im Ungarischen Tieflande. *Jahrbuch der Kaiserlich und Königlich Geologischen Reichanstalt* 16, 427-446.

Nyilas, I., Sümegi, P. (1991). *The Mollusc fauna of Hortobágy at the end of the Pleistocene (Würm3) and in the Holocene.* Proceeding of 10th International Malacological Congress, Tübingen, pp. 481-486.

Lóczy, L. (1910). Magyarország felsőpleisztocén és holocén korszakának klímájáról. *Magyar Királyi Földtani Intézet Népszerű kiadványai* 2, 69 - 76.

Maher, L.J. Jr (1972). Nomograms for computing 0.95 confidence limits of pollen data. *Review of Palaeobotany and Palynology* 13, 85-93.

Mados, L. (1943). A szikesedés és a víz. *Hidrológiai Közlöny* 23, 3-21.

Mádlné Szőnyi, J., Simon, Sz., Tóth, J., Pogácsás, Gy. (2005). Felszíni és felszín alatti vizek kapcsolata a Duna-Tisza közi Kelemenszék és Kolon-tó esetében. *Általános Földtani Szemle* 30, 93-110.

Magyari, E.K. (2002). *Climatic versus human modification of the Late Quaternary vegetation in Eastern Hungary.* PhD Thesis, Debrecen.

Magyari, E.K. (2011). Late quaternary vegetation history in the Hortobágy steppe and Middle Tisza floodplain, NE Hungary. *Studia Botanica Hungarica* **42**, 185-203.

Magyari, E., Chapman, J.C., Passmore, D. G., Allen, J. R. M., Huntley, J.P., Huntley, B. (2010). Holocene persistence of wooded steppe in the Great Hungarian Plain. *Journal of Biogeography* 37, 915-935.

Miháltz, I. (1947). *A Duna-Tisza csatorna geológiai viszonyainak tanulmányozása.* "A Duna-Tisza csatorna" Egyetemi Nyomda, Budapest.

Molnár, B. (1960). Pliocén és pleisztocén lehordási területek az Alföldön. *Földtani Közlöny* 89, 403-413.

Molnár, B. (1964). A magyarországi folyók homoküledékeinek nehézásvány-összetétel vizsgálata. *Hidrológiai Közlöny* 44, 347-355.

Molnár, B. (1965). Changes in Area and Directions of Stream Erosion in the Eastern Part of the Hungarian Basin (Great Plain) during the Pliocene and Pleistocene. *Acta Minerologica et Petrographica* 17, 39-52.

Molnár, B. (1967). A Dél-Alföld feltöltődésének ritmusai és vízföldtani jelentőségük. *Hidrológiai Közlöny* 47, 537-552.

Molnár, B. (1973). Az Alföld harmadidőszak-végi és negyedkori feltöltődési ciklusai. *Földtani Közlöny* 103, 294-310.

Molnár, B., Szónoky, M. (1976). On the origin and geohistorical evolution of Natron-lakes of the Bugac Region. *Móra Ferenc Múzeum Évkönyve*, 1975-76/1, 257-270.

Molnár, V. A. (2007). *Kitaibel Pál élete és öröksége*. Kitaibel Kiadó, Biatorbágy.

Molnár, Zs. (1996). A Pitvarosi-puszták és környékük vegetáció- és tájtörténete a Középkortól napjainkig. *Natura Bekesiensis* 2, 65-97.

Molnár, Zs. (1997). Vegetation history of the Kardoskút area (S.E. Hungary) II: The lake Fehér – tó int he last 200 years. *Tiscia* 30, 27-34.

Molnár Zs. (1999). Ősi és másodlagos (szikes) puszták a Tiszántúlon. In: Füleky Gy. (ed.), *A táj változásai a Kárpát-medencében*, pp. 231-233. Szent István Egyetem Kiadványa, Gödöllő.

Molnár, Zs. (2003). Tájtörténeti adatok a hazai szikesek növényzetének ismeretéhez. In: Tóth, A. (ed.), *Ohattól Farkas-szigetig. Ökológiai kultúra – ökológiai nevelés*, pp. 71-95. Alföldkutatásért Alapítvány, Budapest – Kisújszállás.

Molnár, Zs. (2007). *Történeti tájökológiai kutatások az Alföldön*. PhD thesis. University of Pécs, Pécs.

Molnár, Zs. (2008). A Duna – Tisza köze és a Tiszántúl fontosabb vegetáció típusainak holocén kori története: irodalmi értékelés egy vegetációkutató szemszögéből. *Kanitzia* 16, 93-118.

Molnár, Zs. (2009). *A Hortobágy hazánk egyik legősibb növényzetű tája*. Válogatás az MTA Ökológiai és Botanikai Kutatóintézet Kutatási eredményeiből, ÖBKI Műhelyfüzetek, 2. Vácrátót, pp. 143-148.

Molnár, Zs., Bíró, M. (1997). Vegetation history of the Kardoskút area (SE Hungary) I: History of the steppes fro the Middle Ages to the present. *Tiscia* 30, 15-25.

Molnár, Zs., Borhidi, A. (2003). Continental alkali vegetation in Hungary: syntaxonomy, landscape history, vegetation dynamics, and conservation. *Phytocoenologia* 21, 235-245.

Moore, P.D., Webb, J.A., Collinson, M.E., (1991). *Pollen Analysis*. Blackwell Scientific Publications, Oxford.

Muraközy, K. (1902). A talajról. *Természettudományi Közlöny* 34, 593-667.

Nyilas, I., Sümegi, P. (1991). *The Mollusc fauna of Hortobágy at the end of the Pleistocene (Würm3) and in the Holocene*. Proceeding of 10th International Malacological Congress, Tübingen, pp. 481-486.

Pálfai, I. (1989). Az Alföld aszályossága. *Alföldi Tanulmányok* 13, 7-25.

Pálfai, I. (1993). Talajvízsüllyedés a Duna-Tisza közén. *Vízügyi Közlemények* 75, 431-434.

Pálfai, I. (1994). *A Duna-Tisza közi Hátság vízgazdálkodási problémái*. A Nagyalföld Alapítvány kötetei 3. Békéscsaba.

Pálfai, I., Boga, T.L., Sebesvári, I. (2003). *Vízháztartási változások a Duna-Tisza közi Homokhátságon*. MHT XXI: vándorgyűlésén elhangzott előadások, Szolnok http://www.vizugy.hu.

Prentice, I.C. 1985. Pollen representation, source area, and basin size: toward an unified theory of pollen analysis. *Quaternary Research* 23, 76–86.

Prentice, I.C., Guiot, J., Huntley, B., Jolly, D., Cheddadi, R. (1996). Reconstructing biomes from palaeoecological data: a general method and its application to European pollen data at 0 and 6 ka. *Climate Dynamics* 12, 185–194.

Rakonczai, J. (2006). Klímaváltozás – aridifikció - változó tájak. In: Kiss, A., Mezősi, G., Sümeghy, Z. (eds.), *Táj, környezet és társadalom. Ünnepi tanulmányok Keveiné Bárány Ilona professzor asszony tiszteletére,* pp. 593-601. Szeged Tudományegyetem, Szeged.

Rakonczai, J., Kovács, F. (2006). A padkás erózió folyamata és mérése az Alföldön. *Agrokémia és Talajtan* 55, 329-346.

Rapaics, R. (1916). A Hortobágy növényföldrajza. *Gazdasági Lapok* 68, 88-89; 102-103; 115-116; 124-126.

Reille, M., (1992). *Pollen et Spores d'Europe et d'Afrique du Nord.* Laboratoirede Botanique Historique et Palynologie, Marseille.

Reille, M., (1995). *Pollen et Spores d'Europe et d'Afrique du Nord.* Supplement 1. Laboratoirede Botanique Historique et Palynologie, Marseille.

Reille, M., (1998). *Pollen et Spores d'Europe et d'Afrique du Nord.* Supplement 2. Laboratoirede Botanique Historique et Palynologie, Marseille.

Réthly, A. (1933). Kísérlet Magyarország klímatérképének szerkesztésére a Köppen-féle klímabeosztás értelmében. *Időjárás* 37, 105-115.

Scherf, E. (1935). Alföldünk pleisztocén és holocén rétegeinek geológiai és morfológiai viszonyai és ezeknek összefüggése a talajkutatással, különösen a sziktalajképződéssel. *Magyar Királyi Földtani Intézet Jelentése 1925-1928 évekről*, 265-273.

Scherf, E. (1946). Szénhidrogének és sósvizek felkutatásának lehetősége a Duna - Tisza – közében. *Jelentés a kincstári sókutatás 1946. évi munkálatairól*, *Pénzügyminisztérium kiadványa, Budapest.*

Scherf, E. (1948). A szabolcsmegyei sósvizek (Tiszagyulaháza stb.) geológiai, hidrológiai és chemiai viszonyai (Adatok az Alföld töréses szerkezetének geomechanikai értelmezéséhez). *Jelentés a jövedéki mélykutatás 1947-1948. évi munkálatairól, Pénzügyminisztérium* kiadványa, Budapest.

Sigmond, E. (1906). Alföldünk szikeseinek válfajairól. *Földtani Közlöny* 36, 389-403.

Sigmond, E (1923). A hidrológiai viszonyok szerepe a szikesek képződésében. *Hidrológiai Közlöny* 3, 5-9.

Sigmond, E. (1934). A Magyar Alföld szikeseinek jellemzése. In: Sajó, E., Trummer, A. (eds.), *A magyar szikesek*, pp. 1-20. Patria Press, Budapest.

Soepboer, W., Sugita, S., Lotter, A.F., van Leeuwen, J.F.N., van der Knaap, W.O. (2007). Pollen productivity estimates for quantitative reconstruction of vegetation cover on the Swiss Plateau. *The Holocene* 17, 65–77.

Stegena, L., Szebényi, L. (1948). Tiszagyulaháza és környékének sós talajvizei. *Jelentés a jövedéki mélykutatás 1947-1948. évi munkálatairól*, Pénzügyminisztérium kiadványa, Budapest.

Stefanovits, P. (1956). *Magyarország talajai.* Akadémiai Kiadó, Budapest.

Stockmarr, J. (1971). Tablets with spores used in absolute pollen analysis. *Pollen et Spores* 13, 614-621.

Strömpl, G. (1931). A szik geomorfológiája. *Földrajzi Közlemények* 4-5, 62-74.

Stute, M., Deák, J. (1989). Environmental isotope study (14C, 13C, 18O, D, noble gases) on deep groundwater circulation systems in Hungary with reference to paleoclimate. *Radiocarbon* 31, 902-918.

Stute, M., Sonntag, C., Deák, J., Schlosser, P. (1992). "Helium in Deep Circulating Groundwater in the Great Hungarian Plain: Flow Dynamics and Crustal and Mantle He Fluxes" *Geochemica et Cosmochemica Acta* 56, 2051-2067.

Stuiver M., Reimer, P.J., Braziunas, T.F. (1998). High-precision radiocarbon age calibration for terrestrial and marine samples. *Radiocarbon* 40, 1127-1151.

Sugita, S. (1994). Pollen representation of vegetation in Quaternary sediments: theory and method in patchy vegetation. *Journal of Ecology* 82, 881–897.

Sümeghy, J. (1937). A Nagykunság felszíni képződményei. *Magyar Királyi Földtani Intézet Jelentése 1929-1932-ről*, 409-436.

Sümeghy, J. (1944). *Tiszántúl.* Magyar Tájak földtani leírása, 6. Budapest.

Sümegi, P. (1989). *Hajdúság felső-pleisztocén fejlődéstörténete finomrétegtani (üledékföldtani, őslénytani, geokémiai) vizsgálatok alapján.* University doctoral thesis, p. 96. Debrecen.

Sümegi, P. (1997). *The Process of Sodification on Hortobágy in Space and Time according to geopedological investigation.* Proceeding and Field Trip of Hydro-Petro-Geology and Hungary, Budapest. pp. 237-242.

Sümegi, P. (2001). *A negyedidőszak földtanának és őskörnyezettanának alapjai.* JATEPress, Szeged.

Sümegi, P. (2003). *Régészeti geológia és történeti ökológia alapjai.* JATEPress, Szeged,

Sümegi, P. (2004). Hortobágy fejlődéstörténete. In: Ecsedi, Z. (ed), *Hortobágy madárvilága*, pp. 33-38. Hortobágyi Természetvédelmi Egyesület, Balmazújváros-Szeged.

Sümegi, P. (2005). *Loess and Upper Paleolithic environment in Hungary.* Aurea Kiadó, Nagykovácsi.

Sümegi, P., Szilágyi, G. (2010). A Hortobágy új felszínfejlődési modellje és a szikesedés eredete. *Acta Biologica Debrecina* 22, 27-36.

Sümegi, P., Szilágyi, G. (2011). A quarter-malacological inventory of Hungarian kurgans. In: Pető, Á., Barczi, A. (eds.), *Kurgan Studies: An environmental and archaeological multiproxy study of burial mounds in the Eurasian steppe zone.* British Archaeological Reports, 2238, Oxford.

Sümegi, P., Magyari, E., Daniel, P., Hertelendi, E., Rudner, E. (1999). A kardoskúti Fehér-tó negyedidőszaki fejlődéstörténetének rekonstrukciója. *Földtani Közlöny* 129, 479-519.

Sümegi, P., Molnár, A., Szilágyi, G. (2000). Szikesedés a Hortobágyon. *Természet Világa* 131, 213-216.

Sümegi, P., Bodor, E., Törőcsik, T. (2005). The origins of sodification in the Hortobágy region in the light of the palaeoenvironmental studies at Zám–Halasfenék. In: Gál, E., Juhász, I., Sümegi, P. (eds.), *Environmental Archaeology in North-Eastern Hungary.* Varia Archaeologica Hungarica, XIX. MTA Régészeti Intézet, Budapest.

Sümegi, P., Bodor, E., Törőcsik, T. (2006). A hortobágyi szikesedés eredete. In: Kiss, A., Mezősi, G., Sümeghy, Z. (eds.), *Táj, környezet és társadalom. Ünnepi tanulmányok Keveiné Bárány Ilona professzor asszony tiszteletére*, pp. 633-641. SZTE Kiadványa, Szeged, Hungary.

Sümegi, P., Molnár, M., Jakab, G., Persaits, G., Majkut, P., Páll, D.G., Gulyás, S., Jull, A.J.T., Törőcsik, T. (2011). Radiocarbon-dated paleoenvironmental changes on a lake and peat sediment sequence from the central part of the Great Hungarian Plain (Central Europe) during the last 25.000 years. *Radiocarbon* 52, 85-97.

Sümegi, P., Persaits, G., Gulyás, S. (2012). Woodland-Grassland Ecotonal Shifts in Environmental Mosaics: Lessons Learnt from the Environmental History of the Carpathian Basin (Central Europe) During the Holocene and the Last Ice Age Based on

Investigation of Paleobotanical and Mollusk Remains. In: Myster, R.W. (ed.) *Ecotones Between Forest and Grassland,* pp. 17-57. Springer Press, New York.

Svensson, A., Andersen, K.K., Bigler, M., Clausen, H.B., Dahl-Jensen, D., Davies, S.M., Johnsen, S.J., Muscheler, R., Parrenin, F., Rasmussen, S.O. (2008). A 60 000 year Greenland stratigraphic ice core chronology. *Climate of the Past* 4, 47-57.

Szabolcs, I. (1954). *Hortobágy talajai.* Mezőgazdasági Kiadó, Budapest.

Szabolcs, I. (1957). *Szikes talajaink genetikája és jelentősége hasznosításuk nézőpontjából.* Akadémiai Kiadó, Budapest.

Szabolcs, I. (1961). *A vízrendezések és öntözések hatása a tiszántúli talajképződési folyamatokra.* Akadémiai Kiadó, Budapest.

Szabó, J. (1864). Hevesmegye földtani leírása. In: Montedégói, A.F. (ed) *Heves és Külső Szolnok törvényesen egyesült vármegyéknek leírása.* Érseki Lyceum Könyv- és Kőnyomda, Eger.

Szendrei, G., Tóth, T. (2006). A magyarországi szikes talajok felszíni sóásványai. *Topographia Mineralogica Hungariae* 9, 47-60.

Szelepcsényi, Z., Breuer, H., Ács, F., Kozma, I. (2009a) Biofizikai klímaklasszifikációk (1. rész: a módszerek bemutatása). *Légkör* 54, 21–26.

Szelepcsényi, Z., Breuer, H., Ács, F., Kozma, I. (2009b) Biofizikai klímaklasszifikációk (2. rész: magyarországi alkalmazások). *Légkör* 54, 21–26.

Székyné Fux, V., Szepesi, K. (1959). Az "alföldi" lösz szerepe a szikes talajképződésben. *Földtani Közlöny* 89, 53-64.

Mádlné Szőnyi, J. Simon, Sz., Tóth J., Pogácsás, Gy. (2005). Felszíni és felszín alatti vizek kapcsolata a Duna-Tisza közi Kelemenszék és Kolon-tó esetében. *Általános Földtani Szemle* 30, 93-110.

Szöőr, G., Sümegi, P., Balázs, É. (1991). Sedimentological and geochemical facies analysis Upper Pleistocene fossil soil zones discovered in the Hajdúság region, NE Hungary. In: Pécsi, M., Schweitzer, F. (eds.) *Quaternary environment in Hungary,* pp. 47-59. Studies in geography in Hungary 26. Akadémiai Kiadó, Budapest.

Szöőr, G., Sümegi, P., Balázs, É. (1992). A Hajdúság területén feltárt felső pleisztocén fosszilis talajok szedimentológiai és geokémiai fácieselemzése. In: Szöőr, Gy. (ed.) *Fáciesanalitikai, paleobiogeokémiai és paleoökológiai kutatások,* pp. 81-91. MTA Debreceni Bizottsága, Debrecen.

Tarasov, P. E., Webb, T. III, Andreev, A. A., Afanaseva, N. B., Berezina, N. A., Bezusko, L. G., Blyakharchuk, T. A., Bolikhovskaya, N. S., Cheddadi, R., Chernavskaya, M. M., Chernova, G. M., Dorofeyuk, N. I., Dirksen, V. G., Elina, G. A., Filimonova, L. V., Glebov, F. Z., Guiot, J., Gunova, V. S., Harrison, S. P., Jolly, D., Khomutova, V. I., Kvavadze, E. V., Osipova, I. M., Panova, N. K., Prentice, I. C., Saarse, L., Sevastyanov, D. V., Volkova, V. S., Zernitskaya, V. P., (1998). Present-day and mid-Holocene biomes reconstructed from pollen and plant macrofossil data from the Former Soviet Union and Mongolia. *Journal of Biogeography* 25, 1029–1053.

Tarasov, P.E., Volkova, V.S., Webb, T. III., Guiot, J., Andreev, A.A., Bezusko, L.G., Bezusko, T.V., Bykova, G.V., Dorofeyuk, N.I., Kvavadze, E.V., Osipova, I.M., Panova, N.K., Sevastyanov, D.V. (2000). Last glacial maximum biomes reconstructed from pollen and plant macrofossil data from northern Eurasia. *Journal of Biogeography* 27, 609-620.

Teschedik, S. (1804). Über die Kultur und Benützung der sogenannte Székes-. Felder in der Gegend an der Theiss. In: *Patriotisches Wochenblatt für der Gegend an der Theiss*. In: Patriotisches Wochenblatt für Ungarn. Jon K. Lübeck, Pest.

Timár, G., Sümegi, P., Horváth, F. (2005). Late Quaternary dynamics of Tisza River: Evidence of climatic and tectonic controlls. *Tectonophysics* 410, 97-110.

Timkó, I. (1934). A borsodi nyílt ártér szikesei. In: Sajó, E., Trummer, A. (eds.), *A magyar szikesek*, pp. 109-126. Patria Nyomda, Budapest.

Timmermann,A., Menviel, L., Okumura,Y., Schilla, A., Merkel,U., Timm,O., Hu,A., Otto-Bliesner, B., Schulz,M. (2010). Towards a quantitative understanding ofmillennial-scale Antarctic warming events. *Quaternary Science Reviews* 29, 74-85.

Tóth, J. *(1963)*. A theoretical analysis of groundwater flow in small drainage basins. *Journal of Geophysical Research* 68, 4795-4812.

Tóth, T., Kuti, L., Fórizs, I., Kabos, S. (2001). A talaj sótartalmának változása a hortobágyi Nyírőlapos mintaterületen. *Agrokémia és Talajtan*, 50, 409-426.

Treitz, P. (1898). Szikes talajok Magyarországon. *Természettudományi Közlöny*, 45-48, 121-128.

Treitz, P. (1917). Jelentés az 1916. évben végzett agrogeológiai munkálatról. II. rész: A szikes talajok szerkezete Kiskunfélegyháza határában. *Magyar Királyi Földtani Intézet Jelentése 1916-ról*, 499-510.

Treitz, P. (1925). Az agrogeológia múltja és feladatai hazánkban. *Földtani Közlöny* 55, 20-25.

Treitz, P. (1934). Csonka-magyarország sós és szikes talajai. In: Sajó, E., Trummer, Á. (eds.), *A magyar szikesek (különös tekintettel vízgazdálkodás útján való hasznosításukra),* pp. 177-206. Magyar Királyi Földművelésügyi Minisztérium Kiadványai, 2. Vízügyi Műszaki Csoport, Budapest.

Troels-Smith, J. (1955). *Karakterisering af lose jordater*. Danmarks Geologiske Undersogelse, ser.IV. (10).

Towson, R. (1796). *Travels in Hungary, with a short account of Vienna in the year 1793*. Robinson, London.

Yiou, P., Jouzel, J., Johnsen, S., Rögnvaldsson, Ö. E. (1995). Rapid oscillations in Vostok and GRIP Ice Cores. *Geophyical Research Letters* 22, 2179–2182.

Várallyay, G. (1967). A Duna-völgyi talajok sófelhalmozódási folyamatai. *Agrokémia es Talajtan* 16, 327-356.

Várallyay, G. (1993). Soils in the region between the rivers Danube and Tisza (Hungary). In: Szujkó-Lacza, J., Kováts, D. (eds), *The flora of the Kiskunság National Park.,* pp. 21-42. Magyar Természettudományi Múzeum Kiadványa, Budapest

Várallyay, G. (1999). Szikesedési folyamatok a Kárpát-medencében. *Agrokémia es Talajtan* 48, 399-418.

Walker, M.J.C., Björck, S., Lowe, J.J., Cwynar, L.C., Johnsen, S., Knudsen, K. L., Wohlfarth, B., and INTIMATE GROUP. (1999). Isotopic "events" in the GRIP ice core: A stratotype for the Late Pleistocene, *Quaternary Science Reviews* 18, 1143–1150.

Willis, K.J. (2007). Impact of the early Neolithic Körös culture on the landscape: evidence from palaeoecological investigations of Kiri-tó. In: Whittle A. (ed.), *The Early Neolithic on the Great Hungarian Plain: investigations of the Körös culture site of Ecsegfalva 23, Co. Békés,* pp. 83-99. Varia Archaeologica Hungarica, 21. Budapest.

Willis, K.J., Sümegi, P., Braun, M., Tóth A. (1995). The Late Quaternary environmental history of Bátorliget, N.E. Hungary. *Palaeogeography, Palaeoclimatology, Palaeoecology* 118, 25-47.

Willis, K.J., Braun, M., Sümegi, P., Tóth A. (1997). Does soil change cause vegetation change or vice-versa? A temporal perspective from Hungary. *Ecology* 78, 740-750.

Willis, K.J., Sümegi, P., Braun, M., Bennett, K.D., Tóth, A. (1998). Prehistoric land degradation in Hungary: who, how and why? *Antiquity* 72, 101-113.

Wright, J. (2003). *Environmental Chemistry*. Routledge, London.

Zipser, C. A. (1817). *Versuch eines topographish-mineralogischen Handbuches von Ungarn.* Carl Friedrich Wiegand, Odenburg.

Zólyomi, B. (1952). Magyarország növénytakarójának fejlődéstörténete az utolsó jégkorszaktól. *MTA Biológiai Osztályának Közleményei* 1, 491–544.

Conservation and Management of Steppe Ecosystems

In: Steppe Ecosystems
Editors: M. B. Morales Prieto and J. Traba Diaz
ISBN: 978-1-62808-298-2

Chapter 9

THE CONSERVATION OF HIGH-INTEREST PLANT SPECIES OFFERS THE CHANCE TO PRESERVE UNIQUE AND VULNERABLE REPRESENTATIVES OF GYPSUM STEPPES

***Màrcia Eugenio*[1,*], *Carlos Molina*[2] *and Gonzalo Montamarta*[2]**
[1]Área de Botánica, EU de Ingenierías Agrarias de Soria, Universidad de Valladolid
[2]Actividades, Estudios y Proyectos en el Medio Ambiente S.L., Spain

ABSTRACT

Senecio auricula Bourgeau (Asteraceae) is a hemicriptophyte endemic to the Iberian Peninsula and North Africa (Ascaso and Pedrol, 1991) that inhabits gypsum steppes. Three subspecies have been distinguished, all of them showing narrow distribution ranges, restricted to particular, well-defined territories: *S. auricula* subsp. *sicoricus* (-Bolòs and Vigo-, Ascaso and Pedrol, 1991) is found mainly in the Ebro valley, Spain. At the westernmost extreme of its distribution range, populations in the Spanish province of Soria constitute the only representation of the species across the whole region of Castilla y León. Thus, *S. auricula* subsp. *sicoricus* is protected by Castilian regional legislation ("Priority Attention"; Decree 63/2007, 14th June), besides being included in the Red List of Spanish Vascular Flora as "Near Threatened Species" (Moreno, 2008). In, 2010, the government of Castilla y León promoted the creation of a flora micro-reserve to guarantee the long-term preservation of such populations. Micro-reserves consist in legally-defined areas under, 200 ha aiming to protect selected population samples of one or several high-interest plant species (Laguna et al.., 2004). Therefore, they constitute an adequate conservation strategy for narrow endemics such as S. auricula subsp. sicoricus, given that they are planned at the local scale and based specifically on the spatial distribution of the species of interest (Laguna, 2001).

Here, we present an overview of the technical works and an assessment related with the process of definition of the "Flora Micro-reserve of Gypsum Outcrops of Monteagudo and Fuentelmonge" in the southeastern part of Soria province, and particularly in the

* Corresponding Author address: Área de Botánica, EU de Ingenierías Agrarias de Soria, Universidad de Valladolid. 42004 Soria, Spain. Email: m.eugenio@agro.uva.es.

municipalities of Monteagudo de las Vicarías and Fuentelmonge. Local landscapes are constituted by gentle hills intermingled with cultivated, flat-bottomed valleys. *S. auricula* subsp. sicoricus appears mainly at gypsum glacis, where the upwards movement of sulfate-saturated water under high evaporation rates results in the formation of sulfate crystals onto soil surface.

Glacis are particularly vulnerable to different threats, remarkably mainly to the use of agricultural machinery at nearby cultivated valleys, which frequently entails trampling or destruction of fragments of gypsophile habitats. The species appears together with other strict edaphic endemics such as *Herniaria fruticosa*, *Reseda stricta*, *Lepidium subulatum* or *Launaea pumila*, that together conform the syntaxa Lepidion subulati, included in the syntaxonomic class Gypsophiletalia, classified as high-priority habitat 1520* according to the European Habitats Directive. Overall, the effective creation of the proposed micro-reserve to protect *S. auricula* subsp. *sicoricus* would entail the preservation of very restricted and isolated areas that constitute excellent habitat representatives of Iberian gypsum steppes, and that are highly vulnerable to the occurrence of local extinctions.

Keywords: Endemic flora, gypsophile habitats, gypsophyte species, gypsum outcrops, micro-reserve, protection, *Senecio auricula*

INTRODUCTION

Gypsum steppes, given their spoiled appearance of low open shrublands surrounded by a matrix of whitish bare soils (Figure 1), are frequently considered by local human populations as areas of low environmental quality and little conservation interest, i.e., as marginal lands.

Figure 1. Gypsum steppes at Monteagudo de las Vicarías y Fuentelmonge; the optimum habitat type for *Senecio auricula* subsp. *sicoricus* is shown in the forefront.

Traditional land uses such as cultivation of flat-bottomed valleys and livestock grazing are still common, and entail both habitat loss and degradation. Besides, practices such as littering or afforestation with non-native species are widespread and further damage them. However, Iberian gypsum steppes (syntaxonomic class *Gypsophiletalia*) account among the habitat types compiled by Annex I of the Habitats Directive (Council Directive 92/43/EEC) as requiring the designation of special areas for their conservation. Furthermore, they are considered as high-priority habitats in the European Union (habitat 1520*). Iberian gypsum steppes are rare in the European geographical frame, since their distribution is restricted to gypsum-rich substrates occurring in the Mediterranean part of the Iberian Peninsula, under arid or semiarid climate conditions.

As a consequence of their particular environmental conditions, particularly the physical and chemical rigors of gypsosoils, the general water scarcity, and the existence of an intense summer drought, a conspicuous set of endemic plants has evolved (Braun-Blanquet and Bolòs, 1957; Rivas-Martínez and Costa, 1970), which constitutes a main value for the conservation of these habitats.

The presence of endemic species is dependent on the preservation of particular ecological processes leading to such conditions, mainly soil physical and chemical conditions. Lastly, it is worth underlining that Iberian gypsum steppes are particularly vulnerable and threatened habitats.

Gypsosols are characterized by gypsum (hydrous calcium sulfate) contents over 5%. Soil chemical conditions have been for long time considered to be the main limiting factor for plant life in these environments, since ion concentration is unbalanced, commonly characterized by an excess of sulphur and calcium, and by scarcity of phosphorous, potassium, and nitrogen (Guerrero-Campo et al.., 1999b), which has been explained by gypsum-saturated water moving through the soil and exchanging calcium for other ions retained in the soil complex, thus leading to a nutritional impoverishment (Meyer et al.., 1992).

Meyer (1986) hypothesized that physical soil crusts (PSCs) could be even more restrictive for plant life in these environments than soil chemical characteristics themselves. In gypsum outcrops, the high evaporation rates associated to arid conditions result in the formation of sulfate crystals on soil surface, particularly in ground-water discharge areas, where such process is intensified.

There has been subsequent evidence supporting this hypothesis, such as the capability of several gypsophyte species (*sensu* Meyer, 1986) to recruit in extremely hard gypsum surfaces (e.g. *Helianthemum squamatum*, Escudero et al.., 1999; and *Lepidium subulatum*, Escudero et al.., 2000), and the inability of emerging seedlings of some gypsovag species (*sensu* Meyer, 1986) to surpass gypsum crusts (e.g. *Teucrium capitatum*, Romão and Escudero, 2005). Overall, it seems that gypsum edaphic specialists base their behavior to a great extent on their ability to surpass extremely hard gypsum surface crusts, although this seems a marginally adaptive trait, as shown by the general negative effect that crusts show in the early development life stages of plants (Romão and Escudero, 2005).

Biological soil crusts (BSCs) composed of eukaryotic algae, cyanobacteria, mosses, liverworts, fungi and lichens, also appear on the surface of gypsum soils, and show relevant interactions with other organisms, including vascular plants (see the review by Maestre et al.., 2011 for semiarid areas of Spain). Most studies on plant-BSCs interactions have focused on seed germination and early phases of seedling establishment, and have revealed contradictory effects that likely result from the fact that such interactions are highly species-specific (Escudero et al.., 2007). Last but not least, topography plays a main role in explaining vegetation patterns in these habitats, as revealed by numerous studies (Guerrero-Campo et al.., 1999a, 1999b; Pueyo and Alados, 2007a, 2007b).

Topography deeply conditions relevant ecological processes and their resulting conditions, thus enhancing or modulating the rigors of both gypsum soils and the regional climate. Remarkably, the slope and aspect profoundly condition the result of evaporation processes, thus determining the formation of PSCs.

Iberian gypsum steppes are inhabited by numerous high-interest plant species, some of which are included in the Red List of Spanish Vascular Flora, 2008, and protected by different regional legislations. *S. auricula* Bourgeau (*Asteraceae*, Figure 2), a hemicriptophyte endemic to the Iberian Peninsula and North Africa (Ascaso and Pedrol, 1991, Figure 3a), is one of such species.

Three subspecies have been distinguished; all of them are characterized by narrow distribution ranges restricted to several well-defined territories. Whereas the subespecies *castellanus* and *sicoricus* behave as strict gypsophytes, the subspecies *auricula* shows a preference for gypsum substrates, although it is also found on a narrow range of related substrates, notably on salt-rich soils ("Subgipsophyte Gipsocline", *sensu* Mota et al.., 2009), which might not be attributed to a different habitat preference, but to the fact that it occurs in drier geographic areas, where evapotranspiration processes are more intense, and halophile habitats more patent (C. Molina, *pers. obs.*). The subsp. *sicoricus* (Ascaso and Pedrol, 1991) occurs mainly in the Ebro Valley. At the westernmost extreme of its distribution, populations in Soria province constitute the only representation of the species in the region of Castilla y León (Figure 3b), where it is protected by the regional legislation ("Priority Attention"; Decree 63/2007, 14th June).

Figure 2. *Senecio auricula* subsp. *sicoricus*.

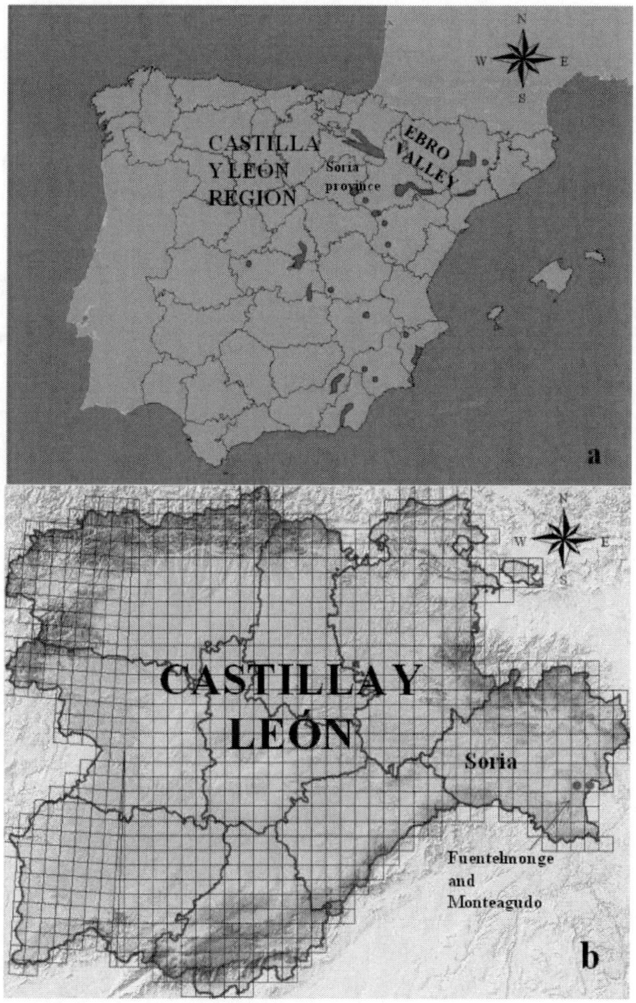

Figure 3. Distribution of *Senecio auricula* (a) in the Iberian Peninsula, and (b) in the region of Castilla y León (corresponding exclusively to *S. auricula* subsp. *sicoricus*).

Castilla y León legally incorporated the designation micro-reserves as a means to guarantee the conservation of the 284 plant species included in the Catalogue of Protected Flora of Castilla y León (Decree 63/2007, 14th June). In general, micro-reserves are considered a suitable management tool for the conservation of high-interest plant species, since they are planned at the local scale and defined following their particular distribution (Laguna et al.., 2004). In this case, micro-reserves were defined as areas under 200 ha where selected samples of one or several catalogued species are found. For the purpose of identifying potential micro-reserves, the regional government established a specific agreement with the University of León in, 2008. In that frame, a study was conducted in, 2010 on populations of *S. auricula* subsp. *sicoricus* in Soria province, which provided with the two documents required for the definition of a micro-reserve: a technical card compiling all available scientific information on the species (own unpublished data), and the proposal itself, including a detailed description of the area (landscapes, geology, habitat types, vegetation dynamics), flora inventories, an assessment on the state of conservation and current threats to the habitat and the species of interest (thus justifying the need for their conservation), the spatial definition proposed, and management recommendations for the future protected area.

Here we present a summary on such technical works, and an assessment on the process of micro-reserve definition. Our ultimate aim is to underline how, in this particular case, the conservation of a high-interest plant species offers the chance to preserve unique and vulnerable representatives of a high-priority habitat, Iberian gypsum steppes.

TERRITORIAL AND GEOMORPHOLOGICAL FRAMEWORK

In the region of Castilla y León, *S. auricula* subsp. *sicoricus* is only found at the municipalities of Fuentelmonge and Monteagudo de las Vicarías, at the southeastern part of the province of Soria. Gypsum outcrops of Monteagudo and Fuentelmonge are located at 800-900 m a.s.l. at one of the driest areas of Castilla y León, where annual precipitation values of around 300 mm are recorded. The unit encompasses a series of paleogene formations of lacustrine origin interspersed with dendritic alluvial successions where calcareous materials such as limestones, marls, and loams alternate with gypsum outcrops. Landscapes resemble to bad-lands, and are constituted by gentle hills arranged as dendritic alignments, sometimes isolated from one another by erosive processes, whose maximum height corresponds to the old structural surface; and between them, erosive plains and river beds whose flat topography is originated by filling and favors cultivation. These croplands usually consist in extensive dry cereal fields.

In this environment, *S. auricula* is mainly found at gypsum glacis, i.e. on smooth slopes or on flat relief at the base of the hills, along terrain strips located between hillsides and valleys, in direct contact with dry cereal croplands that in fact occupy part of the potential areas of strict gypsophile communities and most of the potential areas of nearby sub-halophile communities. At gypsum glacis, physical soil crusts are formed as a result of sulfate-saturated water rising to soil surface under arid climatic conditions ensuring high evaporation rates. Such crusts remain wet through the winter until spring. Remarkably, they constitute particularly vulnerable areas, threatened mainly by the use of agricultural machinery at nearby, cultivated valleys, which too often results in the extension of arable lands into the foot of the hills. *S. auricula* appears together with other strict gypsophyte species of low stature such as *Herniaria fruticosa*, *Reseda stricta*, *Lepidium subulatum* or *Launaea pumila*, all corresponding to *Lepidion subulati*, one of the characteristic syntaxa included in the high-priority habitat *Gypsophiletalia*, together with *Gypsophilion hispanicae*, and *Thymo-Teucrion verticillati* (European Comission, 2007). Other frequent species are gypsovag endemic prostrated shrubs such as *Thymus mastigophorus* and *Teucrium gnaphalodes*, and widely distributed grasses such as *Koeleria vallesiana*.

VEGETATION CHARACTERIZATION

At the gypsum outcrops of Monteagudo and Fuentelmonge, gypsophile steppes at different conservation states constitute the most characteristic vegetation type. Notably, one of the best regional representatives of the European high-priority habitat 1520* (Iberian gypsum steppes (*Gypsophiletalia*)) is found here (Table 1). Where gypsum substrates are not massive, other communities included in the European habitat 4090 appear (Endemic oro-Mediterranean heaths with gorse). Thus, on sharp hillsides where gypsum marls alternate with calcareous substrates and soils are either poorly developed or eroded, vegetation mosaics constituted by *Lepidion subulati* steppes and *Sideritido incanae-Salvion lavandulifoliae* xeric calcareous shrublands are common. Such shrublands usually present an open structure, are rich in species of the families *Labiatae* and *Fabaceae*, and tend to appear together with xeric grasslands dominated by *Brachypodium retusum*, which substitute them on deeper and better-structured soils. On bare areas, open grasslands of the syntaxa *Brachypodion distachyi* and *Thero-Brachypodion retusi* are found (both included in the European high-priority habitat 6220*, Pseudo-steppe with grasses and annuals (*Thero-Brachypodietea*)). At gypsum glacis, when soils suffer from nitrification or physical alteration due to sheep grazing, gypsophile steppes are substituted or appear together with xeric nitrophilous grasslands (corresponding to different syntaxa, such as *Thero-Brachypodetalia, Astragalo sesamei-Poion bulbosae,* etc.), where plant species adapted to gypsum or salt-rich substrates such as *Convolvulus lineatus*, *Plantago albicans*, *Bromus rubens* or *Aegilops triuncialis*, are dominant. *Thero-Brachypodetalia* grasslands appear at the first stages of succession on turned-over soils and abandoned croplands, and are constituted by small annual species characterized by a quick spring development such as *Medicago minima* or *M. trucatula*, and by annual grasses such as *Bromus rubens* or *B. hordeaceus*. *Astragalo sesamei-Poion bulbosae* grasslands appear at latter successional stages, when nitrophilous conditions are maintained. On bare gypsum soils, ephemeral gypsophile lawnlands of *Campanula fastigiata* (*Sedo-Ctenopsion gypsophillae*), corresponding to the European high-priority habitat 6220*, constitute micro-communities that intermingle with other vegetation types (*Lepidion subulati*, *Thero-Brachypodietalia*, *Astragalo sesamei-Poion bulbosae* and *Salsolo vermiculatae-Peganion harmalae*). Halo-nitrophilous steppes dominated by *Artemisia herba-alba*, corresponding to *Salsolo vermiculatae-Peganion harmalea*, appear on moderately steep gypsum hillsides when sheep grazing or cultivation has resulted in soil nitriphication, turn over and trampling. Nitrophilous grasslands and gypsophile species are intermingled with such halo-nitrophilous steppes, although the latter dominate when the regeneration of soil crusts has occurred. Lastly, in areas such as gullies, ravines, etc., where soils retain water in winter, the dominant species is *Elymus campestris*. Other common species are *Brachypodium phoenicoides* or *Dactylis glomerata* subsp. *hispanica*, which usually appear in a mosaic-like pattern with grasslands dominated by *Brachypodium retusum* (*Thero-Brachypodion*).

In relation to dynamics, *Lepidion subulati* steppes constitute stable communities where gypsum physical soil crusts are well-developed, as it is the case at the lower half of gypsum hills in the Nágima river basin at the municipalities of Fuentelmonge and Monteagudo de las Vicarías. When superficial soil crusts are broken and erosion occurs, calcareous chamaephytes such as *Lithodora fruticosa*, *Genista scorpius*, *Helichrysum stoechas* or *Santolina chamaecyparissus* (characteristic of the European habitat type 4090) appear. When soils are nitrified or physically altered, the presence of formations of *Artemisia herba-alba* and *Salsola vermiculata*, corresponding to the European habitat type 1430 (Halo-nitrophilous scrubs *Pegano-Salsoletea*) is increased. Conversely, when anthropogenic or zoogenic disturbances disappear from gypsum areas dominated by xeric calcareous shrublands, these communities move back and gypsophile vegetation partially recovers on flat microrelief, a process which occurs along with regeneration of superficial physical soil crusts. Similarly, when disturbances disappear from areas occupied by *Salsolo vermiculatae-Peganion harmalae,* the development of either gypsophile steppes or calcareous shrublands is promoted.

Flora inventories that were conducted as part of the technical works directed to defining the micro-reserve of Monteagudo and Fuentelmonge allowed identifying two other high-interest species besides *S. auricula*, namely *Campanula fastigiata* Dufour and *Thymus mastigophorus* Lacaita. *Campanula fastigiata* is a terophyte that grows on gypsum marls in the central, southern and eastern Iberian Peninsula. It is one of the three protected gypsophyte plant species in Castilla y León ("Preferential Concern"; Decree 63/2007, 14th June), where it shows two main populations (in the provinces of Palencia and Soria), one of which corresponds to the gypsum outcrops of Monteagudo and Fuentelmonge. It is a tiny plant difficult to detect, except during its flowering in July. *Campanula fastigiata* is scarce in the area, where it appears in small isolated groups; remarkably, it is absent from many potential locations. *Thymus mastigophorus* is a prostrated chamaephyte that inhabits base-rich soils on limestones, marls, and gypsum marls. It is endemic to the Mediterranean part of the Iberian Peninsula, and its distribution range encompasses the high Ebro valley, the northern Iberian plateau, and the Alcarria region.

Table 1. Summary of the main habitat types compiled by Annex I of the Habitats Directive (Council Directive 92/43/EEC) that are found in the area of interest, together with their characteristic species

Habitat Code	Habitat Type	Characteristic Species
1520*	Iberian gypsum steppes (*Gypsophiletalia*)	*Senecio auricula, Herniaria fruticosa, Reseda stricta, Lepidium subulatum, Ononis tridentata, Launaea pumila, Astragalus alopecuroides, Thymus mastigophorus*
6220*	Pseudo-steppe with grasses and annuals (*Thero-Brachypodietea*)	*Brachypodium distachyon, Bupleurum baldense, Desmazeria rigida, Galium parisiense, Neatostema apulum, Wangenheimia lima, Helianthemum salicifolium, Vulpia unilateralis* (*Brachypodion distachyi*)
		Campanula fastigiata (*Sedo-Ctenopsion gypsophillae*)
		Avenula bromoides, Brachypodium retusum, Dactylis hispanica, Koeleria vallesiana, Stipa sp., Phlomis lychnitis, Dipcadi serotinum (*Thero-Brachypodion retusi*)
1430	Halo-nitrophilous scrubs (*Pegano-Salsoletea*)	*Artemisia herba-alba*
4090	Endemic oro-Mediterranean heaths with gorse	*Genista scorpius, Linum suffruticosum, Thymus vulgaris, Bupleurum fruticescens, Salvia lavandulifolia, Santolina chamaecyparissus, Helianthemum appeninum*

DEFINITION OF THE MICRO-RESERVE

A total of 31,78 hectares were finally selected to integrate the "Flora Micro-reserve of the Gypsum Outcrops of Monteagudo and Fuentelmonge". Notably, this area was not continuous, but the result of considering together a network of areas of varying size, and belonging to two different groups based on their degree of interest and their current state of conservation. On the one hand, a total of 9 areas of maximum interest (MI) were defined following the best population representations of *S. auricula* and *C. fastigiata* (8 and 1 over the total, respectively). On the other hand, a total of 11 areas requiring management measures were also identified and included, most of them corresponding to small populations of *S. auricula* which were threatened by nearby land cultivation, vehicle trails, and littering. It has been repeatedly underlined that micro-reserves need to be precisely delimited, on the basis of the accurate delimitation of species' current distribution, as well as on their population dynamics, in such a way that not only present conservation but also future development are ensured (Laguna et al.., 2004). It is also generally accepted that a micro-reserve is meant to be as large as possible whenever land surface is available around the current area of population occupation (Draper et al.., 2001). In the case of the gypsum outcrops of Monteagudo and Fuentelmonge, the current distribution of *S. auricula* was accurately defined based on field work, which allowed delimiting the different areas that made up the micro-reserve. The approach focused on the preservation of those locations where the species of interest currently exists; thus, gypsum hills were included, and surrounding cultivated valleys were excluded. Given the historical tendency to extend arable lands into the foot of the hills using agricultural machinery, it is nowadays impossible to precisely define the limits of potential territories of *S. auricula* at the contact with croplands.

Moreover, it is unknown whether including (and thus preserving from agricultural practices) the outermost cropland strips into the micro-reserve would result in their colonization by *S. auricula*, and how many years or decades might such process take. Only recently scientific attention has been paid to secondary successional processes occurring either after quarrying (Mota et al.., 2004; Dana and Mota, 2006) or after agriculture abandonment (Martínez-Duro et al.., 2010; Eugenio et al., 2012) in semi-arid gypsum habitats of the Iberian Peninsula. These processes seem to differ from the general amelioration mechanism described for other Mediterranean-type ecosystems, in which the structure and composition of vegetation change along with the improvement of soil physical and chemical properties that occurs with time since abandonment (Debussche et al.., 1996; Haase et al.., 1996, Ne'eman and Izhaki, 1996). Contrastingly, in gypsum environments of central Spain, the most relevant predictor of vegetation composition along succession after agriculture cessation was soil chemical status, independently of time since abandonment (Martínez-Duro et al.., 2010). In relation to individual plant species, responses varied along succession; notably, some gypsophyte species competing better on genuine gypsum soils with well-developed superficial soil crusts needed more time to reappear after agriculture cessation (Martínez-Duro et al.., 2010).

According to our field observations, such would presumably be the case for *S. auricula,* and thus, the recovery of soil crusts may constitute a key process determining the potential colonization of the outermost cropland strips by this high conservation-interest species. After disturbance cessation, physical crusts are firstly recovered, followed by biological crusts.

Studies along succession after agriculture cessation are missing both for PSCs and BSCs. In relation to BSCs dynamics after other types of disturbances, observations in southeastern Spain point out to decadal recovery times after removal and under low erosion levels (Lázaro et al.., 2008). Nevertheless, recovery rates are considered quite variable both at local and regional scales, depending on crust and soil types, severity of disturbance, climatic variability and rainfall erosivity (Belnap and Eldridge, 2003, Lázaro et al.., 2008). In general, and since most gypsophytes are narrow endemic species, it is a priority to ascertain whether allowing secondary succession after agriculture abandonment involves the recovery of their populations in the medium term (years to decades), or contrarily, it may constitute a drawback for their future conservation.

CONCLUSION

The "Flora Micro-reserve of the Gypsum Outcrops of Monteagudo and Fuentelmonge" has been proposed and defined specifically to preserve the only population representative of the endemism *S. auricula* subsp. *sicoricus* in Castilla y León and, secondarily, to preserve the presence of the rare *C. fastigiata*, which is other gypsophyte species protected in the region by means of the Decree 63/2007, 14th June. In the studied area, *S. auricula* subsp. *sicoricus* appears together with other strict edaphic endemics such as *Herniaria fruticosa*, *Reseda stricta*, *Lepidium subulatum* or *Launaea pumila*, where superficial soil crusts occur, commonly on gypsum glacis. Thus, its presence indicates flat relief or micro-relief where water rises by evaporation, resulting in the formation of soil crusts that remain wet until spring or, more generally speaking, environments where the syntaxa *Lepidion subulati*, included in the European high-priority habitat type 1520*, develop. Thus, in this case, the conservation of a high-interest plant species has offered the chance to preserve unique and vulnerable representatives of a high-priority habitat, Iberian gypsum steppes.

Iberian gypsum steppes are particularly vulnerable: cultivation of valleys involves the use of agricultural machinery, which has tended to extend arable lands to potential areas of *S. auricula*, and livestock pressure causes physical (by trampling) and chemical (nitriphication) soil alteration that results in a progressive substitution of gypsophile communities. Overall, their desolated appearance has favored an erroneous perception by local people as marginal lands, which in turn has involved detrimental uses such as littering and stocking of hay bales. Besides their vulnerability, it is necessary to underline the rarity of Iberian gypsum steppes in the region of Castilla y León, where they appear mainly in the central Duero river basin and, to a lesser extent, in the Ebro valley, like the study community of Monteagudo de las Vicarías. Thus, the flora micro-reserve created to protect *S. auricula* subsp. *sicoricus* offers the chance to preserve as well one of the best representatives of this threatened habitat.

ACKNOWLEDGMENTS

We are very grateful to the Department of Biodiversity and Environmental Management of the Faculty of Environmental Sciences, University of León, and to the Consejería de Medio Ambiente, Junta de Castilla y León, for inviting "Actividades, Estudios y Proyectos en el

Medio Ambiente S.L." to collaborate in the "Convenio para la realización de trabajos científicos vinculados al desarrollo del Decreto 63/2007, de 14 de junio, por el que se crean el Catálogo de Flora Protegida de Castilla y León y la figura de protección denominada Microreserva de Flora".

REFERENCES

Ascaso, J., Pedrol, J. (1991). De plantis vascularibus praesertim ibericis. *Fontqueria* 31, 135-140.

Belnap, J. Eldridge, D. (2003). Disturbance and recovery of biological soil crusts. In: J. Belnap, Lange, O. L. (eds.), *Biological soil crusts: structure, function, and management*, pp. 363-383. Springer-Verlag, Berlin.

Braun-Blanquet, J., Bolòs, O. (1957). Les groupements vegetaux du Bassin Moyen de l'Ebre et leur dynamisme. *Anales de la Estación Experimental de Aula Dei* 5, 1-226.

Dana, E.D., Mota, J. F. (2006). Vegetation and soil recovery on gypsum outcrops in semi-arid Spain. *Journal of Arid Environments* 65, 444-459.

Debussche, M., Escarré, J., Lepart, J., Houssard, C., Lavorel, S. (1996). Changes in Mediterranean plant succession: old-fields revised. *Journal of Vegetation Science* 7, 519-526.

Draper, D., Marques, I., Roselló-Graell, A. Iriondo, J.M. (2001). Definition of a micro-reserve for endangered *Narcissus cavanillesii* in Portugal. http://www.nerium.net/plantaeuropa/Download/Procedings/Draper_et_al.pdf.

Escudero, A., Somolinos, R. C., Olano, J.M., Rubio, A. (1999). Factors controlling the establishment of *Helianthemum squamatum*, and endemic gypsophile of semi-arid Spain. *The Journal of Ecology* 87, 290-302.

Escudero A., Iriondo, J.M., Olano, J.M., Rubio, A., Somolinos, R.C. (2000). Factors affecting establishment of a gypsophyte: the case of *Lepidium subulatum* (Brassicaceae). *American Journal of Botany* 87, 861-871.

Escudero, A., Martínez, I., de la Cruz, A., Otálora, M.A.G., Maestre, F.T. (2007). Soil lichens have species-specific effects on the seedling emergence of three gypsophile plant species. *Journal of Arid Environemnts* 70, 18-28.

Eugenio, M., Olano, J.M., Ferrandis, P., Martínez-Duro, E., Escudero, A. (2012). Population structure of two dominant gypsophyte shrubs through a secondary plant succession. *Journal of Arid Environments* 76, 30-35.

European Comission, DG Environment, Nature and Biodiversity. (2007). Interpretation Manual of European Union Habitats –EUR27. http://ec.europa.eu/environment/nature/legislation/habitatsdirective/docs/2007_07_im.pdf.

Guerrero-Campo, J., Alberto, F., Hodgson, J., García-Ruiz, J.M., Montserrat-Martí, G. (1999a). Plant community patterns in a gypsum area. I. Interactions with topographic factors and soil erosion. *Journal of Arid Environments* 41, 401-410.

Guerrero-Campo, J., Alberto, F., Hodgson, J., García-Ruiz, J.M., Montserrat-Martí, G. (1999b). Plant community patterns in a gypsum area. II. Effects of ion washing on topographic distribution of vegetation. *Journal of Arid Environments* 41, 411-419.

Haase, P. Pugnaire, F.I., Clarck, S.C., Incoll, L.D. (1996). Spatial patterns in a two-tiered semi-arid shrubland in southeastern Spain. *Journal of Vegetation Science* 7, 527-534.

Laguna, E. (2001). The micro-reserves as a tool for conservation of threatened plants in Europe. *Nature and the Environment nº 121*. Convention on the Conservation of European Wildlife and Natural Habitats (Bern Convention). Council of Europe Publishing, Strasbourg.

Laguna, E., Deltoro, V.I., Pérez-Botella, J., Pérez-Rovira, P., Serra, Ll., Olivares, A., Fabregat, C. (2004). The role of small reserves in plant conservation in a region of high diversity in Eastern Spain. *Biological Conservation* 119, 421-426.

Lázaro, R., Cantón, Y., Solé-Benet, A., Bevan, J., Alexander, R., Sancho, L.G., Puigdefábregas, J. (2008). The influence of competition between lichen colonization and erosion on the evolution of soil surfaces in the Tabernas badlands (SE Spain) and its landscape effects. *Geomorphology* 102, 252-266.

Maestre, F.T., Bowker, M.A., Cantón, Y., Castillo-Monroy, A.P., Cortina, J., Escolar, C., Escudero, A., Lázaro, R., Martínez, I. (2011). Ecology and functional roles of biological soil crusts in semi-arid ecosystems of Spain. *Journal of Arid Environments* 75, 1282-1291.

Martínez-Duro, E.; Ferrandis, P., Escudero, A., Luzuriaga, A.L., Herranz, J.M. (2010). Secondary old-field succession in an ecosystem with restrictive soils: does time from abandonment matter? *Applied Vegetation Science* 13, 234-248.

Meyer, S.E. (1986). The ecology of gypsophile endemism in the Eastern Mojave Desert. *Ecology* 67, 1303-1313.

Meyer, S.E., García-Moya, E., Lagunes-Espinoza, L.C. (1992). Topographic and soil surface effects on gypsophile plant community patterns in central Mexico. *Journal of Vegetation Science* 3, 429-438.

Moreno, J.C., coord. (2008). *Lista Roja 2008 de la flora vascular española*. Dirección General de Medio Natural y Política Forestal (Ministerio de Medio Ambiente, y Medio Rural y Marino, and Sociedad Española de Biología de la Conservación de Plantas), Madrid.

Mota, J.F.; Sola, A.J., Jiménez-Sánchez, M.L., Pérez-García, F J., Merlo, M.E. (2004). Gypsicolous flora, conservation and restoration of quarries in the southeast part of the Iberian Peninsula. *Biodiversity and Conservation* 13, 1797-1808.

Mota, J.F., Sánchez-Gómez, P., Merlo, M.E., Catalán, P., Laguna, E., de la Cruz, M., Navarro, F.B., Marchal, F., Bartolomé, C., Martínez, J.M., Sainz-Ollero, H., Valle, F., Serra, L., Martínez, F., Garrido, J.A., Pérez-García, F.J. (2009). Aproximación a la checklist de los gipsófitos ibéricos. *Anales de Biología (Universidad de Murcia)* 1, 71-80.

Ne'eman, G., Izhaki, I. (1996). Colonization in an abandoned east Mediterranean vineyard, *Journal of Vegetation Science* 7, 465-472.

Pueyo, Y., Alados, C.L. (2007a). Abiotic factors determining vegetation patterns in a semi-arid Mediterranean landscape: Different responses on gypsum and non-gypsum substrates. *Journal of Arid Environments* 69, 490-505.

Pueyo, Y., Alados, C.L., Maestro, M., Komac, B. (2007b). Gypsophile vegetation patterns under a range of soil properties induced by topographical position. *Plant Ecology* 189, 301-311.

Rivas-Martínez, S., Costa, M. (1970). Comunidades gipsícolas del centro de España. *Anales del Instituto Botánico A. J. Cavanilles* 27, 193-224.

Romão, R.L., Escudero, A. (2005). Gypsum physical soil crusts and the existence of gypsophytes in semi-arid central Spain. *Plant Ecology* 181, 127-137.

In: Steppe Ecosystems
Editors: M. B. Morales Prieto and J. Traba Diaz
ISBN: 978-1-62808-298-2

Chapter 10

THE COVER OF STEPPE VASCULAR PLANTS BY THE PROTECTED AREA NETWORK OF THE REPUBLIC OF MORDOVIA (RUSSIA)

Anatoliy A. Khapugin[1,2,*] ***and Tatyana B. Silaeva***[1]
[1]Mordovian State University, Saransk, Republic of Mordovia, Russia
[2]Mordovian State Nature Reserve, Pushta, Temnikov district, Republic of Mordovia, Russia

ABSTRACT

The Republic of Mordovia is located in Central Russia in the boundary of the forest and forest-steppe zones. Here are the limits of distribution of many steppe species. At present, most steppe plots are used as arable lands and pastures, some of which are abandoned lands. Elements of steppe vegetation remained in plots that are unsuitable for agriculture.

As a consequence of this fragmentation process, steppe plants present a vulnerable conservation status in the Republic of Mordovia. Many species are included in the regional Red Data Book. The protected area (PA) network of Mordovia was established for the conservation of populations of rare species in the region. However, many populations of rare steppe species of the Republic of Mordovia remained outside these protected areas.

In this chapter we analyze the distribution of populations of rare steppe species in the protected area network of the Republic of Mordovia. Changes of cover of the regional conservation network after taking into account the new PAs recommended in recent years are presented and discussed.

Keywords: Republic of Mordovia, steppe, protected area, rare plant species, population

* Adress: Russia, Republic of Mordovia, Saransk, Bolshevistskaya Street, 68, Mordovian State University, 430005. Email: hapugin88@yandex.ru.

INTRODUCTION

Conservation of biodiversity becomes one of the major problems of today and is essential for the maintenance of the ecological balance in the biosphere. Inventory of flora, detecting rare and endangered species, creation of Red Data Books, and the establishment of protected areas (PAs) are among the primary objectives in conservation management and policy. Among other conservation priorities, creation of PAs networks serves to the establishment of links between populations of rare plants across regions (Silaeva, 2003; Bilz et al., 2011).

The territory of the Republic of Mordovia extends over the transition between forest and forest steppe biomes of central Russia (from 53°38' N to 55°11' N latitude and 42°11' E to 46°45' E longitude) (Figure 1). Its eastern part lies on the Volga Uplands, while the western portion covers the Oka-Don Lowland. Steppe vegetation is distributed irregularly in Mordovia. It is most widely distributed in the eastern part, on the western spur of the Volga Uplands. Here, where large areas have been covered by herbaceous steppe, its elements occur on a small number of "islands" of land unsuitable for agricultural use (on slopes of ravines, on hills and on indigenous banks in the river valleys). Therefore, it is very difficult to determine the area currently covered by steppe vegetation. Nevertheless, it has been estimated at about 0.3 % of the total area (26,200 square km) of the region (Yamashkin, 1998). However the number of steppe areas may increase due to vegetation overgrowth on sedimentary deposits, especially near steep slopes covered by steppe formations. This is associated with a reduction of grazing impact, reflected in the overgrowth of feather grass, especially *Stipa pennata*. On the slopes, however, this overgrowth corresponds mainly to mesophytic grasses. This occurs due to the reduction of grazing impact, frequently due to stabling of cattle. These areas are usually colonized by shrubs and trees. A significant number of populations of steppe plants are located here near the northern borders of their ranges. Many of these steppe species are included in the Red Data Book of the Republic of Mordovia (Silaeva, 2003).

Figure 1. Locality of the Republic of Mordovia in Russian Federation (Map with modifications from web site BankGorodov.RU: www.bankgorodov.ru).

The natural conditions of the Republic of Mordovia help separate several species subgroups of steppe plant communities (Silaeva et al., 2010a). Calciphilic-steppe plant species grow in areas with close bedding of limestone rocks. Psammophilic steppe species are associated to slopes on poor sandy soils; sometimes these species are also found in dry pine forests. Halophilic-steppe plants are associated to habitats characterized by soil salinity. Finally, species, which grow on plots of meadow steppes on leached chernozem, are attributed to the meadow-steppe subgroup.

In this chapter, we present a summary of inventory results on rare steppe species of flora in the Republic of Mordovia, and discuss the representativeness of the existing PA network in the region studied. In addition, we compare the level of protection achieved by the current PA network, and the level of protection that an extended network including new PA recommended in the period 2003–2012 would provide.

SPECIES DISTRIBUTION BY CATEGORIES OF THE RED DATA BOOK

The Red Data Book of the Republic of Mordovia (Silaeva, 2003) includes 170 species of vascular plants, including 65 representatives of the steppe flora element. Results of the regional Red Data Book have been published in previous years (Silaeva et al., 2004–2012). They included information about new findings of rare species in the Republic of Mordovia and about plants that have been recommended for inclusion in the regional Red Data Book for the first time (*Galatella angustissima, Astragalus asper, Valeriana rossica, Echinops ritro, Carex pediformis*), increasing the number of rare steppe flora elements in the Republic of Mordovia from 65 to 70 species, which are listed in Table 1.

Steppe vascular plants of the regional Red Data Book have categories of rarity from “0” to “4” that is shown in the Table 1.

Rare Steppe Plants on Protected Area Network in the Republic of Mordovia

There are 102 protected areas of different levels in the Republic of Mordovia: 2 protected areas have federal status: Mordovian State Nature Reserve (since 1936) and the National Park “Smolny” (since 1995). In addition, there are 98 natural monuments and 1 botanical garden of the Mordovian State University. In spite of this, the plant populations of the regional Red Data Book are poorly represented in the current protected area network.

Of the 1397 populations of rare plants, 1133 are located out the PAs network of the Republic of Mordovia (Silaeva et al., 2010a). This result shows that the existing protected area network in the region does not adequately protect the remaining populations of these species, and that rare steppe species of the Republic of Mordovia are clearly under-represented in this network. For example, all 15 known populations of *Artemisia pontica* are located outside it.

Similarly, only 3 out of 34 known populations of *Iris aphylla* (included in the Red Data Book of the Russian Federation, (Bardunov, and Novikov, 2008)) are outside the PAs network in the Republic of Mordovia (Silaeva et al., 2010a).

Table 1. Vascular plants of the Republic of Mordovia included in the Red Data Book rarity categories (asterisks indicate species proposed for inclusion in the main list in 2004–2012)

Category	Species	Number of species
0 – Probably extinct	*Stipa dasyphylla* (Lindem.) Trautv.	4
	S. pulcherrima C. Koch	
	Silene multiflora Pers.	
	Astragalus sulcatus L.	
1 – Endangered	*Helictotrichon desertorum* (Less.) Nevski	27
	H. schellianum (Hack.) Kitag.	
	Melica transsilvanica Schur	
	Stipa sareptana A. Beck. s. str.	
	S. zalesskii Wilensky	
	Allium globosum Bieb. ex Redoute	
	Krascheninnikovia ceratoides (L.) Gueldenst.	
	Arenaria procera Spreng.	
	Silene amoena L.	
	S. baschkirorum Janisch.	
	Clematis recta L.	
	Spiraea litvinovii Dobrocz.	
	Amygdalus nana L.	
	**Astragalus asper* Jacq.	
	Lathyrus pallescens (Bieb.) C. Koch	
	Linum perenne L.	
	Euphorbia rossica P. Smirn.	
	Helianthemum nummularium (L.) Mill.	
	Silaum silaus (L.) Schinz et Thell.	
	Scutellaria supina L.	
	Thymus cimicinus Blum ex Ledeb.	
	Scabiosa isetensis L.	
	Senecio integrifolius (L.) Clairv.	
	Inula germanica L.	
	**Galatella angustissima* (Tausch) Novopokr.	
	G. villosa (L.) Reichenb. fil.	
	Hieracium virosum Pall.	
2 – Vulnerable	*Elytrigia lolioides* (Kar. et Kir.) Nevski	30
	Stipa capillata L.	
	S. pennata L.	
	S. tirsa Stev.	
	Carex supina Wahlenb.	
	Allium flavescens Bess.	
	Iris aphylla L.	
	Silene sibirica (L.) Pers.	
	Delphinium cuneatum Stev. ex DC.	
	Anemone sylvestris L.	
	Adonis vernalis L.	
	Spiraea crenata L.	
	Potentilla arenaria Borkh.	
	Cotoneaster niger (Wahlenb.) Fries	
	Astragalus austriacus Jacq.	
	A. onobrychis L.	
	Linum flavum L.	
	Polygala sibirica L.	
	Onosma simplicissima L.	

Category	Species	Number of species
	Verbascum phoeniceum L.	
	Veronica spuria L.	
	Senecio schvetzovii Korsh.	
	Artemisia armeniaca Lam.	
	A. latifolia Ledeb.	
	A. pontica L.	
	A. sericea Web. ex Stechm.	
	Aster amellus L.	
	Galatella linosyris (L.) Reichenb. fil.	
	**Echinops ritro* L.	
	Centaurea ruthenica Lam.	
3 – Rare	**Carex pediformis* C.A. Mey.	5
	Rosa rubiginosa L.	
	Hypericum elegans Steph. ex Willd.	
	Orobanche coerulescens Steph.	
	O. elatior Sutt.	
4 – Indeterminate	*Hierochloë repens* (Host) Beauv.	4
	Carex diluta Bieb.	
	Dianthus campestris Bieb.	
	**Valeriana rossica* P. Smirn.	

0 – Probably extinct species. Populations of these plants have probably disappeared from the territory of the Republic of Mordovia. These plants have not been recorded in the wild during the past 50 years, either in points where the species were known to be formerly present, or at any other potential locations. Nevertheless, the possibility that some individuals or populations have been overlooked due to dormancy cannot be completely excluded. Under favorable conditions, plants develop vegetative and /or generative organs and become detectable. In this context, it is important to note that 3 species of this group (*Stipa dasyphylla, S. pulcherrima, Astragalus sulcatus*) changed category from 0 to 1, because new localities with presence were found after 80–90 years.

1 – Endangered species. Species whose populations have reached critically small sizes and / or their habitats have changed in such a way that their survival is unlikely if the impact of threat factors persists.

2 – Vulnerable species. Species characterized by steadily declining populations in the region, which can quickly fall into the category of endangered species if impacts of unfavorable factors persist.

3 – Rare species. Species of high vulnerability because of their small population size in the region. They are distributed over a limited area or a large scale, but in a very low density.

4 – Indeterminate species. Species whose populations could be classified into one of the previous categories, but information about their present state is insufficient to accurately determine their status.

The inventory of regional flora and the monitoring of the populations of rare and endangered species have been carried out since 2003, when the Red Data Book of the Republic of Mordovia was published (Silaeva et al., 2004–2012). In the course of these works, new localities with populations of rare steppe plants were found. Therefore, the establishment of new PAs was proposed, based on the analysis of species richness of steppe areas. The purpose of these measures is providing protection for the largest possible number of known populations of rare plants, mosses, fungi and lichens. Consequently, 75 new protected areas have been recommended to date, including 50 PAs with populations of rare steppe plants. All known data on the representation of rare steppe plants (i.e. included in the steppe ecological-cenotic group considered in the Flora of the Republic of Mordovia (Silaeva et al., 2010a)) in the existing PA network in the region are summarized in Table 2.

Table 2. Protected steppe species of vascular plants in the existing Protected Area (PA) Network of the Republic of Mordovia. Plots that have been recommended for protection in the 2004–2012 years are also considered (Asterisks indicate species proposed to be included in the main list of the re-issued regional Red Data Book)

Nº	Species of the Red Data Book of the Republic of Mordovia (Silaeva, 2003)	Number of steppe species populations			
		At the present stage		Recommendations of 2004–2012 years	
		Within PA network	Outside PA network	Within PA network	Outside PA network
1	*Helictotrichon desertorum*	0	6	6	0
2	*Helictotrichon schellianum*	0	8	8	0
3	*Hierochloe repens*	0	2	0	2
4	*Elytrigia lolioides*	0	8	4	4
5	*Melica transsilvanica*	0	2	2	0
6	*Stipa capillata*	1	26	17	10
7	*Stipa zalesskii*	0	4	4	0
8	*Stipa pulcherrima*	0	3	3	0
9	*Stipa dasyphylla*	0	3	3	0
10	*Stipa pennata*	4	93	32	65
11	*Stipa sareptana*	1	5	5	1
12	*Stipa tirsa*	0	13	9	4
13	*Carex supina*	0	12	12	0
14	*Carex diluta*	0	1	0	1
15	**Carex pediformis*	0	1	1	0
16	*Allium flavescens*	0	10	6	4
17	*Allium globosum*	0	1	1	0
18	*Iris aphylla*	4	32	24	12
19	*Krascheninnikovia ceratoides*	0	2	2	0
20	*Arenaria biebersteinii*	0	3	0	3
21	*Silene baschkirorum*	0	1	1	0
22	*Silene multiflora*	0	1	0	1
23	*Silene amoena*	0	4	2	2
24	*Silene sibirica*	0	8	3	5
25	*Dianthus campestris*	0	12	5	7
26	*Delphinium cuneatum*	2	21	7	16
27	*Anemone sylvestris*	2	33	15	20
28	*Clematis recta*	0	1	1	0
29	*Adonis vernalis*	6	52	26	32
30	*Spiraea crenata*	1	15	6	10
31	*Spiraea litvinovii*	1	0	1	0
32	*Cotoneaster niger*	2	3	2	3
33	*Potentilla arenaria*	0	6	0	6
34	*Rosa rubiginosa*	0	3	1	2
35	*Amygdalus nana*	1	3	4	0
36	**Astragalus asper*	0	1	1	0
37	*Astragalus austriacus*	0	18	9	9
38	*Astragalus sulcatus*	0	4	3	1
39	*Astragalus onobrychis*	0	11	3	8

N°	Species of the Red Data Book of the Republic of Mordovia (Silaeva, 2003)	Number of steppe species populations			
		At the present stage		Recommendations of 2004–2012 years	
		Within PA network	Outside PA network	Within PA network	Outside PA network
40	*Lathyrus pallescens*	1	1	1	1
41	*Linum flavum*	4	20	12	12
42	*Linum perenne*	0	3	3	0
43	*Polygala sibirica*	2	12	8	6
44	*Euphorbia rossica*	0	3	2	1
45	*Hypericum elegans*	3	14	11	6
46	*Helianthemum nummularium*	0	1	1	0
47	*Silaum silaus*	0	2	1	1
48	*Onosma simplicissima*	0	8	6	2
49	*Scutellaria supina*	0	1	1	0
50	*Thymus cimicinus*	2	1	3	0
51	*Verbascum phoeniceum*	1	21	10	12
52	*Veronica spuria*	0	10	6	4
53	*Orobanche elatior*	0	3	3	0
54	*Orobanche coerulescens*	0	1	1	0
55	**Valeriana rossica*	0	1	1	0
56	*Scabiosa isetensis*	0	1	1	0
57	*Senecio integrifolius*	0	4	2	2
58	*Senecio schvetzovii*	1	19	10	10
59	*Inula germanica*	1	1	1	1
60	*Artemisia armeniaca*	3	13	16	0
61	*Artemisia pontica*	0	18	11	7
62	*Artemisia sericea*	1	6	6	1
63	*Artemisia latifolia*	3	24	21	6
64	*Aster amellus*	4	27	16	15
65	**Galatella angustissima*	0	2	2	0
66	*Galatella villosa*	0	1	1	0
67	*Galatella lynosyris*	0	9	9	0
68	**Echinops ritro*	1	33	10	24
69	*Centaurea ruthenica*	1	7	6	2
70	*Hieracium virosum*	0	4	3	1
Total		53	703	414	342

As can be noticed, only 7 % of the known populations of steppe species are registered within the current PA network. Moreover, only 25 species occur in areas included in the network. Therefore, it can be concluded that it does not properly protect populations of steppe plants.

The ratio of number of populations in the PAs Network to the total number of populations of species in the region is very small for most rare steppe plant species in the Republic of Mordovia. For comparison, this relation for *Inula germanica* is 1:2, for *Thymus cimicinus* is 2:3, but it is even smaller for *Iris aphylla* – 4:36 and for *Linum flavum* – 4:24. Three out of 6 steppe plant species of the Red Data Book of the Russian Federation (Bardunov, and Novikov, 2008) are also located out of the existing protected area network in the Republic of Mordovia. These species are *Stipa dasyphylla*, *S. pulcherrima*, *S. zalesskii*.

Therefore, the problem of improving the existing protected area network in the Republic of Mordovia comes to the forefront. In the first place, it requires the designation of new protected areas, which should cover most of the known populations of rare plants in the Republic of Mordovia.

Fifty steppe plots have been recently recommended for in the PA Network of Mordovia (Silaeva et al., 2004–2012). Here we estimated the change of effectiveness of the PA network of the region taking these new (50) PAs into account. The proportion of populations of rare steppe species protected by the PAs network in the region will increase from 7.0 to 54.8 %.

According to Table 2, all species may be classified into three groups after their degree of protection under the PA Network:

- First group: 75–100 % of known populations of the species are located within the recommended protected area network in the Republic of Mordovia: *Helictotrichon desertorum*, *H. schellianum*, *Melica transsilvanica*, *S. dasyphylla*, *S. pulcherrima*, *S. sareptana*, *S. zalesskii*, *Carex supina*, *C. pediformis*, *Allium flavescens*, *A. globosum*, *Krascheninnikovia ceratoides*, *Silene baschkirorum*, *Clematis recta*, *Spiraea litvinovii*, *Amygdalus nana*, *Astragalus asper*, *A. sulcatus*, *Linum perenne*, *Helianthemum nummularium*, *Onosma simplicissima*, *Scutellaria supina*, *Thymus cimicinus*, *Orobanche coerulescens*, *O. elatior*, *Valeriana rossica*, *Scabiosa isetensis*, *Artemisia armeniaca*, *A. latifolia*, *A. sericea*, *Galatella angustissima*, *G. lynosyris*, *G. villosa*, *Centaurea ruthenica*, *Hieracium virosum*. It is worth noting that this group includes 4 species of the Red Data Book of the Russian Federation (Bardunov, and Novikov, 2008): *Stipa dasyphylla*, *S. pulcherrima*, *S. zalesskii*, *Thymus cimicinus*. All known populations of these species are in the recommended PA network, which implies the highest level of protection in the territory of Mordovia.
- Second group: it includes species that presented less than 75 % of all known populations within the recommended PA network: *Elytrigia lolioides*, *Stipa capillata*, *S. pennata*, *S. tirsa*, *Iris aphylla*, *Silene amoena*, *S. sibirica*, *Dianthus campestris*, *Delphinium cuneatum*, *Anemone sylvestris*, *Adonis vernalis*, *Spiraea crenata*, *Rosa rubiginosa*, *Astragalus austriacus*, *A. onobrychis*, *Linum flavum*, *Polygala sibirica*, *Euphorbia rossica*, *Hypericum elegans*, *Silaum silaus*, *Verbascum phoeniceum*, *Veronica spuria*, *Senecio integrifolius*, *S. schvetzovii*, *Inula germanica*, *Artemisia pontica*, *Aster amellus*, *Echinops ritro*. This group contains 2 species of the Red Data Book of the Russian Federation, (Bardunov, Novikov, 2008) (*Stipa pennata* and *Iris aphylla*) (Bardunov and Novikov, 2008), a significant number of whose populations (32 and 24 respectively) are located within the recommended protected areas in the Republic of Mordovia. They are assigned to the second group, and not to the first, because a significant number of populations of these species are registered in the region. Therefore, the number of locations of *Stipa pennata* and *Iris aphylla* within the recommended protected area network is relatively small.
- Third group: it includes species whose protection status will not change as a result of the improvement of protected areas recommended: *Hierochloe repens*, *Carex diluta*, *Arenaria biebersteinii*, *Silene multiflora*, *Spiraea litvinovii*, *Potentilla arenaria*, *Cotoneaster niger*, *Lathyrus pallescens*. The reasons for which the protection status

of these species will not change are varied. However the main reason is that locations of these steppe plants are single; they were recorded a long time ago, or in areas or communities where other steppe plants are absent.

Community Aspects of Rare Steppe Vascular Plants in Mordovia

At the community level, the steppe subgroup contains the greatest number of species (44, Silaeva et al., 2010a). The calciphilic-steppe subgroup ranked second with 19 species. Other minor subgroups are the meadow steppe (four species: *Clematis recta*, *Veronica spuria*, *Valeriana rossica*, *Senecio integrifolius*), the psamophilic-steppe (2 species: *Arenaria procera*, *Potentilla arenaria*) and the halophilic-steppe (1 species: *Carex diluta*) subgroups.

Unfortunately, halophilic-steppe and psammophilous-steppe plants will remain outside of protected area network, even if all the recommended PAs are implemented. The proportion of plant populations of the steppe, calciphilic-steppe and meadow-steppe community groups that would be within the PA network will increase to 75–100 %.

Therefore, the representativeness of protected area network of the Republic of Mordovia will increase by 8-fold (from 7.0 to 54.8 %), subject to the implementation of all the proposals on the organization of Protected Areas (Silaeva, 2003; Silaeva et al., 2004–2012). All populations of 4 out of 6 steppe species of the Red Data Book of the Russian Federation (Bardunov, and Novikov, 2008) that are present in the Republic of Mordovia (*Stipa dasyphylla*, *S. pulcherrima*, *S. zalesskii*, *Thymus cimicinus*) will thereby be within its protected area network.

REFERENCES

Bardunov, L.V., Novikov, V.S. (eds.). (2008). *Red Book of Russian Federation (plants and fungi)*. (2008). KMK Scientific Press Ltd., Moscow. [in Russian].

Bilz, M., Kell, S.P., Maxted, N., Lansdown, R. V. (2011). *European Red List of Vascular Plants*. Luxembourg, Publications Office of the European Union. Luxembourg.

Silaeva, T.B. (ed.). (2003). *Red Book of the Republic of Mordovia: Rare species of plants, lichens and fungi*. Mordovian Book Publishing House, Saransk. [in Russian].

Silaeva, T.B., Ageeva, A.M., Barmin, N.A., Kiryukhin, I.V., Pismarkina, E.V., Smirnov, V.M., Chugunov, G.G. (2004). *Rare plants and fungi: materials for maintenance of the Red Book of the Republic of Mordovia for 2004*. Publisher of the Mordovian State University, Saransk. [in Russian].

Silaeva, T.B., Kiryukhin, I.V., Pismarkina, E.V., Ageeva, A.M., Barmin, N.A., Vargot, E.V., Smirnov, V.M., Chugunov, G.G. (2005). *Rare plants and fungi: materials for maintenance of the Red Book of the Republic of Mordovia for 2005*. Publisher of the Mordovian State University, Saransk. [in Russian].

Silaeva, T.B., Kiryukhin, I.V., Pismarkina, E.V., Barmin, N.A., Chugunov, G.G., Ageeva, A.M., Vargot, E.V., Grishutkina, G.A., Smirnov, V.M. (2006). *Rare plants and fungi: materials for maintenance of the Red Book of the Republic of Mordovia for 2006*. Publisher of the Mordovian State University, Saransk. [in Russian].

Silaeva, T.B., Kiryukhin, I.V., Pismarkina, E.V., Barmin, N.A., Chugunov, G.G., Ageeva, A.M., Vargot, E.V., Grishutkina, G.A., Smirnov, V.M. (2007). *Rare plants and fungi: materials for maintenance of the Red Book of the Republic of Mordovia for 2007*. Publisher of the Mordovian State University, Saransk. [in Russian].

Silaeva, T.B., Kiryukhin, I.V., Pismarkina, E.V., Chugunov, G.G., Vargot, E.V., Ageeva, A.M., Smirnov, V.M., Khapugin, A.A. (2008). *Rare plants and fungi: materials for maintenance of the Red Book of the Republic of Mordovia for 2008*. Publisher of the Mordovian State University, Saransk. [in Russian].

Silaeva, T.B., Kiryukhin, I.V., Pismarkina, E.V., Chugunov, G.G., Ivoylov, A.V., Ageeva, A.M., Vargot, E.V., Smirnov, V.M. (2009). *Rare plants and fungi: materials for maintenance of the Red Book of the Republic of Mordovia for 2009*. Publisher of the Mordovian State University, Saransk. [in Russian].

Silaeva, T.B., Kiryukhin, I.V., Chugunov, G.G., Levin, V.K., Mayorov, S.R., Pismarkina, E.V., Ageeva, A.M., Vargot, E.V. (2010a). *Vascular plants of the Republic of Mordovia (synopsis of flora)*. Publisher of the Mordovian State University, Saransk. [in Russian].

Silaeva, T.B., Kiryukhin, I.V., Chugunov, G.G., Ageeva, A.M., Vargot, E.V., Pismarkina, E.V., Khapugin, A.A., Bolshakov, S.Yu., Ivoylov, A.V., Smirnov, V.M. (2010b). *Rare plants and fungi: materials for maintenance of the Red Book of the Republic of Mordovia for 2010*. Publisher of the Mordovian State University, Saransk. [in Russian].

Silaeva, T.B., Vargot, E.V., Khapugin, A.A., Chugunov, G.G., Ageeva, A.M., Bolshakov, S.Yu., Ivoylov, A.V., Grishutkin, O.G., Kiryukhin, I.V. (2011). *Rare plants and fungi: materials for maintenance of the Red Book of the Republic of Mordovia for 2011*. Publisher of the Mordovian State University, Saransk. [in Russian].

Silaeva, T.B., Vargot, E.V., Bolshakov, S.Yu., Khapugin, A.A., Chugunov, G.G., Ivoylov, A.V., Grishutkin, O.G., Kiryukhin, I.V. (2012). *Rare plants and fungi: materials for maintenance of the Red Book of the Republic of Mordovia for 2012*. Publisher of the Mordovian State University. Saransk, [in Russian].

Yamashkin, A.A. (1998) *Physical and geographical conditions of the Republic Mordovia*. Publisher of the Mordovian State University, Saransk. [in Russian].

In: Steppe Ecosystems
Editors: M. B. Morales Prieto and J. Traba Diaz
ISBN: 978-1-62808-298-2

Chapter 11

Recovery of Degraded Steppe Soils of Northern China after Grazing Exclusion

Martin Wiesmeier*
Lehrstuhl für Bodenkunde,
Technische Universität München, Germany

Abstract

In the last decades, land use changes led to severe degradation of semi-arid steppe ecosystems in Northern China. In particular, overgrazing due to rising stocking rates of grazing animals resulted in a considerable deterioration of steppe soils. Besides soil erosion, degraded steppe soils are subject to compaction of the topsoil, destruction of soil aggregates, and a deterioration of chemical and physical soil properties. In order to restore heavily degraded soils in Northern China, cessation of grazing was proposed. In this study, we present results from short- and long-term experimental sites in Inner Mongolia, where grazing was excluded. Grazing cessation resulted in a significant improvement of topsoil conditions already after few years that continued after 30 years. This recovery was related to a higher input of particulate organic matter (POM) and to an increased physical stabilization of soil oragnic matter (SOM) due to increased formation of soil aggregates. Enhanced soil aggregation can be attributed to a greater input of organic matter acting as binding agents in combination with an exclusion of animal trampling that physically protects SOM from degradation. An analysis of the spatial distribution of SOM showed the formation of a heterogeneous soil pattern after grazing exclusion at different spatial scales. Apparently, the recovery of degraded steppe soils starts with the formation of "islands of fertility", where a higher input of water and deposition of windblown material promotes the development of vegetation and associated SOM patches.

However, it has to be noted that increased amounts of SOM after grazing cessation are very sensitive to a resumption of land use and sequestered mainly in readily decomposable POM fractions. This generally questions a sustainable management of

* Corresponding Author address: Lehrstuhl für Bodenkunde, Wissenschaftszentrum Weihenstephan für Ernährung, Landnutzung und Umwelt, Technische Universität München, Emil-Ramann-Str. 2, 85354 Freising, Germany. Email: wiesmeier@wzw.tum.de.

intensively grazed steppes and emphasizes the high sensitivity of this semi-arid ecosystem to land use intensification.

Keywords: Soil organic carbon (SOC), overgrazing, semi-arid grassland, soil aggregation, soil organic matter (SOM) stabilization

INTRODUCTION

Degradation of semi-arid and arid environments is a global environmental problem, particularly in Northern China, where around 2.5 million km² are affected by desertification (Yang et al., 2005). In the autonomous province of Inner Mongolia in Northern China, the change from a sustainable nomadic culture to a static grazing management over the last 50 years has resulted in severe degradation of the semi-arid steppes (Akiyama and Kawamura, 2007). Moreover, due to a large increase of the population in China, agricultural use further expands in the boundary areas of semi-arid steppe regions where intensification of land use can lead to a severe deterioration of this ecologically sensitive environment (Christensen et al., 2004; Wiesmeier et al., 2011; 2012b). In some regions of Inner Mongolia, up to 72% of the total area are classified as degraded steppe (Tong et al., 2004). Detrimental implications for physical and chemical topsoil properties of the steppes in Northern China due to heavy grazing have been documented in several publications (e.g. Steffens et al., 2008; Su et al., 2005; Xie and Wittig, 2004). In particular, semi-arid grasslands are of global importance for carbon sequestration as they store approximately 15% of total soil organic carbon (SOC) (Anderson, 1991; Lal, 2004). Intensification and changes of land use caused large losses of SOC in these sensitive ecosystems during past decades (Guo and Gifford, 2002; Houghton, 1999).

In order to prevent grasslands from considerable ecological and economical damages and to establish a sustainable land-use management, a Sino-German research project (MAGIM, Matter fluxes in grasslands of Inner Mongolia as influenced by stocking rate) was conducted in the semi-arid steppes within the Xilin River catchment in Inner Mongolia. Besides a detailed investigation of the processes leading to degradation of the grasslands induced by intensive grazing, several management options were assessed to restore degraded steppe soils. As most effective and simple method, grazing exclusion was proposed that seems to be a promising approach for the recovery of intensively grazed steppe soils. However, information about a beneficial effect of grazing cessation on chemcical and physical soil properties and underlying processes is limited. The amount of soil organic matter (SOM) is assumed to increase due to higher input of organic matter (OM) into the soil as grazing exclusion led to a increase of above- and belowground net primary productivity in the study area (Gao et al., 2008). Furthermore, a high surface roughness as a consequence of increased height and density of the vegetation led to an enhanced deposition of the C-rich material (Hoffmann et al., 2008a, 2008b).

The recovery of degraded steppe soils after grazing exclusion could further be attributed to an improvement of the soil structure and related physical soil properties as a consequence of decreased animal templing. In particular, a possible enhanced soil aggregation would significantly contribute to the recovery as a large proportion of SOM is stabilized in soils by physical protection within aggregates (Jastrow, 1996; Six et al., 2002).

At ungrazed sites, soil aggregation is assumed to be promoted by a higher input of OM acting as a binding agent (Jastrow et al., 1998; Oades, 1984) and due to the absence of mechanical stress.

Moreover, little attention has been drawn to the importance of the spatial distribution of soil properties for ecosystem functioning of semi-arid grasslands. Under semi-arid and arid climatic conditions, vegetation is arranged in patches with relatively dense cover separated by more or less bare soil (Aguiar and Sala, 1999; Burke et al., 1998).

These vegetation patches change the microenvironment, as they act as surface obstructions that increase water infiltration, reduce radiation and surface temperatures and accumulate water and windborne sediments, seeds, litter and other organic materials (Bhark and Small, 2003; Garner and Steinberger, 1989). It is unclear whether heterogeneous vegetation patterns also lead to a spatial variability of SOM and other soil properties in undisturbed steppes of Northern China. A detailed knowledge of the spatial distribution of soil organic matter is important for SOC inventories (Wang et al., 2002). Moreover, the maintenance of heterogeneous vegetation patterns is of particular importance for the stability and productivity of the steppe ecosystem (Noy-Meir, 1973; Tongway and Ludwig, 1990). Relatively small amounts of water from single rainfall events are insufficient for plant growth in semi-arid environments.

Only the redistribution of run-off water from bare soil with low infiltration to vegetation patches with a higher infiltration enables primary production there and creates a heterogeneous pattern of vegetation (Tongway and Ludwig, 2005). Therefore, heterogeneous vegetation patterns are a crucial precondition for the productivity and stability of semi-arid environments (Aguiar and Sala, 1999; Tongway and Ludwig, 2001).

The aim of this study was to investigate the impact of grazing exclusion on the amount, the physical protection due to aggregation and the spatial distribution of SOM in the semi-arid grasslands of Northern China. To investigate short-term effects of grazing cessation, a controlled grazing/grazing exclusion experiment was conducted. Basic soil parameters such as SOC, total nitrogen (N_{tot}) and bulk density (BD) were determined at the beginning of the experiment in 2005 and again after three years in 2008. At long-term ungrazed sites which were fenced in 1979, soil properties were investigated in detail both at *Stipa grandis* dominated grasslands that represent drier soil conditions and at wetter *Leymus chinensis* dominated areas which account together for 60% of the study area (Chen et al., 2005b; Tong et al., 2004).

Besides basic soil parameters, soil fractions, aggregate stability and carbon mineralization as indicators for the aggregation and stabilization of SOM were determined. As grazing intensity presumably will increase in Northern China in the future and grazing management further extends into undisturbed boundary areas of the semi-arid grasslands, we also investigated the effect of grazing induced mechanical stresses exceeding the natural soil stability (precompression stress) on SOC mineralization. High grazing intensities were simulated by artificial mechanical compaction of undisturbed soil samples. Furthermore, the spatial distribution of chemical and physical soil parameters was analyzed at the plant scale, as studies in arid and semi-arid environments worldwide emphasized the importance of small-scale investigations at distances of cm to several m (Ludwig et al., 1999a; Schlesinger et al., 1996).

METHODS

Study Area

The study area is a 3600 km² large section of the Xilin River basin (43°24′ to 44°40′ N and 115°20′ to 117°13′ E) located in the Xilingol steppe at the southeast Mongolian plateau in the autonomous province Inner Mongolia, Northern China (Figure 1). The Chinese Academy of Sciences maintains the Inner Mongolia Grassland Ecosystem Research Station (IMGERS) within the study area that is located about 60 km southeast of the city of Xilinhot.

Elevation in the area ranges between 1010 and 1609 m above sea level. The region is part of the continental semi-arid grasslands of the Central Asian steppe ecosystem, with a dry and cold mid-latitude climate (Kawamura et al., 2005). Mean annual temperature at IMGERS is 0.7 °C and mean annual precipitation is around 350 mm, with the highest values in the summer from June to August. The vegetation period from May to September is relatively short (<150 days). The prevalent vegetation is grassland, which is typically used for semi-nomadic or static grazing with sheep and goats. Dominant geological units are quarternary deposits in the center of the catchment, shales from the Tertiary and Cretaceous period in the north and a wide basalt plateau in the southwest of the basin. Totally, nine reference soil groups (RSGs) were identified in the Xilin River catchment according to the World Reference Base (WRB) for Soil Resources (IUSS Working Group WRB, 2006). The largest part of the catchment is covered by the typical steppe soils Phaeozems and Chernozems. Further RSGs are Gleysols, Kastanozems, Arenosols, Calcisols, Regosols, Cryosols and Cambisols. Further information on the study area can be found in Wiesmeier et al. (2011) and Barthold et al. (2013).

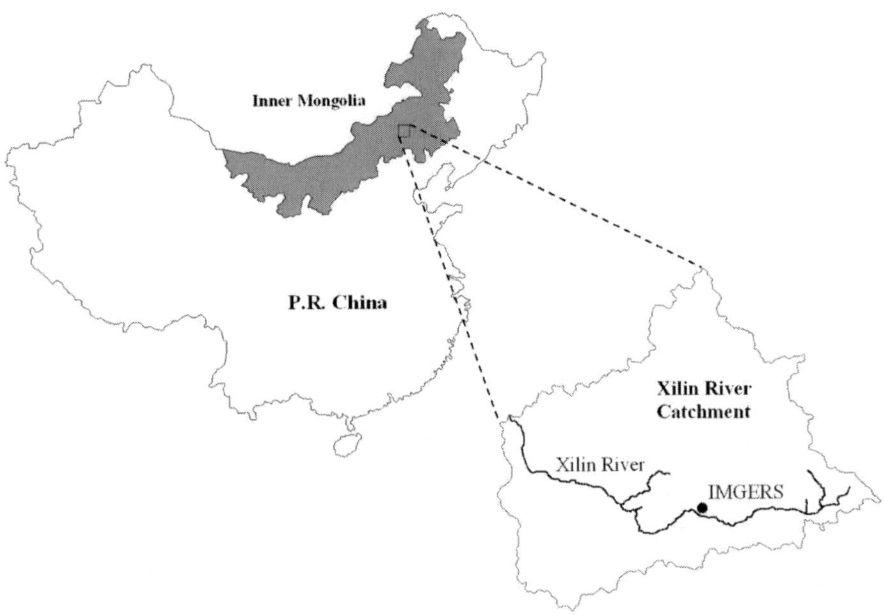

Figure 1. Location of the Xilin River Catchment with the Inner Mongolia Grassland Ecosystem Research Station (IMGERS) in Inner Mongolia, Northern China.

Short-term ungrazed experimental sites

In order to investigate the impact of short-term grazing cessation, a controlled grazing/grazing exclusion experiment was established in 2005 on a grazed typical steppe next to a farm (Schönbach et al., 2009, 2010; Wiesmeier et al., 2012a). Until 2003, grazing intensity was moderate and thereafter, the area recovered from grazing for two years before the experiment started in June 2005. Approximately 160 ha were fenced and divided into plots of 2 ha with different grazing intensities. In this study, ungrazed plots (UG05) were compared to heavily grazed plots with stocking rates of 4.5 (HG4.5) and 7.5 (HG7.5) sheep units per hectare, respectively. Grazing animals were 2-year-old female sheep, averaging 35 kg initial live-weight, and were kept continuously on the plots throughout the grazing season from June to September. Soil samples from the upper 4 cm of the soil were taken from eight UG05 plots, and four HG plots in June 2005 before the grazing experiment started and again after three grazing seasons in June 2008. At each plot three undisturbed soil samples were taken from five randomly selected locations using steel cylinders with a volume of 100 cm^3 and bulked to obtain a composite sample. Soil material was dried at 105°C for 24 h, weighed, sieved (2 mm) and analysed for soil texture, pH, BD, SOC and N_{tot} concentration.

Long-Term Ungrazed Experimental Sites

To reveal the long-term effect of grazing exclusion on amount and spatial distribution of SOM, continuously grazed (CG) and ungrazed plots, which were fenced in 1979 (UG79), were sampled at both *Leymus chinensis* and *Stipa grandis* dominated steppe types (Steffens et al., 2008; Wiesmeier et al., 2009). Before 1979, all areas were exposed to light gazing intensity (< 1 sheep unit ha^{-1} a^{-1}). While the stocking rates of sheep and goats rose at the continuously grazed sites after 1979 to a moderate level (approximately 1.2 sheep units ha^{-1} a^{-1}), grazing was excluded at the fenced sites. At all experimental sites, grids with 100 sampling points as well as soil pits were sampled to determine several soil properties, such as soil texture, BD, SOC, N_{tot}, pH values, contribution of soil fractions to SOM and aggregate stability. To elucidate the small-scale spatial variability of SOM, 40 randomly selected points (20 points at bare soil areas and 20 points under vegetation patches) were sampled inside an area of 2 × 2 m at CG and UG79 sites from both steppe types (Wiesmeier et al., 2009). At each point, three undisturbed soil samples were taken from the upper 4 cm of the topsoil using a steel cylinder with a volume of 100 cm^3 and bulked to obtain a composite sample. To determine the thickness of the Ah horizon and the depth of the occurrence of calcium carbonate ($CaCO_3$), a soil auger was used at each sampling point.

In order to investigate the effect of grazing exclusion on soil respiration, an incubation experiment was conducted (Wiesmeier et al., 2012b). Undisturbed soil cores were taken from the upper 10 cm of the topsoil at 10 randomly selected locations within a radius of 5 m around the soil pits from all experimental sites using steel cylinders with a volume of 235 cm^3. Samples were air dried and four randomly selected soil cores (at least one core from each pit location) were used for the determination of water holding capacity (WHC) and basic soil parameters and for the selection of 20 soil aggregates (10 – 20 mm) from each experimental site in order to quantify aggregate stability. From the remaining six undisturbed samples for each site, three soil cores were mechanically compacted and precompression stress (σ_p) was measured. The three compacted and the remaining three undisturbed soil cores from each soil pit location were further used for the incubation experiment.

Soil Chemical and Physical Analyses

Bulk density was quantified with the mass of the oven-dried soil (105° C) divided by the core volume (Hartge and Horn, 1989). Soil pH values were measured in 0.01 M $CaCl_2$ at room temperature. Water holding capacity was measured in triplicate by calculating the water content after draining the soil samples over a porous plate for 24 h at 10 kPa.

Soil organic carbon and N_{tot} were determined in duplicate by dry combustion on a Vario Max CNS elemental analyser (Elementar Analysensysteme, Hanau, Germany). The measured C concentrations of the samples that were free of carbonate represent the SOC concentration. Samples that contained $CaCO_3$ were heated to 500 °C for 4 hours to remove organic carbon and the concentration of inorganic C of the residual material was determined by dry combustion. The content of inorganic C was subtracted from the C_{tot} concentration of the untreated material and represents the SOC content.

For the analyses of soil texture, soil samples sieved to <2 mm were oxidized with H_2O_2 to remove organic matter. The remaining material was dispersed with $Na_4P_2O_7$ and shaken for 16 to 24 hours, followed by wet sieving to isolate sand fractions >63 µm. To determine silt and clay fractions, approximately 3 g of the <63 µm fraction was suspended in deionized water with $Na_4P_2O_7$ and an ultrasonication for 3 minutes with 75 J ml^{-1} was conducted. Afterwards, the distribution of silt and clay fractions was obtained by measuring the X-ray absorption of the soil-water suspension during sedimentation of the soil particles with a Micromeritics Sedigraph 5100 (Micromeritics, Norcross, USA) (Spörlein et al., 2004).

Bulk soil samples (*Stipa*) and aggregate size classes (*Leymus*) were separated by a physical fractionation procedure in order to obtain meaningful soil fractions (Steffens et al., 2009b). The free particulate organic matter (fPOM) was separated using a Na-polytungstate solution (ρ = 1.8 g cm^{-3}). The floating fPOM was extracted by sucking via a water jet pump. For the determination of POM occluded in aggregates (oPOM), the heavy fraction (>1.8 g cm^{-3}) was treated by ultrasonication. An energy input of 150 J ml^{-1} was applied to disrupt all macroaggregates and to obtain highest similarity of clay yields compared to standard particle size analysis, but to minimise the production of artefacts following heavy ultrasonication (Schmidt et al., 1999). With a subsequent density fractionation step (Na-polytungstate solution, ρ = 1.8 g cm^{-3}), the oPOM floating on the suspension was obtained after centrifugation (10 min at 3500 rpm). To remove the Na-polytungstate from the POM fractions, the samples were washed with bi-distilled water over a 20 µm sieve and by subsequent filtration of the remaining solution to 0.22 µm a small oPOM fraction (oPOMs) was obtained.

The remaining mineral associated organic matter (MOM) was divided into coarse and medium sand (cS), fine sand (fS) and coarse silt (cSi) fractions by sieving at 200 µm, 63 µm and 20 µm, respectively. Finer particles were separated by sedimentation (Atterberg) into medium silt (20-6.3 µm; mSi), fine silt (6.3-2.0 µm; fSi), coarse and medium clay (0.2-2.0 µm; cC) and by centrifugation into fine clay (<0.2 µm; fC). All particle size fractions were washed with bi-distilled water in order to remove excessive salt, freeze-dried, ground and analyzed in duplicate for organic carbon and N_{tot} concentrations by dry combustion on a Vario EL elemental analyser (Elementar Analysensysteme, Hanau, Germany). In order to compare the soil fractions of bulk soil material from both steppe types, the proportions of the soil fractions from the three aggregate size classes determined for *Leymus* sites were mathematically recombined in relation to their relative proportion to the bulk soil to bulk soil fractions.

In order to quantify macroaggregate stability, tensile strength (TS) was measured using the crushing method (Dexter and Kroesbergen, 1985). The aggregates were crushed individually between two flat parallel plates using a loading frame (Instron GmbH, Pfungstadt, Germany) and the force (*F*) that was needed to break up the aggregate was recorded. The TS was calculated using following equation:

$$TS = 0.576\left(\frac{F}{D^2}\right) \quad \text{(Equation 1)}$$

where 0.576 is a coefficient for the ratio between the compressive and the tensile stress within the aggregate and *D* is the mean aggregate diameter:

$$D = \frac{d_x + d_y + d_z}{3} \quad \text{(Equation 2)}$$

where d_x is the longest, d_y is the intermediate and d_z is the smallest diameter of each aggregate. Mechanical compaction of undisturbed, air-dried soil samples was conducted with a standard oedometer device and precompression stress was determined from the obtained compression curve (Krümmelbein et al., 2006). Stresses were applied with a pneumatic piston under confined compression. During the whole compaction process the vertical displacement (settlement) of the soil sample was recorded by a potentiometric displacement transducer. To avoid compacted and disturbed zones around the tensiometer tip, a hole of the size of the tensiometer tip was prepared with a hollow drill. The soil samples were compacted under static loading with stepwise increasing loads of 1, 10, 20, 30, 40, 50, 60, 80, 100, 120, 150, 300 and 400 kPa for 10 min per loading step. Precompression stress values were determined graphically according to Casagrande (1936). All soil physical investigations were conducted at the Christian-Albrechts-University, Kiel.

For determination of carbon mineralization, three undisturbed and three compacted soil cores from each experimental site were incubated for a period of 75 days at a constant temperature of 20° C with an automated incubation system according to Heinemeyer et al. (1989) and Mueller (2009). The compaction of soil cores was conducted with a standard oedometer device and stresses were stepwise applied with a pneumatic piston under confined compression with a maximum loading of 400 kPa in order to destroy soil aggregates (Krümmelbein et al., 2006). Before incubation all samples were pre-incubated for two weeks in order to obtain constant respiration rates for each sample. Undisturbed as well as compacted soil samples were transferred into plastic sample tubes with a volume of 165 cm^3 with several holes capped with polyester fibre of 23 µm mesh size (PET 1500, Sefar AG, Heiden, Switzerland) in order to improve gas diffusion and placed in airtight sealed glass bottles. Soil material from compacted samples was transferred stepwise to incubation tubes so that the original BD was obtained in order to ensure the comparability of all samples in terms of pore volume. Immediately before incubation, all samples were wetted to 70% WHC and inoculated with 100 µl of a soil solution derived from field-moist topsoil samples from the studied sites in order to provide comparable starting conditions. The inoculum was uniformly injected within the soil sample with a syringe. The samples were continuously flushed with ambient air that was free of CO_2 due to absorption with soda lime. In order to maintain a constant water content, the air was purged with deionized water acidified by H_2SO_4 adjusting a pH of 2.5. The CO_2 concentration of the incubated samples was measured hourly by an infrared CO_2-analyzer (Type 225 Mk3, ADC BioScientific Ltd., England) and the air flow rate by a flow meter (Tylan FM 260, Mykrolis, USA). The calculated release of CO_2-C was related to the OC content of the soil samples.

Statistical and Geostatistical Analyses

All statistical analyses were carried out using SPSS 15.0 (SPSS Inc., Chicago, USA). Descriptive statistical methods were used to describe the central trend and the spread of the data set, which included extreme maximum and minimum values, mean, median, standard deviation, skewness, kurtosis, variance and coefficient of variation. To test the significance of grazing exclusion effects on the examined parameters, a one-way analysis of variance (ANOVA) combined with post hoc tests (LSD, Tukey and Scheffé) and a Student`s T-test were applied. The significance of correlations between various parameters was expressed by Pearson`s correlation coefficient.

For all geostatistical analyses R 2.6.1 (Vienna, Austria) in combination with the package gstat (Pebesma, 2004) was used and experimental semivariograms were computed to characterize the spatial pattern of SOC, N_{tot} and BD at the plant scale (Webster and Oliver, 2001). This method describes the spatial variation of a regionalized variable by measuring the degree of spatial dependence as a function of the distance between the sampling points. Prior to geostatistical analysis, the distributions of all the parameters were tested for normality using the Kolmogoroff-Smirnoff test. The semivariance γ is calculated as half the average of squared differences between measurements at points separated by the distance interval h:

$$\gamma(h) = \frac{1}{2N(h)} \sum_{i=1}^{N(h)} \left[A_i(x_i) - A_i(x_i + h)\right]^2 \qquad \text{(Equation 3)}$$

where $N(h)$ is the number of pairs of points separated by the distances h, $A_i(x_i)$ is a measured property at point x_i, and $A_i(x_i + h)$ are measurements of the same property at various distances h. Models were used to fit the form of the semivariograms on the basis of visual accordance and minimized sum of squares. Most semivariograms were fitted with a spherical model (Webster and Oliver, 2001). When no spatial dependence was detected over the sampling area, a pure nugget model was fitted. In order to visualize the spatial patterns of SOC, N_{tot} and BD, ordinary kriging was applied, which interpolates values for the investigated sites on the basis of the calculated semivariograms. The maps were generated with the geostatistical analyst of the ArcMap 9.2 software (ESRI, Redlands, USA).

Results and Discussion

Impact of Short-Term Grazing Exclusion on Steppe Soils

In 2005 elemental concentrations and BD were similar at all sites with 20 mg g^{-1} SOC, 2.1 mg g^{-1} N_{tot} and a BD of 1.27 g cm^{-3} (Figure 3). In 2008 after 3 years of grazing exclusion, UG05 plots showed significantly ($P<0.05$) increasing SOC and N_{tot} concentrations compared with 2005. Bulk density was significantly lower ($P<0.05$) and decreased to 1.15 g cm^{-3}. After three seasons of heavy grazing in 2008, elemental concentrations tended to decrease while BD remained almost unchanged at HG sites. Soil organic carbon and N_{tot} were 10–18% lower compared with the results of 2005, but differences were only significant for HG4.5 ($P<0.05$) and not for HG7.5 ($P = 0.06–0.22$). Although, differences between UG05 and HG sites were significant ($P<0.05$) in 2008 after three years of grazing but not in 2005 before the start of the grazing experiment.

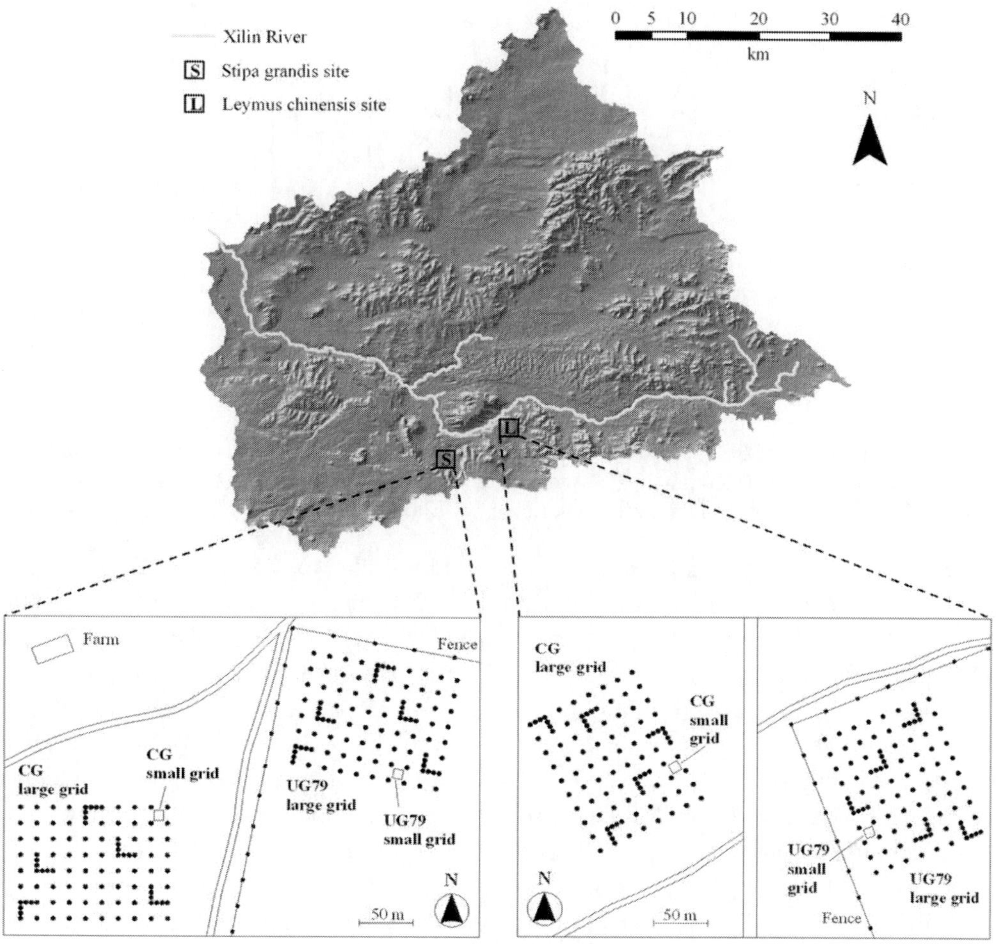

Figure 2. Location of the *Stipa grandis* and *Leymus chinensis* dominated grassland locations with the sampling grids at continuously grazed (CG) and long-term ungrazed (UG79) sites.

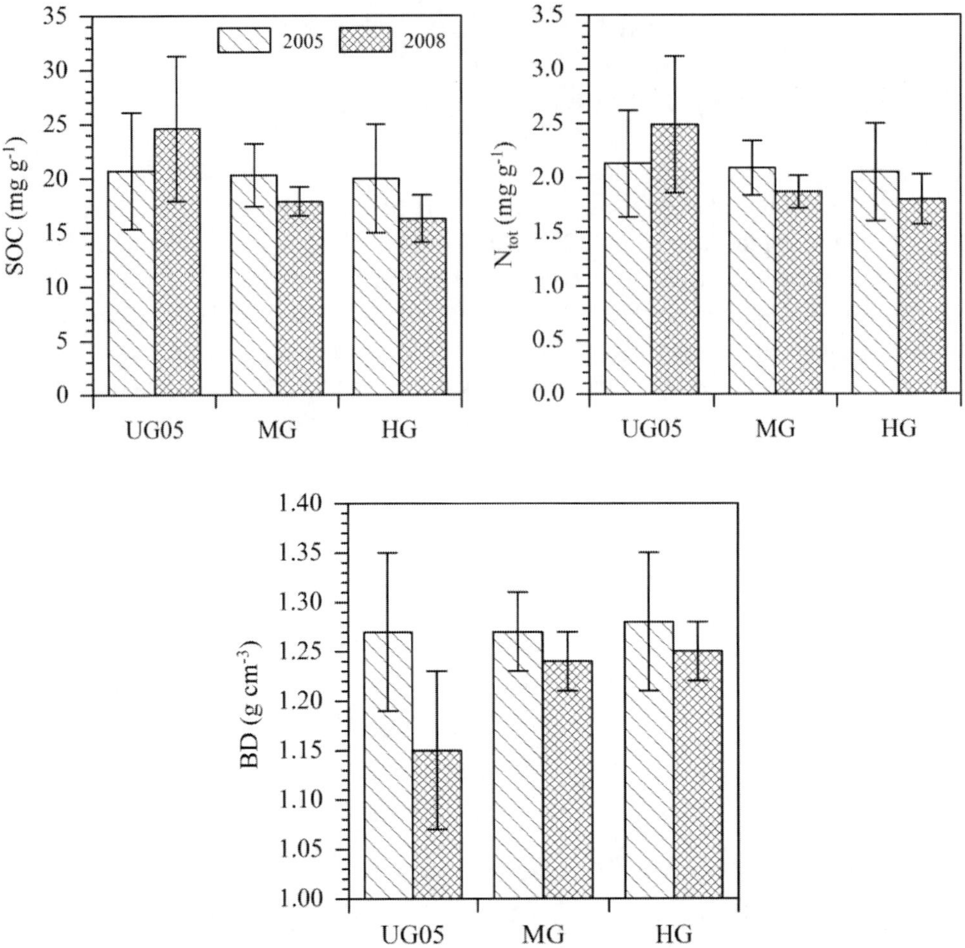

Figure 3. Concentrations of SOC, N_{tot} and BD determined in 2005 and 2008 of short-term ungrazed (UG05) and heavily grazed (HG4.5, HG7.5) sites. Error bars represent standard deviation (n = 40 for UG05, n = 10 for HG4.5 and HG7.5, (Wiesmeier et al., 2012a).

Elemental stocks calculated for the first 4 cm of the topsoil from all plots were comparable in 2005 with amounts of 1.0 kg m^{-2} for SOC and 0.1 kg m^{-2} for N_{tot}. In 2008, UG05 plots revealed significantly ($P<0.05$) higher SOC and N_{tot} stocks whereas HG4.5 and HG7.5 sites tended to show lower elemental stocks, but differences were not significant.

Results from the initial soil sampling in 2005 showed that the starting conditions for all studied plots of the grazing experiment were comparable in terms of physical and chemical soil properties (Figure 3). Grazing exclusion for three years resulted in a considerable increase of SOM associated with a decrease in soil compaction, indicated by significantly higher concentrations and stocks of SOC and N_{tot} and significantly lower BD. Obviously, the recovery of continuously grazed steppe soils after cessation of grazing takes place in short time periods. A short-term improvement of grassland soils after grazing exclusion was also indicated for the Alxa steppe in Inner Mongolia, where SOC and N contents were 22% and 14% higher after six years of grazing cessation (Pei et al., 2008). Other investigations in semi-arid grasslands of Northern China also reported a significant increase of SOM and a decrease of BD when grazing was excluded for 3 – 5 years, but evidence for these findings is missing as the starting conditions were not investigated (Su et al., 2003; Su et al., 2004; Wu et al., 2008).

In contrast to the positive effects at UG05 sites, heavy grazing for three years resulted in a considerable reduction of SOM in the topsoil of HG4.5 and HG7.5 plots, although the differences compared with 2005 were only significant for HG4.5 plots. The processes leading to degradation of topsoils of long-term grazed grasslands in Inner Mongolia are well documented and understood (e.g. (Steffens et al., 2008; Wiesmeier et al., 2009). Nevertheless, the observed short-term changes three years after intensified grazing pressure were not expected. Zhou et al. (2008) investigated a degraded grassland area with different desertification gradients in Northern China and assumed that most of SOC and N_{tot} is lost in early stages of the desertification process but no results are available from controlled grazing experiments with known starting conditions.

The loss of SOM can be ascribed to a limited input of above- and belowground biomass, destruction of soil aggregates due to increased animal trampling associated with mineralization of formerly protected organic matter and increased soil erosion, particularly of SOM-rich fine material (Hoffmann et al., 2008b; Steffens et al., 2009b). In general, a negative correlation of SOC and BD was found at all plots (Wiesmeier et al., 2012a). However, HG4.5 and HG7.5 sites showed almost unchanged BD but decreasing SOC contents in the third year of the grazing experiment (Figure 3). Continuous grazing for several decades has probably caused a relatively high pre-compression level (Krümmelbein et al., 2006) and the number of additional sheep at HG4.5 and HG7.5 sites did not result in further topsoil compaction. The observed rapid decrease of SOM after short-term intensive grazing reveals that semi-arid grasslands are very sensitive to land use changes and intensification.

Steppe Soils as Affected by Long-Term Grazing Cessation

Soil Organic Matter and Bulk Density

The concentrations of SOC, N_{tot} and BD determined at 100 grid points from the topsoil of all sites are given in Figure 4. In general, all elemental concentrations were significantly higher ($P<0.01$) at long-term ungrazed sites from both *Stipa* and *Leymus* steppe types compared to grazed sites. SOC contents of UG79 sites were 37% higher at the ungrazed site from the *Stipa* steppe and 35% higher at *Leymus* steppe. For N_{tot}, 28% and 31% higher concentrations were determined for the long-term ungrazed sites, respectively.

C/N ratios were significantly different ($P<0.01$) between CG and UG79 sites, but were in a narrow range of 9.1 to 9.8. Bulk density decreased under long-term grazing exclusion and was 19% lower compared to the grazed site at the *Stipa* steppe type and 20% lower at the *Leymus* dominated steppe type.

Beyond differences in elemental concentrations and BD as a result of grazing exclusion, there were also regional differences between *Stipa* and *Leymus* dominated steppe types under the same management. Both *Stipa* sites, which are supposed to be drier and less productive, showed significantly ($P<0.01$) lower contents of SOC, N_{tot} and higher BD than the *Leymus* sites with a higher soil moisture and primary production. At the fenced site of the *Stipa* steppe type, concentrations of SOC and N_{tot} were 28% and 27% lower compared to the ungrazed *Leymus* dominated site.

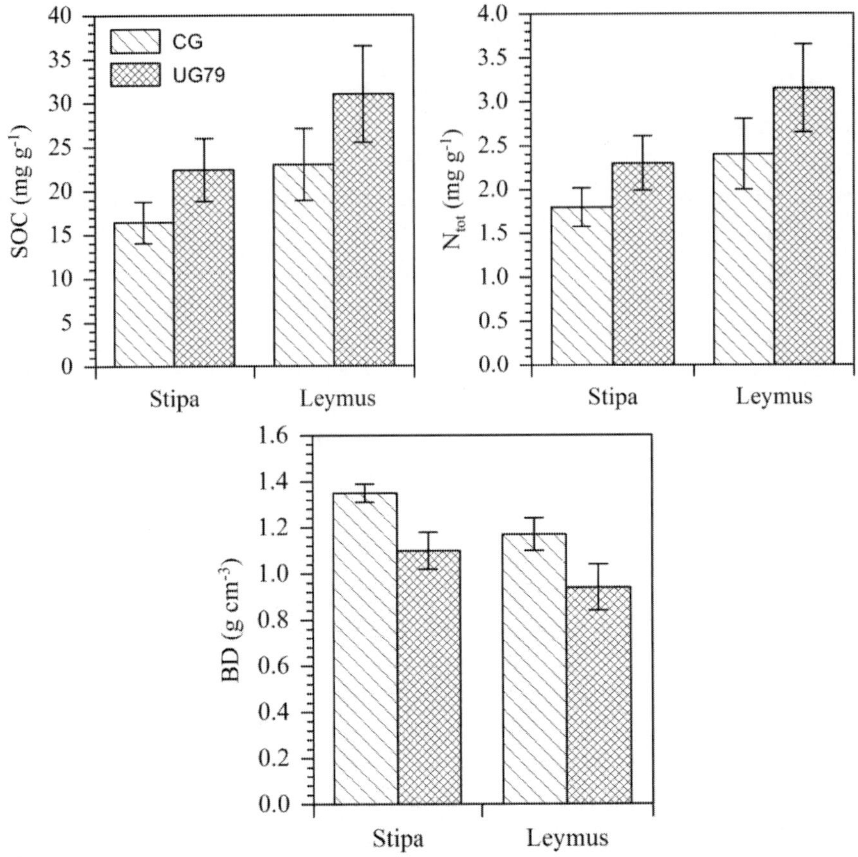

Figure 4. Concentrations of SOC, N_{tot} and BD at continuously grazed (CG) and long-term ungrazed (UG79) sites. Error bars represent standard deviation (n = 100, (Wiesmeier et al., 2009).

Under continuous grazing management, the respective contents of the *Stipa* site were 29% and 25% below the concentrations of the *Leymus* site. An opposite trend was found for BD, with significantly ($P<0.01$) higher values for the continuously grazed (+ 13%) and ungrazed (+ 15%) *Stipa* dominated site compared to CG and UG79 of the *Leymus* steppe type.

Continuous grazing was responsible for a significant reduction of SOM in the topsoils of the grazed sites compared to ungrazed sites at both steppe types. Steffens et al. (2008) discussed several processes to account for the degradation of SOM at intensively grazed grasslands. Primarily, grazing reduces above- and belowground biomass of grassland species (Gao et al., 2008), associated with decreasing incorporation of organic matter into the soil, as consumed biomass is not redistributed to the grazed areas, because sheep are held in paddocks overnight and the dung collected there is used as heating material. Furthermore, animal trampling results in soil compaction indicated by increased BD, which is followed by a mineralization of SOC released from destroyed macroaggregate structures (Six et al., 2002). A shifted negative correlation between BD and SOC content at CG sites compared to UG79 sites (Wiesmeier et al., 2009) supports the assumption of soil compaction and enhanced decomposition of SOC. On the other hand, the degree of soil aggregation is generally lower under limited input of organic materials that act as binding agents, and thus stabilization of

SOC in aggregates is probably lower at CG sites with reduced litter input. Besides, with decreasing height and density of vegetation, soil erosion is also increased (Hoffmann et al., 2008b), leading to a depletion of SOM.

In general, *Stipa* dominated sites showed lower concentrations of SOM and higher BD compared to *Leymus* dominated sites. Obviously there are considerable differences between *Leymus chinensis* and *Stipa grandis* dominated steppe types. Both fenced sites had been ungrazed for the same period of time since 1979, grazing intensity was not different before fencing, and climatic conditions were similar due to the short distance of approximately 10 km between the sites, but the *Stipa* dominated site had significantly lower contents and stocks of SOC and N_{tot}. These differences are most likely explained by a lower biomass production at the Stipa site. For the period between 1980 and 1989, Xiao et al. (1995) found 20% lower peak aboveground live biomass for *Stipa grandis* steppe compared to *Leymus chinensis* steppe. Bai et al. (2004) reported 35% lower aboveground biomass for a 24 year period between 1980 and 2004. These findings are consistent with our results of lower contents of SOC and N_{tot} at *Stipa* dominated sites compared to *Leymus* dominated sites. Species richness and plant cover are also lower at *Stipa grandis* steppe compared to *Leymus chinensis* steppe (Xiao et al., 1995). The lower primary production is probably due to drier soil conditions at the Stipa sites. In the Xilin River catchment, soil water content was lower at a *Stipa grandis* steppe compared to a *Leymus chinensis* steppe and, thus, the authors concluded that *Leymus chinensis* is found on relatively wet and fertile areas, whereas *Stipa grandis* is restricted to dryer and leaner regions (Chen et al., 2005a). This is consistent with our observation of a distinct $CaCO_3$ layer under the soil surface at the *Stipa grandis* site. At the *Leymus chinensis* site, $CaCO_3$ is less accumulated and occurred at greater soil depths (Wiesmeier et al., 2009). Because of drier soil conditions, capillary rise of soil water is more pronounced at *Stipa grandis* sites, associated with a higher accumulation of $CaCO_3$ in the topsoil. As water is the limiting factor in this environment, primary production is reduced at *Stipa grandis* dominated sites and, thus, lower input of organic materials into the soil causes lower amounts of SOM.

Soil Fractions, Aggregate Stability and SOM Mineralization

The distribution of bulk SOM between soil fractions derived from the physical fractionation is given in Figure 5. As soil texture is comparable between CG and UG79 sites from both steppe types (Wiesmeier et al., 2012b), one can assume that differences in the proportions of SOM fractions between these experimental sites can be attributed to the different land use management (grazed/ungrazed).

In general, significant ($P<0.05$) differences were found for light fractions (fPOM, oPOM) for both steppe types between CG and UG79. Mass contributions of both light fractions were 26 – 35% lower at CG sites compared with UG79 sites. As also the OC concentration of fPOM and oPOM was considerably lower at CG sites, these fractions had much lower OC contributions (31 – 58%) to the bulk soil. For MOM fractions, no significant differences were determined between grazed and ungrazed sites. Generally, *Leymus* dominated experimental sites showed considerably higher amounts of OC of all fractions compared to *Stipa* dominated sites. Grazing exclusion led to almost twice the amount of OC of fPOM and oPOM compared with grazed sites from *Stipa* and *Leymus*. Although the absolute mass contribution of light fractions is relatively small, their high OC concentrations result in considerable contributions of these fractions to bulk OC at the UG79 sites.

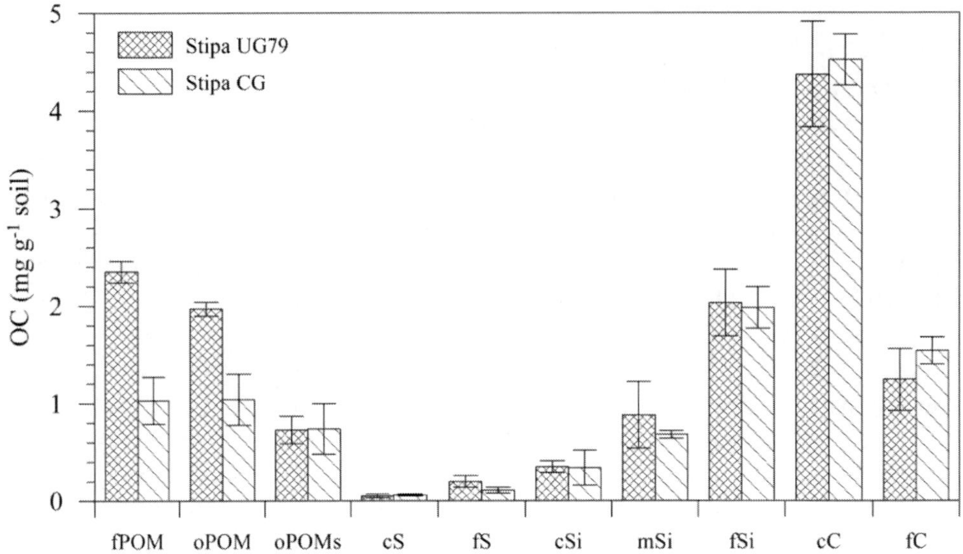

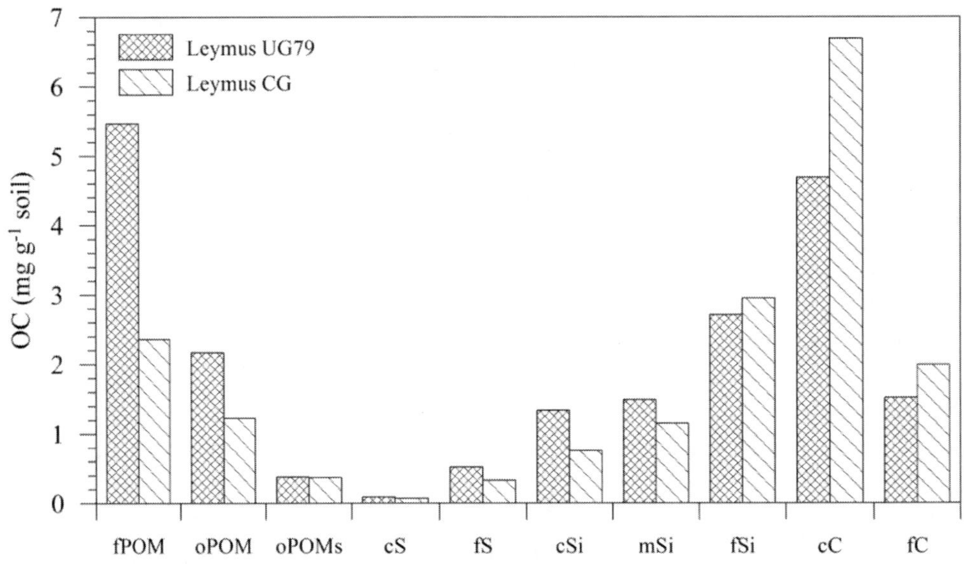

Figure 5. Contribution of OC of soil fractions to the bulk soil from continuously grazed (CG) and ungrazed (UG79) Stipa grandis (n = 3) and Leymus chinensis (n = 1) dominated experimental sites (error bars represent standard deviation; fPOM = free particulate organic matter, oPOM = occluded particulate organic matter, oPOMs = small occluded particulate organic matter, cS = coarse and medium sand, fS = fine sand, cSi = coarse silt, mSi = medium silt, fSi = fine silt, cC = coarse and medium clay, fC = fine clay).

At *Stipa* sites, fPOM and oPOM have OC contributions of 16.5 and 19.1% to bulk OC, and at *Leymus* sites, contributions of 20.1 and 14.2% were calculated, respectively. For CG sites, significantly ($P<0.05$) lower amounts of OC of fPOM (8.5 – 10.3%) and oPOM (8.3 – 14.8%) contribute to total OC as relatively more SOM is stored in mineral associated fractions.

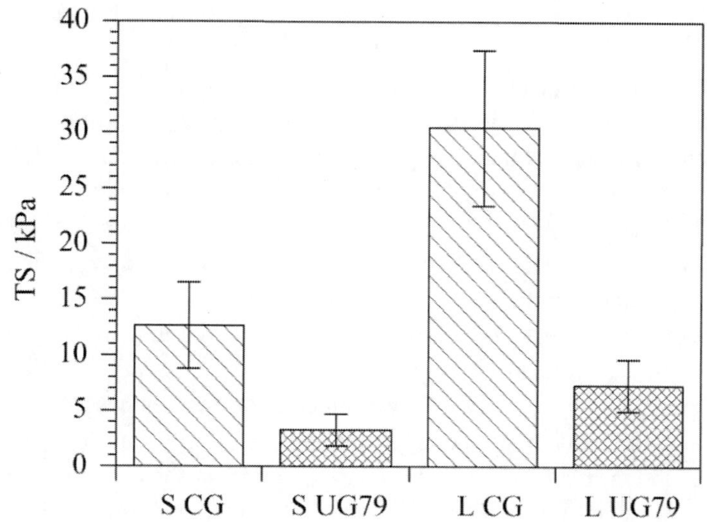

Figure 6. Tensile strength (TS) of macroaggregates (10 – 20 mm) from continuously grazed (CG) and ungrazed (UG79) *Stipa grandis* (S) and *Leymus chinensis* (L) dominated sites (error bars represent standard deviation, n = 20, (Wiesmeier et al., 2012b).

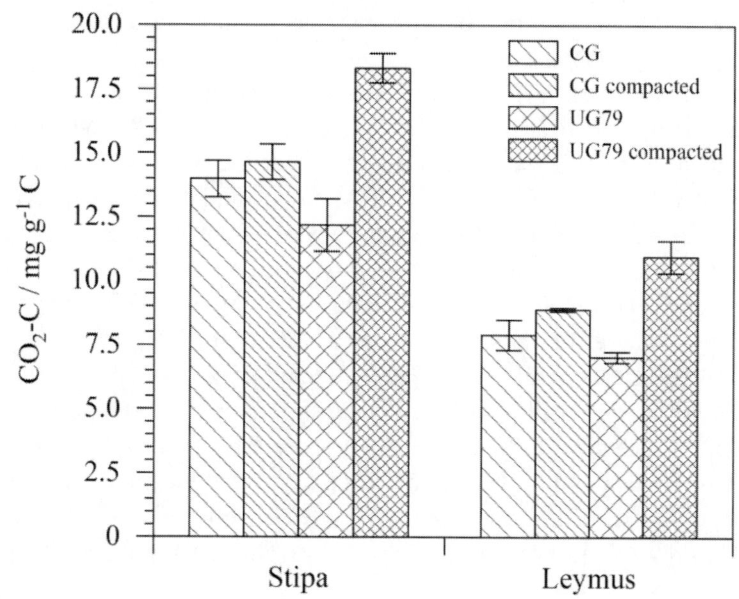

Figure 7. Cumulative release of CO_2-C per g C of the bulk soil over an incubation period of 75 days of undisturbed and mechanically compacted topsoil material from continuously grazed (CG) and ungrazed (UG79) *Stipa grandis* and *Leymus chinensis* dominated experimental sites (error bars represent standard deviation, n = 3, (Wiesmeier et al., 2012b).

Determination of aggregate stability, measured as tensile strength (TS), showed an approximately four times higher stability for grazed sites compared to ungrazed sites (Figure 6). Tensile strength for soil aggregates from CG and UG79 sites of *Stipa* dominated grassland was 12.7±3.9 and 3.3±1.4 kPa, respectively. Significantly ($P<0.05$) higher values of TS were measured for CG and UG79 sites of *Leymus* dominated grassland with 30.4±7.0 and 7.3±2.3 kPa, respectively.

The cumulative amount of carbon in the form of carbon dioxide (CO_2-C) released from undisturbed as well as mechanically compacted topsoils from all experimental sites during the incubation experiment are presented in Figure 7. The undisturbed samples showed generally the lowest degree of CO_2 release for UG79 sites with amounts of 7.0±0.2 mg CO_2-C g^{-1} C for *Leymus* and 12.2±1.0 mg CO_2-C g^{-1} C for *Stipa* after 75 days of incubation. Mineralization of SOM was significantly ($P<0.05$) higher for grazed *Leymus* and *Stipa* dominated sites with total releases of 7.9±0.6 mg CO_2-C g^{-1} C and 14.0±0.7 mg CO_2-C g^{-1} C, respectively. Mechanical compaction of soil samples from CG sites resulted in a slightly higher release of CO_2-C for *Stipa* and a significantly ($P<0.05$) higher amount for *Leymus*. The effect of soil compaction on the mineralization of LF was much more pronounced for UG79 sites that showed 50–57 % higher total releases of 11.0±0.6 mg CO_2-C g^{-1} C and 18.3±0.6 mg CO_2-C g^{-1} C for *Leymus* and *Stipa* when compared to the undisturbed samples. In general, all investigated *Stipa* samples showed 64 – 77% higher amounts of respired CO_2-C compared to the respective *Leymus* samples.

Results of the incubation and the physical fractionation of topsoil samples from UG79 confirmed our assumption that grazing exclusion leads to an increased physical protection of SOC due to enhanced aggregation. This was indicated by a considerably lower release of CO_2-C (Figure 7) and a higher proportion of SOM occluded in aggregates (oLF) for UG79 sites compared to CG sites (Figure 5).

However, it should be pointed out that there is no evidence for aggregate protection as aggregation was not measured directly but was deduced from SOC mineralization and soil fractions. Further, one has to bear in mind that a proportion of oLF was not released from aggregates during ultrasonication but from MOM, for example. But due to a low energy input, we assume that the majority of oLF was released from disrupted soil aggregates. In contrast to the low respiration, quantity and quality of potential mineralizable SOM is significantly higher for UG79 sites. This was indicated by higher SOC concentrations of the bulk soil (Figure 4) as well as higher amounts of OC of fLF (Figure 5), that comprises relatively undecomposed, labile organic matter (Golchin et al., 1994). Nevertheless, a large proportion of SOM is physically protected against microbial decomposition within soil aggregates that are inaccessible for microbes due to pore sizes exclusion (Jastrow and Miller, 1997; Oades, 1984; Tisdall and Oades, 1982). We assume that the higher input of particulate organic matter that acts as binding agent for aggregates (Jastrow et al., 1998; Oades, 1984) in combination with prevention of mechanical stress promotes the formation of soil aggregates at UG79 sites.

In contrast, the degree of aggregate-associated protection of SOC was considerably lower at CG sites. This was indicated by much lower proportions of oLF (Figure 5) and a higher amount of mineralized SOC at the grazed sites (Figure 7). This high release of CO_2-C seems to be contrary to lower SOC concentrations (Figure 4) and a lower proportion of the easily degradable fLF at grazed sites (Figure 5). On the one hand, grazing probably improves the availability of fLF for microbial mineralization as the continuous sheep trampling may promote the breakup of plant residues and thus their accessibility for microbes. The OC concentrations of fLF were much lower at CG sites compared to UG79 sites and that is probably a hint to the assumption that this fraction is more decomposed after sheep trampling leading to an increased release of CO_2-C. On the other hand, the potential for aggregate formation may be lower at CG sites as there is probably a decreased availability of aggregate binding agents indicated by generally lower SOC concentrations and proportions of LF (Figure 5). Moreover, sheep trampling results in the destruction of soil aggregates associated with a release of formerly protected SOM that is subsequently mineralized. A deterioration of soil structure due to grazing was observed in several semi-arid grasslands and other environments of the world (e.g. Chanasyk and Naeth, 1995; Greenwood and McKenzie, 2001; Villamil et al., 2001). The hoof pressure of grazing sheep is in the range of 48 – 83 kPa (Greenwood and McKenzie, 2001) but can reach up to 200 kPa when the sheep are moving (Willatt and Pullar, 1984). These high mechanical stresses result in a deterioration of the soil structure by aggregate destruction and associated soil compaction (Hamza and Anderson, 2005) that was also indicated by higher BD and higher precompression stresses at least for *Leymus* CG (Figure 4).

Considerable differences for the mineralization of SOM and aggregation were found between *Stipa* and *Leymus* dominated sites (Figure 5 and 7). In general, *Leymus* sites showed a higher release of CO_2-C for all incubated soils as well as higher contributions of OC in all soil fractions, particularly in fLF, than *Stipa* dominated sites. The degree of aggregation is probably higher at *Leymus* sites due to higher SOC concentrations (Figure 4). Although the amount of OC stored within aggregates is only slightly higher for *Leymus*, the stability of soil aggregates from *Leymus* dominated sites was much higher compared to *Stipa* sites (Figure 6). Consequently, the physical protection of SOM is lower at *Stipa* sites due to a lower aggregate stability leading to a higher mineralization of SOM.

Evidence was found for our assumption that grazing exclusion considerably promotes the aggregation of the grassland soils and thus the physical protection of SOC. A supposed detrimental effect of grazing on this stabilization process could also be confirmed. However, contrary results were found for the stability of soil aggregates. At ungrazed sites, aggregates were not as stable as expected but showed a much lower stability towards mechanical pressure compared to grazed sites (Figure 6) which was also found by Reszkowska et al. (2010), who conducted TS measurements at various moisture conditions. This seems to be in contradiction to the finding that the input of OM and formation of soil aggregates is strongly enhanced after cessation of grazing. Moreover, Steffens et al. (2009b) found higher aggregate stabilities for UG79 compared to CG at *Leymus* dominated sites for smaller aggregates (<2 mm) that was explained by greater amounts of LF and neutral sugars. At CG sites, tensile strength of aggregates was much higher for both steppe types although soil aggregation was less developed.

In order to explain these apparently inconsistent findings we developed a conceptual model that links the physical protection and mineralization of SOC with the stability of aggregates (Figure 8). At ungrazed sites, lower soil aggregate stabilities can be related to the "porosity exclusion principle" (Dexter, 1988) that is based on the existence of an aggregate hierarchy. In grassland soils where OM is the major binding agent, soil aggregates constitute not a random arrangement of all soil components but are often built up in a hierarchical order (Oades and Waters, 1991).

Mineral particles are bound together with fungal, bacterial and plant debris to microaggregates (<250 μm) that build up larger macroaggregates (>250 μm) together with intra-macroaggregate binding agents like roots and hyphae (Jastrow and Miller, 1997; Oades, 1984; Oades, 1993; Six et al., 2002; Tisdall and Oades, 1982).

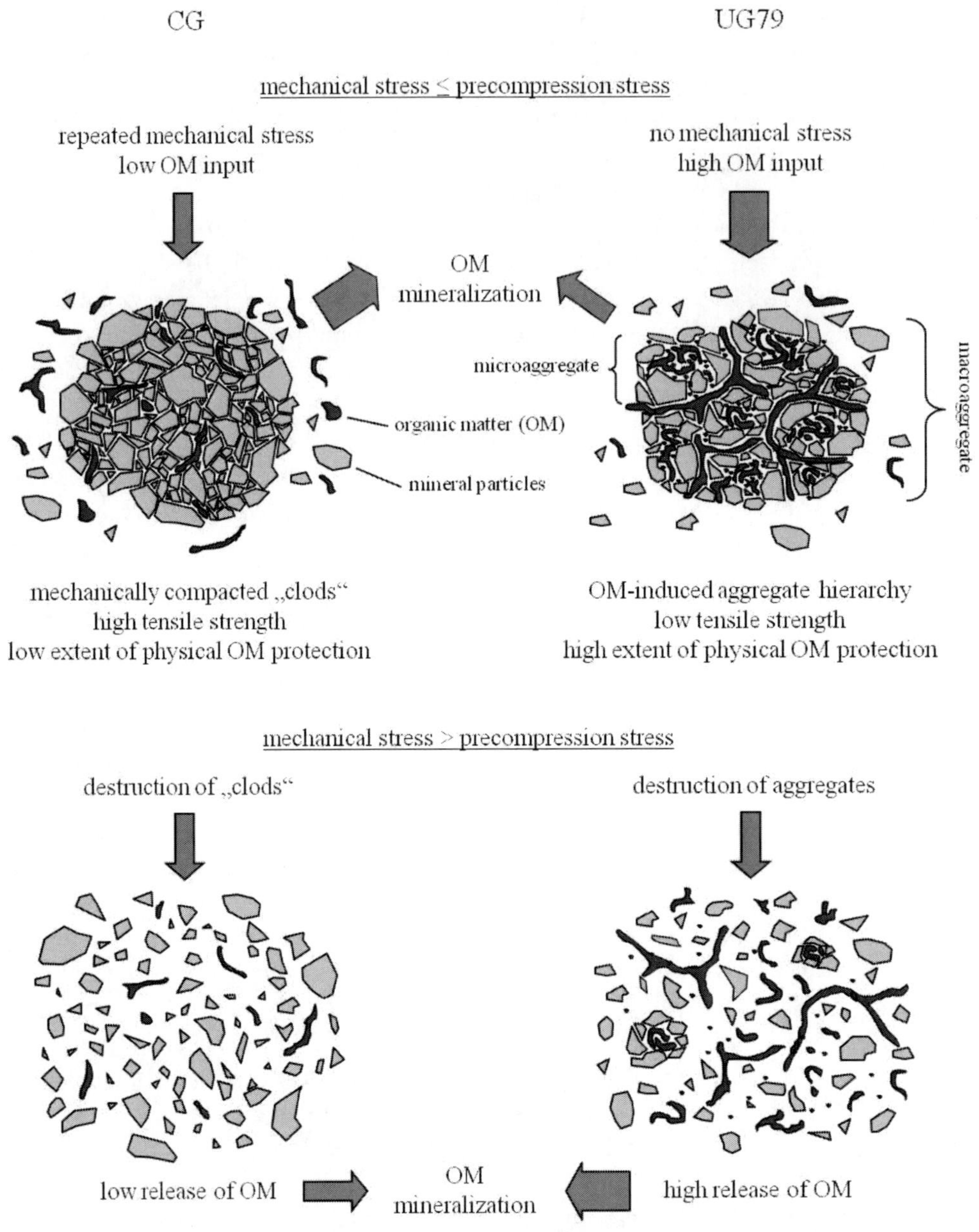

Figure 8. Conceptual model of the physical protection of SOC in semiarid grassland soils, its relation to aggregate stability and the sensitivity to mechanical stress (Wiesmeier et al., 2012b).

As a consequence microaggregates at a lower hierarchical order exclude the pore spaces between the particles and aggregates that comprise a higher macroaggregate order (Dexter, 1988). Thus, macroaggregates are generally less stable compared to microaggregates. Moreover, roots and hyphae are transient intra-macroaggregate binding agents that are decomposed relatively fast resulting in a further decrease of macroaggregate stability (Tisdall and Oades, 1982). The high input of OM at UG79 sites probably promotes the formation of soil aggregates in a hierarchical order (Steffens et al., 2009b) associated with relatively low macroaggregate stabilities. However, aggregate hierarchy was not directly measured and the low stabilities were only determined for large macroaggregates (10–20 mm) in this study. For smaller aggregates from UG79, particularly microaggregates, stabilities are expected to be higher compared to CG, as it was indicated by Steffens et al. (2009b).

At grazed sites, a low input of OM and continuous trampling prevent the biologically-induced formation of soil aggregates in a hierarchical order. We suppose that soil components, particularly mineral particles of different sizes, are mechanically pressed together by grazing animals to dense stable groupings, called "clods" by Dexter (1988). These compacted units have much higher tensile strengths compared to soil aggregates of UG79 sites but protect considerably lower amounts of SOC against microbial mineralization.

To summarize, at ungrazed sites a high proportion of SOC is physically protected against mineralization within soil aggregates with a low stability whereas at grazed sites a lower amount of SOC is protected in compacted stable clods. Results of the incubation of artificially compacted topsoils beyond the precompression stresses of the soils provided evidence for these assumptions. For CG sites, where high mechanical pressures resulted in a destruction of clods, only a slight increase of SOC mineralization was observed (Figure 7). This indicates that only a low proportion of SOC is stored within the clods that can be mineralized after the disruption. The impact of artificial soil compaction on SOC mineralization was much more pronounced at UG79 sites. Here, a very high release of CO_2-C indicated that a considerable amount of SOC is stored within aggregates, which are easily destroyed after applying pressures beyond the relatively low precompression stresses. The amount of mineralized CO_2-C of compacted samples from UG79 was 50–57% higher compared to undisturbed samples of the same site. Only small additional loadings are required at ungrazed sites to disrupt aggregates followed by a high mineralization of released C-rich oLF. However, it should be noted that relatively stable aggregates <6.3 mm probably are not completely destroyed as it was indicated by the findings of Steffens et al. (2009b) who determined higher aggregate stabilities at UG79 compared to CG after ultrasonication.

Therefore, our assumption that sites where grazing was excluded are more resistant against high grazing pressures than grazed sites due to the formation of the stable aggregates cannot be maintained. Whereas already degraded CG sites revealed only a slight increase of SOC mineralization, UG79 sites showed a high sensitivity towards additional mechanical loadings.

Spatial Distribution of SOM as Affected by Grazing Cessation

Analyses of the spatial distribution of SOM at the field scale for both steppe types indicated a high spatial variability at smaller scales <5 m (Kölbl et al., 2011; Steffens et al., 2009a; Wiesmeier et al., 2009). Therefore, the small-scale spatial distribution of SOM at the plant scale was investigated by calculating semivariances of SOC and N_{tot} concentrations and stocks and BD derived from the small grids of 2 × 2 m. All parameters were normally distributed and fitted with spherical models, except N_{tot} stocks of *Stipa* CG that were fitted with a nugget model. All semivariograms of both ungrazed sites revealed semivariances approximately one order of magnitude higher compared to grazed sites, pointing towards a much stronger spatial dependence. The differences in terms of semivariance values between grazed and ungrazed sites were more pronounced at the plant scale than the field scale. Both CG sites showed similar ranges of autocorrelation, with distances on average of around 40 cm. In contrast, ranges of spatial dependence at ungrazed sites differed between *Stipa* and *Leymus* dominated steppe types. *Stipa* UG79 showed relatively small ranges of about 23 cm, whereas *Leymus* UG79 had ranges of 90 cm on average. Semivariances of all parameters were higher for *Leymus* sites (CG and UG79) compared to *Stipa* sites at the plant scale. Furthermore, mean concentrations of SOC, N_{tot} and BD were calculated for CG sites from 40 sampling points and for UG sites from 20 points under bare soil areas and from 20 points under vegetation patches. Highly significant ($P<0.001$) differences for all parameters were found between vegetation patches and bare soil at both ungrazed sites (Wiesmeier et al., 2009). In contrast, CG sites and bare soil areas from UG79 sites showed comparable elemental concentrations and BD at *Leymus* dominated sites and only slightly higher contents of SOC, N_{tot} and lower BD for bare soil areas at *Stipa* dominated sites compared to *Stipa* CG.

A strong spatial dependence of SOM properties and BD was determined at both ungrazed sites, pointing towards a heterogeneous spatial pattern of these parameters. In contrast, topsoil parameters of continuously grazed steppe types showed a more homogeneous distribution, indicated by lower total semivariances (Wiesmeier et al., 2009). The more patchy or heterogeneous pattern of SOM and BD can be explained by the formation of resource patches also referred to as "islands of fertility" under grass tussocks after grazing exclusion at UG79 sites (Garner and Steinberger, 1989).

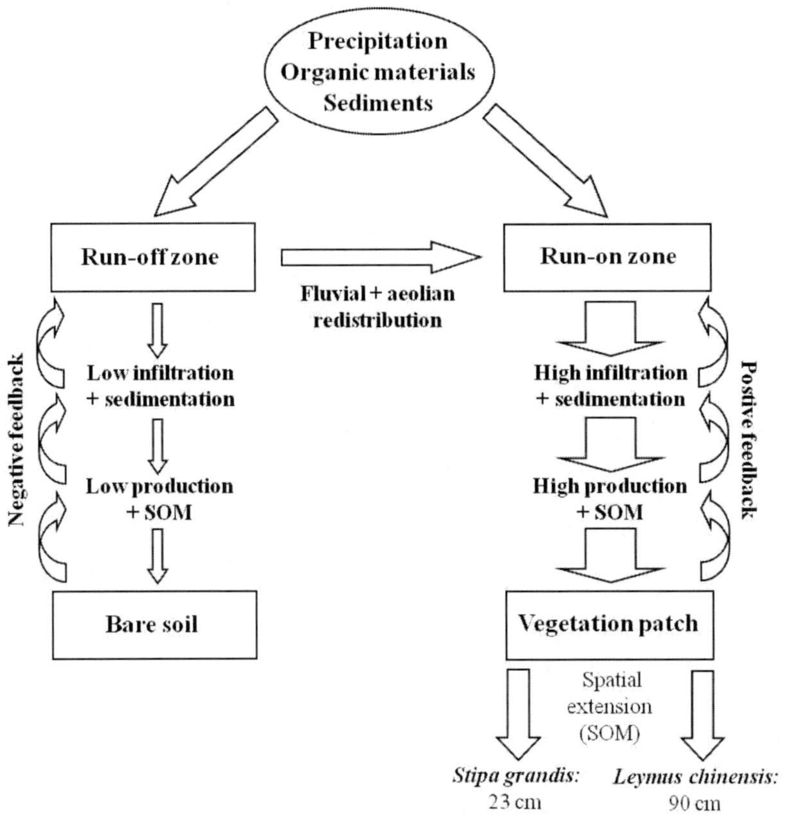

Figure 9. Processes leading to a spatially heterogeneous landscape of vegetation patches separated by bare soil in semi-arid *Stipa grandis* and *Leymus chinensis* dominated grasslands (Wiesmeier et al., 2009).

Figure 9 gives a schematic overview of the processes leading to spatial heterogeneity at ungrazed sites. Precipitation water as well as organic materials and sediments derived from aeolian and fluvial erosion are redistributed from bare soil areas (run-off zones) with low infiltration to vegetation patches (run-on zone) with higher infiltration. The higher input of water, which is the limiting factor in semi-arid steppes, and the accumulation of eroded materials enhance primary production at run-on zones and create a heterogeneous pattern of vegetation and associated chemical and physical soil properties (Ludwig et al., 1999b).

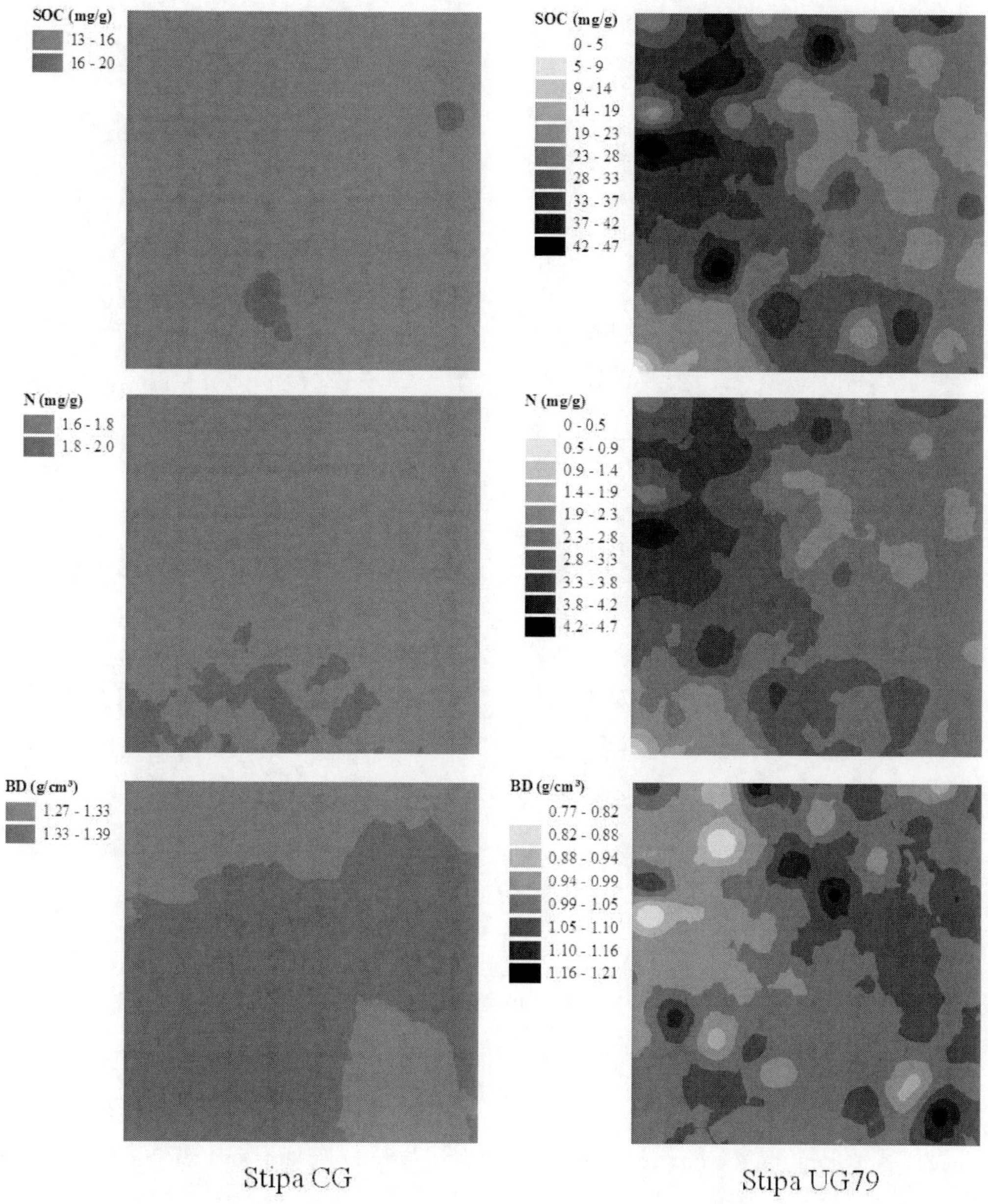

Figure 10. Maps derived from ordinary kriging of SOC, N_{tot} and BD from 2 m × 2 m plots of continuously grazed (CG) and ungrazed (UG79) *Stipa* dominated sites (Wiesmeier et al., 2009).

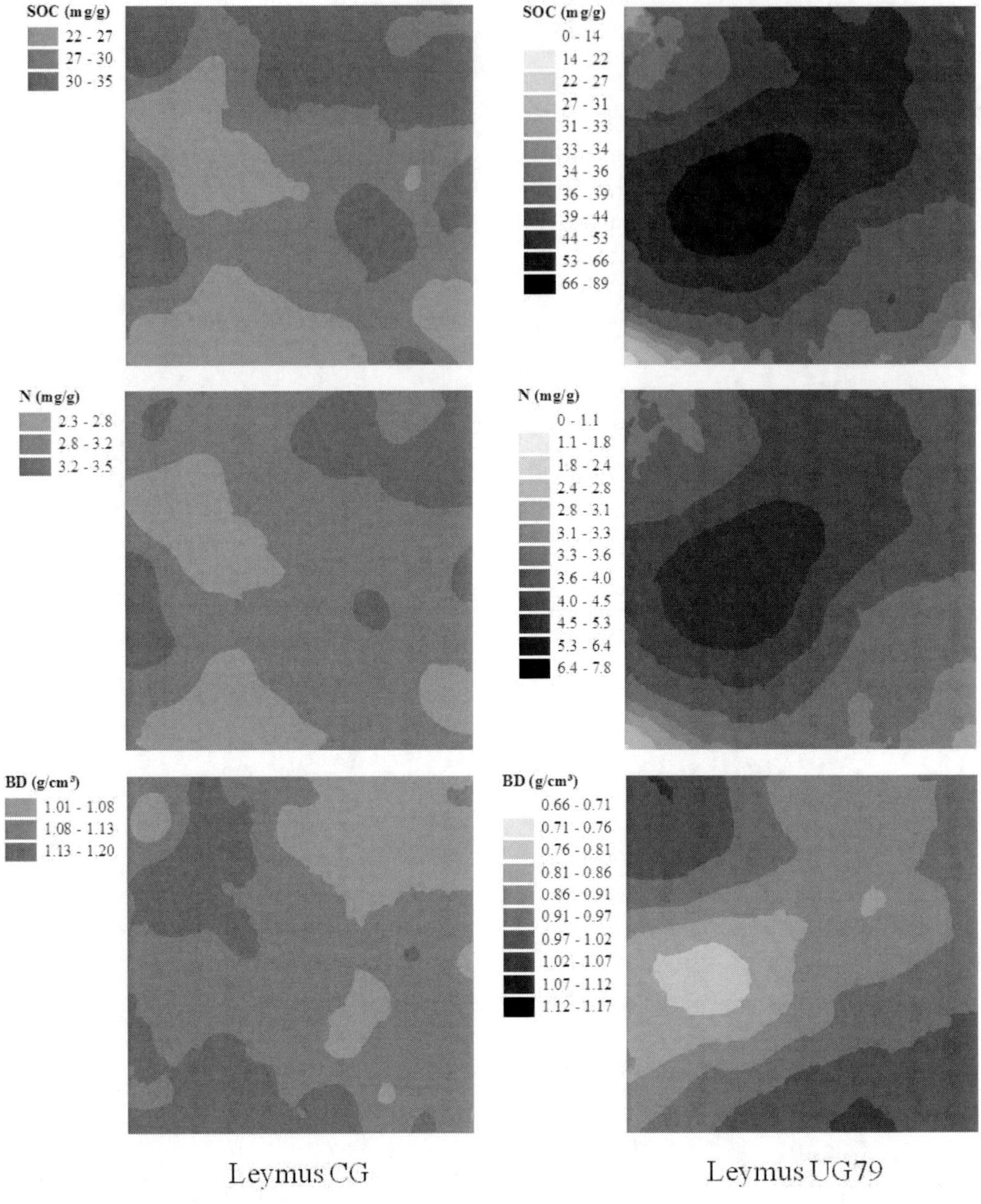

Figure 11. Maps derived from ordinary kriging of SOC, N_{tot} and BD from 2 m × 2 m plots of continuously grazed (CG) and ungrazed (UG79) *Leymus* dominated sites (Wiesmeier et al., 2009).

These heterogeneous patterns of topsoil properties are readily identifiable on maps of the 2 × 2 m plots of UG79 sites generated by ordinary kriging (Figure 10 and 11). At larger scales, vegetation patches are normally created by trees and shrubs that act as "ecosystem engineers" ameliorating the environmental conditions for other plants in their vicinity (Jones et al., 1997).

But there is also evidence for the formation of resource islands at smaller scales in semi-arid grasslands, where even small clumps of grasses can initiate the accumulation of water and nutrients beneath their canopy (Hook et al., 1991). In a semi-arid Mediterranean steppe in

Spain, *Stipa* tussocks were regarded as ecosystem engineers that favoured the growth of other species in their vicinity (Maestre, 2003). Xie and Wittig (2003) investigated growth parameters of characteristic species of *Stipa* steppes in Northern China and found diameters for *Stipa grandis* tussocks of 26 cm, which corresponds well with ranges of 23 cm for the spatial dependences determined at the plant scale of the ungrazed *Stipa* site.

At *Leymus* UG79, a range of about 90 cm refers to larger vegetation patches created not only by grass tussocks but also by shrubs such as *Carragana microphylla*, which had invaded the fenced *Leymus* site. In desert ecosystems of the south-western United States, ranges of autocorrelation were smaller than 20 cm in grasslands and 1 – 3 m in adjacent shrublands (Schlesinger et al., 1996). The locations and extensions of patches of SOC, N_{tot} and BD on the maps of the 2 × 2 m plots (Figure 10 and 11) were congruent with observed vegetation patches of grass tussocks and shrubs at *Stipa* and *Leymus* dominated ungrazed sites. Distinct differences of mean concentrations of SOC, N_{tot} and mean BD between vegetation patches and bare soil confirmed the congruence of vegetation and soil patches (Wiesmeier et al., 2009). Besides, comparable elemental concentrations at CG small plots and at bare soil areas from UG79 small plots clearly show that the increase of SOM after grazing exclusion is spatially restricted to vegetation patches.

Intensive grazing can change the spatial distribution of vegetation and topsoil properties and both increasing and decreasing heterogeneity were observed. Adler et al. (2001) proposed that increased spatial heterogeneity following grazing is due to a stronger spatial heterogeneity of grazing than of vegetation which they called "patch grazing". Several studies determined a more heterogeneous pattern as a consequence of grazing (Schlesinger et al., 1990).

Conversely, if the spatial heterogeneity of grazing is weaker than of vegetation, referred to as "homogeneous grazing", spatial distribution of vegetation and topsoil parameters will decrease. This scenario often occurs where randomly distributed grazing takes place in fine-scale heterogeneous environments, such as the grasslands of Inner Mongolia. Hoof action of grazing animals can destroy heterogeneous patterns by lateral movement of accumulated micromounds of soil under the vegetation patches into the intermound depressions (Nash et al., 2004). At the same time, aboveground biomass is reduced by herbivores and thus high-cover vegetation patches are removed. Formerly protected islands of fertility become susceptible to erosion, input of organic material and water infiltration are strongly reduced and higher radiation and surface temperatures as a result of lower soil protection probably favour the microbial degradation of SOC. Moreover, increased wind erosion at grazed steppes may support the homogenization of the topsoil. In arid grasslands of southern New Mexico, elimination of grass cover at ungrazed plots resulted in a homogenization of soil variables after three years, which was explained by strongly increased wind erosion (Li et al., 2008). Beside a homogenization of SOM, BD of the upper 4 cm also becomes more uniformly distributed by homogeneous trampling of grazing animals. This is indicated by the spatial correlations of SOM properties and bulk density evident in crossvariograms and semivariograms (Wiesmeier et al., 2009). A spatial homogenization of heterogeneous semi-arid environments as a result of grazing has also been observed in other semi-arid environments, e.g. Australia (Ludwig and Tongway, 1995), Africa (Allsopp, 1999; Lechmere-Oertel et al., 2005) and North America (Adler and Lauenroth, 2000).

The determined homogenization of chemical and physical topsoil properties at small spatial scales probably contributes to the degradation of semi-arid steppes in Northern China.

Under semi-arid climatic conditions, the heterogeneous pattern of vegetation is essential for the productivity of the grasslands. Low amounts of precipitation water are redistributed from bare soil areas to vegetation patches and enable primary production there, referred to as "functional heterogeneity" (Ludwig et al., 1999b). Thus, productivity of patchy landscapes is greater than that of non-patchy landscapes. A homogenization of vegetation and associated soil properties would limit productivity and the landscape would become "disfunctional" (Ludwig and Tongway, 1995). Valentin et al. (1999) concluded a decline of soil fertility and water infiltration as a result of the loss of landscape patchiness. In semi-arid woodlands of Australia, the loss of landscape heterogeneity due to overgrazing accounted for a reduction of 25% of soil water and 40% lower annual net primary productivity (Ludwig et al., 1999b). We assume that the determined loss of spatial heterogeneity at intensively grazed *Stipa* and *Leymus* dominated steppes in Inner Mongolia causes a decrease of productivity associated with a decline of SOM. Therefore, spatial homogenization of vegetation and associated topsoil properties due to intensive grazing may be another process leading to grassland degradation. Moreover, the spatial distribution of soil properties at small spatial scales (< 2 m) depending on tussock diameter may be an appropriate indicator for steppe degradation and desertification, as it points to the status of ecosystem function (Imeson and Cammeraat, 2000). To apply the amount of vegetation or soil properties exclusively as indicators is ambiguous, because of the strong heterogeneity reported in this study. The spatial patterns of soil properties are probably better indicators for degradation of steppe ecosystems than vegetation parameters, as they are not influenced by short-term climatic variations such as droughts (Imeson and Prinsen, 2004; Tongway and Hindley, 2000).

Summary and Conclusion

Our study provided evidence that exclusion of grazing is an effective management option to restore degraded semi-arid steppe soils. Even short-term grazing cessation of three years resulted in a notable recovery of grassland soils in Northern China in terms of chemical and physical soil properties. At long-term ungrazed sites, where grazing was excluded for 30 years, SOM increased up to 40% compared to adjacent continuously grazed sites and soil compaction, measured as BD, significantly decreased. The main process of SOM stabilization after grazing exclusion is the physical protection of a considerable amount of SOM due to improved soil aggregation as it was indicated by increased amounts of oPOM as well as lower mineralization rates in ungrazed steppe soils. The recovery is accompanied by the developed of a heterogeneous structure of SOM-rich spots under highly productive vegetation patches separated by depleted bare soil areas. These patches can be regarded as "Islands of fertility", where precipitation water as well as organic materials and sediments are redistributed from bare soil areas (run-off zones) with low infiltration to patches (run-on zone) with higher infiltration that favours plant growth and input of OM into the soil.

In contrast, intensive grazing resulted in a considerable deterioration of chemical and physical soil properties. A low input of OM and the continuously applied mechanical stress of grazing animals prevent the development of OM-induced soil aggregation but lead to the formation of mechanically compacted stable clods, which do not provide an effective physical protection for SOC. The destruction of soil aggregates is associated with the release of physically protected SOM that is mineralized. Beside direct detrimental consequences of grazing such as reduced input of organic materials, a change in the spatial distribution of topsoil parameters may also adversely affect productivity of the grasslands. Grazing led to a homogenization of heterogeneous vegetation and soil patterns. Patches of high SOM stocks around grass tussocks were removed and thus the ability of the naturally heterogeneous landscape to accumulate OM under vegetation patches was diminished. We conclude that the ability of the steppe ecosystem to capture, store and recycle materials is reduced by grazing, which contributes to the deterioration of the ecosystem. The maintenance of heterogeneous vegetation and associated topsoil structures is essential for the accumulation of SOM in semi-arid grassland ecosystems.

Our study clearly demonstrated that grazing exclusion is an effective management option to restore degraded grassland soils and increase C sequestration as soil aggregation was promoted and SOM started to accumulate in a patchy pattern. However, our findings also indicate that development of soil aggregation after grazing cessation is not accompanied by the formation of stable aggregates. A possible hierarchical arrangement of aggregates in the ungrazed grassland soils is associated with a relatively low stability of soil aggregates that results in a high sensitivity towards newly applied grazing pressures beyond the precompression stresses. On the one hand, degraded semi-arid grasslands have a high potential for sequestration of atmospheric C after grazing cessation but on the other hand these sensitive ecosystems are very susceptible to intensive grazing management. Therefore, we recommend that an intensification of grazing management or a further expansion into undisturbed boundary areas of the semi-arid grasslands should be prevented.

A sustainable land use management in the grasslands of Inner Mongolia should incorporate first of all significantly reduced stocking rates in order to initialize the regeneration of vegetation and soil fertility. The extensification of grazing should be accompanied by a change of property rights of land tenure and the existing structure of the pastoral economy in Inner Mongolia. A sustainable grazing management should be based on the traditional nomadic pastoralism and organized in small associations of herders who manage their own piece of grassland. These areas should be large enough to allow a seasonal change of pastures in order to utilize the heterogeneously distributed resources and to prevent overexploitation of the sensitive steppes. The economic status of the herding families is thereby not worsened as the sedentarized life style in houses can be maintained if seasonal camps are used for the supervision of the herds. In small, manageable groups, all involved persons feel responsible for their environment and detrimental developments of land use practices can be rapidly recognized and prevented.

REFERENCES

Adler, P.B., Lauenroth, W.K. (2000). Livestock exclusion increases the spatial heterogeneity of vegetation in Colorado shortgrass steppe. *Applied Vegetation Science* 3, 213-222.

Adler, P.B., Raff, D.A., Lauenroth, W.K. (2001). The effect of grazing on the spatial heterogeneity of vegetation. *Oecologia* 128, 465-479.

Aguiar, M.R., Sala, O.E. (1999). Patch structure, dynamics and implications for the functioning of arid ecosystems. *Trends in Ecology and Evolution* 14, 273-277.

Akiyama, T., Kawamura, K. (2007). Grassland degradation in China: methods of monitoring, management and restoration. *Japanese Society of Grassland Science* 53, 1-17.

Allsopp, N. (1999). Effects of grazing and cultivation on soil patterns and processes in the Paulshoek area of Namaqualand. *Plant Ecology* 142, 179-187.

Anderson, J.M. (1991). The Effects of Climate Change on Decomposition Processes in Grassland and Coniferous Forests. *Ecological Applications* 1, 326-347.

Bai, Y.F., Han, X.G., Wu, J.G., Chen, Z.Z., Li, L.H. (2004). Ecosystem stability and compensatory effects in the Inner Mongolia grassland. *Nature* 431, 181-184.

Barthold, F.K., Wiesmeier, M., Breuer, L., Frede, H.G., Wu, J., Blank, F.B. (2013). Land use and climate control the spatial distribution of soil types in the grasslands of Inner Mongolia. *Journal of Arid Environments* 88, 194-205.

Bhark, E.W., Small, E.E. (2003). Association between plant canopies and the spatial patterns of infiltration in shrubland and grassland of the Chihuahuan Desert, New Mexico. *Ecosystems* 6, 185-196.

Burke, I.C., Lauenroth, W.K., Vinton, M.A., Hook, P.B., Kelly, R.H., Epstein, H.E., Aguiar, M.R., Robles, M.D., Aguilera, M.O., Murphy, K.L., Gill, R.A. (1998). Plant-soil interactions in temperate grasslands. *Biogeochemistry* 42, 121-143.

Casagrande, A. (1936). The determination of preconsolidation load and its practical significance disc. In: *Proceedings of the 1st Int. Conference on Soil Mechanics and Foundation*, pp. 137-151, Cambridge.

Chanasyk, D.S., Naeth, M.A. (1995). Grazing impacts on bulk density and soil strength in the foothills fescue grasslands of Alberta, Canada. *Canadian Journal of Soil Science* 75, 551-557.

Chen, S.P., Bai, Y.F., Lin, G.H., Han, X.G. (2005a). Variations in life-form composition and foliar carbon isotope discrimination among eight plant communities under different soil moisture conditions in the Xilin River Basin, Inner Mongolia, China. *Ecological Research* 20, 167-176.

Chen, S.P., Bai, Y.F., Zhang, L.X., Han, X.G. (2005b). Comparing physiological responses of two dominant grass species to nitrogen addition in Xilin River Basin of China. *Environmental and Experimental Botany* 53, 65-75.

Christensen, L., Coughenour, M.B., Ellis, J.E., Chen, Z.Z. (2004). Vulnerability of the Asian typical steppe to grazing and climate change. *Climatic Change* 63, 351-368.

Dexter, A.R. (1988). Advances in Characterization of Soil Structure. *Soil and Tillage Research* 11, 199-238.

Dexter, A.R., Kroesbergen, B. (1985). Methodology for determination of tensile strength of soil aggregates. *Journal of Agricultural Engineering Research* 31, 139-147.

Gao, Y.Z., Giese, M., Lin, S., Sattelmacher, B., Zhao, Y., Brueck, H. (2008). Belowground net primary productivity and biomass allocation of a grassland in Inner Mongolia is affected by grazing intensity. *Plant and Soil* 307, 41-50.

Garner, W., Steinberger, Y. (1989). A proposed mechanism for the formation of fertile islands in the desert ecosystem. *Journal of Arid Environments* 16, 257-262.

Golchin, A., Oades, J.M., Skjemstad, J.O., Clarke, P. (1994). Study of Free and Occluded Particulate Organic-Matter in Soils by Solid-State C-13 CP/MAS NMR-Spectroscopy and Scanning Electron Microscopy. *Australian Journal of Soil Research* 32, 285-309.

Greenwood, K.L., McKenzie, B.M. (2001). Grazing effects on soil physical properties and the consequences for pastures: a review. *Australian Journal of Experimental Agriculture* 41, 1231-1250.

Guo, L.B., Gifford, R.M. (2002). Soil carbon stocks and land use change: a meta analysis. *Global Change Biology* 8, 345-360.

Hamza, M.A., Anderson, W.K. (2005). Soil compaction in cropping systems - A review of the nature, causes and possible solutions. *Soil and Tillage Research* 82, 121-145.

Hartge, K.H., Horn, R. (1989) *Die physikalische Untersuchung von Böden.* Enke Verlag. Stuttgart.

Heinemeyer, O., Insam, H., Kaiser, E.A., Walenzik, G. (1989). Soil microbial biomass and respiration measurements - an automated technique based on infrared gas-analysis. *Plant and Soil* 116, 191-195.

Hoffmann, C., Funk, R., Li, Y., Sommer, M. (2008a). Effect of grazing on wind driven carbon and nitrogen ratios in the grasslands of Inner Mongolia. *Catena* 75, 182-190.

Hoffmann, C., Funk, R., Wieland, R., Li, Y., Sommer, M. (2008b). Effects of grazing and topography on dust flux and deposition in the Xilingele grassland, Inner Mongolia. *Journal of Arid Environments* 72, 792-807.

Hook, P.B., Burke, I.C., Lauenroth, W.K. (1991). Heterogeneity of soil and plant N and C associated with individual plants and openings in North-American shortgrass steppe. *Plant and Soil* 138, 247-256.

Houghton, R.A. (1999). The annual net flux of carbon to the atmosphere from changes in land use 1850-1990. *Tellus Series B-Chemical and Physical Meteorology* 51, 298-313.

Imeson, A., Cammeraat, E. (2000) Scaling up from field measurements to large areas using the Desertification Response Unit and Indicator Approaches. In: Arnalds, O., Archer, S. (eds.), *Rangeland Desertification*, pp. 99-114. Kluwer Academic Press, London.

Imeson, A.C., Prinsen, H.A.M. (2004). Vegetation patterns as biological indicators for identifying runoff and sediment source and sink areas for semi-arid landscapes in Spain. *Agriculture Ecosystems and Environment* 104, 333-342.

IUSS Working Group WRB. (2006). *World reference base for soil resources 2006.* World Soil Resources Reports No. 103. FAO. Rome.

Jastrow, J.D. (1996). Soil aggregate formation and the accrual of particulate and mineral-associated organic matter. *Soil Biology and Biochemistry* 28, 665-676.

Jastrow, J.D., Miller, R.M. (1997). Soil aggregate stabilization and carbon sequestration: feedbacks through organomineral associations. In: Lal, R., Kimble, J.M., Follett, R.F., Stewart, B.A. (eds.), *Soil Processes and the Carbon Cycle*, pp. 207-223. CRC Press, Boca Raton, (FL), USA.

Jastrow, J.D., Miller, R.M., Lussenhop, J. (1998). Contributions of interacting biological mechanisms to soil aggregate stabilization in restored prairie. *Soil Biology and Biochemistry* 30, 905-916.

Jones, C.G., Lawton, J.H., Shachak, M. (1997) Positive and negative effects of organisms as physical ecosystem engineers. *Ecology* 78, 1946-1957.

Kawamura, K., Akiyama, T., Yokota, H., Tsutsumi, M., Yasuda, T., Watanabe, O., Wang, S.P. (2005) Quantifying grazing intensities using geographic information systems and satellite remote sensing in the Xilingol steppe region, Inner Mongolia, China. *Agriculture Ecosystems and Environment* 107, 83-93.

Kölbl, A., Steffens, M., Wiesmeier, M., Hoffmann, C., Funk, R., Krümmelbein, J., Reszkowska, A., Zhao, Y., Peth, S., Horn, R., Giese, M., Kögel-Knabner, I. (2011). Grazing changes topography-controlled topsoil properties and their interaction on different spatial scales in a semi-arid grassland of Inner Mongolia, P.R. China. *Plant and Soil* 340, 35-58.

Krümmelbein, J., Wang, Z.Y., Zhao, Y., Peth, S., Horn, R. (2006). Influence of various grazing intensities on soil stability, soil structure and water balance of grassland soils in Inner Mongolia, PR China. *Soil Management for Sustainability* 38, 93-101.

Lal, R. (2004). Carbon sequestration in dryland ecosystems. *Environmental Management* 33, 528-544.

Lechmere-Oertel, R.G., Cowling, R.M., Kerley, G.I.H. (2005). Landscape dysfunction and reduced spatial heterogeneity in soil resources and fertility in semi-arid succulent thicket, South Africa. *Austral Ecology 30*, 615-624.

Li, J., Okin, G.S., Alvarez, L., Epstein, H. (2008). Effects of wind erosion on the spatial heterogeneity of soil nutrients in two desert grassland communities. *Biogeochemistry* 88, 73-88.

Ludwig, J.A., Tongway, D.J. (1995). Spatial organization of landscapes and its function in semiarid woodlands, Australia. *Landscape Ecology* 10, 51-63.

Ludwig, J.A., Tongway, D.J., Eager, R.W., Williams, R.J., Cook, G.D. (1999a). Fine-scale vegetation patches decline in size and cover with increasing rainfall in Australian savannas. *Landscape Ecology* 14, 557-566.

Ludwig, J.A., Tongway, D.J., Marsden, S.G. (1999b). Stripes, strands or stipples: modelling the influence of three landscape banding patterns on resource capture and productivity in semi-arid woodlands, Australia. *Catena* 37, 257-273.

Maestre, F.T. (2003). Small-scale spatial patterns of two soil lichens in semi-arid Mediterranean steppe. *Lichenologist* 35, 71-81.

Mueller, C. (2009). *Carbon sequestration and soil organic matter composition affected by altered soil aggregate dynamics due to land-use changes.* Ph Dissertation, Technische Universität München, Munich.

Nash, M.S., Jackson, E., Whitford, W.G. (2004). Effects of intense, short-duration grazing on microtopography in a Chihuahuan Desert grassland. *Journal of Arid Environments* 56, 383-393.

Noy-Meir, I. (1973). Desert ecosystems: environment and producers. *Annual Reviews of Ecology, Evolution, and Systematics* 4, 25-51.

Oades, J.M. (1984). Soil organic matter and structural stability - mechanisms and implications for management. *Plant and Soil* 76, 319-337.

Oades, J.M. (1993). The Role of Biology in the Formation, Stabilization and Degradation of Soil Structure. *Geoderma* 56, 377-400.

Oades, J.M., Waters, A.G. (1991). Aggregate Hierarchy in Soils. *Australian Journal of Soil Research* 29, 815-828.

Pebesma, E.J. (2004). Multivariable geostatistics in S: the gstat package. *Computers and Geosciences* 30, 683-691.

Pei, S.F., Fu, H., Wan, C.G. (2008). Changes in soil properties and vegetation following exclosure and grazing in degraded Alxa desert steppe of Inner Mongolia, China. *Agriculture Ecosystems and Environment* 124, 33-39.

Reszkowska, A., Krummelbein, J., Peth, S., Horn, R., Zhao, Y., Gan, L. (2010). Influence of grazing on hydraulic and mechanical properties of semiarid steppe soils under different vegetation type in Inner Mongolia. *China. Plant and Soil* 340, 59-72.

Schlesinger, W.H., Raikes, J.A., Hartley, A.E., Cross, A.E. (1996). On the spatial pattern of soil nutrients in desert ecosystems. *Ecology* 77, 364-374.

Schlesinger, W.H., Reynolds, J.F., Cunningham, G.L., Huenneke, L.F., Jarrell, W.M., Virginia, R.A., Whitford, W.G. (1990). Biological feedbacks in global desertification. *Science* 247, 1043-1048.

Schmidt, M.W.I., Rumpel, C., Kögel-Knabner, I. (1999). Evaluation of an ultrasonic dispersion procedure to isolate primary organomineral complexes from soils. *European Journal of Soil Science* 50, 87-94.

Schönbach, P., Wan, H., Schiborra, A., Gierus, M., Bai, Y., Müller, K., Glindemann, T., Wang, C., Susenbeth, A., Taube, F. (2009). Short-term management and stocking rate effects of grazing sheep on herbage quality and productivity of Inner Mongolia steppe. *Crop and Pasture Science* 60, 963-974.

Schönbach, P., Wan, H.W., Gierus, M., Bai, Y.F., Müller, K., Lin, L.J., Susenbeth, A., Taube, F. (2010). Grassland responses to grazing: effects of grazing intensity and management system in an Inner Mongolian steppe ecosystem. *Plant and Soil* 340, 103-115.

Six, J., Conant, R.T., Paul, E.A., Paustian, K. (2002). Stabilization mechanisms of soil organic matter: Implications for C-saturation of soils. *Plant and Soil* 241, 155-176.

Spörlein, P., Dilling, J., Joneck, M. (2004). Pilot study to test the equivalence or comparability of soil-particle-size analysis according to E DIN ISO 11277: 06.94 (pipette method) and by the use of the sedigraph. *Journal of Plant Nutrition and Soil Science* 167, 649-656.

Steffens, M., Kölbl, A., Giese, K.M., Hoffmann, C., Totsche, K.U., Breuer, L., Kögel-Knabner, I. (2009a). Spatial variability of topsoil and vegetation in a grazed steppe ecosystem in Inner Mongolia (PR China). *Journal of Plant Nutrition and Soil Science* 172, 78-90.

Steffens, M., Kölbl, A., Kögel-Knabner, I. (2009b). Alteration of soil organic matter pools and aggregation in semi-arid steppe topsoils as driven by organic matter input. *European Journal of Soil Science* 60, 198-212.

Steffens, M., Kölbl, A., Totsche, K.U., Kögel-Knabner, I. (2008). Grazing effects on soil chemical and physical properties in a semiarid steppe of Inner Mongolia (PR China). *Geoderma* 143, 63-72.

Su, Y.Z., Li, Y.L., Cui, H.Y., Zhao, W.Z. (2005). Influences of continuous grazing and livestock exclusion on soil properties in a degraded sandy grassland, Inner Mongolia, northern China. *Catena* 59, 267-278.

Su, Y.Z., Zhao, H.L., Zhang, T.H. (2003). Influences of grazing and exclosure on carbon sequestration in degraded sandy grassland, Inner Mongolia, north China. *New Zealand Journal of Agricultural Research* 46, 321-328.

Su, Y.Z., Zhao, H.L., Zhang, T.H., Zhao, X.Y. (2004). Soil properties following cultivation and non-grazing of a semi-arid sandy grassland in northern China. *Soil and Tillage Research* 75, 27-36.

Tisdall, J.M., Oades, J.M. (1982). Organic matter and water-stable aggregates in soils. *Journal of Soil Science* 33, 141-163.

Tong, C., Wu, J., Yong, S., Yang, J., Yong, W. (2004). A landscape-scale assessment of steppe degradation in the Xilin River Basin, Inner Mongolia, China. *Journal of Arid Environments* 59, 133-149.

Tongway, D., Hindley, N. (2000). Assessing and monitoring desertification with soil indicators. In: Arnalds, O., Archer, S. (eds.), *Rangeland Desertification*, pp. 88-89. Kluwer Academic Press, London.

Tongway, D.J., Ludwig, J.A. (1990). Vegetation and soil patterning in semiarid Mulga lands of Eastern Australia. *Australian Journal of Ecology* 15, 23-34.

Tongway, D.J., Ludwig, J.A. (2001). Theories on the Origins, Maintenance, Dynamics, and Functioning of Banded Landscapes. In: Tongway, D.J. Valentin, C.and Seghieri, J. (eds.). *Banded Vegetation Patterning in Arid and Semiarid Environments*, pp. 20-31. Springer, New York.

Tongway, D.J., Ludwig, J.A. (2005). Heterogeneity in arid and semiarid lands. In Lovett, G.M. Jones, C.G. Turner, M.G.and Weathers, K.C. (eds.), *Ecosystem Function in Heterogeneous Landscapes*, pp 189-205. Springer. New York.

Valentin, C., d'Herbes, J.M., Poesen, J. (1999). Soil and water components of banded vegetation patterns. *Catena* 37, 1-24.

Villamil, M.B., Amiotti, N.M., Peinemann, N. (2001). Soil degradation related to overgrazing in the semi-arid southern Caldenal area of Argentina. *Soil Science* 166, 441-452.

Wang, S.Q., Zhou, C.H., Liu, J.Y., Tian, H.Q., Li, K.A., Yang, X.M. (2002). Carbon storage in northeast China as estimated from vegetation and soil inventories. *Environmental Pollution* 116, S157-S165.

Webster, R., Oliver, M. (2001). *Geostatistics for environmental scientists*. John Wiley and Sons, Chichester.

Wiesmeier, M., Barthold, F.K., Blank, F.B., Kögel-Knabner, I. (2011). Digital mapping of soil organic matter stocks using Random Forest modeling in a semi-arid steppe ecosystem. *Plant and Soil* 340, 7-24.

Wiesmeier, M., Kreyling, O., Steffens, M., Schönbach, P., Wan, H.W., Gierus, M., Taube, F., Kölbl, A., Kögel-Knabner, I. (2012a). Short-term degradation of semiarid grasslandsuresults from a controlled-grazing experiment in Northern China. *Journal of Plant Nutrition and Soil Science* 175, 434-442.

Wiesmeier, M., Steffens, M., Kölbl, A., Kögel-Knabner, I. (2009). Degradation and small-scale spatial homogenization of topsoils in intensively grazed steppes of Northern China. *Soil and Tillage Research* 104, 299-310.

Wiesmeier, M., Steffens, M., Mueller, C.W., Kölbl, A., Agnieszka, R., Peth, S., Horn, R., Kögel-Knabner, I. (2012b). Aggregate stability and physical protection of soil organic carbon in semi-arid steppe soils. *European Journal of Soil Science* 63, 22-31.

Willatt, S.T., Pullar, D.M. (1984). Changes in soil physical-properties under grazed pastures. *Australian Journal of Soil Research* 22, 343-348.

Wu, L., He, N., Wang, Y., Han, X. (2008). Storage and dynamics of carbon and nitrogen in soil after grazing exclusion in Leymus chinensis grasslands of northern China. *Journal of Environmental Quality* 37, 663-668.

Xiao, X.M., Wang, Y.F., Jiang, S., Ojima, D.S. and Bonham, C.D. (1995). Interannual variation in the climate and aboveground biomass of Leymus chinensis steppe and Stipa grandis steppe in the Xilin river basin, Inner-Mongolia, China. *Journal of Arid Environments* 31, 283-299.

Xie, Y.Z., Wittig, R. (2003). Growth parameters of characteristic species of Stipa steppes in Northern China as indicators of the grazing intensity. *Journal of Applied Botany* 77, 68-74.

Xie, Y.Z., Wittig, R. (2004). The impact of grazing intensity on soil characteristics of Stipa grandis and Stipa bungeana steppe in northern China (autonomous region of Ningxia). *Acta Oecologica* 25, 197-204.

Yang, X., Zhang, K., Jia, B., Ci, L. (2005). Desertification assessment in China: An overview. *Journal of Arid Environments* 63, 517-531.

Zhou, R.L., Li, Y.Q., Zhao, H.L., Drake, S. (2008). Desertification effects on C and N content of sandy soils under grassland in Horqin, northern China. *Geoderma*, 145, 370-375.

In: Steppe Ecosystems
Editors: M. B. Morales Prieto and J. Traba Diaz

ISBN: 978-1-62808-298-2

Chapter 12

EFFECTS OF UNGULATE GRAZING ON SOIL MICROBIAL DIVERSITIES IN INNER MONGOLIAN STEPPES OF NORTHERN CHINA

***Xiaoqi Zhou*[1,2,3]* *and Yanfen Wang*[1]**

[1]College of Life Sciences,
Graduate University of Chinese Academy of Sciences, Beijing, China
[2]Scion (New Zealand Forest Research Institute),
Fendalton, Christchurch, New Zealand
[3]Environmental Futures Centre, Griffith University,
Nathan, Brisbane, Australia

ABSTRACT

Ungulate grazing plays a crucial role in regulating nutrient cycling and controlling plant diversity and productivity in grassland ecosystems. However, little is known about the effects of grazing on soil bacterial and methanotrophic community structure and diversity. We selected two experimental sites, i.e., different grazing intensities and long-term grazing grasslands, in the Inner Mongolia of northern China. There were four grazing intensity treatments, i.e., non-grazed (CK), low-intensity grazing (LG), moderate-intensity grazing (MG), and high-intensity grazing (HG) at the experimental field 1. On the experimental field 2, there were fenced site since 1979 (UG79), fenced site since 1999 (UG99) and outside them a freely grazing site (FG). Community structure and diversities of soil bacteria and methanotrophs were assessed with PCR amplification of genomic DNA and following denaturing gradient gel electrophoresis (DGGE) separation and clone library analyses.

Principal component analysis of DGGE patterns showed that the LG and MG treatments were different from the CK and HG treatments. Soil bacterial diversities in the LG and MG treatments were significantly higher than those in the CK and HG treatments. The relationships between environmental variables and soil bacterial community structure were assessed using redundancy analysis, and we found that soil

* Corresponding author address: Environmental Futures Centre, Griffith University, Nathan, 4111, Brisbane, Australia, Email: xiaoqi.zhou@griffith.edu.au.

moisture content, *Artemisia frigida* biomass, and pH were the best indicators of the changes in soil bacterial community structure among all the treatments. Additionally, the UG99 soil had significantly higher bacterial diversity as compared with the FG and UG79 soils. Moreover, 55 clones from the UG79 soil and 56 clones from the FG soil were selected and sequenced. Phylogenetic analysis of all clone sequences indicated that bacterial communities were dominated by the groups of *Actinomycetes*, *Proteobacteria* and *Bacteroidetes*, but there were no significant differences in bacterial diversity between the UG79 and FG sites, consistent with the results obtained from DGGE.

Additionally, there were no significant differences in methane flux among four grazing intensity treatments, but methanotrophic communities showed significant differences. Diversity indices of type II methanotrophs were higher at the LG and MG treatments than at the CK and HG treatments. Though grazing intensities had an effect on the structure of the methanotrophic community, changes in the community structure and diversity of methanotrophs did not reflect methane consumption capacity across the treatments. Cluster analysis indicated that the methanotrophic communities structure of UG99 and FG soils were similar to each other but different from that of UG79 soil. Sequence analysis showed that most bands belonged to the cluster of USCγ. This is the first report showing that USCγ cluster is dominant in grassland soils.

Overall, our results indicated that intermediate grazing intensities (LG and MG) increased soil bacterial diversities and short-term rather than long-term exclusion of grazing increased bacterial diversity, which will be helpful for managements of grassland ecosystems. On the other hand, we also suggest that methane consumption is a complex process in soil, and we should be cautious when speculating on methane consumption rates based on changes in methanotrophic community structure.

Keywords: Grazing, Bacteria, Methanotroph, Methane flux, DGGE, Inner Mongolia, Grassland

INTRODUCTION

Ungulate grazing plays a vital role in carbon (C) and nitrogen (N) cycling in grassland ecosystems (Wardle et al. 2004). Overgrazing may lead to degradation of grassland (Steffens et al. 2008). Many studies have documented the effects of grazing on aboveground plant communities (Wardle et al. 2008) and C and N mineralization (Wang et al. 2006, Xu et al. 2007). Grazing has been found to affect soil microbial biomass irrespective of changes in both the quantity and quality of litter (Bardgett et al. 2005). However, knowledge on the effects of grazing on microbial community structure and diversity in grassland ecosystems is limited.

Soil microorganisms represent the world's largest reservoir of biological diversity with enormous diversities of 16S rRNA gene sequences amplified from soil DNA and RNA (McCaig et al. 1999, Sun et al. 2004). It is well recognized that plate count cannot reflect the whole community composition of soil microbes since the majority of them are not culturable (Amann et al. 1995). Molecular techniques provide a way to circumvent this problem (Muyzer et al. 1993). Usually these techniques are based on the extraction of genomic DNA from soil, followed by PCR amplification of 16S rRNA genes and separation of amplicons with procedures such as denaturing gradient gel electrophoresis (DGGE). The DGGE method has successfully described the structure and diversity of microbial community in waster water sludge (Boon et al. 2002), grassland soil (Clegg 2006, Zhou et al. 2008b) and agricultural soils (Seghers et al. 2004).

Responses of bacterial communities to grazing are more complicated than we expected (Sankaran and Augustine 2004). McCaig et al. (2001) did not observe obvious changes in soil bacterial community structure between the grazed and non-grazed sites using DGGE method. However, Jangid et al. (2010) found that community structure rather than diversity of soil bacteria significantly differed along a chronosequence of restoration of grassland using phospholipid fatty acid and clone libraries.

Recently, Klumpp et al. (2009) have reported that compared with low level grazing, intensive grazing leads to a change in soil microbial communities and a proliferation of Gram(+) bacteria in a controlled experiment. Some studies have characterized soil bacterial communities, attempting to draw links between plants and below-ground bacterial diversity (Miethling et al. 2003). A number of factors were found to influence soil microbial community composition and diversity and these factors included plant species (Smalla et al. 2001), soil organic C content and soil C:N ratio (Wieland et al. 2001). Given the importance of soil microorganisms relating to C and N cycling, it is necessary to investigate the effect of grazing on microbial community structure and diversity, which is helpful for improving sustainable management of grassland ecosystems.

Atmospheric methane is one of the most important greenhouse gases and contributes approximate 20% of global warming (IPCC 2007). Aerobic soils also oxidize a significant amount (20-60 Tg y^{-1}) of the atmospheric methane (Hutsch 2001). Different from general bacteria, methane-oxidizing bacteria (methanotrophs) are a group of functional microorganisms that can utilize atmospheric methane as sole source of carbon and energy. The methanotrophs can be divided into two distinct physiological group, type I and type II, based on their phylogenetic affiliation, intracellular membrane arrangement, carbon assimilation pathway and phospholipid fatty acid composition (Hanson and Hanson 1996, Bowman 2000).

Type I is low-affinity methanotrophs and can grow in the high methane concentration and low oxygen concentration, while Type II is high-affinity methanotrophs and can grow in the low concentration of atmospheric methane (Bender and Conrad 1992). Notably, cultivated isolates of methanotrophs were not able to grow on the low mixing ratio of atmospheric methane (Benstead et al. 1998).

The new independent culture technique, however, proved that a kind of methanotroph exists in soils which could be the predominant methanotrophs consuming atmospheric methane (Bull et al. 2000, Crossman et al. 2005, Zhou et al. 2008a).

The enzyme methane monooxygenase catalyzes the first step in the process of methane oxidation (Figure 1). With the exception the genus of *Methylocella* (Dedysh et al. 2004), all methanotrophs possess the particulate methane monoxygenase (pMMO). The *pmoA* gene which encodes for an active subunit of the pMMO is therefore a specific and almost universal marker gene for methanotrophs of the upland soils. Molecular studies have not only detected strains of most recognized genera of methanotrophs in upland soils (Bourne et al. 2001, Knief et al. 2003), but indicated that uncultivated methanotrophs are present in these soils. Uncultivated organisms that harbor *pmoA* sequences of the 'forest soil cluster' or upland soil cluster α (USCα), and upland soil cluster γ (USCγ) are most likely involved in atmospheric methane oxidation in upland soils (Figure 2) (Holmes et al. 1999).

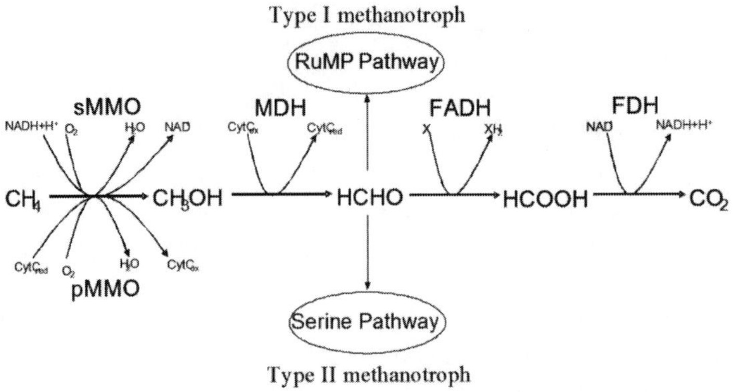

Figure. 1 Pathways of methane oxidation by methanotrophs. Methane (CH_4) is first oxidized into methanol (CH_3OH) via particulate (pMMO) and soluble methane monooxygenase (sMMO), following oxidizing into formaldehyde (HCHO), formic acid (HCOOH), and finally carbon dioxide (CO_2). The methanotrophs can be divided into type I and type II methanotrophs based on different carbon assimilation pathways.

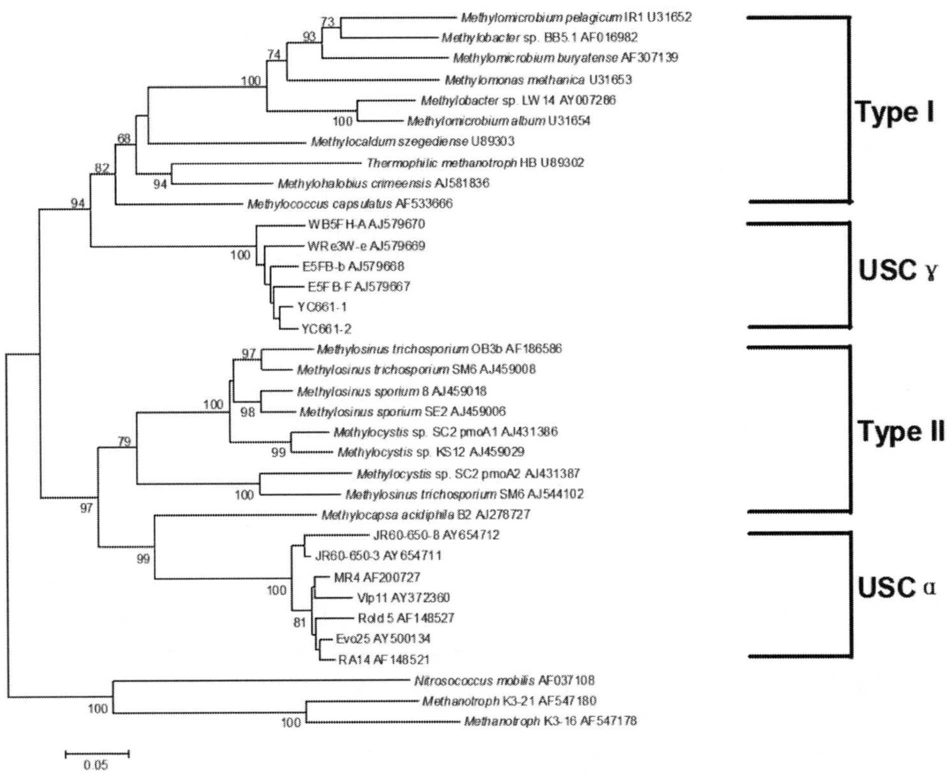

Figure 2. Phylogenetic analysis of methanotroph based on the conserved sequence of *pmoA* gene.

Among upland soils, grassland soils are the second terrestrial methane sinks and its capacity of methane oxidation is less than that of the forest soils (IPCC 2007). Recently, many studies have documented the structure of the methanotrophic community of diverse forest soils (Bourne et al. 2001, Knief et al. 2005). Compared with the forest soils, a few studies are available for the community structure of methanotrophs in grassland soils (Horz et al. 2005, Zhou et al. 2008a).

Typical grassland of Inner Mongolia is one of the representative parts of Eurasian grasslands and regarded as an important sink for atmospheric methane (Zhou et al. 2008b). There is about 3.13×10^8 km^2 of grassland in China (Bai et al. 2004). Grazing is one of the most important land-uses on the Inner Mongolia steppe and has been so far for thousands of years. Nowadays these grasslands have undergone considerable changes over the last decades because of grassland managements. This has led to a more intensive grazing regime which may eventually result in grassland degradation finally (Chen and Wang 2000). Only a few studies were available for the effects of grazing intensities on the microbial communities using traditional plate counts and enzyme activities (e.g., Liu and Liao 1997).

The goal of this study was to evaluate the effect of the different grazing intensities and long-term exclusion of grazing on community structure and diversity of bacteria and methanotrophs with specific primers of 16S ribosomal RNA (rRNA) and *pmoA* genes. We combined a kinetic description of the methane consumption capacity of these experimental treatments with a molecular analysis of the bacterial populations involved. The main objectives of this study were to (1) investigate the effects of different grazing intensities and long-term exclusion of grazing on bacterial community structure and diversity using PCR-DGGE, (2) quantify the relationships between environmental variables and community composition of bacteria in soils, and (3) determine which grazing intensities were more suitable for local grazing management in the Inner Mongolian steppe. For methanotrophs, we hypothesized that management could induce changes in the structure and diversity of the methanotrophic community, which might in turn lead to changes in methane consumption capacity of soils.

METHODS

Site Description

The two experimental sites were nearby the Inner Mongolia Grassland Ecosystem Research Station of Chinese Academy of Sciences, located in the central part of Inner Mongolia Autonomous Region (43°50′N, 116°34′E), with an average elevation of 1,100 m above sea level. The two experimental sites were about 60 km away from each other. The climate is temperate and semi-arid with cold, dry winters and mild, wet summers. The mean annual temperature is -0.4 °C with mean monthly temperatures ranging from 17.9 °C in July to -23 °C in January. The mean annual precipitation is 350 mm with most rain events occurring from June to August. During the sampling years, 2004 exhibited normal annual rainfall of 364 mm, whereas 2005 was a dry year with only 126 mm recorded.

Experimental Field 1

Prior to the experimental design, *Artemisia frigida,* a native grass with an area of 16 ha, was dominant plant species in the experimental field 1 (Figure 3).

Figure 3. Landscape view of the grazing intensity site (Experimental field 1) which started in 1989.

Vegetation within the experimental plots is dominated by *Artemisia frigida*, *Kochia prostrate* and *Potentilla acaulis*. Soils are coarse textured with a mean of 71% sand, 20% silt, and 9% clay across all experimental plots (Zhou et al. 2008b, Zhou et al. 2010).

Grazing management experimental plots were established in 1989 using a randomized complete block design with six grazing intensities with three replicates. There were 0, 4, 8, 12, 16 and 20 Inner Mongolia fine wool sheep grazing rotationally in three replicated 1-ha plots, representing the grazing intensities of 0, 1.33, 2.67, 4.00, 5.33 and 6.67 sheep ha^{-1}, respectively (Chen and Wang 2000). Four grazing intensity were selected in this study: non-grazed (CK; 0 sheep ha^{-1}), light-grazing intensity (LG; 1.33 sheep ha^{-1}); moderate-grazing intensity (MG; 4.00 sheep ha^{-1}) and heavy-grazing intensity (HG; 6.67 sheep ha^{-1}) (Zhou et al. 2010). Every year, grazing started on May 20th and ended on October 5th. Each plot was rotationally grazed freely three times per year, each time for 15 days with a rotation interval of 30 days. The total grazing period of each grazed plot was thus 45 days per year. During grazing periods, the sheep were driven to the enclosures at 5:00 am and back home at 8:00 pm. The sheep drank water twice daily from the nearby Xilin River, once before being driven to the enclosures and once after being driven out of the enclosures. No fertilizer was used in any plot. The plant biomass and soil properties (soil texture, structure and chemical characteristics) were relatively uniform among all experimental plots before the initiation of the grazing treatments. The grazing intensity treatments started in 1989.

Experimental Field 2

Prior to 1979, the whole experimental area of 75 ha was generally grazed under a low intensity (Figure 4). The area being grazed by herds consisted of 70-90% sheep and 10-30% goats. In 1979, one site (24 ha) was fenced and excluded from grazing (UG79) and grazing intensity in the remaining region increased to a moderate level.

Figure 4. Landscape view of the long-term grazing exclusion site which started in 1979 (Experimental field 2).

In 1999, a second site (25 ha) was fenced and excluded from grazing (UG99). Outside the two sites was freely grazed grassland (FG). The FG site is assumed to represent a grazing intensity of 2 sheep units (1 SU= 1 ewe and 1 lamb) ha^{-1} yr^{-1} (Zhou et al. 2012c). The aboveground plant biomass at the UG79, UG99 and FG sites was 218.7, 215.7 and 120.0 g m^{-2}, respectively (Wang et al. 2006). The UG79 site is dominated by perennial rhizome grasses such as *Leymus chinensis* and *Stipa grandis* which represent a widely distributed grassland community in the Eurasian grasslands. However, the FG site is dominated by herbaceous species such as *Artemisia frigid* and *Pedicularis acaulis* (Wang et al. 2006). Soils are a sandy loam texture classified as calcic Chernozems with a mean of 50% sand, 35% silt, and 15% clay across all experimental sites (Steffens et al. 2008, Zhou et al. 2013).

CH_4 Consumption Analyses

Methane consumption rates were measured in situ in July and August in 1998 and 1999 as well as in August 2004 and in July 2005 using a closed static chamber method (Zhou et al. 2008b). The chambers were equipped with a battery driven ventilator and a small vent for pressure equilibration was used for gas flux measurements. Three replicate chambers, 0.5 m×0.5 m×0.5 m, were used in every plot. During the measurement, the chambers were 10 to 15 m away from each other. After chamber closure, gas samples were taken from the chambers five times at a 10-min interval within 40 minutes using 10ml gas tight syringes with stopcocks at the top. All gas samples were analyzed within 6 h in a laboratory at the Inner Mongolia Grassland Ecosystem Research Station on a gas chromatrograph HP 5890 II equipped with a flame ionization detector. Linear regressions were performed on the five samples for CH_4 uptake from each flux chamber. Finally, the CH_4 flux rates were adjusted using air pressure and air temperate within the chamber headspace.

Soil Sampling and Plant Aboveground Biomass and Diversity Measurements

For all experimental sites, within each replicated plot, five 1×1 m quadrats were randomly selected in August 2004 and aboveground vegetation was clipped at ground level to determine plant community composition and diversity. Soil samples were collected at a depth of 0 to 10 cm with a 3 cm diameter auger within each plot in August 2004 and in August 2005.

All soil cores were hand-sorted to remove stones and coarse root, cooled on ice in the field and transported into the laboratory. After passing through a 2-mm sieve, soil cores within each plot were mixed to form a composite sample. All the samples were stored at 4 °C and soil biochemical properties were determined within two weeks. For aboveground plant biomass, the plant samples were separated into different species and the weights of each species were determined prior to oven-drying at 65 °C for 48 h. The Shannon index of general diversity (H) was used to measure the plant diversity (Degens et al. 2000); H was calculated as follows: $H = -\sum(n_i/N)\log(n_i/N)$, where n_i is the number of individual plant species and N is the sum of all plant species.

Soil Biochemical Property Measurements

Moisture content of each soil sample was determined gravimetrically by oven-drying at 105 °C for 24 h. Air-dried soil samples were used for measuring pH, organic C content and organic N content. The pH values were determined in water suspension (water/soil=2.5:1). Soil organic C contents were analyzed using a H_2SO_4-$K_2Cr_2O_7$ oxidation method. Soil organic N contents were measured using the Kjeldahl digestion method (Kjektec System 1026 Distilling Unit, Sweden). Soil NH_4^+-N and NO_3^--N concentrations were determine using KCl extraction method (Zhou et al. 2012a). Briefly, a 5 g field soil sample was extracted with 30 ml of 2 M KCl on an end-to-end shaker for 1 h. The soil suspension was filtered through a Whatman no. 42 filter paper and the filtrate analyzed for on a flow injection auto-analyzer (FIAstar 5000 Analyzer, Foss Tecator, Denmark) (Zhou et al. 2011, Zhou et al. 2012b).

Extraction and PCR Amplification of total DNA from Soil

At first, we tried two extraction methods of frozen-thawing and bead-beating to determine which one can better reflect bacterial communities in grassland soils (Zhou et al. 2007). Nucleic acids were extracted and purified from 0.5 g aliquots of each soil sample following the procedures (Zhou et al. 2007, Zhou et al. 2010). Briefly, soil samples were placed in 2 ml screw-cap tubes containing a mixture of ceramic and silica particles (Bio101, Carlsbad, Calif.); the mixture was homogenized for 30 s in a FastPrep bead beater cell disrupter (Bio101).

The nucleic acids were then precipitated and washed twice using 75% (v/v) ethanol, and the final DNA was re-suspended in 100 μl TE buffer. The crude extracts were purified with Qiagen gel extraction Kit (Qiagen, Inc.) and DNA concentrations were determined by nucleic acid analysis (Shimadzu, Kyoto, Japan).

Table 1. PCR primers used in this study

Target	Primers	PCR conditions							References
		Numbers of cycles	Denaturation		Annealing		Elongation		
			°C	min	°C	min	°C	min	
First PCR round									
Type I methantrophs	methT1dF, methT1bR	30	95	1	55	1	72	2	(Wise et al. 1999)
Type II methanotrophs	27f, methT2R	30	95	1	55	1	72	2	(Lane 1991, Wise et al. 1999)
Second PCR									
Bacteria	P338F[a], P518r	30	94	1	56	1	72	1	

Before cycles started, the temperature was held at 95 °C for 7 min and after each run the temperature was kept 72 °C for 10 min for final template elongation.

[a]A 5'GC-clamp was added for DGGE analysis (Muyzer et al. 1993)/

PCR amplifications of bacterial 16S rRNA were carried out using the primers P338F (5' ACTTCTACGGGAGGCAGCAG 3') with 40 GC clamps and P518R (5' ATTAC CGCGG CTGCT GG) (Muyzer et al. 1993, Zhou et al. 2007). All the primers and PCR conditions of methanotrophs are listed in Table 1. In order to increase the sensitivity and to facilitate the DGGE by analyzing fragments of the same length, a nested PCR technique was applied. In the first round different group-specific primers and one set of bacterial primers were used, each with their own corresponding PCR protocol. During the second PCR round, the obtained fragments were re-amplified by using the bacterial primers P338F and P518r in the same PCR protocol (Table 1). PCR amplification of pmoA genes was performed via a semi-nested PCR approach using 5' primer A189 and 3' primer A682 (Holmes et al. 1995). Aliquots of the first round of PCR (0.25μl) were used as the template in the second round of PCR using the 5' primer A189 with a GC clamp (Henckel et al. 1999) and 3' primer mb661which was designed for the detection of type I and type II methanotrophs (Costello and Lidstrom 1999).

Amplification was performed in a 25μl (total volume) reaction mixture. The final concentration of different components in the mixture was as follows: ~ 50ng of purified DNA, 0.4 μM of each primer, 200 μM of each deoxynucleoside triphosphate, 1.5 μM $MgCl_2$, 1× thermophilic DNA polymerase 10×reaction buffer ($MgCl_2$-free), 1.25U per 50 μl of Taq DNA polymerase (Promega, Madison, WI, USA), and DNAse and RNAse free filter sterilized water (LiYuan Apparatus Com., Beijing, China). The reaction was performed in a PTC-200 thermal cycler (MJ Research Com., USA). After each PCR, the size of the amplification products was verified on a 1.2% agarose gel.

DGGE Analysis

The DGGE analysis was performed using a Dcode system (Bio-Rad, Hercules, CA, USA) (Zhou et al. 2008b, Zhou et al. 2010). Polyacrylamide gels (8% acrylamide, 0.5 × TAE (20 mM Tris, 20 mM acetate, 1 mM EDTA [pH 8.0]), 37:1 acrylamide: bisacrylamide) were cast using 40-60% denaturant; 100% denaturant was defined as 7 M urea with 40% (vol/vol) formamide. Electrophoresis was run overnight for 12 h at a constant temperature of 60 °C at 110 V. After electrophoresis, the gels were soaked in SYBR green II nucleic acid gel stain (1:10000 dilution; FMC Bioproducts, Rockland, ME, USA) for 30 min. The stained gels were immediately photographed on a UV trans-illumination table with a Hewlett Packard Scanjet 5370C. The DGGE patterns were determined using the Labworks software (Labworks TM software version 4.0. UVP, UK). Band intensities of each lane were read based on peak heights in the densitometric curve. The Shannon index of general diversity (H) was used to measure the bacterial diversity (Degens et al. 2000); H was calculated based on the band intensity in the gel lanes as follows: $H = -\sum(n_i/N)\log(n_i/N)$, where n_i is the height of the peak and N the sum of all peak heights of the densitometric curve. Evenness (E) was calculated by the equation of $E=H/H_{max}$, where H_{max} was the largest H within a special sample (Zhou et al. 2012d).

Sequencing for Specific Bands and Clone Library Analysis

The specific bands of the DGGE pattern were excised from the gel and re-amplified with the same reaction procedure as described above. The PCR products were excised from an agarose gel, purified with Qiagen gel extraction kit (Qiagen Inc.) to remove primers and short oligonucleotides and then cloned into plasmids with a pGEM-T easy vector kit (Promega) following the manufacturers' procedure. Ligations were transformed into Escherichia coli DH5α supercompetent cells. Plasmid DNAs were isolated from randomly selected clones and screened for inserts of the expected sizes. Following confirmation of their DGGE positions under the conditions as described before, the plasmid DNAs were sequenced with primers (T7/SP6 forward/reverse) on an ABI PRISM 3700 DNA analyzer (ABI Biosystems, Inc.).

The PCR products of bacterial communities from the UG79 and FG soils were also cloned into pGEM-T easy vector (Promega) as well. After plasmid DNAs were screened for inserts of the expected sizes, then 55 and 56 clones were at last randomly selected and sequenced from UG79 and FG soils, respectively. The plasmid DNAs were sequenced using primers T7 and SP6 on an ABI PRISM 3700 DNA analyzer (ABI Biosystems, Inc.) as well. The identities of the 16S rRNA gene sequences were confirmed by searching the international sequence NCBI database using the BLAST programs.

Statistical Analysis

One-way analysis of variance (ANOVA) was employed to determine the effect of grazing on the soil biochemical properties, plant biomass and diversity, bacterial and methanotrophic diversity among all treatments for each field. For DGGE patterns of soil bacteria, principal components analysis (PCA) was carried out with a correlation matrix. A dendrogram analysis based on correlation matrix was used to cluster soil methantrophic DGGE patterns across the long-term grazing exclusion sites. All ANOVA, PCA and regression analyses were performed using SPSS 12.0 software (SPSS Inc., USA).

The sequences from the two sites were compared to a current database of genetic sequences (GenBank) with the Blast (basic local alignment search tool) program of the Ribosomal Database Project to determine their approximate phylogenetic affiliation. Sequence assembly and manual refinement of alignments were carried out using the multiple-alignment algorithm as implemented in CLUSTALX 1.81. Phylogenetic analysis was carried out using Neighbor Joining (NJ) method with Jukes-Cantor model in the PHYLIP 3.5 package. Nodal robustness on the NJ tree was estimated by the nonparametric bootstrap (1,000 replicates). The clones were clustered into operational taxonomic units (OTUs) at a level of sequence similarity of >97%. Finally, collector's curves or species abundance curves were constructed to compare bacterial diversities between the two sites.

Redundancy analysis (RDA) was used to study the relationships between environmental variables (soil properties and plant biomass and diversity) and bacterial community structure based on DGGE profiles. Data were analyzed by detrended correspondence analysis using Canoco Software 4.5 (Microcomputer Power, USA), which revealed that the data exhibited a linear, rather than unimodal, response to the environmental variables. Redundancy analysis of the DGGE profiles in all the treatments were related to environmental variables by selecting the linear combination of environmental variables that gave the smallest residual sum of squares (Kennedy et al. 2004, Zhou et al. 2012d).

Nucleotide Sequence Accession Numbers

The 16S rRNA sequences of clone libraries were deposited in the Genbank Data Library under accession numbers of DQ414821 to DQ414826, DQ414832 to DQ414844, DQ420618 to DQ420621 and DQ 973190 to DQ973281. The partial pmoA gene fragments of DGGE bands were deposited in the GenBank database under accession numbers EF165624 to EF165630.

RESULTS

Comparison of the Two DNA Extraction Methods

Both the Frozen-thawing and bead-beating methods yield 30-50 μg of microbial DNA in 0.5 g soil and there were no significant differences between them (Figure 5). However, DNA bands were much less in DGGE profiles using the frozen-thawing method than using the bead-beating method for DNA extraction (Figure 6). Moreover, with the former method, DNA bands were concentrated in a small area of the gel. Such pattern did not change with the variation of gel denaturant gradient (50%–60%). Dendrogram analysis of DGGE patterns indicated that there was some, not significantly different, variation (<10%) in DGGE patterns between duplicates in each method, while the difference between the two methods was remarkable (Figure 6). In this study, the bead-beating method performed better than the frozen-thawing method in terms of genomic DNA extraction efficiency (Figure 6). Because of better performance of bead-beating methods, we used this method to investigate the bacterial and methanotrophic community structure and diversity.

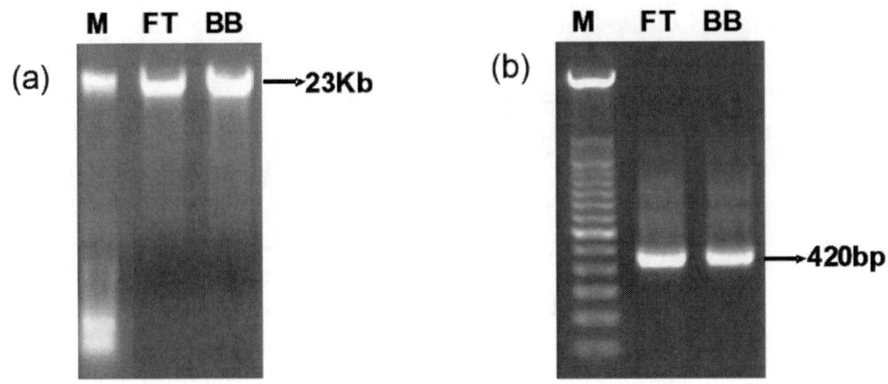

Figure 5. Comparison of genomic DNA and subsequent bacterial PCR results from *Leymus Chinensis* field using two extraction methods: FT, frozen-thawing method (a); b, BB, bead beating method (b).

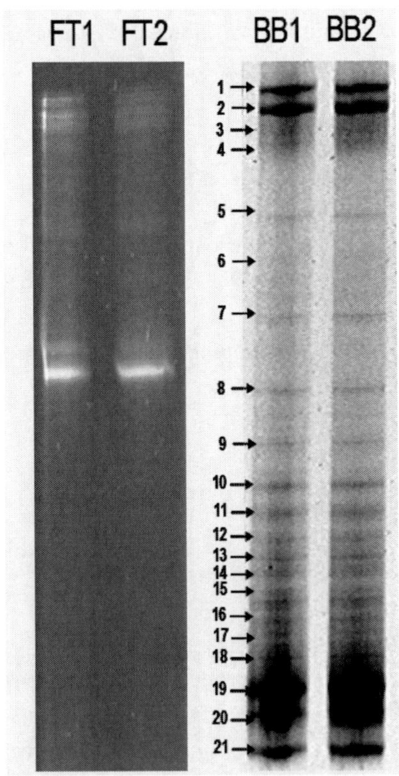

Figure 6. Comparison of DGGE profiles with duplicates for two different bacterial 16S rDNA extraction methods in UG79 soils (FT, Frozen-thawing method; BB, bead-beating method).

Bacterial Diversity under Grazing Intensities (Experimental Field 1)

Soil Characteristics and Plant Biomass and Diversity

At the experimental field 1, the CK soil had significantly higher moisture contents, organic C, NH_4^+-N and NO_3^--N concentrations than the grazed soils (Table 2), while the MG soil had the highest pH. There were no significant differences in the soil organic N contents among all the treatments.

Table 2. Soil properties under different grazing intensities at the experimental field 1 in the Inner Mongolia steppe. CK, non-grazed; LG, low-intensity grazing; MG, moderate-intensity grazing; HG, heavy-intensity grazing. The CK, LG, MG and HG sites were grazed by 0.00, 1.33, 4.00 and 6.67sheep ha^{-1}, respectively. Each plot was rotationally grazed three times per year, each time for 15 days with a rotation interval of 30 days. Each value represents the mean with standard errors. Different letters within columns showed significant differences at P<0.05

Treatment	Soil moisture (%)	pH	soil organic C (g kg^{-1})	NH_4^+-N (μg g^{-1})	NO_3^--N(μg g^{-1})	soil organic N (g kg^{-1})
CK	18.7±0.05a	6.65±0.02b	1.79±0.02a	4.34±0.07a	3.10±0.06a	0.71±0.05
LG	11.16±0.04b	6.65±0.03b	1.47±0.02b	3.33±0.02b	1.84±0.09b	0.69±0.05
MG	12.13±0.07b	6.73±0.02a	1.47±0.05b	3.34±0.03b	1.34±0.03b	0.66±0.08
HG	12.65±0.09b	6.58±0.04b	1.56±0.03b	3.25±0.08b	0.89±0.05c	0.55±0.09

Table 3. Dominant species, total aboveground biomass and plant diversity Shannon index (H) under different grazing intensities at the experimental field 1. CK, non-grazed; LG, low-intensity grazing; MG, moderate-intensity grazing; HG, heavy-intensity grazing. Each value represents the mean with standard errors. Different letters within rows showed significant differences at P<0.05

Treatment		CK	LG	MG	HG
Dominant species aboveground biomass (g m^{-2})	*Artemisia frigida*	27.5±0.3c	36.5±0.2b	49.1±0.5a	12.1±0.6d
	Kochia prostrata		14.2±2.92b	21.1±1.8a	12.6±3.3b
	Potentilla acaulis	19.9±2.8b			29.5±2.1a
	Leymus chinensis	44.4±2.3			
	Carex duriuscula		16.6±5.49	23.7±6.09	
Total aboveground biomass (g m^{-2})		127.1±12.9a	98.3±12.1a	102.1±13.2a	56.7±12.6b
Plant community Shannon diversity H		1.48±0.11a	1.33±0.07a	1.23±0.12ab	1.10±0.06b

The dominant species, plant biomass and diversity index are shown in Table 3. The CK site was dominated by *Leymus chinensis*, *Artemisia frigida* and *Potentilla acaulis*, while the grazed sites were dominated by *Artemisia frigida*, *Kochia prostrate* and *Carex duriuscula*. The plant biomass and diversity in the HG treatment were significantly lower than those in the other treatments.

PCR-DGGE Analysis of Bacterial Community

Before PCR-DGGE analysis, DNA yields after extraction were examined and no significant differences were found among all treatments. DGGE patterns of soil bacterial community under different grazing intensities are shown in Figure 7. A great number of bands in DGGE patterns were detected and there were also many bands forming a smear. Clear differences were observed across sites with many bands absent in the CK and HG soils and some bands present in the LG and MG soils.

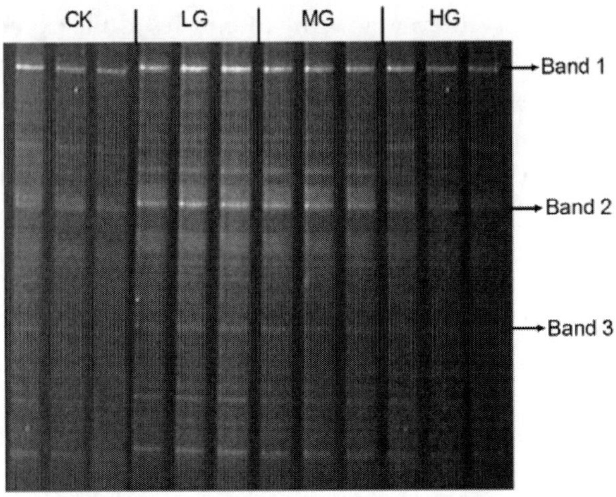

Figure 7. DGGE patterns of soil bacterial community under different grazing intensities. Bands 1, 2 and 3 were excised from the gel and were further sequenced to indentify the specific bacterial taxon. CK, non-grazed; LG, low-intensity grazing; MG, moderate-intensity grazing; HG, heavy-intensity grazing.

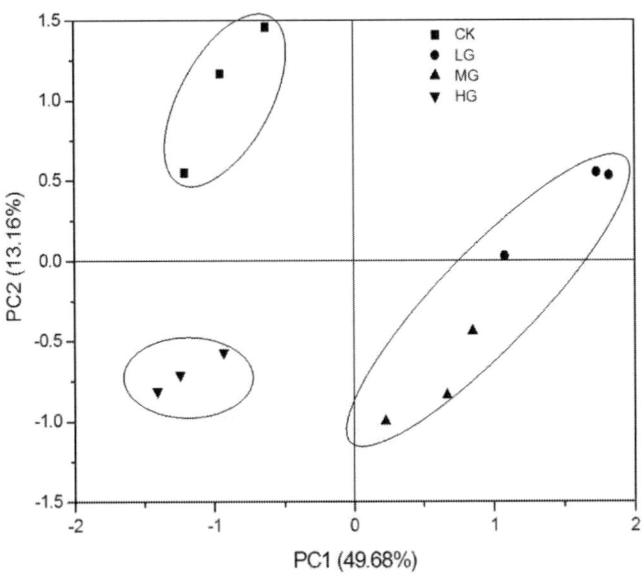

Figure 8. Ordination plot of principal components analysis (PCA) of DGGE patterns in soils under different grazing intensities. Numbers in parenthesis are percentage variance by each principal component (PC). CK, non-grazed; LG, low-intensity grazing; MG, moderate-intensity grazing; HG, heavy-intensity grazing.

To identify the specific dominant bacterial species in the DGGE pattern under different grazing intensities, three main bands were excised from the gel (Figure 7), re-amplified and sequenced.

Table 4. Shannon diversity index (H) and Evenness (E) of soil bacterial diversity based on the DGGE profiles under different grazing intensities in the Inner Mongolian steppe. CK, non-grazed; LG, low-intensity grazing; MG, moderate-intensity grazing; HG, heavy-intensity grazing. Different letters within columns showed significant differences at $P<0.05$

Treatment	H	E
CK	1.88±0.04b	0.84±0.01b
LG	2.39±0.05a	0.89±0.02a
MG	2.31±0.07a	0.89±0.01a
HG	1.76±0.09b	0.79±0.03b

The sequences of bands 1, 2 and 3 showed 98% similarity to *Intestinal bacterium D22* (AY374116), 97% to *Uncultured Acidobacteria bacteria* (AKYH520*)* and 98% to *Bosea minatitlanensis* (AF273081), respectively. These sequences fell into subtaxa of *Gamma-proteobacteria*, *Acidobacteria* and *Alpha-proteobacteria,* respectively.

The bacterial community structure among all the treatments was analyzed by PCA (Figure 8). The PC1 and PC2 explained 49.68% and 13.16% of the variance of the data, respectively. The LG and MG soils differed from the CK and HG soils with high ordinate scores on the PC1. There was a clear separation between the CK and HG soils along the PC2. The LG and MG soils had higher bacterial diversities and evenness than the CK and HG soils (Table 4).

Relationships between Environmental Variables and Bacterial Community Structure

The relationships between environmental variables and soil bacterial community composition are shown in Figure 9. Eigenvalues of RDA indicated that axes 1 and 2 explained 0.21 and 0.19 of the overall variance within the soil bacterial community in all the treatments, respectively. Species-environment correlation for both axes was 0.98, indicating that soil bacterial community composition was strongly correlated with environmental parameters. The plot can be interpreted quantitatively by noting that the length of an arrow indicates how much variance is explained by that factor. The direction of arrows for each environmental variable indicates an increasing concentration of that variable (the longer the arrow, the stronger the relationship; Kennedy et al. 2004). The arrows for soil moisture content, *A. frigida* biomass and pH were longer than those of the other variables, indicating that these factors accounted for the greatest proportion of variance in the soil bacterial community composition.

The patterns of soil bacterial community structure in the treatments with approximately the same direction as the environmental variable arrow indicate a high positive correlation. The patterns of community composition of bacteria in the CK and HG treatments showed a positive relationship with soil moisture and organic C contents, while those in the LG and MG treatments were positively influenced by *A. frigida* biomass and pH, followed by soil organic N contents.

Ammonium-N and NO_3^--N concentrations, and plant biomass and diversity contributed to some extent to the changes in community composition of bacteria under different grazing intensity treatments.

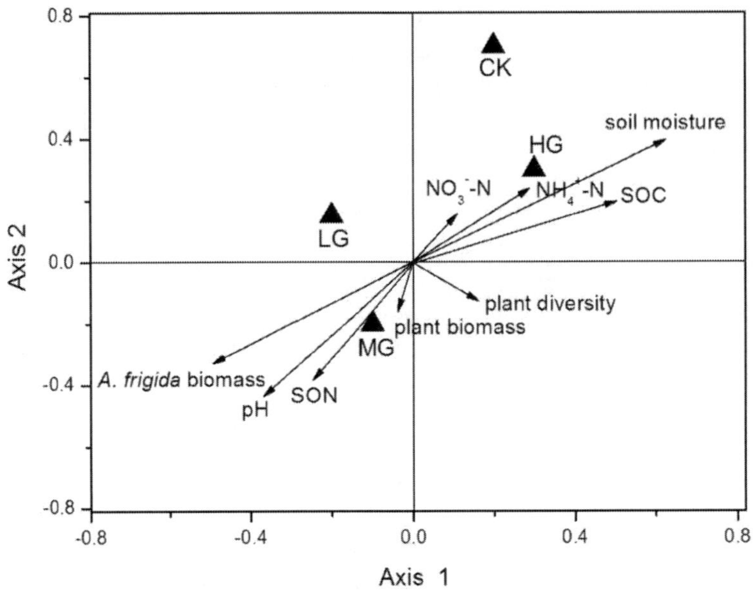

Figure 9. Biplots of redundancy analysis (RDA) of the relationships between environmental variables and soil bacterial community structure based on DGGE profiles under different grazing intensities. SOC, soil organic C and SON, soil organic N. CK, non-grazed; LG, low-intensity grazing; MG, moderate-intensity grazing; HG, heavy-intensity grazing.

Bacterial Diversity under Long-Term Grazing Exclusion (Experimental Field 2)

At the experimental field 2, the UG99 soil had significantly higher moisture content, organic C, organic N, and NH_4^+-N concentrations than the other two soils, while the UG79 soil had the highest NO_3^--N concentration and the FG soil had the lowest organic C, organic N, NH_4^+-N and NO_3^--N concentrations (Table 5).

Table 5. Soil properties at the three grasslands at the experimental field 2 for 2004 sampling. Each value represents means and standard errors. Different letters within columns showed significant differences at P<0.05. FG, freely grazing site; UG99, site fenced since 1999; UG79, site fenced since 1979

Site	Moisture (%)	Moisture (%) (2005 sampling)	pH	Organic C (g kg^{-1})	Organic N (g kg^{-1})	NH_4^+-N (mg kg^{-1})	NO_3^--N (mg kg^{-1})
FG	17.8±0.1b	7.1±0.1b	7.15±0.02b	18.7±0.1c	1.95±0.05c	4.96±0.11c	1.01±0.06c
UG99	23.1±0.1a	11.1±0.1a	7.31±0.03a	24.3±0.2a	28.5±0.05a	14.76±0.09a	1.71±0.09b
UG79	18.1±0.1b	10.3±0.2a	7.10±0.02b	21.7±0.2b	23.1±0.08b	8.24±0.12b	2.12±0.03a

PCR-DGGE Analysis of Bacterial Community

DNA yields after extraction showed no significant differences across the treatments. DGGE patterns of bacterial communities in soils sampled in 2004 and 2005 at the three grassland sites are shown in Figure 10. A large number of bands were detected in the DGGE patterns with many bands forming a smear. The bacterial community structure across the treatments was analyzed by PCA (Figure 11). The PC1 and PC2 explained 32.5% and 20.4% of the variance of the data, respectively.

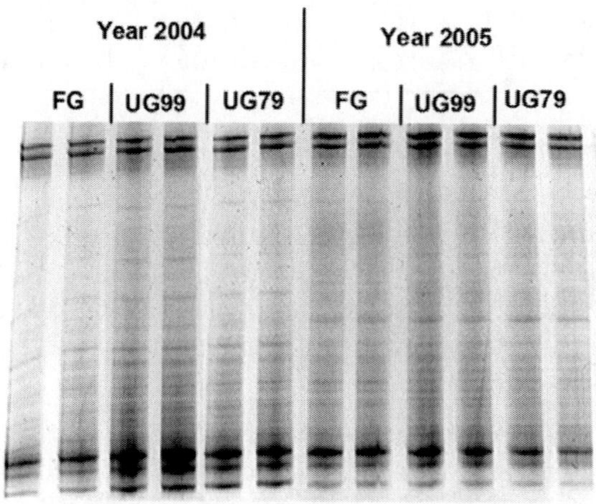

Figure 10. Representative DGGE pattern of bacterial communities in soils sampled in two consecutive years at the three grasslands at the experimental field 2. FG, freely grazing site; UG99, site fenced since 1999; UG79, site fenced since 1979.

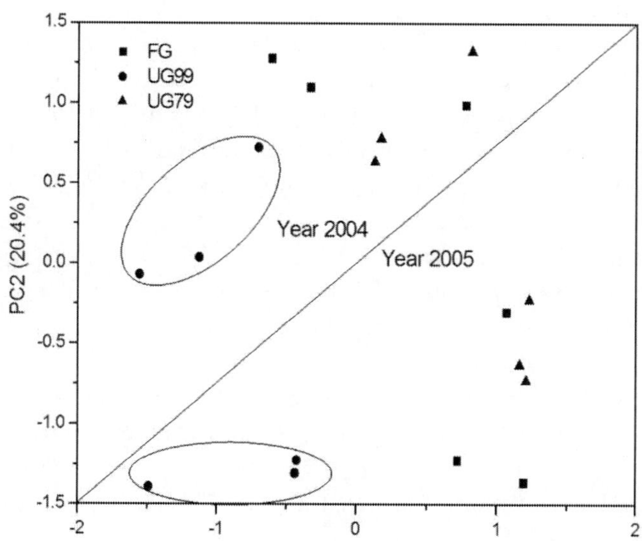

Figure 11. Ordination plot of principal components analysis (PCA) of DGGE patterns in soils at three grasslands. Numbers in parenthesis are percentage variance by each principal component (PC). The soils sampled in 2004 lie above the diagonal of the square, whereas the soils sampled in 2005 lie below the diagonal. FG, freely grazing site; UG99, site fenced since 1999; UG79, site fenced since 1979.

Table 6. Bacterial diversity Shannon-Wiener index H and Evenness (E) calculated based on DGGE patterns of bacterial community in soils of two consecutive years at the three grasslands at the experimental field 2. Different letters within rows showed significant differences at $P<0.05$. FG, freely grazing site; UG99, site fenced since 1999; UG79, site fenced since 1979

		Year 2004			Year 2005	
	FG	UG99	UG79	FG	UG99	UG79
H	2.75±0.02c	2.99±0.02b	2.95±0.02c	3.03±0.02b	3.19±0.02a	3.08±0.03b
E	0.54±0.03c	0.66±0.01b	0.60±0.03b	0.65±0.02b	0.76± 0.03a	0.67±0.02b

There was a clear separation of the bacterial community between the years 2004 and 2005. The UG99 treatment differed significantly from the FG and UG79 treatments with lower ordinate scores on PC1. The three treatments in 2004 had lower soil bacterial diversities than those of 2005 and the UG99 soil had higher bacterial diversity than the other soils (Table 6). The differences of evenness among the treatments showed the same pattern as those of soil bacterial diversity (Table 6).

Bacterial Diversities at UG79 and FG Sites

Prior to phylogenetic analysis, two clone libraries were identified chimeras which showed not only typical bell-shaped histograms from the CHECK-CHIMERA program but also low percentages of matching with sequences in the database. The UG79 and FG clone libraries contained 55 and 56 sequences, respectively. Based on comparison of the two clone libraries, we found that YC53 from the UG79 soil and LW32 from the FG soil were identical based on BLAST analysis (Figure 12).

Table 7 lists the phylogenetic relationships of all clone sequences from the two sites. Many sequences belonged to characterized groups such as groups of *Proteobacteria* and *Actinomycetes*, while some sequences fell into recently recognized groups such as groups of *Acidobacteria*, *Firmicutes* and *Gemmatimonadetes*. Soil bacterial diversity did not differ between the two sites based on the collector's curves, consistent with the DGGE results. However, the groups of *Proteobacteria, Acidobacteria* and *Firmicutes* showed large differences between the two sites (Table 7). For example, *β- Proteobacteria* and *Firmicutes* groups only occurred in the FG soil and *γ- Proteobacteria* only occurred in the UG79 soil.

The results of phylogenetic analysis of 111 sequences are shown in Figure 12. Seven individual clones were not related to cultured or uncultured representatives of the sequence databases and were considered novel groups. Cluster analysis of clone sequences revealed six major clusters with numerous diverse groups. Clones that exhibited >97% sequences similarity were clustered on the trees. In all, 11 clusters containing two or three sequences with >97% sequence homology were observed, the UG79 and FG soils contained 8 and 7 clusters, respectively (Figure 12).

Among all clone sequences, the largest major group was the *Actinomycetes* (50.9% and 50%, respectively). Four clones (YC49, YC52, LW45 and LW48) of this group exhibited 100% similarity to the culturable bacteria *Arthrobacter globiformis* (X80736), *Streptosporangium roseum* (X89949), *Amycolatopsis sp.* (AF453718) and *Bradyrhizobium sp.* (AF230720), respectively.

Figure 12. Neighbour-Joining tree showing relationship of soil bacterial communities from the UG79 (YC) site fenced since 1979 and the FG (LW) site subjected to continually grazing all along, based on analysis of 111 clone sequences of aligned 16S rRNA sequences. Clones exhibiting >97% sequence similarity are included in numbered clusters. The scale bar indicates an estimated change of 5%. The bootstrap values (>50%) are shown below the branches.

The *Proteobacteria* was the second most dominant group (12.8% and 19.7%). In this group, LW24 had a similarity of 100% to uncultured bacterium (AY218681) in the GenBank. The third major group belonged to *Bacteroidetes* (14.5% and 14.3%). Two clones (YC4 and LW9) in this group exhibited 99% similarity to the culturable bacteria *Pseudomonas sp.* (DQ339153) and *Sphingomonas aquatilis* (AF131295), respectively.

Table 7. Relative abundances of soil bacterial clones from the UG79 site fenced since 1979 and the FG site subjected to continually grazing all along

	Relative clone abundances (%)	
	UG79	FG
Proteobacteria		
Alfa-proteobacteria	5.5	14.3
Beta-proteobacteria	0	3.6
Gamma-proteobacteria	1.8	0
Delta-proteobacteria	5.5	1.8
Acidobacteria	10.9	1.8
Actinomycetes (Gram+)	50.9	50
Gemmatimonadetes	3.6	5.4
Firmicutes (Gram+)	0	3.6
Bacteroidetes	14.5	14.3
Unclassfied bacteria	7.3	5.4

The group of *Acidobacteria* is a recently recognized bacterial division with only several cultivated representatives; the majority of sequences in this division come from environmental clones. Phylogenetic analysis indicated that bacterial YC16, YC34 and YC43 fell into the representative subdivision (Figure 12).

Methanotrophic Diversity under Grazing Intensities (Experimental Field 1)

Methane Consumption Rates

Methane fluxes at different stocking rates are shown in Figure 13.

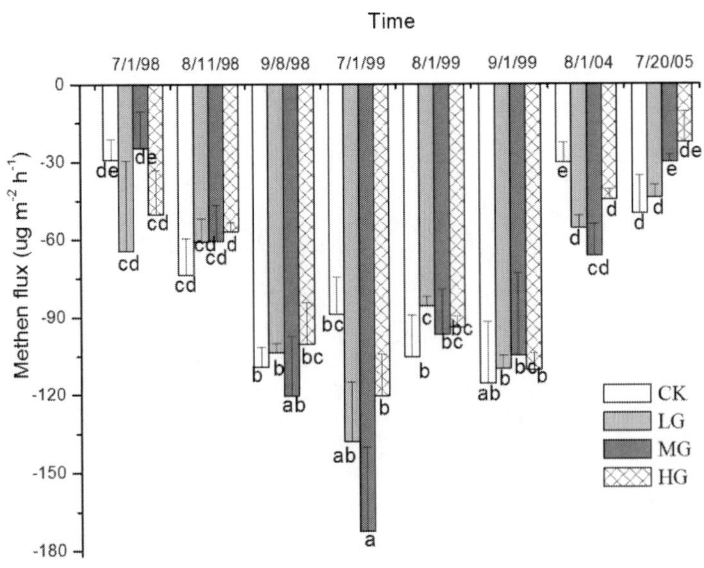

Figure 13. The methane consumption rates under different grazing intensities. CK, non-grazing; LG, low-intensity grazing; MG, moderate- intensity grazing; HG, heavy- intensity grazing. Different letters indicated significant differences at P<0.05.

The methane consumption rates at all sites were negative, which showed Inner Mongolia grassland as a sink for atmospheric methane, consistent with previous studies in this area (Wang et al. 2005). The large variations of methane uptake flux observed at one site at the different sampling times confirm flux rates observed by Wang et al. (2005). In total, methane uptake rates were slightly higher on 9/8/98, 7/1/99, 8/1/99 and 9/1/99 than at the other sampling times (Figure 13). The mean annual precipitation in 1998, 1999, 2004 and 2005 was 357, 214, 364 and 162 mm, respectively. Methane uptake flux at one site at one sampling time also showed large variations (error bars, Figure 13). When the data of different sampling time at the same site were averaged, however, no significant differences were observed among different grazing sites (75, 83, 85 and 75 μg C m^{-2} h^{-1} at the CK, LG, MG and HG sites, respectively).

DGGE Analysis

Before PCR-DGGE analysis, DNA yields after extraction were examined, and no significant differences were found between the different samples. Figure 14 shows DGGE patterns of the type I methanotrophs and type II methanotrophs. After further PCA analysis of these patterns, the structure of the type I methanotroph community in the grazing treatments was clearly different from the CK (Figure 15a). No pronounced differences were observed between the other grazing intensity treatments. Grazing management greatly affected the structure of the type II methanotrophs community as well.

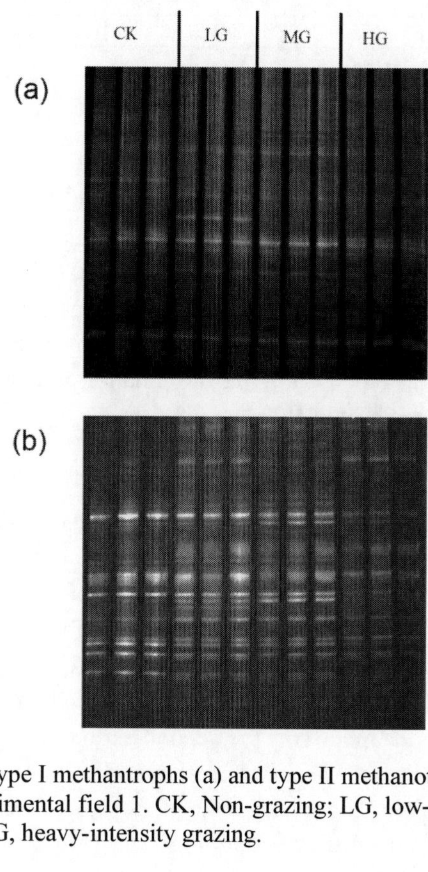

Figure 14. DGGE patterns of type I methantrophs (a) and type II methanotrophs (b) under different grazing intensities at the experimental field 1. CK, Non-grazing; LG, low- intensity grazing; MG, moderate-intensity grazing; HG, heavy-intensity grazing.

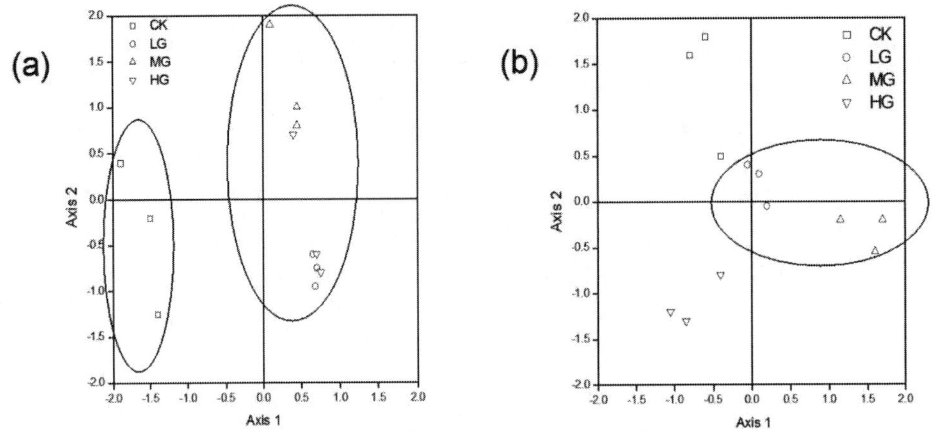

Figure 15. Principal components analysis (PCA) analysis DGGE patterns of type I methanotrophs (a) and type II methanotrophs (b) under different grazing intensities.

Table 8. Shannon diversity (H) and Evenness (E) of type I and type II methanotrophs under different grazing intensities at the experimental field 1. CK, Non-grazing; LG, low-intensity grazing; MG, moderate-intensity grazing; HG, heavy-intensity grazing. Different letters within rows showed significant differences at P<0.05

	CK	LG	MG	HG
Type I methanotrophs	0.98±0.05b	1.03±0.02a	1.06±0.09a	0.97±0.03b
Type II methanotrophs	1.13±0.04b	1.26±0.02a	1.21±0.03a	1.09±0.02b

Type II methanotroph communities at the LG and MG sites were clearly different from the other two sites (Figure 15b). Additionally, the Shannon indices (Table 8) indicated that type I and type II methanotrophs communities at the LG and MG sites were higher than at the CK and HG sites. Each band in the DGGE pattern was assumed to represent one unique ribotype. In the pattern of type II methanotrophs, a phylotype (band) was only detected in LG and MG soil. Sequence analysis revealed that the band was an unclassified *Rhodospirillaceae* which had the highest similarity 99% with *unidentified bacterium* (AY344369). The Shannon indices (Table 8) indicated that type I and type II methanotrophs communities were higher at the LG and MG sites than sites CK and HG.

Methanotrophic Diversity under Long-Term Grazing Exclusion (Experimental Field 2)

When primer set A189F-A682R was used, a large number of non-expected bands in the agarose gel emerged. When primer set A189F- mb661R was used, little or no *pmoA* amplicons produced. Nested PCR, however, using the A189F- 692R primer set in the first round and either reverse primer mb661 in the second round, generated high yields of *pmoA* amplicons.

The DGGE patterns revealed that clear differences were observed in the composition of methanotrophic community at the different sampling sites (Figure 16).

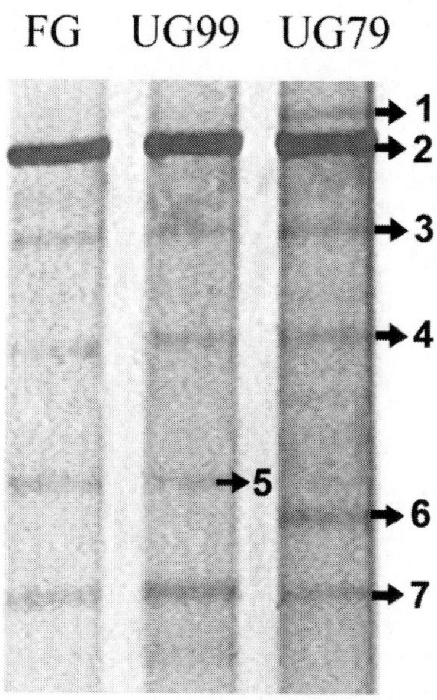

Figure 16. DGGE pattern of methanotrophs in soils under long-term grazing exclusion at the experimental field 2. Specific bands were cut from the gel and partial *pmoA* sequences for each band were identified.

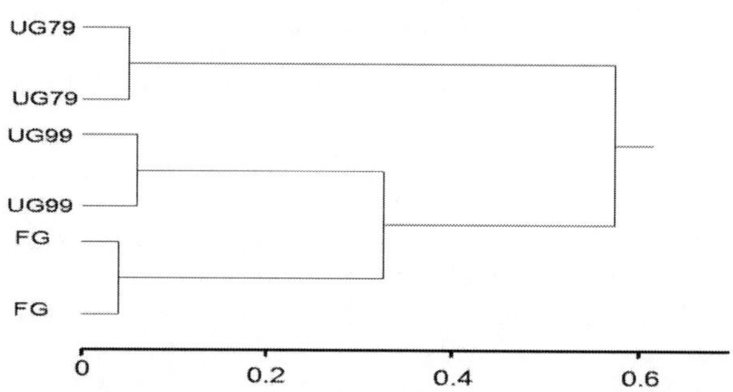

Figure 17. Dendrogram analysis of methanotrophic DGGE profiles under long-term grazing exclusion. The dendrogram is obtained from the DGGE banding patterns presented in Figure 16.

The duplicate analyzed from each sampling site were similar to each other (Figure 17). Each band of the DGGE pattern was assumed to represent one unique ribotype. Highest number of ribotypes was detected in the UG79 soils (13 bands), whereas the UG99 and FG soils represented 8 and 7 bands, respectively. The dendrogram analysis of DGGE patterns indicated that UG99 site firstly clustered with FG, and then clustered with UG79 site (Figure 17).

Moreover, seven main bands in the DGGE pattern were selected, cloned and sequenced. The results of phylogenetic analysis revealed that all these sequences belonged to methanotrophs based on comparison the conserved amino acid sequences in the particulate methane monooxygenase. A phylogenetic tree of all seven sequences from DGGE pattern along with some sequences published was constructed with neighbour-joining method (Figure 18).

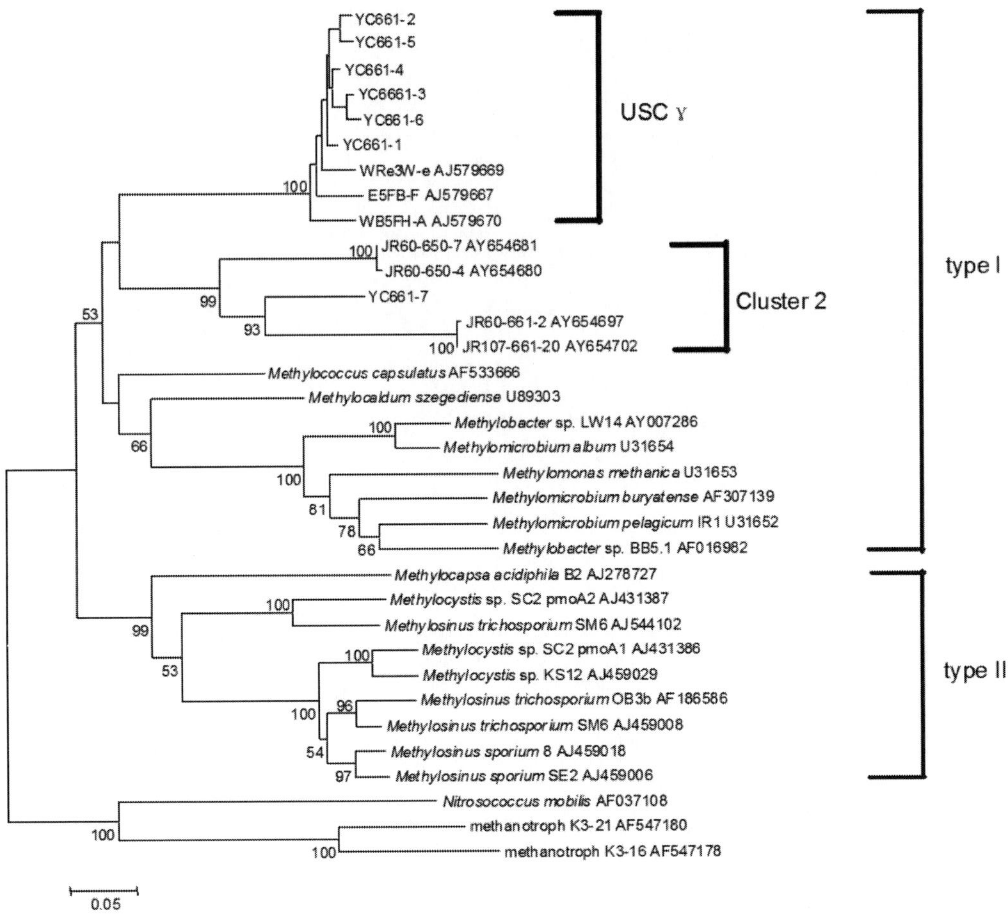

Figure 18. Phylogenetic tree based on deduced amino acid sequences of partial *pmoA* sequences, showing the relationship of sequences detected in Inner Mongolia grassland sampling sites to the sequences detected in other cultivation-independent studies. The tree was calculated using neighbour joining method with Jukes-Cantor model. The positions of *pmoA* and *amoA* sequences in the tree were confirmed by the topology of a Maximum likelihood tree. *AmoA* sequences of ammonia-oxidizing bacteria was used as an outgroup. The scale bar indicates an estimated change of 5%. The bootstrap values (>50%) are shown below the branches.

The recovery of *pmoA* gene sequences were distantly related to sequences of those well-known cultivated genera of methanotroph and formed two groups in the phylogenetic tree. The major group which almost contained all sequences belonged to USCγ which represented uncultivated methanotrophs.

Sequences of YC661-1 and YC661-6 were not found in the UG99 and FG soils, and sequence of YC661-5 was not found in the UG79 soil. The levels of amino acid identity of *pmoA* sequences in this group were 93% to each other, 87 to 95% to sequences of the uncultivated bacteria represented by WRe3W-e in Figure 18. Cluster 2 was only a sequence which was closely similar to JR2 and JR3 which recently were detected in Californian grassland (Horz et al. 2005). Sequence of YC661-7 which occurred at all sites exhibited level of amino acid identity of 87% to sequence of JR60-661-2.

DISCUSSION

Effects of Grazing Intensities on the Bacterial Community Structure

Grazing represents the most important land-use in the Inner Mongolian steppe and can determine the grassland productivity and lifestyles of local residents (Chen and Wang 2000). The decrease in plant biomass along the grazing gradient is consistent with a previous study (Chen and Wang 2000). In addition, there were considerable changes in dominant plant species with the increasing grazing intensities (Table 3). It has been shown that plant species had a major selective influence on the soil bacterial community of their rhizosphere (Smalla et al. 2001, Jangid et al. 2010), which could result in the differences in the soil bacterial community. In this study, we found that intermediate grazing intensity treatments (LG and MG) considerably changed soil bacterial community structure and significantly increased bacterial diversities as compared to the CK and HG treatments (Figure 8 and Table 4). Our results were in contrast to what reported by McCaig et al. (1999, 2001) and Clegg (2006), who have conducted their studies in the unimproved and improved grassland ecosystems in England and reported that there were no significant differences in soil bacterial diversity between both ecosystems using the methods of DGGE and clone library. However, Zhou et al. (2010 and 2012c) found that grazing intensity was a very important land use and largely affect the bacterial diversity in the Inner Mongolian steppe.

These large variations of bacterial diversity in grazed and non-grazed treatments could be attributed to grazing intensity. In fact, grazing by ungulate animal affects soil bacterial diversity through various factors such as faecal and urine deposition (Ritz et al. 2004), shifts in rhizosphere exudation (Guitian and Bardgett 2000), or changes in aboveground plant community composition (Dorrough et al. 2004). These factors could exert direct or indirect influences on soil bacterial community composition and diversity (Nunan et al. 2005). In general, intermediate grazing intensities can greatly affect or even benefit grassland ecosystems in terms of dry matter production, nutrient cycling and C and N storage, possibly due to increased nutrient availability and facilitated vegetation regeneration (Han et al. 2008). Previous studies revealed that intermediate grazing intensities increased plant productivity (plant biomass + biomass removal by grazing) (Chen and Wang 2000) and stimulated the cumulative net N mineralization, nitrification and ammonification (Xu et al. 2007), which might support higher soil bacterial diversities. However, it has been shown that long-term heavy grazing significantly decreased the storage of soil C and N and caused grassland degradation (Cui et al. 2000, Han et al. 2008), which might result in a decrease in bacterial diversity. Along with previous studies at the same experimental site (Chen and Wang 2000; Xu et al. 2007), we suggested that the MG treatment was the most suitable management practice in this area, not only supporting greater livestock amounts and rapid N cycling but also harbouring greater soil bacterial diversity.

What Are the Factors Driving Changes in Bacterial Community Structure?

The RDA seems to be a suitable method for quantifying the relationships between environmental variables and changes in soil bacterial community structure (Macdonald et al.

2009), aiming to determine the main factors affecting soil bacterial community structure among all the treatments. It has been shown that the soil microbial community were predominately affected by abiotic environmental factors, such as soil moisture content (Macdonald et al. 2009), pH (Fierer et al. 2009), soil organic C content (Degens et al. 2000), and C:N ratio (Graham and Haynes 2005).

In grassland soils, soil bacterial community can be driven by plant community composition and diversity (Kowalchuk et al. 2002). On the contrary, Nunan et al. (2005) reported that soil bacterial community structure in the natural grassland soil was subject to a wide range influences such as plant species, plant community composition and soil characteristics, and so on, none of which was found to be dominant. In this study, we found that soil moisture content, *A. frigida* biomass and pH were the best indicators of the changes in soil bacterial community structure under grazing intensity treatments. The *A. frigida* biomass contributed more to the changes in bacterial community structure compared to overall plant biomass and diversity. This result indicated that keystone plant species had a more important influence on bacterial community structure than plant biomass and diversity. The patterns of soil bacterial community structure in the LG and MG treatments were positively influenced by *A. frigida* biomass and pH (Figure 9). The *A. frigida* species in the LG and MG treatments could affect soil bacterial communities through increasing root exudation, possibly due to the stimulation by intermediate grazing (Chen and Wang 2000).

Effects of Long-Term Grazing Exclusion on Bacterial Diversity

The lower concentrations of soil organic C and N, NH_4^+-N and NO_3^--N in the FG treatment suggested negative effects of grazing disturbance on soil N pools, which was consistent with the results of previous studies (Wang et al. 2006). The reasons for this could be attributed to (a) enhanced N loss through leaching and gaseous emission from denitrification (Wang et al. 2006), (b) increased biomass N removal by animals, and (c) changes in N transformation rates due to altered soil physical and chemical characteristics (Steffens et al. 2008).

Zhou et al. (2010) reported that bacterial diversity changed monotonically with increasing grazing intensities and there were no significant differences in bacterial diversity between non-grazed and heavy-intensity grazing treatments. In light of this statement, it should have similar bacterial diversities between exclusion of grazing and FG sites. However, changes in soil bacterial diversity under ungulate grazing are more complicated than we expected (Ritz et al. 2004, Sankaran and Augustine 2004). Effects of long-term grazing exclusion could exert direct or indirect influences on soil bacterial community composition and diversity (Sankaran and Augustine 2004). Previous studies indicated that the UG99 soil had higher cumulative net N mineralization and N availability (Xu et al. 2007), which might support more bacterial species involved in this process.

In the Inner Mongolian grassland, there is great synchrony between plant growth and high temperature and moisture. Water availability in soils is an important factor limiting aboveground plant biomass (Bai et al. 2004). The aboveground plant biomass in the UG79 treatment was measured with five replicates in each of the two years and we found that the plant biomass in 2004 was significantly higher than that of 87.8±4.2 g m^{-2} in 2005. It is well known that soil microbial communities are closely associated with plant communities. The majority of microorganisms are heterogeneous and require nutrients and C sources provided by plants, but in turn soil microorganisms can decompose organic matter and litter and then provide available nutrients for plant growth (Zhou et al. 2012d). The lower plant biomass in 2005 may decrease the nutrients and C sources available to microorganisms and subsequently bacterial diversity (McCaig et al. 2001, Nunan et al. 2005, Zhou et al. 2010). Tilman et al. (1997) have put forward a theory of niche complementary to explain the relationships between plant diversity and limiting factors, such as a gradual decrease in a particular factor leading to a gradual increase in plant diversity. Such a relationship is present in these results, whereby as water availability became a limiting factor, bacterial diversity increased. Our results complement this theory (Tilman et al. 1997) although further experiments are needed to test whether this theory applies in the other fields of microbial ecology.

Effects of Grazing on Methanotrophic Diversity

We examined the effects of different grazing intensities on methane consumption capacity and methanotrophic community structure. The soil atmosphere exchange of methane comprises two parts: atmospheric methane oxidation by soil and methane produced by the soil methanogenesis (Wager et al. 1996, Wuebbles and Hayhoe 2002). Much of this methane produced by methanogenesis is oxidized by aerobic methanotrophs in the oxidic cover layer of these anaerobic soils before it is released into the atmosphere.

Methane production in the Inner Mongolia steppe can be negligible because the amount of methane produced by methanogenesis is very little in the aerated soil in this semi-arid grassland. The mean that methane uptake flux observed agrees with previous studies in this area (Wang et al. 2005). These uptake rates are slightly higher than those reported in the Colorado shortgrass steppe (25-45 μg C m^{-2} h^{-1}) (Mosier et al. 1991), in an English grassland (10-19 μg C m^{-2} h^{-1}) (Kammann et al. 2001) and in a burned native and annual prairie (20 - 32μg C m^{-2} h^{-1}) (Tate and Striegl 1993).

Compared with type I methanotrophs, the type II methanotrophs are likely to be the main group in the grassland soil of Inner Mongolia. These organisms can oxidize atmospheric methane at low concentrations (Lane 1991). Compared with the CK and HG sites, methane uptake at the LG and MG sites was slightly higher. The type II methanotrophic phylotype was obvious at the LG and MG sites and disappeared at the CK and HG sites. This implies that the phylotype seems to play an important role in oxidizing atmospheric methane in this region. The further sequence analysis indicated that it is a kind of uncultured Methylobacteriaceae bacterium. Previous studies indicated that long-term agricultural practices may shift the structure of the methanotrophic community, which results in the change of methane oxidation (Seghers et al. 2003). Are the changes in methane oxidation related with the shifting of structure and diversity of methanotrophic community induced by management practices? Previous research indicated that fire significantly increased methane oxidation rates, but the community structure did not change through comparing *pmoA* recovery products (Jaatinen et al. 2004). It is possible that shifts occurred on a fine phylogenetic scale not distinguishable by *pmoA* products.

In this study, comparing the change of methane uptake flux and the change of methantroph communities structure at different stocking rates sites, we found that although grazing intensity had an impact on the methanotrophic community structure, it did not greatly impact the methanotrophic activity. A range of specific factors, such as fecal and urine deposition, shifts in plant rhizosphere exudation, shifts in plant composition and the change in soil structure and aerobicity, changes soil characteristics (Clegg 2006). Methanotrophic activities are affected by many factors such as ammonium concentrations (Bodelier and Laanbroek 2004). The higher NH_4^+-N concentrations at the CK site cannot account for no significant differences in methane consumption rates across the sites (Figure 13).

For almost all the sequences similar to each other, the changes of DGGE patterns at the different sites were considered the variation of methanotrophic community structure under the condition of grazing. On the other hand, the results of methane uptake at the different sampling site showed that there were no significant differences between each other based on the data of July 2004 and August 2005 which were confirmed by Wang et al. (2005). Because Inner Mongolia steppe soil is well-drained, methane productions by methanogenesis are so weak that they can be negligible, thus methane uptake must be greatly mediated by the activity of methanotrophs in soils which also showed the variation of the structure of methanotrophic community in the different soils. Recently more methanotrophic groups that participate in consuming atmospheric CH_4 have been found in terrestrial ecosystems (Bull et al. 2000, Henckel et al. 2000, Zhou et al. 2008a). Studies have indicated that atmospheric methane oxidization can also be mediated by a kind of unknown methanotrophs, which are distant from type I and type II methanotrophs in the phylogenetic tree (Crossman et al. 2005). The type II methanotroph primers used in this study, which targeted all of the *methylosinus/methylocystis* groups, are degenerate primers and can detect some unknown methanotrophs (Wise et al. 1999).

At the experimental field 2, we used *pmoA* sequences to investigate methanotrophic community structure and diversity and found a specific group of methanotroph, different from those of the well-known cultivated methanotrophs. The recovery of the *pmoA* sequences in the DGGE patterns diverged from sequences of the well-known cultivated methanotrophic bacteria. The cluster of USCγ is the main group in the Inner Mongolia grassland soil, probably capable of oxidizing methane at atmospheric concentrations, which had been detected previously in several upland soils (Auman et al. 2000). The group of USCγ was previously detected in other upland soils (Bourne et al., 2001; Knief et al., 2003). However, to the best of our knowledge, it is the first study to demonstrate the dominance of this group of methanotrophs in the Inner Mongolia steppe soil (Zhou et al. 2008a).

CONCLUSION

In summary, grazing intensities over long-term periods (ca. 16 years) greatly influenced soil bacterial community structure and diversity. Among the grazing intensity treatments, the HG treatment had the lowest plant biomass and diversity. Intermediate grazing intensity treatments (LG and MG sites) considerably changed soil bacterial community structure and significantly increased bacterial diversities as compared with the CK and HG treatments. The relationships between environmental variables and soil bacterial community structure were analyzed using RDA and we found that soil moisture content, *A. frigida* biomass and pH were the best indicators of the changes in soil bacterial community structure among all the treatments. Additionally, bacterial diversity was higher in the UG99 treatment than in the FG and UG79 treatments. Principal components analysis of DGGE patterns showed that soil bacterial community in 2004 clearly separated from those in 2005, and the UG99 soil was significantly different from the FG and UG79 soils over the two consecutive years. Phylogenetic analysis of the clone libraries from the two sites indicated that the grassland soils were dominated by the groups of *Actinomycetes, Bacteroidetes* and *Proteobacteria*. Overall, our results indicated that intermediate grazing intensities (LG and MG) increased soil bacterial diversities and short-term rather than long-term exclusion of grazing increased bacterial diversity, which will be helpful for managements of grassland ecosystems.

Comparing the trends of methane consumption rates and the changes of methanotrophic communities at the different grazing intensity sites, we found that although grazing intensities had an impact on the structure of the methanotrophic community, management-induced changes in the structure of methanotrophic community did not reflect methane consumption capacity across sites. These results suggest that methane consumption is a complex process in soil, and we should be cautious when speculating on the change of methane consumption rates based on changes in methanotrophic community structure. Additionally, methanotrophic communities at the long-term grazing exclusion site were different from those at the short-term grazing exclusion and freely-grazing sites. The *pmoA* sequence analysis showed that most methanotrophs at the long-term grazing exclusion site belonged to the USCγ cluster. This is the first report showing that the USCγ cluster is dominant in grassland soils.

ACKNOWLEDGMENTS

This work was jointly supported by the President Foundation of the Chinese Academy of Sciences and National Science Foundation of China (No. 90411001), Griffith University Research Fellowship and Postdoctoral Fellowship at Scion, New Zealand. We thank Prof. Juan Traba at the Universidad Autónoma de Madrid for careful revision of the early version of the chapter.

REFERENCES

Amann, R.I., Ludwig, W., Schleifer, K.H. (1995). Phylogenetic identification and in situ detection of individual microbial cells without cultivation. *Microbiology Review* 59, 143-169.

Auman, A.J., Stolyer, S., Costello, A.M., Lidstrom, M.E. (2000). Molecular characterization of methantorophic isolates from freshwater lake sediment. *Applied and Environmental Microbiology* 66, 5259-5266.

Bai, Y.F., Han, X.G., Wu, J.G., Chen, Z.Z., Li, L.H. (2004). Ecosystem stability and compensatory effects in the Inner Mongolia grassland. *Nature* 431, 181-184.

Bardgett, R.D., Bowman, W.D., Kaufmann, R., Schmidt, S.K. (2005). A temporal approach to linking aboveground and belowground ecology. *Trends in Ecology and Evolution* 20, 634-641.

Bender, M., Conrad, R. (1992). Kinetics of CH_4 oxidation in oxic soils exposed to ambient air or high CH_4 mixing ratios. *FEMS Microbiology Letters* 101, 261-269.

Benstead, J., King, G.M., Williams, H.G. (1998). Methanol promotes atmospheric methane oxidation by methanotrophic cultures and soils. *Applied and Environmental Microbiology* 64, 1091-1098.

Bodelier, P.L.E., Laanbroek, H.J. (2004). Nitrogen as a regulatory factor of methane oxidation in soils and sediments. *FEMS Microbiology Ecology* 47, 265-277.

Boon, N., De Windt, W., Verstraete, W., Top, E.M. (2002). Evaluation of nested PCR-DGGE (denaturing gradient gel electrophoresis) with group-specific 16S rRNA primers for the analysis of bacterial communities from different wastewater treatment plants. *FEMS Microbiology Ecology* 39, 101-112.

Bourne, D.G., McDonald, I.R., Murrell, J.C. (2001). Comparison of *pmoA* PCR primer sets as tools for investigating methanotroph diversity in three Danish soils. *Applied and Environmental Microbiology* 67, 3802-3809.

Bowman, J.P. (2000). The methanotrophs: the families *methylococcacea* and *methylocystaceae.* In: Dworkin, M. (Ed.), *The Prokaryotes*. Springer, New York.

Bull, I.D., Parekh, N.R., Hall, G.H., Ineson, P., Evershed, R.P. (2000). Detection and classification of atmospheric methane oxidizing bacteria in soil. *Nature* 405, 175-178.

Chen, Z.Z., Wang, S.P. (2000). *Chinese Typical Grassland Ecosystem* (Chen ZZ and Wang SP Eds.) Science Press, Beijing, China.

Clegg, C.D. (2006). Impact of cattle grazing and inorganic fertilizer additions to managed grasslands on the microbial community composition of soils. *Applied Soil Ecology* 31, 73-82.

Costello, A.M., Lidstrom, M.E. (1999). Molecular Characterization of Functional and Phylogenetic Genes from Natural Populations of Methanotrophs in Lake Sediments. *Applied and Environmental Microbiology* 65, 5066-5074.

Crossman, Z.M., Ineson, P., Evershed, R.P., 2005. The use of ^{13}C labelling of bacterial lipids in the characterisation of ambient methane-oxidising bacteria in soils. *Organic Geochemistry.* 36, 769-778.

Cui, X.Y., Chen, S.P., Chen, Z.Z. (2000). CO_2 release from typical *stipa grandis* grassland soil. *Chinese Journal of Applied Ecology* 11, 390-394.

Dedysh, S.N., Berestovskaya, Y.Y., Vasylieva, L.V., Belova, S.E., Khmelenina, V.N., Suzina, N.E., Trotsenko, Y.A., Liesack, W., Zavarzin, G.A. (2004). *Methylocella tundrae* sp. nov., a novel methanotrophic bacterium from acidic tundra peatlands. *International Journal of Systematic Evolutionary Microbiology* 54, 151-156.

Degens, B.P., Schipper, L.A., Sparling, G.P., Vojvodic-Vukovic, M. (2000). Decreases in organic C reserves in soils can reduce the catabolic diversity of soil microbial communities. *Soil Biology and Biochemistry* 32, 189-196.

Dorrough, J., Ash, J., McIntyre, S. (2004). Plant responses to livestock grazing frequency in an Australian temperate grassland. *Ecography* 27, 798-810.

Fierer, N., Strickland, M.S., Liptzin, D., Bradford, M.A., Cleveland, C.C. (2009). Global patterns in belowground communities. *Ecology Letters* 12, 1238-1249.

Graham, M.H., Haynes, R.J. (2005). Catabolic diversity of soil microbial communities under sugarcane and other land uses estimated by Biolog and substrate-induced respiration methods. *Applied Soil Ecology* 29, 155-164.

Han, G.D., Hao, X.Y., Zhao, M.L., Wang, M.J., Ellert, B.H., Willms, W., Wang, M.J. (2008). Effect of grazing intensity on carbon and nitrogen in a meadow steppe in Inner Mongolia. *Agricultute Ecosystems and Environment* 125, 21-32.

Hanson, R., Hanson, T. (1996). Methanotrophic bacteria. *Microbiology Review* 60, 439-471.

Henckel, T., Jäckel, U., Schnell, S., Conrad, R. (2000). Molecular analyses of novel methanotrophic communities in forest soil that oxidize atmospheric methane. *Applied and Environmental Microbiology* 66, 1801-1808.

Holmes, A.J., Costello, A., Lidstrom, M.E., Murrell, J.C. (1995). Evidence that participate methane monooxygenase and ammonia monooxygenase may be evolutionarily related. *FEMS Microbiology Letters* 132, 203-208.

Holmes, A.J., Roslev, P., McDonald, I.R., Iversen, R., Henriksen, K., Murrell, J.C. (1999). Charaterization of methanotrophic bacterial populations in soils showing atmospheric methane uptake. *Applied and Environmental Microbiology* 65, 3312-3318.

Horz, H.-P., Rich, V., Avahami, S., Bohanna, J.M. (2005). Methane- oxidizing bacteria in a california upland grassland soil: diversity and response to simulated global change. *Applied and Environmental Microbiology* 71, 2642-2652.

Hutsch, B.W. (2001). Methane oxidation in non-flooded soils as affected by crop production - invited paper. *European Journal of Agronomy* 14, 237-260.

IPCC Working Group 1 Climate Change (2007): *The physical Science Basis* (Cambridge Uni. Press, 2007).

Jaatinen, K., Knief, C., Dunfield, P.F., Yrjala, K., Fritze, H. (2004). Methanotrophic bacteria in boreal forest soil after fire. *FEMS Microbiology Ecology*50, 195-202.

Jangid, K., Williams, M.A., Franzluebbers, A.J., Blair, J.M., Coleman, D.C., Whitman, W.B. (2010). Development of soil microbial communities during tallgrass prairie restoration. *Soil Biology and Biochemistry* 42, 302-312.

Kammann, C., Grünhage, L., Jäger, H.J., wachinger, G. (2001). Methane fluxes from differentially managed grassland study plots: the important role of CH_4 oxidation in grassland with a high potential for CH_4 production. *Environental Pollution* 115, 261-273.

Kennedy, N., Brodie, E., Connolly, J., Clipson, N. (2004). Impact of lime, nitrogen and plant species on bacterial community structure in grassland microcosms. *Environmental Microbiology* 6, 1070-1080.

Klumpp, K., Fontaine, S., Attard, E., Roux, X.L., Gleixner, G., Soussana, J. (2009). Grazing triggers soil carbon loss by altering plant roots and their control on soil microbial community. *Journal of Ecology* 97, 876-885.

Knief, C., Lipski, A., Dunfield, P.F. (2003). Diversity and activity of methanotrophic bacteria in different upland soils. *Applied and Environmental Microbiology* 69, 6703-6714.

Knief, C., Vanitchung, S., Harvey, N.W., Conrad, R., Dunfield, P.F., Chirdthaisong, A. (2005). Diversity of methanotrophic bacteria in tropical upland soils under different land uses. *Applied and Environmental Microbiology* 71, 3826-3831.

Kowalchuk, G.A., Buma, D.S., de Boer, W., Klinkhamer, P.G.L., van Veen, J.A. (2002). Effects of above-ground plant species composition and diversity on the diversity of soil-borne microorganisms. *Antonie Leeuwenhoek* 8, 509-520.

Lane, D.J. (1991). 16S/23S rRNA sequencing, p.115-175. In: Stackebrandt, E., Goodfellow, M. (ed.), *Nucleic acid techniques in bacterial systematics*. John Wiley and Sons Ltd., Chichester, United Kingdom.

Liu, L.P., Liao, Y.N. (1997). Biological charateristics and biodiversity of the soil microorganisms in Leymus Chinensis steppe and Stipa Grandis steppe under different grazing intensities In: *Inner Mongolia Grassland Ecosystem Research Station Research on Grassland Ecosystem*. vol 5, pp. 13-22. China Scientific Press, Beijing, China.

Macdonald, C.A., Thomas, N., Robinson, L., Tate, K.R., Ross, D.J., Dando, J., Singh, B.K. (2009). Physiological, biochemical and molecular responses of the soil microbial community after afforestation of pastures with Pinus radiata. *Soil Biology and Biochemistry* 41, 1642-1651.

McCaig, A.E., Glover, L.A., Prosser, J.I. (1999). Molecular analysis of bacterial community structure and diversity in unimproved and improved upland grass pastures. *Applied and Environmental Microbiology* 65, 1721-1730.

McCaig, A.E., Glover, L.A., Prosser, J.I. (2001). Numerical analysis of grassland bacterial community structure under different land management regimens by using 16S ribosomal DNA sequences data and denaturing gradient gel eletrophoresis banding patterns. *Applied and Environmental Microbiology* 67, 4554-4559.

Miethling, R., Ahrends, K., Tebbe, C.C. (2003). Structural differences in the rhizosphere communities of legumes are not equally reflected in community-level phsiological profiles. *Soil Biology and Biochemistry*35, 1405-1410.

Mosier, A., Schimel, D., Valentine, D., Bronson, K., Parton, W. (1991). Methane and nitrous oxide fluxes in native, fertilized and cultivated grasslands. *Nature* 350, 330-332.

Muyzer, G., de Waal, E.C., Uitterlinden, A.G. (1993). Profiling of complex microbial populations by denaturing gradient gel electrophoresis analysis of polymerase chain reaction-amplified genes coding for 16S rRNA. *Applied and Environmental Microbiology* 59, 695-700.

Nunan, N., Daniell, T.J., Singh, B.K., Papert, A., McNicol, J.W., Prosser, J.I. (2005). Links between plant and rhizosplane bacterial communities in grassland soils, chracterized using molecular techniques. *Applied and Environmental Microbiology* 71, 6784-6792.

Ritz, K., McNicol, J.W., Nunan, N., Grayston, S., Millard, P., Atkinson, D., Gollotte, A., Habeshaw, D., Boag, B., Clegg, C.D., Griffiths, B.S., Wheatley, R.E., Glover, L.A., McCaig, A.E., Prosser, J.I. (2004). Spatial structure in soil chemical and microbiological properties in an upland grassland. *FEMS Microbiology Ecology* 49, 191-205.

Sankaran, M., Augustine, D.J. (2004). Large herbivores suppress decomposer abundance in a semiarid grazing ecosystem. *Ecology* 85, 1052-1061.

Seghers, D., Top, E.M., Reheul, D., Bulcke, R., Boeckx, P., Verstraete, W., Siciliano, S. (2003). Long-term effects of mineral versus organic fertilizers on activity and structure of the methanotrophic community in agricultural soils. *Environental Microbiology* 5, 867-877.

Seghers, D., Wittebolle, L., Top, E.M., Verstraete, W., Siciliano, S.D. (2004). Impact of Agricultural Practices on the Zea mays L. Endophytic Community *Applied and Environmental Microbiology* 70, 1475-1482.

Smalla, K., Wieland, G., Buchner, A., Zock, A., Parzy, J., Kaiser, S., Roskot, N., Heuer, H., Berg, G. (2001). Buck and rhizosphere soil bacterial communities studied by denaturing gradient gel eletrophoresis: plant-dependent enrichment and seasonal shifts revealed. *Applied and Environmental Microbiology* 67, 4742-4751.

Steffens, M., Koelbl, A., Totsche, K.U., Koegel-knabner, I. (2008). Grazing effects on soil chemical and physical properties in a sem-arid steppe of Inner Mongolia (P.R. China). *Geoderma* 143, 63-72.

Sun, H.Y., Deng, S.P., Raun, W.R. (2004). Bacterial community structure and diversity in a century- old manure- treated agroecosystem. *Applied and Environmental Microbiology* 70, 5868-5874.

Tate, C.M., Striegl, R.G. (1993). Methane consumption and carbon dioxide emission in tallgrass prairie: effects of biomass burning and conversion to agriculture. *Globlal Biogeochemical Cycles* 7, 735-748.

Tilman, D., Lehman, C.L., Thomson, K.T. (1997). Plant diversity and ecosystem productivity: theoretical considerations. *Proceedings of the National Academy of Sciences of United States of America* 94, 1857-1861.

Wager, C., GriessHammer, A., Drake, H.L. (1996). Acetogenic capacities and the anaerobic turnover of carbon in kansas prairie soil. *Applied and Environmental Microbiology* 62, 494-500.

Wang, C.H., Wan, S.Q., Xing, X.R., Zhang, L., Han, X.G. (2006). Temperature and soil moisture interactively affected soil net N mineralization in temperate grassland in Northern China. *Soil Biology and Biochemistry* 38, 1101-1110.

Wang, Y., Xue, M., Zheng, X., Ji, B., Du, R., Wang, Y. (2005). Effects of environmental factors on N_2O emission from and CH_4 uptake by the typical grasslands in the Inner Mongolia. *Chemosphere* 58, 205-215.

Wardle, D.A., Bardgett, R.D., Klironomos, J.N., Setälä, H., Van der Putten, W.H., Wall, D.H. (2004). Ecological linkages between aboveground and belowground biota. *Science* 304, 1629-1633.

Wardle, D.A., Bardgett, R.D., Walker, L.R., Peltzer, D.A., Lagerstrom, A. (2008). The response of plant diversity to ecosystem retrogression: evidence from contrasting long-term chronosequences. *Oikos* 117, 93-103.

Wieland, G., Neumann, R., Backhaus, H. (2001). Variation of microbial communities in soil, rhizosphere, and rhizoplane in response to crop species, soil type, and crop development. *Applied and Environmental Microbiology* 67, 5849-5854.

Wise, M.G., McArthur, J.V., Shimkets, L.J. (1999). Methanotroph diversity in landfill soil: isolation of novel type I and type II methanotrophs whose presence was suggested by culture-independent 16S ribosomal DNA analysis. *Applied and Environmental Microbiology* 65, 4887-4897.

Wuebbles, D.J., Hayhoe, K. (2002). Atmospheric methane and global change. *Earth-Science Review* 57, 177-210.

Xu, Y.Q., Li, L.H., Wang, Q.B., Chen, Q.S., Cheng, W.X. (2007). The pattern between nitrogen mineralization and grazing intensities in an Inner Mongolian typical steppe. *Plant and Soil* 300, 289-300.

Zhou, X.Q., Chen, C.R., Wang, Y.F., Smaill, S., Clinton, P. (2013). Warming rather than increased precipitation increases soil recalcitrant organic carbon in a semiarid grassland after 6 years of treatments. *Plos One* 8, e53761. DOI: 10.1371/journal.pone.0053761.

Zhou, X.Q., Chen, C.R., Wang, Y.F., Xu, Z.H., Hu, Z.Y., Cui, X.Y., Hao, Y.B. (2012a). Effects of warming and increased precipitation on soil carbon mineralization in an Inner Mongolian grassland after 6 years of treatments. *Biology and Fertility of Soils* 48, 859-866.

Zhou, X.Q., Chen, C.R., Wu, H.W., Xu, Z.H. (2012b). Dynamics of soil extractable carbon and nitrogen under different cover crop residues. *Journal of Soils and Sediments* 12, 844-853.

Zhou, X.Q., Liu, X., Rui, Y.C., Chen, C.R., Wu, H.W., Xu, Z.H. (2011). Symbiotic nitrogen fixation and soil N availability under legume crops in an arid environment. *Journal of Soils and Sediments* 11, 762-770.

Zhou, X.Q., Wang, J.Z., Hao, Y.B., Wang, Y.F. (2010). Intermediate grazing intensities by sheep increase soil bacterial diversities in an Inner Mongolian steppe. *Biology and Fertility of Soils* 46, 817-824.

Zhou, X.Q., Wang, Y.F., Cai, Y., Huang, X.Z., Hao, Y.B., Tian, J.Q., Chai, T.Y. (2007). PCR-DGGE detection of the bacterial community structure in the Inner Mongolia steppe with two different DNA extraction methods. *Acta Ecologica Sinica* 27, 1684-1689.

Zhou, X.Q., Wang, Y.F., Hao, Y.B. (2012c). Short-term rather than long-term exclusion of grazing increases soil bacterial diversity in an Inner Mongolian steppe. *Acta Ecologica Sinica* 32, 180-183.

Zhou, X.Q., Wang, Y.F., Huang, X.Z., Hao, Y.B., Tian, J.Q., Wang, J.Z. (2008a). Effects of grazing by sheep on the structure of methane-oxidizing bacterial community of steppe soil. *Soil Biology and Biochemistry* 40, 258-261.

Zhou, X.Q., Wang, Y.F., Huang, X.Z., Tian, J.Q., Hao, Y.B. (2008b). Effect of grazing intensities on the activity and structure of methane-oxidizing bacterial community in the grassland of Inner Mongolia. *Nutrient Cyclyng in Agroecosystems* 80, 145-152.

Zhou, X.Q., Wu, H.W., Koetz, E., Xu, Z.H., Chen, C.R. (2012d). Soil labile carbon and nitrogen pools and microbial metabolic diversity under winter crops in an arid environment. *Applied Soil Ecology* 53, 49-55.

In: Steppe Ecosystems
Editors: M. B. Morales Prieto and J. Traba Diaz
ISBN: 978-1-62808-298-2

Chapter 13

THE PSEUDO-STEPPE OF LA CRAU (SOUTHEASTERN FRANCE): ORIGIN, MANAGEMENT AND RESTORATION OF A MEDITERRANEAN RANGELAND

Thierry Dutoit**, Elise Buisson, Sylvain Fadda, Frédéric Henry, Clémentine Coiffait-Gombault, Renaud Jaunatre, Jean-François Alignan, Solène Masson and Adeline Bulot
Mediterranean Institute of Biodiversity and Ecology.
UMR CNRS-IRD IMBE. Université d'Avignon. France

ABSTRACT

The Crau plain in the Bouches-du-Rhône county, between the cities of Arles, Salon-de-Provence and Fos-sur-Mer in southeastern France, is considered as a Mediterranean steppe in Western Europe. In a landscape covered by short herbaceous vegetation (*Coussouls* de Crau), this ecosystem displays an exceptional range of biodiversity with regard to vegetation and fauna. Resulting from particular climatic and soil conditions, but also from the centuries-old practice of itinerant sheep grazing, this original ecosystem is also considered such as the grazed pastures in Western Iberia as outposts in the northern part of the Mediterranean Basin of North African semi-arid steppe. Its area has been reduced by more than 80% as the result of military, industrial and agricultural intensive land-use since the 16th century. Since 2000, major research programs have been undertaken with the aim of, (i) identifying the impact of land-use changes (intensification of agricultural practices) on biodiversity; (ii) identifying the original and reference ecosystems; (iii) setting up ecological restoration operations. Our results based on field and lab experiments show that despite the progress made in ecological rehabilitation, the

* Corresponding Author address: Mediterranean Institute of Biodiversity and Ecology. UMR CNRS-IRD IMBE. Université d'Avignon. Site Agroparc BP 61207. 84911 Avignon cedex 09, France, Email: thierry.dutoit@imbe.fr.

existing knowledge and techniques in ecological engineering are still inadequate to restore the integrity of the destroyed ecosystems.

Keywords: Herbaceous plant communities, coleoptera, orthoptera, biodiversity, palaeoecology, plant ecology, entomology, ecological engineering

INTRODUCTION

The Mediterranean basin has been identified as one of the 34 'hotspots' of biodiversity worldwide (Médail & Quézel, 1997). This high level of biodiversity may be correlated with a particular development of Mediterranean flora and fauna, but the diversity of the landscapes and ecosystems found there cannot be explained without reference to thousands of years of interaction with man. These interactions have been the cause of the creation and maintenance of highly original ecosystems (Grove & Racham, 2001) whose natural capital has been in some cases maintained to a greater or lesser extent by millennia of sustainable agricultural practices (Blondel et al., 1996) but have also sometimes generated severe degradation in terms of loss of biodiversity and functions of some ecosystems.

This is particularly the case for the Plain of Crau in the county of the Bouches-du-Rhône in Provence, in the southeast of France (Buisson et al., 2009). This plain, with a surface area of 600 km², forms an isosceles triangle whose side measure is 30 km, between the cities of Arles, Salon de Provence and Fos-sur-Mer. It is situated between the Alpilles to the north, the Berre Lagoon to the east, the Camargue wetlands to the west and the Mediterranean Sea to the south (Figure 1). The vegetation that characterizes the plain over more than 100 km², is a short herbaceous plant community, with numerous patches of bare soil, with neither trees nor bushes. Called *Coussouls* or *Coussous* (from lat. *cursorium*, the place one tramples underfoot), this vegetation is made up of more than 50% of annual species and may be related to the vegetation of the semi-arid steppe of North Africa, of which La Crau constitutes an isolated 'outpost' on the northern shore of the Mediterranean.

The Coussoul is composed of the unique phytosociological association *Asphodelletum fistulosii* (Devaux et al., 1983) found nowhere else in Europe, and also holds a quite remarkable avifauna, including the little bustard (*Tetrax tetrax*), the pin-tailed sandgrouse (*Pterocles alchata*) and the lesser kestrel (*Falco naumanni*), as well as numerous insect species, some of them endemic such as *Prionotropis hystrix rhodanica* and *Acmaeoderella cyanepennis perroti*.

The biological spectrum of this vegetation is also related to the vegetal formations of the *Dehesas* in Spain (3 million hectares) and the *Montado* in Portugal (700,000 hectares), but with differences in floristic composition. The absence of trees throughout the centre of the plain, the maintenance of herbaceous vegetation leaving bare patches of soil and the ubiquity of sheep grazing over a period of at least 2000 years (Badan et al., 1995) have favored the use of the term 'steppe' to refer to this vegetation (Devaux et al., 1983). We prefer however the term 'pseudo-steppe' since in the phyto-geographical sense, the climatic conditions and in particular the average rainfall (545 mm/year) do not justify classifying this dry grassland vegetation with the semi-arid steppe of North Africa which is rather characterized by average rainfall values between 100 and 400 mm/year (Le Houérou, 1995).

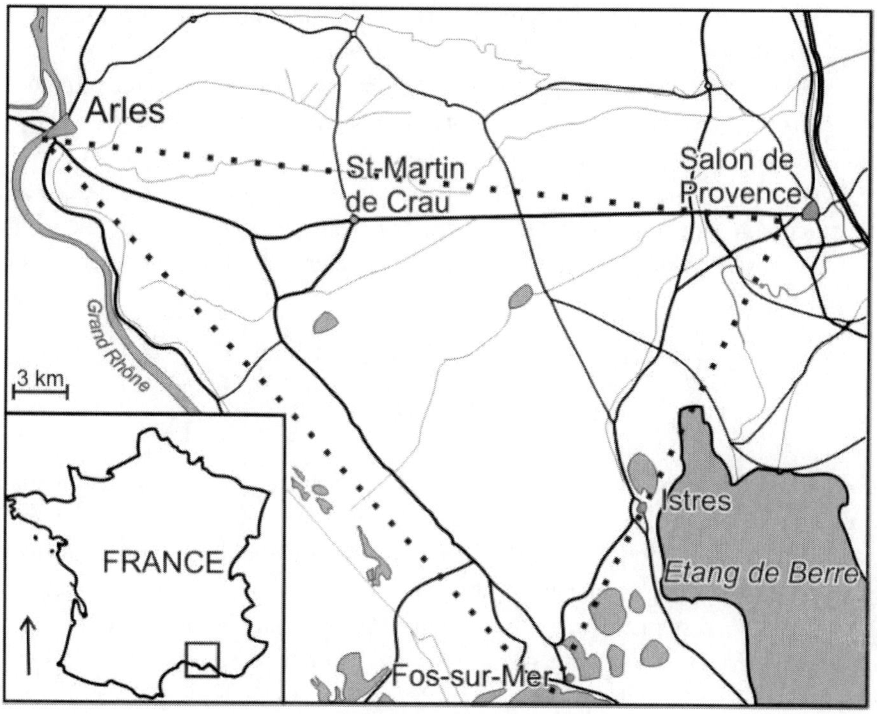

Figure 1. Localization of the La Crau plain in Southeastern France.

It is only over the past thirty years or so that awareness has developed of the exceptional biodiversity of this area and of the threats that it has faced since the conversion from the 16th century of its steppe-type vegetation to hay meadows, vegetable crops, cereal production or orchards after the construction of submersion irrigation networks. To these changes in agricultural uses were added military and industrial installations favored by the low demography of the area. This development resulted in the destruction of over 80% of the pseudo-steppe in four centuries, with a marked acceleration in the 19th and 20th centuries (Buisson & Dutoit, 2006).

Although there has been research on Crau since the beginning of the 19th century, this was initially of naturalistic and agronomic character, and it is only since 1975 that there has been research applied to the conservation of species and ecosystems in parallel with the introduction of different levels of protection. Protection was established in 2001 (74 km^2) to preserve the last remaining patches left over from the fragmentation of the pseudo-steppe. The first research program was undertaken to rehabilitate and/or restore this ecosystem in 2001, although the impact of the changes in land-use on the flora and fauna was the focus of research since the early 1980s (Devaux et al., 1983).

Since 2001, the relict and degraded areas of pseudo-steppe in the Crau Plain have been a real workshop for research in restoration ecology. Ecological restoration was initially undertaken at experimental spatial and time scales. Nevertheless, only the acquisition of knowledge and of techniques in ecological engineering has made possible the application of operational ecological restoration. These operations concern several hundred hectares scientifically monitored over a period of at least 30 years (Dutoit et al., 2013).

A survey of our results from experimental researches in community ecology (vegetation and insects), population ecology (plants) and palaeoecology (wood charcoal analyses) will be presented here on the basis of the three axes which have provided the basic structure of our research in this area over the past ten years: (1) identification of the human-generated degradation of biodiversity and the pseudo-steppe dynamics; (2) research on the original and reference ecosystem or ecosystems with a view to (3), designing operations aimed at rehabilitating certain functions of the ecosystem or orienting it towards a trajectory that will enable it in the shorter or longer term to recover the state prior to disturbances and its place in the traditional sheep breeding sector.

METHODS

Since the beginning of the 21st century, our research has been carried out mainly on patches of pseudo-steppe and degraded areas nearby, but where the level of ecological disturbance had not been enough to cross thresholds of irreversibility, which can be defined as levels where ecosystems cannot be restored from their natural resilience, precluding the rehabilitation and/or restoration of the reference pre-existing ecosystem. If many such thresholds have been crossed, it becomes no longer viable from the biological or socio-economic points of view to return to an earlier ecosystem. In such case it may be preferable the neo-formation of a new ecosystem, also known as re-allocation (Muller et al., 1998).

In order to select areas to apply restoration management, we have consulted aerial photographs taken since 1944. The maintenance of steppe patches can be confirmed on more than 10 000 ha of the 55 000 ha potentially existing originally, and for more than 1 500 ha of fallow land resulting from the abandonment of vegetable crops between the 1960s and the1980s. These fallow areas were chosen for the restoration of the pseudo-steppe since they presented no major changes in the substratum (preservation of the layer of conglomerate composed of pebbles embedded in a matrix of calcium carbonate at an average depth of 40 cm) and were all to a greater or lesser extent implanted in a landscape matrix still including patches of pseudo-steppe that had not been destroyed by cultivation.

In order to measure the impacts of land abandonment on vegetation and coleoptera assemblages, we select 6 sites distributed at random throughout the plain, in which we performed soil analyses (particle sizes, major chemical elements, organic matter, pH, etc.), vegetation sampling (10 x 4m^2 quadrats sampled on a presence-absence basis in 25 40 x 40 cm sub-quadrats) and coleopterans catching with pit fall traps (3 pit fall traps arranged in a triangle with 1m sides at random in the fallow land with repetitions in three representative fallow areas; traps were sampled every 15 days from April to November). These experimental designs were undertaken to measure the impact of the early cultivation phases on the composition and richness of the plant communities and of the assemblages of terrestrial coleopterans compared to the non-destroyed neighbouring patches of pseudo-steppe. This monitoring procedure was also applied every meter on 10 m transects placed perpendicularly (3 transects replicated within each plot with a 10 m space between each on three plots of different fallow areas with contrasting cultivation histories) at the edges of the fallow areas limiting with patches of pseudo-steppe in order to measure possible edge effects (Buisson & Dutoit, 2004).

There is a consensus on the choice of the pseudo-steppe as reference ecosystem, based on the strong socio-cultural references linked to migratory sheep raising (transhumance), the maintenance of the local breed (*Mérinos d'Arles*) and the tourism attraction of this area. Nevertheless, knowledge on the original ecosystem (prior to the first significant human impact) was insufficient as a basis for determining whether the restoration should be passive (*via* natural reconstitution or resilience) or whether it should be active (with human intervention, Etienne et al., 1998 ; Molinier & Tallon, 1950).

In order to identify the origin of the ecosystem at a scale compatible with the existence of steppe vegetation, we have chosen to sample wood carbon contained in the soil (Thinon, 1992). In order to improve our chances of finding carbon in soils that had experienced at least 2000 years of sheep grazing, major groundwater and wind induced erosion, we carried out sampling within the undisturbed steppe vegetation and also beneath the foundations of old sheepfolds. In all, 14 pits were dug, enabling the sampling of 60 layers of soil between 0 and 40 cm depth, or a total of 900 litres of soil sampled, from which were extracted 597 wood carbon samples, of which only 23 could be identified and dated (Henry et al., 2010).

Restoration operations have consisted in forcing of the processes of dispersal by transfer of hay in exclosures (areas fenced off to prevent sheep grazing) installed in patches of pseudo-steppe without destruction in order to maximise the harvesting of seeds produced (Coiffait et al., 2011b). Because of the presence at the surface of large stones, this harvesting of hay is carried out by aspiration in undisturbed patches of the reference steppe and the hay thus harvested is spread over the soil at a rate of 700g/m² in quadrats (3 repetitions of 4 pseudo-repetitions of 40 x 40 cm) in presence or absence of grazing. As control, this operation was also carried out in the pseudo-steppe vegetation. In all, there were thus 48 quadrats with spreading of hay and 48 quadrats without spreading of hay that were monitored over three years (2007 – 2009). Since the seeds of the perennial species which dominate the steppe, namely the ramose false brome (*Brachypodium retusum)* and the common thyme *(Thymus vulgaris*), are very rarely produced and are thus scarcely represented in the hay, they were directly sown mixed with common commercial seeds of subterranean clover (*Trifolium subterraneum*). The mixture is composed of 73% of ramose false brome (47kg/ha), 20% of subterranean clover (13 kg/ha) and 7% of Common thyme (4 kg/ha). The hypothesis behind sowing these species is that they will produce a drop in the density of the weeds and an increase in safe sites favourable for the establishment of species typical of the pseudo-steppe when this mixture is grazed.

Twelve repetitions of 12.5 x 25 m were sown with this mixture, half of them accessible for grazing and the others not accessible. Monitoring was carried out over a period of two years (2008 – 2009) using 3 quadrat (1m²) pseudo-replicas in each treatment, 72 quadrats per year. In addition, comparisons were made with three quadrats in the control pseudo-steppe and the three quadrats in the fallow area without treatment but with traditional grazing for these last two lots (for more details on the experimental design, see Coiffait-Gombault et al., 2012b.

In order to measure the effects of restoration treatments on Coleoptera assemblages, small areas (1,5 x 3m) randomly distributed were again covered with stones with or without the exclusion of grazing in order to assess their importance with regard to the distribution of Coleoptera in a fallow area (12 repetitions of each treatment). This protocol was also carried out in the relictual pseudo-steppe in order to measure the impact of withdrawal of the stones and of grazing (also 12 repetitions of each treatment). Ninety six quadrats were monitored over four years after finishing the rehabilitation protocol in order to avoid the negative impact of the disturbance due to its application. The Coleoptera were trapped using the same method as that described for measuring the impact of degradations.

Finally, as part of a conservation operation following the construction of a logistical platform, the steppe vegetation and its soil were removed from the surface layer to a depth of between 10 and 20 cm, depending on the technical capacity of the mechanical digger. In spring 2009, the soil was transported and spread the same day onto the soil of fallow area resulting from the uprooting of an intensive orchard area abandoned in 2006. In all, 3000 m^2 were removed and transferred from the donor site and spread over 9 000 m^2 on the receiver site.

In all, 600 m^3 of soil were spread, on a ratio of one to three, which corresponds to what is generally recommended for the restoration of dry grassland by soil spreading (Kiehl et al., 2010). Monitoring has been done one year after soil spreading on three pseudo-replica quadrats of 4m^2 sampled on a presence-absence basis with attribution of a cover coefficient for each species. They were compared with 18 quadrats positioned in the control pseudo-steppe and 18 in the fallow area without soil spreading.

All the data were analysed using ANOVA tests after verification of the standardisation of the data (Shapiro-Wilk test) and the homogeneity of variance (Levene test, calculation statistics version R 3.0.1.2005). Different post-hoc tests were used as *a posteriori* tests. More details on data analyses can be found in our references cited in the bibliography.

RESULTS

Our results showed that even 30 years after its cessation, cultivation still has a significant impact on the plant species composition, structure and specific richness in plant species of fallow areas compared to the pseudo-steppe (Figure 2). In particular, a significant drop in the specific richness of fallow areas was apparent, whether they were cultivated for the production of melons or of cereals, along with a drastic change in floristic composition, which became commoner than the heterogeneous composition of the steppe vegetation in relation with a north-south climatic gradient (Römermann et al., 2005). The most typical species of the pseudo-steppe (dominant species and characteristic species) disappeared in favour of a flora composed of weeds from cultivated areas (Römermann et al., 2005). This strong persistence of the effects of cultivation was more marked when the cultivation phase was long, but the date of abandonment of the fallow land had little influence on the current floristic composition (Römermann et al., 2005). This impact on the vegetation may also be explained by changes in the soil chemistry of the formerly cultivated areas which still possessed levels of total phosphorus, potassium, calcium and calcium carbonate contents that were significantly higher than in the soils of the pseudo-steppe (Römermann et al., 2005). These changes should be related to the impact of cultivation practices, including the presence of relictual fertilization (phosphorus, potassium in particular) and the raising of carbonates resulting from deep ploughing, that also caused an increase in pH, admittedly of only 0.5 of a unit, but which was enough to transform the pH of the pseudo-steppe from slight acidity to neutrality.

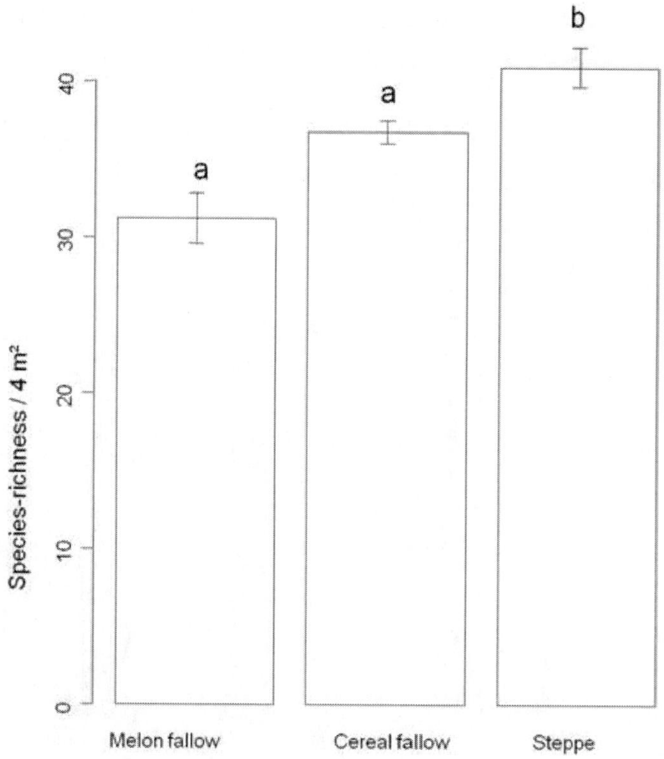

Figure 2. Species richness (mean ± SE) of the original steppe area, cereal fallow and melons fallow in the La Crau plain, Southeastern France. Different letters for ANOVAs with subsequent post-hoc LSD tests (least significant difference) indicate a significant difference for $p < 0.001$ ($N = 60$) (Römermann et al. 2005, modified).

The sampling of the seed rain and the permanent and transitory seed bank of the soil showed that the seeds of species typical of the pseudo-steppe were rarely wind-dispersed and are of poor viability in the soil, in contrast with the results obtained for the seeds of weed species (Buisson et al., 2006).

For the Coleoptera, the results obtained were quite the opposite, showing greater richness of Coleoptera in the old abandoned cultivated areas compared to the relictual pseudo-steppe (Fadda et al., 2007 ; Fadda et al., 2008b).

For the edges, we did not record any significant increase in plant species richness and terestrial coleopterans assemblages along the transect in relation with the distance from the edge (Buisson & Dutoit, 2004; Fadda et al., 2008a). For the vegetation, an edge effect was only apparent in the first three meters of the transect.

Considering the origin of the steppe, of the 23 carbon samples identified, none corresponded clearly to the dominant species of old forests (> 200 years) under Mediterranean climate (*Quercus ilex*) in situations without trace of human occupation. At the end of the Neolithic Age, an increase in the frequency of fires may be noted, probably correlated with more widespread human exploitation (Figure 3).

Considering restoration experiments, from the first year of monitoring, our results showed a significant increase in the species richness of the quadrats (Coiffait et al., 2011a) where the hay was spread. A change in composition was also measured that allowed the establishment of numerous pseudo-steppe species (Figure 4).

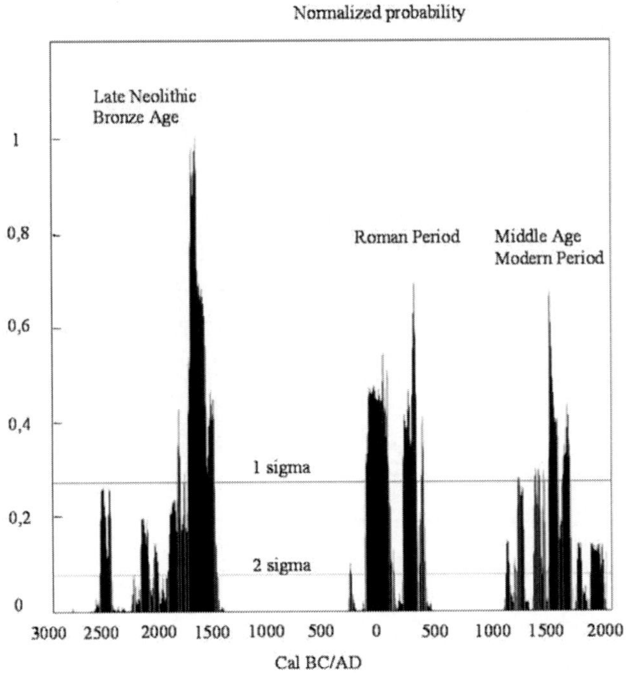

Figure 3. Temporal distribution of 14C dates for the 23 charcoals sampled in the steppe of La Crau, Southeastern France (Henry et al. 2010, modified).

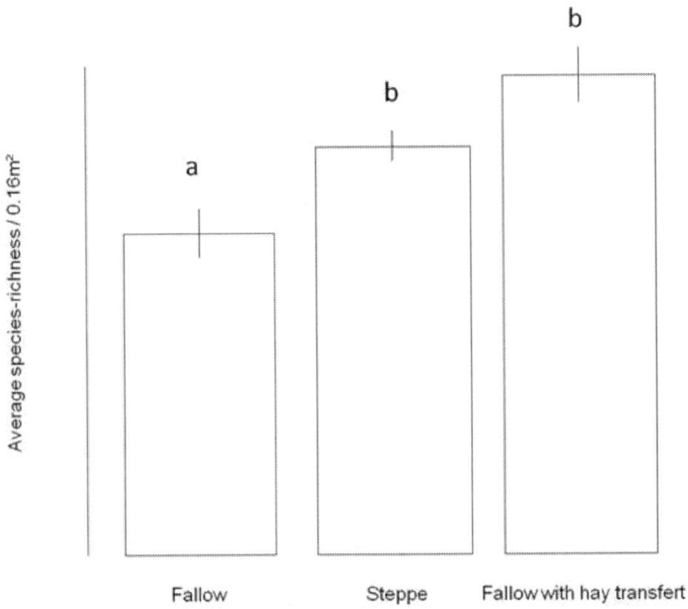

Figure 4. Species-richness (mean ± SE) one year after steppe hay transfer in a former cultivated land for its restoration in the La Crau plain, Southeastern France Different letters for ANOVAs with Bonferroni tests indicate a significant difference for $p < 0.001$ ($N = 12$) (Coiffait et al., 2012b, modified).

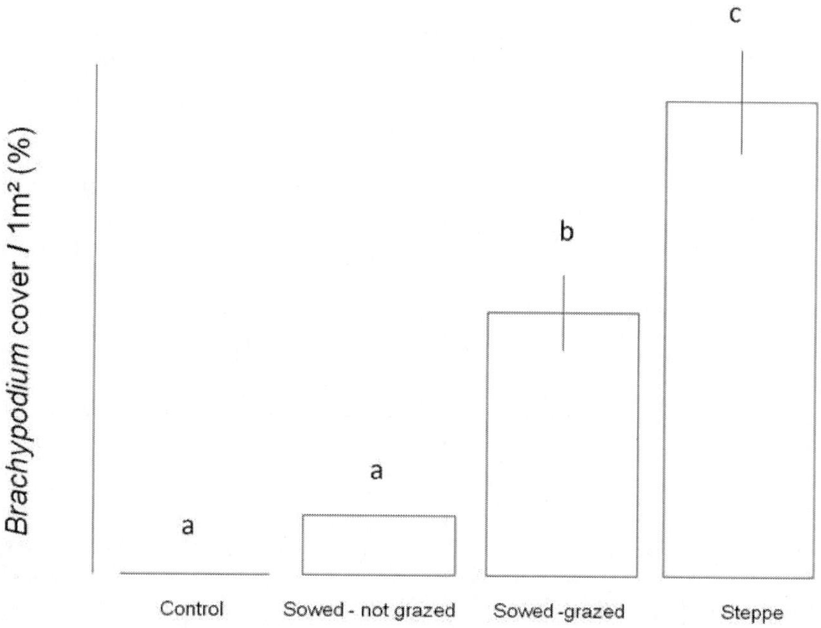

Figure 5. Cover of *Brachypodium retusum* (mean ±SE) cover one year after it has been sowing in a former cultivated area for its restoration in the La Crau plain, Southeastern France. Different letters for ANOVAs with Bonferroni tests indicate a significant difference for $p < 0.001$ ($N = 30$) (Coiffait et al., 2012b, modified).

However, if these results were particularly encouraging for the annual species, this was not the case for the dominant perennial species (ramose false brome and common tyme in particular) which were not transplanted by this method because of insufficient production of seeds accessible at the time of collection by the aspirators (beginning of July 2006 during the peak fructification of the steppe species).

Consequently, these two dominant species were sown mixed with the subterranean clover. From 2008 to 2009, our results showed that these sown species were indeed implanted in the fallow areas. In addition, grazing had a favourable impact on ramose false brome that covered more soil compared to the non-grazed exclosures (Figure 5). The action of mixture sowing did not have any significant effect of on the richness of species present spontaneously in fallow land as compared with the ploughed area without sowing treatment. However, a significant increase in plant species richness was recorded when mixture sown areas were compared to control areas with grazing only.

With regard to the composition, the areas grazed and sown with the mixture differed from other treatments. In particular, they presented plants from the family Poaceae also present in the pseudo-steppe (*Lolium perenne, Catapodium rigidum, Bromus hordeaceus, Taenatherum caput-medusae*), which thus proved capable of spontaneously colonising these areas under the influence of structuring sown species and of grazing (Coiffait et al., 2012b).

In relation to coleopterans, the reestablishment of the stone cover and the exclusion of grazing in the fallow land resulted in a significant increase in the density of the Coleoptera especially *Asida sericea* (Figure 6) but not in their overall species-richness (Blight et al., 2011; Fadda et al., 2008a).

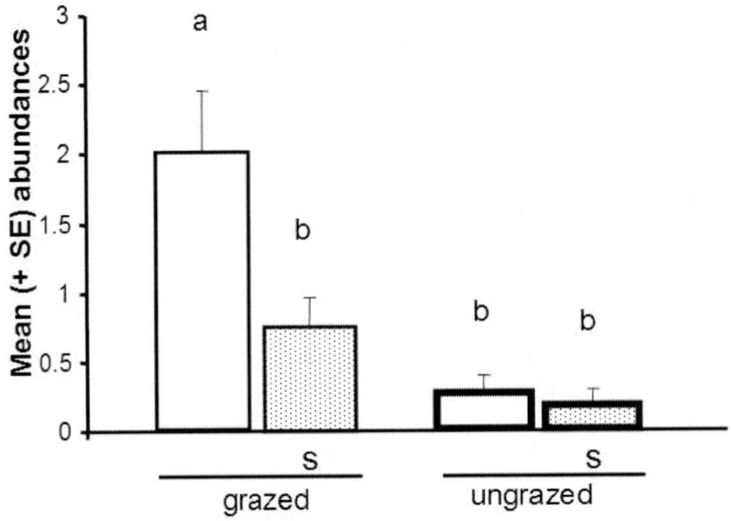

Figure 6. Abundance of *Asida sericea* (mean ± SE) on formerly-cultivated field subplots, ungrazed (thick line) and stone (dash area). Different letters for ANOVAS with subsequent Tukey post-hoc test indicate a significant difference for $p < 0.001$(N=12) (Blight et al., 2011, modified).

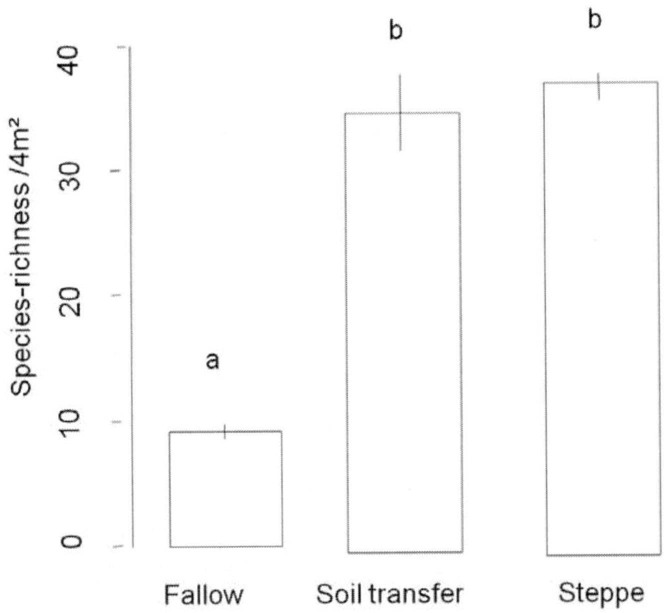

Figure 7. Plant species-richness (mean ± SE) one year after soil spreading of the soil from the original steppe into a former cultivated fallow for its restoration in the La Crau plain, Southeastern France. Different letters for ANOVAS with subsequent Tukey post-hoc test indicate a significant difference for $p < 0.001$ (N = 18) (Jaunatre et al., 2012, modified).

For the transfer of soil carried out in the spring of 2009 (Jaunatre et al., 2012; Jaunatre et al., 2013; Jaunatre et al., *in press a*), the measurements taken a year later showed a significant increase in species richness compared to the control fallow land without spreading and also in relation to the control steppe (Figure 7).

DISCUSSION

Our results have evidenced the undeniable impact of the phases of cultivation on plant species-richness and composition of the steppe vegetation. Thirty years after the abandonment of cultivation and even after the immediate reintroduction of grazing, the vegetation of the formerly cultivated areas remains poor and less heterogeneous than that of the pseudo-steppe. The assemblages of terrestrial coleopterans show the inverse result for the species richness. These results show the interest of monitoring several biodiversity compartments with a view to achieving a better assessment of the responses of living organisms to disturbances.

These results also show the very low natural resilience of the steppe vegetation after a phase of cultivation. Even if such a phase resulted in an increase in the fertility of the soil, that is still apparent 30 years after the abandonment of cultivation (phosphorus and potassium content, in particular). This low resilience might also be explained by the strategies of the species that make up the steppe vegetation (Saatkamp et al., 2010). We would offer here the hypothesis that it is the tolerant species, adapted to stress (water deficiency) and to recurrent intermediate disturbances (itinerant sheep grazing) but are poorly productive that have seeds that remain poorly viable in the soil and are little transported by the wind. This hypothesis however would now need to be tested by experiments involving burial and aerial traps for seeds. The presence of a matrix of pseudo-steppe vegetation and / or patches of relictual pseudo-steppe is of little use for speeding up regeneration of the steppe vegetation and of its entomofauna. It would require at least several tens of years before even half of the steppe species colonise solely the edges of the formerly cultivated areas (Buisson & Dutoit 2004 ; Buisson et al., 2006).

The very low resilience of certain plant communities has however already been identified for the forest communities (Plue et al., 2008) and also for the pseudo-steppe of Crau where the impact of sheep breeding practices during antiquity and in particular the concentration of flocks in sheep folds is still perceptible in the chemical composition of the soil, its seed bank and the vegetation that grows 1500 to 2000 years after they were abandoned (Henry et al., 2010). Our investigations in paleoecology confirm the ancient origin of the herbaceous vegetation of the pseudo-steppe type which may date back at least to 5 000 to 6 000 years BP and result from the exploitation of a Matorral type of vegetation at the end of the Neolithic period. This very early origin exacerbates the fragility of the vegetation with regard to any disturbance other than that with which it has 'co-evolved' (sheep grazing for at least 2 000 years). Although the physiognomy of this vegetation has remained herbaceous over the centuries, its composition has probably changed, enabling the constitution of ancient meadows or mature meadows. These are composed of numerous annual species but dominated by a perennial, the ramose false brome which here takes the place in the dry grassland Mediterranean ecosystem of the trees known as 'climax' (*Q. ilex, Q. pubescens*) for the Mediterranean forest ecosystems (Gaillard et al., 2009).

Like many semi-arid steppe areas in North Africa (Le Houérou, 1995), the vegetation of the plain of Crau is not an original or 'climax' ecosystem, which explains even further its low natural resilience which would not lead spontaneously to the chosen reference ecosystem. This corresponds rather to an old agro-ecosystem which has evolved under the recurrence of sustainable traditional grazing practices, since the current grazing pressure is comparable to that practised by Roman shepherds two thousand years ago (Badan & Brun, 1995). The Crau pseudo-steppe also constitutes a culture-based ecosystem in the sense of Clewell & Aronson (2013) which therefore requires the application of active rehabilitation and / or restoration operations in order to set it on a trajectory that would enable it to recover its former ecological integrity.

The main thresholds of irreversibility for the natural resilience of the steppe vegetation after phases of cultivation consist in the destruction of this vegetation over large areas (1 500 ha in all), an increase in the chemical fertility of the plots but also the colonisation of these areas by numerous common species of weeds. The restoration of the pseudo-steppe vegetation, or of the Coleoptera assemblages, would therefore require the forcing of the process of dispersion in order to bring the seeds and the clones of the steppe species to the formerly cultivated areas (transfer of hay or soil), and also the limitation of the cover of weed species from formerly cultivated areas by the sowing of covering species. The transfer of soil and the transfer of hay give very good results for the restoration of part of the plant composition typical of the pseudo-steppe, in particular for the annual species. These results have however been obtained over a very short term and regular monitoring surveys should now be carried out at least every five years.

The transfer of soil is not however in itself a restoration operation since the pseudo-steppe of the donor site is destroyed. What remains to be studied is the possibility of transferring slabs or mats of different sizes and spacing in order to test their role as 'graft' or 'nucleus' via spontaneous colonisation starting from the edges of the transplanted mats. But it is also necessary to study the 'scarring' at the places where the mats of pseudo-steppe have been removed. This technique is based on the results of the reintroduction of small areas covered with stones at sites that had initially been cleared of stones for cultivation, and which showed an increase in the density of Coleoptera four years after the application of the experimental protocol. It is thus possible to envisage restoration operations on small areas that are spatially functional in order to reduce the cost of this kind of operation.

The operations of soil transfer or of mats of steppe of course involve a high cost, and should only be envisaged for the restoration of small areas of an agro-ecosystem of high patrimonial value, as in the case of a technological accident following the leakage of an oil pipeline, for example. Such operations cannot of course be transposed to the restoration of large areas of the semi-arid steppe of North Africa (Aronson et al., 1993; Pontannier et al., 1995) where as shown by our results, it would be preferable to the transfer of hay from non-degraded areas and the sowing of covering species, if possible the perennial species dominating the non-degraded steppe. Finally, it appears to be the combination of these two practices that would offer the best results, both with regard to rehabilitation of part of the specific composition of the steppe vegetation (in particular annuals) but also to its structuring through the implantation of the dominant species (Coiffait et al., 2012b), and would be capable of creating micro-sites favourable for the establishment of other steppe species via their transport in the fleece or the digestive system of domestic herbivores used for itinerant grazing. Our first results from the sowing of *Brachypodium retusum* confirm the feasibility of its implantation by mechanical means but its impact in terms of facilitating the colonisation of fallow lands by species typical of the pseudo-steppe seems to require several years. In order to speed up this process, operations of soil removal and transfer (to reduce the fertility of the soil and the stock of seeds of weed species) of the soil were carried out in 2009 as well as hay transfer operations where the *Brachypodium* sown seemed to be already well implanted (Jaunatre et al., 2012; Jaunatre et al., *in press b*). Our first results show the undeniable impact of soil transfer on the reduction of the number of weed species issuing from the ground seed bank (e.g. *Chenopodium album*) but the role of the *Brachypodium* facilitating the implantation of seeds transferred by the hay still requires confirmation in the medium term (Coiffait et al., 2011a; Coiffait et al., 2011b).

CONCLUSION

Ten years of research on the pseudo-steppe vegetation of the plain of Crau have enabled us to identify the major impact of its degradation by phases of cultivation and the origin of the development of this ecosystem and provided encouraging results in terms of restoration. In addition to numerous complementary studies, notably on the role of landscape aspects in the spontaneous dynamic of the vegetation of formerly cultivated areas and of the combination of certain restoration practices (sowing of dominant perennial species, then spreading of hay several years later), the validation of these results will require the long term monitoring of the protocols tested and their transfer during operations of rehabilitation or restoration over large areas with the necessary ecological engineering methods adapted to use in developing countries. This should be the role of a workshop site established in the plain of Crau, a real experimental platform for research on the rehabilitation and restoration of Mediterranean steppe rangelands.

ACKNOWLEDGMENTS

We would particularly like to thank the CEN-PACA and the *Réserve Naturelle des Coussouls de la Crau* for granting permission for access to the study sites, and the shepherds, sheep breeders and private landowners. We are also grateful to the companies GRT/GAZ, SAGESS, Gagneraud Constructions and CDC-Biodiversité for funding part of the experiments; the *Région Provence-Alpes-Côte-d'Azur* for funding PhD theses. Our work has also been carried out within the framework of various programmes of the *Ministère de l'Ecologie, de l'Energie, du Développement Durable et de la Mer*, the programme *Espaces Protégés* (2001-2006) and the *Institut d'Ecologie et Environnement du CNRS* (INEE), programmes *Ingéco* (2007-2009) et *Ingécotech* (2010-2011). Thanks to Mr. Paul Michael for correcting the English version.

REFERENCES

Aronson, J., Floret, C., Le Floc'h, E., Ovalle, C., Pontanier, R. (1993). Restoration and rehabilitation of degraded ecosystems of arid and semiarid lands. I. A view of the south. *Restoration Ecology 1*, 8-17.

Badan, O., Brun, J.P., Congès, G. (1995). Les bergeries romaines de La Crau d'Arles. *Gallia* 52, 263-310.

Blight, O., Fadda, S., Orgeas, J., Ponel, P., Buisson, E., Dutoit, T. (2011) Using stone cover patches and grazing exclusion to restore ground-active beetle communities in a pseudo-steppe. *Journal of Insect Conservation* 15, 561-572.

Blondel, J., Aronson, J., Bodiou, J.Y., Bœuf, G. (2010). *The Mediterranean Basin - Biological Diversity in Space and Time*. Oxford University Press, Oxford.

Buisson, E., Dutoit, T, Roche, P, Duhem, C, Campagne, P, Tatoni, T. (2009). A cultural landscape of southern France: the plain of La Crau. In: Krzywinski K, O'Connell M., Küster H. (eds.), *Cultural landscapes in Europe: Fields of Demeter, Haunts of Pan*. Aschenbeck Media, Bremen, Germany.

Buisson, E., Dutoit, T., Torre, F., Römermann, C., Poschlod, P. (2006). The implications of seed rain and seed bank patterns for plant succession at the edges of abandoned fields in Mediterranean landscapes. *Agriculture Ecosystems and Environment* 115, 6-14.

Buisson, E., Dutoit, T. (2004). Colonisation by native species of abandoned farmland adjacent to a remnant patch of Mediterranean steppe. *Plant Ecology* 174, 371-384.

Buisson, E., Dutoit, T. (2006). Creation of the Natural Reserve of La Crau: implications for the creation and management of protected areas. *Journal of Environmental Management* 80, 318-326.

Clewell, A.F., Aronson, J. (2013). *Ecological Restoration. Principles, Values, and Structure of an Emerging Profession*. Second Edition. Island Press, Washington, USA.

Coiffait-Gombault, C., Buisson, E., Dutoit, T. (2011a) Hay transfer and sowing structuring species: two complementary ecological engineering techniques to restore dry grassland communities. *Procedia Environmental Sciences* 9, 33-39.

Coiffait-Gombault, C., Buisson, C., Dutoit, T. (2011b). Hay transfer promotes establishment of Mediterranean steppe vegetation on soils disturbed by pipeline construction. *Restoration Ecology* 19, 214-222.

Coiffait-Gombault, C., Buisson, E., Dutoit, T. (2012a). Are old Mediterranean grasslands resilient to human disturbances? *Acta Oecologica* 43, 86-94.

Coiffait-Gombault, C., Buisson, E., Dutoit, T. (2012b). Using a two-phase sowing approach in restoration – sowing foundation species to restore and subordinate species to evaluate restoration success. *Applied Vegetation Science* 15, 277-289.

Devaux, J.P., Archiloque, A., Borel, L., Bourrelly, M., Louis Palluel, J. (1983). Notice de la carte phytosociologique de la Crau. *Biol. Ecol. Med.* 10, 5-54.

Dutoit T. Jaunatre R., Buisson E. (2013). Mediterranean steppe restoration in France. In: Clewell A., Aronson J. (eds.) *Ecological Restoration Principles, Values, and Structure of an Emerging Profession,* pp. 60-64. Second Edition. Island Press, Washington.

Etienne, M., Aronson, J., Le Floc'h, E. (1998). Abandoned lands and land use conflicts in Southern France. *Ecological Studies* 136, 127-140.

Fadda, S., Henry, F., Orgeas, J., Ponel, P., Buisson, E., Dutoit, T. (2008a). Consequences of the cessation of 3000 years of grazing on dry Mediterranean grassland ground-active beetle communities. *Comptes Rendus de Biologie* 331, 532-546.

Fadda, S., Orgeas, J., Ponel, P., Buisson, E., Dutoit, T. (2008b). Conservation of grassland patches failed to enhance colonization of ground-active beetles on formerly cultivated plots. *Environmental Conservation* 35, 109-116.

Fadda, S., Orgeas, J., Ponel, P., Buisson, E., Dutoit, T. (2007). Past cultivation as a driving factor for the organisation of dry grassland beetle communities (plain of la Crau, Bouches-du-Rhône, France). *Environmental Conservation* 34, 132-139.

Gaillard, M.J., Dutoit, T., Hjelle, K., Koff, T., O'Connel, M. (2009). European cultural landscape - insights into origins and development. In: Krzywinski K, O'Connell M, Küster H, (eds.), *Cultural landscapes in Europe: Fields of Demeter, Haunts of Pan*. Aschenbeck Media, Bremen.

Grove, A.T., Rackham, O. (2001). *The nature of Mediterranean Europe: An Ecological History*. Yale University Press, Yale.

Henry, F., Talon, B., Dutoit, T. (2010). The age and history of the French Mediterranean steppe vegetation revisited by soil wood charcoal analysis. *The Holocene* 20, 25-34.

Jaunatre, R., Buisson, E., Dutoit, T. (2012). First-year results of a multi-treatment steppe restoration experiment in La Crau (Provence, France). *Plant Ecology and Evolution* 145, 13-23.

Jaunatre, R. Buisson, E., Muller, I. Morlon, H., Mesléard, F, Dutoit T. (2013). New synthetic indicators to assess community resilience and restoration success. *Ecological Indicators* 29, 468-477.

Jaunatre, R. Buisson, E., Dutoit, T., Using ecological engineering to restore an abandoned intensive cultivation in a Mediterranean rangeland. *Ecological Engineering*, in press a.

Jaunatre, R. Buisson, E., Dutoit, T., Topsoil removal improves various restoration treatments of a dry Mediterranean grassland (La Crau, southeastern France). *Applied Vegetation Science*, in press b.

Kiehl, K., Kirmer, A., Donath, T.W., Rasran, L., Hölzel. N. (2010). Species introduction in restoration projects – Evaluation of different techniques for the establishment of semi-natural grasslands in Central and Northwestern Europe. *Basic and Applied Ecology* 11, 285-299.

Le Houérou, H.N. (1995). Considérations biogéographiques sur les steppes arides du nord de l'Afrique. *Sécheresse* 6, 167-182.

Médail, F., Quézel, P. (1997). Hot-Spots analysis for conservation of plant biodiversity in the Mediterranean Basin. *Annals of the Missouri Botanical Garden* 84, 112-127.

Molinier, R., Tallon, G. (1950). La végétation de La Crau (Basse-Provence). *Revue Générale de Boanique* 56, 1-111.

Muller, S., Dutoit, T., Alard, D., Grevilliot, F. (1998). Restoration and Rehabilitation of species-rich grassland ecosystems in France: a review. *Restoration Ecology* 6, 94-101.

Plue, J., Hermy, M., Verheyen, K., Thuillier, P., Saguez, R., Decocq, G. (2008). Persistent changes in forest vegetation and seed bank 1,600 years after human occupation. *Landscape Ecololgy* 23, 673-688.

Pontanier, R., M'Hiri, A., Akrimi, N., Aronson, J., Le Floc'h, E. (1995). *L'homme peut-il refaire ce qu'il a défait ?* John Libbey Eurotext, Paris, France.

Römermann, C., Dutoit, T., Poschlod, P., Buisson, E. (2005). Influence of former cultivation on the unique Mediterranean steppe of France and consequences for conservation management. *Biological Conservation* 121, 21-33.

Saatkamp, A., Roëmermann, C., Dutoit T. (2010). Plant functional traits show non-linear response to grazing. *Folia Geobotanica* 45, 239-252.

Thinon, M. (1992). *L'analyse pédoanthracologique. Aspects Méthodologiques et Applications*. Thèse de Doctorat ès-Sciences, Université Paul Cézanne, Marseille.

In: Steppe Ecosystems
Editors: M. B. Morales Prieto and J. Traba Diaz
ISBN: 978-1-62808-298-2

Chapter 14

CRITICAL REVIEW OF THE ALGERIAN EXPERIENCE TO MANAGE THE LAND DEGRADATION

Aziz Hirche[123*], Mostefa Salamani[13], Boughani Abdelmadjid[1], Dalila Nedjraoui[1], Ratiba Hourizi[1] and Youcef Oukil[1]

[1]Laboratory of Vegetal Ecology, Faculty of Biological Sciences, University of Sciences and Technology Houari Boumediene, Bab Ezzouar, Algiers, Algeria

[2]University of Algiers Youcef Benkhedda, Faculté centrale, Algiers, Algeria

[3]Scientific and Technical Research Center on Arid Regions, Front de l'Oued, Biskra, Algeria

ABSTRACT

The arid Algerian High Plateaus, a key area with pastoral activity in Nort Africa, covers 27 million hectares. This steppe underwent significant degradation in the eighties. The combination of a severe drought, the most important of the century, with a peak between 1982 and 1987 and an exponential increase in livestock had a catastrophic impact on pastoral resources. Vegetation cover, which was generally greater than 30% before 1980 is now less than 15%. Phytomass, that exceeded 1t.ha^{-1}.year^{-1} in 1980, is now less than 450 kg.ha^{-1}.year^{-1}. This desertification has in turn greatly undermined the fodder potential of the steppe. The fodder production, that earlier exceeded an average of 150 Uf. ha^{-1}.year^{-1}, can currently provide about 30–40 Uf. ha^{-1}.year^{-1}. In response to this deterioration, the Algerian government, defined a strategy to manage and restore the ecosystems. It had several technical solutions to achieve this. The first was to increase pastoral and agricultural production; the second was to reduce livestock; the third was to combine the first two. Finally, only the first approach was chosen. Several options were considered. The main ones were planting, collecting water for livestock and agriculture and finally installing exclosures. Agro-forestry plantations were carried out on 217,549 ha with different species, particularly the genus *Atriplex, Medicago* and *Opuntia*. Small hydraulic entities were also built to irrigate 410,000 ha and to increase the planted area.

* Corresponding Author address: Email: a_hirche@yahoo.fr.

The exclosures were established on 3 million hectares. We discuss the relevance and success rate of these works from a holistic perspective. The economical and biological impacts (biodiversity, production) are assessed. As a result, the government's strategy is not clear, and sometimes seems at odds with the objectives of the fighting desertification. Worse yet, we show that in some cases, the Algerian government seems paradoxically to be supporting the steppe degradation.

Keywords: Rangelands, Steppe, Overgrazing, Exclosures, Plantations, Desertification

INTRODUCTION

For a little more than half a century, corresponding to the post-independence period, Algeria has been wrestling with the problem of desertification. This desertification is reflected mainly through continued loss of natural resources, especially plant cover and biomass. Many studies have shown that the causes are principally overgrazing and extension of cultivation (Djebaili 1978; CRBT 1978, 1995; Le Houérou 1985; Nedjraoui 1981; Boutonnet 1989; Kadi Hanifi 1998; ROSLET 2010). These two phenomena are also often interrelated.

Overgrazing is a consequence of livestock increases related to population growth, which exacerbates meat demand (Boutonnet 1989; Hirche 2010). Livestock food demand increased and the government tried to respond by increasing forage production through the plantations, and to restore degraded rangelands by multiplying livestock exclosures. In this work, we examine the impact of this approach and propose, when considered necessary, alternative measures.

1. PRESENTATION OF THE STUDY AREA

The study area is located in the so-called Hauts Plateaux corresponding to the Algeria highlands between two mountains ranges, the Atlas Tellian and the Saharan Atlas. It is mainly occupied by a Quaternary polygenic glacis whose average height fluctuates around 1000m in the flat zones to about 2000m for the mountains. There are dug saline depressions, such Chott Chergui, Gharbi and the Zahrez, surrounded by mountains like the Jebel Antar and the Sen Alba (Figure 1).

The climate is harsh. Rainfall is low with an average rainfall between 400 mm to 100 mm. The yearly distribution is very irregular, the coefficient of variability exceeding 30%. Evaporation is high, often higher than 2200 mm.

The area has severe winters and hot summers with an average minimum of -1.5 ° C in January and an average maximum of 38.1 in July. The dry period (*sensu* Bagnouls and Gaussen 1953) lasts 6 months (Table 1). The rainfall evolution during the last century shows an alternation of dry and humid periods with no clear trends during approximately the first half of the century (Figure 2). At the end of the second half, dry periods became notably more frequent and intense, especially during the 1981-1988 period (Hirche et al. 2007), considered to be the driest of the century.

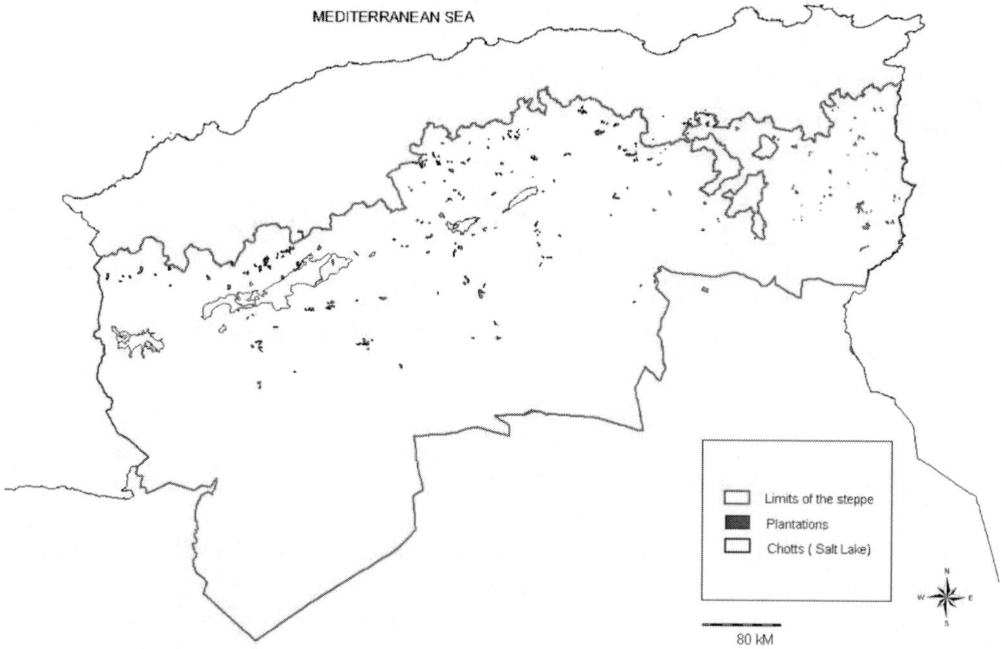

Figure 1. General locatingon of survey the study zonearea.

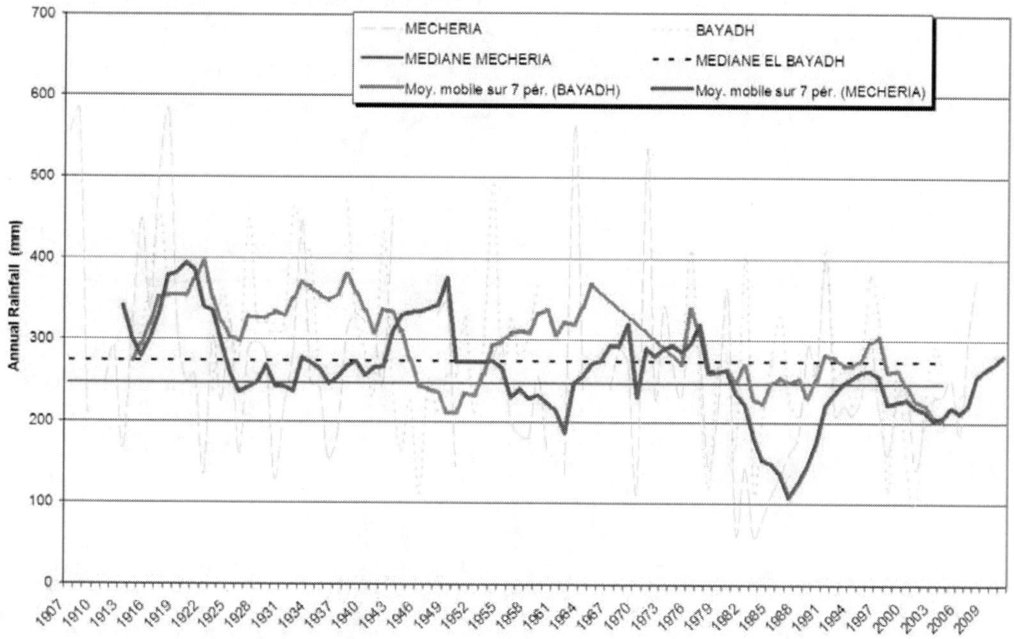

Figure 2. Evolution of annual rainfall in the study area since the beginning of the century.

The former dominant vegetation was Alfa or esparto grass (*Stipa tenacissima*), albardine (*Lygeum spartum*) and wormwood (*Artemisia herba alba*). With regard to the vegetation, several phytoecological and phytosociological studies undertaken by the CRBT (1978), Djebaili (1978), and Aidoud-Lounis (1998), among others, have described the vegetation dynamics of the Hauts Plateaux.

Table1. Main characteristics of the climate

Rainfall (mm)	300
Coefficient of variation (rainfall)	30-45 %
Average minima of the coldest month m (°C)	-1.5° C
Evaporation (mm)	>2200
Bioclimate	Arid cool to cold with 6 dry months

At present, the most widespread vegetation in terms of physiognomy is dominated by *Atractylis serratuloides, Noaea mucronata, Peganum harmala* on rocky soils and *Thymelaea microphylla* on sandy soils.

2. Historical Context of Local Pastoral Economy

If we compare the trends of livestock, population and rainfall since the beginning of the century, two periods appear to be significant. First, the period before independence, while human population increased slowly, and both livestock and rainfall followed an irregular and non-periodic evolution. Additionally, droughts and animal diseases (epizooties) were frequent provoking a severe decline in livestock population that allowed the steppe vegetation to respond favorably with an increase in density. Second, the post-independence period during which both human and livestock populations increased exponentially (Figure 3).

By dint of fodder supply, livestock became a permanent pressure on rangelands. The progress in animal prophylaxis, and sedentary are reasons explaining the sharp increase of livestock.

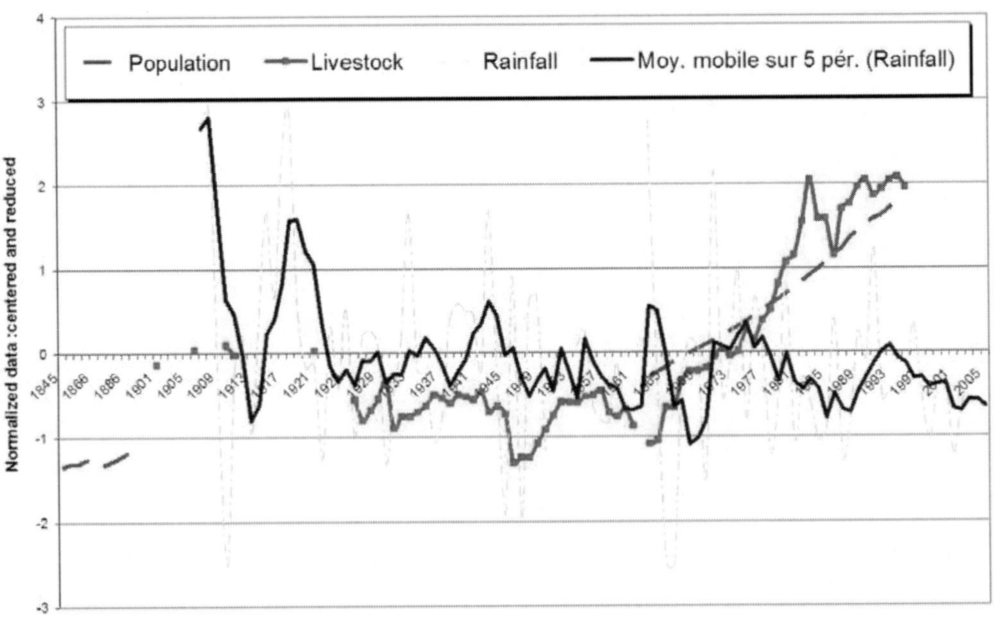

Figure 3. Evolution of population, livestock and rainfall in Méchéria.

Besides, the exceptional drought towards the end of this period deepened the dramatic grazing pressure on rangelands. The traditional balance in fodder offer/demand was broken with the fodder offer in decline and the needs increasing.

The steppe is not a cropland but a rangeland area in which natural fodder produced is insufficient for the current stocking rate. An important resource to fill this gap is represented by stubble from cereal production and fallow. However, this is subject to strong fluctuations and is not enough to cover all the demand. So, fodder must be imported from Northern areas of the country or even from other countries. However, arable land for the whole country represents only 3.5% corresponding to 8.4 million hectares from the 250 million of the country. This means 0.23 ha per capita, much less than the 0.6 ha usually deemed necessary to meet basic food needs.

Average cereal yields since the beginning of the century are generally less than 700 $Kg.ha^{-1}$', one of the lowest in the Mediterranean Basin (FAOSTAT 2009). The fodder production of the steppe, that earlier exceeded an average of 150 $Uf.Ha^{-1}.year^{-1}$ (being Uf the equivalent of one kg of barley; it represents the energy equivalent of the forage) can currently provide only 30-40 $Uf.Ha^{-1}.year^{-1}$ (Nedjraoui, 2006). The spiral of growth in cereals demand only exacerbates the degradation of rangelands, because it means keeping a high grazing pressure on the territory.

As a consequence, the remaining pastoral surface is overgrazed (Cousins et al. 2003). The best fodder resources diminish. Livestock, despite drought, have survived because they received additional energetic food, especially in the form of barley (Aidoud and Touffet 1996). Main sustenance based only on barley keeps animals hungry. As a result, even bad fodder plants are grazed. This is the case of Alfa, a characteristic species of this region, considered as coarse food, with low bromatological interest (low energy content), and not particularly appreciated by livestock is eaten by livestock to ensure their intestinal transit. The degradation occurs mainly on the glacis, flat area, whereas the abrupt lands, less accessible, are clearly more preserved. Wormwood (chih), one of the best appreciated fodder plants by livestock has now completely disappeared (Aidoud 1998).

Protected areas, generally enclosed, were installed in 1975 by the CRBT (the former Algerian research institution) to monitor the main vegetation units represented by Alfa, albardine and wormwood. Field observations showed that protected sites had higher vegetation cover than non-protected ones. This was regardless of the drought that had a decreasing impact on vegetation for both protected and non-protected sites. In consequence, the effect of overgrazing seems to be more important than climate, a situation already underlined by other authors (Slimani 1998; Slimani et al. 2010; Amghar et al. 2012).

Therefore, it appears that the driving forces leading to vegetation cover decline and land degradation are related mainly to socio-economic parameters. These are the principal keys to understand and foresee the ecosystems evolution.

3. Techniques to Manage Rangelands

Two techniques are widely used in Algeria by the administration with the objective of increasing fodder offer. The first one is plantation and the second one is land exclosure.

3.1. Plantations

In 1981, the government created the HCDS (Haut Commissariat au Développement de la Steppe), that is a structure involved in the development of the arid rangelands steppes. Among its activities, 326,489 ha were officially planted (Table 2). These statistics were gathered on administrative documents but no synthetic work was reported to confirm or assess them by field investigation. Actually, the HCDS has launched an inventory of these plantations to be integrated on a GIS platform. The area actually concerned by this ongoing work is 105,161 ha (Table 2).

Using these data, we observed that the main part of the plantations is located in the Center and West of the country. The database is not completely informed but the preliminary results show that 5 species are commonly used: Three *Atriplex, A. canescens, A. nummularia* and *A. halimus*, one Cactaceae, *Opuntia ficus indica* and one Legumineous, *Medicago arborea*. This last species is rarely planted and does not play an important role. Among, the *Atriplex* species, the most commonly planted ones are *A. nummularia, A.halimus subsp. schweinfurthii and A. canescens subsp. linearis.*

The original area of most of these species ranges from central Mexico to Canada. According Le Houerou (1969), they were introduced in North Africa and the Middle East from the United States through Tunisia in the 1980's. These species show different ecological requirements. The choice of species depends on a combination of factors such as:

- Average annual rainfall;
- cold tolerance;
- topographic position,
- accumulation or runoff of water;
- nature of the soil.

Adapted to these constraints, these species could respond, to the following goals of development: Fodder production; production of wood; fixation of sands; and protection against water erosion. Their main inconvenient is their high content in sodium, potassium and mucilages, that may limit their utilization as a fodder.

Table 2. Main actions carried out by the HCDS

Actions	1994/ 1999	2000 /2008	Total
Exclosure (ha)	**200,000**	**2,817,194**	**2,817,194**
Pastoral plantation (ha)	**120,000**	**206,193**	**326,489**
Implementation and development of ceds (Units)	**1**	**1201**	**1202**
Construction and development of water points (Units)	**1108**	**6475**	**7583**
C.E.S Work (m3)	**6519**	**1,975,940**	**1,982,459**
Achieving seguia ML	**3762**	**1,353,459**	**1,357,221**
Rustic Fruit Tree		**8359**	**8359**
Households solar electrification		**4507**	**4507**

ced= small dam.

C.E.S = Work of soil restoration.

In particular, among the most commonly planted species in Algeria, *A. canescens* prefers sand soils and has been used in dune stabilization projects, while *A. halimus*, is rather located on silt soils. Both of them are halophilous and need the presence of salt and water supply. According to Le Houerou (1992), these species shows rapid development, with an excellent performance at low doses of water. Its value of efficiency in the use of water equals to 5-10 Kg. ha^{-1}. mm^{-1}.$year^{-1}$ of dry matter have been observed. Le Houerou and Hoste (1977) report that in the WANA area (North Africa and Middle East) a large area has been planted with forage shrubs, which have been subjected to combined techniques of water harvesting (Boulanouar et al. 2000; Nefzaoui et al. 2000; 2002; Nefzaoui and El Mourid 2010; Redjel and Boukheloua 2000). Dry biomass production depends on the species and plant density.

According to Le Houerou and Franclet (1971), the productivity of *Atriplex* can reach 1800 Uf.Ha^{-1}.$year^{-1}$in favourable conditions. For *A. halimus*, the total green biomass produced annually is estimated at 2911 kg distributed in 1792 kg.$year^{-1}$ for leaves and 1119 kg/year for the stems (Bouzid and Benabdelli 2011). In Algeria, several authors CRBT (1978), Boughani (1995), Amghar (2012) show a biomass that exceeds 1000 Kg of DM and productions that fluctuate between 350 to 1800 UF (CRBT 1978, Amghar 2012, Boughani 1995).

Table 3 shows the area planted in Algeria divided in three regions: West, Centre and East. Plantations ocurred mainly on the western sector with 38236 ha and the central sector with 41188 hectares, on a total planted area of 326489 ha. The success rate of these plantations, measured as the ratio of surviving on the total of planted shrubs, has been rarely assessed. Nevertheless, it has been tentatively estimated by the synthesis of Belala and Meguenni (2011) who reviewed some works related to *A. canescens* (Table 4).

It appears that the average survival rate is quite encouraging with about 83% and the small standard deviation (8%) indicating homogeneity with small differences between measurements. The best results are observed when we have the conjunction of sand and salt (96%) and the worst is found in encrusted glacis (72,05%). So, if we consider in an optimistic estimate suggested by our field observations of 1000 UF.ha^{-1} per plantation, and that the average successful planting rate is 80%, the fodder production is about 800 UF.ha^{-1} The requirements of a zootechnic unit are 400 UF.ha^{-1}. Therefore, each plantation could theoretically support one zootechnic unit composed by a sheep, its lamb and 1/25 of reproductive male. The 217,549 ha can support a livestock of 108,774 sheep. If we compare to the 12,191.302 sheep living in agro-steppic ecosystems, the fodder rate favoured by the plantations is less than 1% (0.88%). The part of plantation fodder added to the natural steppe production is weak, and the rangeland fodder production part fluctuates between 10% in a dry year to 30% in very humid conditions. So, 70 to 90% of the livestock feeding is provided from the Northern cultures (Sub-humid zone) or by the importations.

Table 3. Location of plantations

	Area (Ha)	%
West	38236	36,35
East	25736	24,47
Center	41188	39,16
Total	105160	100

Table 4. Survival rates of planted species linked to geomorphic units

Geomorphology	Mean of survival rates (%)
Glacis	81,21
Glacis with sand	82,44
Encrusted Glacis	72,05
Salt and Sanded Glacis	96
Wadi (River)	82,60
Dune	82,80
Average survival rate	82,85
Standard deviation	7,65

Table 5. Economic costs related to the plantation investment

Plantation fourrageres designation	Unit	Quantity by ha	Unitary cost in da	Cost(da) by ha
Labor				
Tracing and picket	HJ	5	500	2 500
Opening holes	HJ	20	500	10 000
Plantation including watering	HJ	20	500	10 000
S/ total				22 500
Overcome	HJ	1	500	500
S/ total				500
Supply and transportation				
plant (purchase)	unit	1000	7	7 000
transport	HJ	8	50	400
S/total				7 400
SUB TOTAL INVESTMENT				30400
VALUE ADDED TAX (17%)				5168
TOTAL				35 568

HJ: Man Day.

As we have previously seen, Boutonnet (1989) considers that 90% of the cattle food is provided by cereal, in the form of straw and grain. Barley and oats are considered here as forage. So, in spite of all the efforts undertaken by plantation, they have had little impact on cattle feeding supply. If we examine the costs, plantations are prohibitive (Table 5).

If we try to assess the global cost of a plantation compared to the total fund invested by the government (Table 6) for developing the arid rangelands, we find that their costs represent the quarter of the total sum. The cost of plantations is about 217 549 ha x 35568 = 7 737 782 832 Dinars. If we consider that one euro is equal more or less to 100 DA, the total costs reached 7,73 billion DA, about 77 millions Euros

The total invested funds are 30 301 643 000 DA representing 30,30 billions dinars. The part of plantations in the total amount is about 25 %, which is a very important amount.

Table 6. Summary of programs granted, registered and pending registration (State Budget Equipment + State FLDDPS). Unit = 1000 DA (DA: Algerian dinar)

Program		Funds
		13 505 643
	Arid Highlands	3 982 000
	South	50 000
	S/T	17 537 643
FLDDPS	Normal	5 000 000
	Arid Highlands	7 102 000
	South	662 000
	S/T	12 764 000
TOTAL		30 301 643

In summary, we have shown that plantations fodder production is insignificant in relation to its extremely high cost. The question of the profitability of such an approach should be seriously considered.

3.2. Exclosures

It is an ancestral technique well known in North Africa as "Agdal " or "Hima" in the Middle East. The principle is to temporarily subtract a rangeland surface from grazing. The duration is generally 2 years or more. They don't exactly play the same role as experimental exclosures that have been widely used as treatments to exclude the effects of predators, large herbivores, livestock, small mammals or birds on the species richness and recruitment in plant communities (Fraser and Madson 2008; Levick and Rogers 2008; Negussie et al. 2008), or on the abundance of other animals (Aerts et al. 2009). The purpose of these exclosures is to restore the initial ecosystem conditions. This technique has been widely used, especially in Africa, Australia and Middle East, and it is generally efficient and most effective when rainfall is high. Its success is also related to the nature of vegetation, potential forage flora and soil conditions (Figure 4).

One of the main HCDS actions was the preservation and regeneration of degraded rangelands through exclosure (2,817,194 ha). Exclosures are often efficient to clearly raise the fodder production, and in the long term, to improve the physical and chemical properties of the soil, and the richness of flora (Amghar 2012; Zucca et al. 2011).

Exclosures are carried out by the HCDS, the beneficiaries are chosen by the local administration. Indeed, the protection of this exclosures is exclusive to the forest conservation and local administration.

Exclosures are not fenced, but a guardian is present. The boundary is marked by spaced heaps of stones more or less regularly implanted. Installation costs are relatively low, estimated at 500 DA.ha^{-1} and are, at equal surface, incommensurately lower than those derived from plantations. Indeed, although the protected area is huge, the cost is around 408,597,000 DA, about 4 million Euros, much lower than that of 73 million needed for plantations. Guardian costs were not included. According the HCDS (2012), the results of the programs are the following:

- Out of 7,000,000 ha of degraded rangelands, 2,670,000 hectares have been restored by the exclosure, representing 38% of the area.
- On 1,000,000 ha severely degraded, having lost their natural regeneration, 300,000 hectares were restored by planting about 33% of the area.
- Perimeters exclosure and pastoral plantations begin to be a tenancy, which is a basis for the management of these areas.
- This management rises the production of 2.5 billion units, which is an improvement of forage cover for livestock from 25% to 50%
- Improved farming conditions led to the preservation of grazing activity and the increase in livestock numbers from 10 to 15%, improving the meat yield from 13 kg to 21 kg of carcass weight.
- The creation of 142,000 permanent jobs in rural areas with endemic unemployment.
- Improvement of environmental conditions by reducing soil erosion, protecting biodiversity and improving groundwater recharge.

But this idyllic HCDS presentation does not stand comparison with reality. First, from a holistic perspective, by subtracting the exclosure for herbivore, the pressure for grazing outside the exclosure increased. In Algeria, different authors agree that the Algerian steppe grazing pressure is 5-12 larger than normal carrying capacity. In other words, there are 5 to 12 times more sheep than there should be (Nedjraoui and Bedrani 2008; Hirche 2010; Bedrani 1993; Bessaoud 2006; Boughani 1995; Aidoud 1989, 1996; Boutonnet 1989; Boukhobza 1982). Therefore, these rangelands undergo tremendous overgrazing that leads sometimes to the complete disappearance of perennial vegetation.

Figure 4. Three steps of abandoned exclosure: 1985, 1997, 2010.

Finally, we only move the problem, as said Le Houérou (1985), in a metaphoric image: "a hole is dug to raise a mound". The exclosure to be really efficient have to be integrated in a rotation plan. But, this one, like most of the pastoral management principles is based on the concept of an adequately balanced carrying capacity. Under the conditions of current imbalance, no classical pastoral management method could be successful. Normally, these exclosures are installed for 1-2 years and re-opened to the herders for 1-4 months, and leased at about 1000 DA per hectare (about 10 Euros). But, most of them are opened without knowing the corresponding carrying capacity and livestock often exceeds the pastoral possibilities of the exclosure, and sometimes irreversible damages occurs. This problem is intensified by the persistence of livestock for longer periods. In addition, the opening date of enclosures is not always adequate and, in this case, efforts of many years of exclosure may be vain.

It is possible that among these three million hectares, corresponding to at least 8 years of fencing, some of them are now newly degraded. Actually, we have no monitoring nor synthetic evaluation on the effects of these seasonal openings. To say that 3 million hectares are protected, as stated in the HCDS, is questionable, given that a significant portion, if not most of them could be now degraded. Some years ago, our prospection of a large part of the Algerian steppe showed many exclosures with some nice successful examples, especially with *A. canescens*, but a significant part of them have been lost due to the pressure of herbivores. The guarding has not been always successful due to the overlap between both administrations in charge: forest and civil institutions. Furthermore, these agencies have generally had financial difficulties, so the permanent personal is not sufficient and the recruitment of temporary staff is not always possible. Another serious problem concerns breeders. They are extremely powerful in pastoral areas and have a direct influence on the administration. They can easily change the time period of use in relation with rainfall and drought. Finally, this method of renting, which is theoretically similar to a participatory approach is not really such, because these protected areas are generally rented to local powerful people. This apparently explains the recent craze for these exclosures by certain people according to the administration. It must be remembered that they were extremely unpopular few years ago. We can understand it, only if we take into account that the population is here, as is often the case, held in hostage by the notables. Last but not least, in reality the exclosures are not established in priority on degraded areas, as claimed by the HCDS. On the contrary, some of them were chosen, on good state rangelands like the Alfa formations located at the frontier with Morocco. The statistics given by the HCDS are questionable and outdated. In a recent study, led by the same organism, based on remote sensing and field investigation, the degraded rangelands are 17 millions hectares (HCDS, 2011), compared with the 7 millions previously declared. If we consider a total pastoral area of 22 millions hectares of Algerian rangelands, the proportion of exclosures is only 13%.

4. Effect of the Exclosures on the Floristic Composition

Bourbouze and Donadieu (1987) consider that exclosures do not produce spectacular changes on floristic composition. They cite the case of a fenced enclosure during 3 years, with *Noea mucronata* and wormwood as dominant species.

Table 7. Species richness inside (E) and outside (O) exclosures. Data based on Amghar and Kadi Hanifi (2008)

Station	Exclosure (E)	Outside (O)	% (E/O)
El Haouita	59	24	0,41
Ain el Mahdi	50	25	0,50
Ksar el Hirane	15	12	0,80
Taadmit	62	40	0,65
Messaad	31	15	0,48

The results showed that the perennial species did not change. It was just observed the gain of eight new annual herbaceous species with low pastoral interest. In an exclosure of 10 years on Alfa steppe in Tadmit, a little floristic modification, and a small extension-recovery of *Stipa parviflora* and Alfa were observed. But several authors now contradict these preliminary results. Amghar and Kadi Hanifi (2008) showed that the number of species in the enclosure was generally double that in the rangelands without protection. Boughani (1995) and Hirche (2010), among others, found similar ratios (Table 7). In Tunisia the ratio between the number of species inside exclosures in relation to outside was of 2 to 3.5.

5. Effect on Biomass

The consequences on biomass are usually much more spectacular than those on floristic composition. While in the Hodna (North Eastern Algeria) has been found that an exclosure fenced for 3 years tripled the total green and dry biomass in March and April, and annual production was multiplied by 2.5, Amghar (2012) showed that production, after 3 years of protection was doubled. These authors underlined too an annual production higher than 100 UF, that may sometimes reach 300 UF.ha^{-1}. Our exclosures have showed significant differences with nearby rangelands. In the albardine vegetal unit, the biomass was 6 times higher inside the fence that outside (Hirche 1987). The pastoral production was 3.5 times higher inside the fence and reaches 160 UF.ha^{-1}. In the wormwood unit, the biomass production was less significant. The biomass ratio oscillates between 1.5 to 2.5. The production of biomass was about 85-110 UF.ha^{-1}, which is excellent. In alfa unit, the biomass reaches 2000 DM. ha^{-1}. year^{-1} (CRBT 1978; 1990). The forage production was about 90-140 UF. ha^{-1}. year^{-1}.

An agreement exists against the long exclosures (longer than 6 years). They are rarely justified in a pastoral zone. An exclosure must begin in a wet year because:

- The chances of recovery are increased and the effective duration decreases correspondingly.
- The removal of a pastoral surface, important for herds, is less problematic for herders

Knowing that March or September, rainy months correspond to the 2/3 of annual rainfall, the decision of exclosuring will therefore depend on the rainfall of these months. April would be generally preferred to November as closing time because spring growth is higher than autumn.

The opening should be in early summer after a wet spring, to avoid high concentration of animals on the restored surfaces and the rapid disappearance of the beneficial effects of protection.

6. DISCUSSION

If we consider that these fenced 3 millions hectares are perfectly restored, and if we consider too an excellent annual production of 100 UF. ha^{-1}. $year^{-1}$, and that the needs of an ovine equivalent is 400 UF. ha^{-1}.$year^{-1}$, these exclosures could offer fodder for only 0.75 millions sheep. If we confront these figures with the needs of the actual livestock which exceeds 18 millions, we must have 72 millions hectares to be fenced. It is much larger than the real pastoral area, which is about 22 millions hectares (27 millions if we add the presaharan zone). In other words, even if all the area was fenced, it would not be enough to cover the needs for feeding livestock. Increasing the fodder production; neither by plantations, nor by exclosures could not exclusively solve the problem. The solution is more complex, with a combination of livestock reduction and agricultural production increase. Boutonnet (1989) rightly said that breeding was a byproduct of agriculture. We neglect here the agricultural component, which does not produce natural forage, even if it provides the basic nutritional needs of livestock. Forage is produced essentially in the Tell (North Algeria), and the rangelands agricultural production is accessory. The vocation of the steppic area is not to be an agricultural land to intensify the production of fodder. We have seen that these solutions are either too expensive (plantations) or unrealistic (exclosures); and both of them are insufficient to respond to existing needs. But, even if we suppose that the livestock could be reduced, it may cause another problem: large-scale unemployment. Indeed, the steppic economy is based on livestock breeding. If livestock was severely reduced, as it has to be, most breeders will lose their job. This is the reason why the administration would try not to downsize livestock, even if it has to be done, because there is no alternative.

On the contrary, when a dry year occurs, the administration helps the breeders by massively importing barley to counteract the lack of fodder. This allows the continued presence of the entire livestock contributing to a permanent degradation of the steppe. To prevent an eventual popular revolt, the administration is trapped and ultimately reluctant to end up the funding to the rangeland degradation. The inclusion of socio-economic variables must be explicitly addressed (Wang et al. 2001; Ramos and Mulligan 2004; Van Delden et al. 2007; Ibáñez et al. 2008) and the solution belongs to the political and economical component. Technical approaches, as we have seen, could attenuate the problem, but they cannot be the ultimate solution. This passes imperatively by downsizing cattle to reach carrying capacity. This probably will not happen if the government does not use its sovereign power lifting the ban on sheep meat imports to reduce the prices and make breeding a less profitable activity. A part of breeders could sell his cattle if they were in economical difficulty, which requires the installation by the government of several cold rooms. In addition, among these farmers, many will swell the ranks of the unemployed, which implies that the government, to avoid a popular revolt, must promote the development of the rangeland areas. Priority action will include industry and the tertiary sector. Their success depends, among others, on tax reform, such as the exemption of local taxes. These measures should be accompanied by a reinforcement of transport capacity. Ultimately, the type of traditional breeding itself must change, related to a more efficient agriculture. It should be become intensive. "The land for all" is a utopia that can not longer resist the new social and economic reality. The development of these arid zones, could be achieved through promoting industry and services, especially tourism. The construction of a new North-South highway, railway track, a new town (that could be a free zone), with low taxes by the government are in the right direction. But, despite the existence of a national development plan in the long term, it is not obvious that these actions will be coordinated or really successful. The future will tell us.

Conclusion

We have discussed the classical solutions advocated by the Algerian administration to promote rangeland management, like exclosures and plantations. But, the central concept of all techniques of management is based on the idea of a match between needs and supply through the carrying capacity. This condition could be met at a small scale, but rarely at a large one. The prohibitive costs of these measures are another critical factor. Indeed, the most crucial problem today is the imbalance between the increase of agricultural production, including rangelands and the population needs, which grow exponentially. We have shown that neither plantations nor exclosures can be the solutions to increase the fodder production. The plantations contributions to livestock feed is only 1% The exclosures represent less than 13% of Algerian rangelands, and these protected areas lead to increased pressure on contiguous lands. The costs are prohibitive in both cases, especially plantations, with few results. Leaving livestock permanently on these lands, where production and biomass decrease, is not a rational alternative. The solution is, as in developed countries, to increase the agriculture productivity and to transform the traditional "way of life". The breeding has to become intensive and not extensive. A consequence of this gain of productivity is that a notable part of the breeders have to leave their activity and work in the secondary or tertiary sector. But, this condition is linked to the country development, and could not be accomplished in an undeveloped country. Ultimately, the solution to land degradation is not only technical, but is also affected by the socio-political dimension, which seems more crucial, clearly.

References

Aerts, R., Nyssen, J., Haile, M. (2009). On the difference between "exclosures" and "enclosures" in ecology and the environment. *Journal of Arid Environments* 73, 762-763.

Aidoud, A. (1989). *Contribution à l'étude des écosystèmes steppiques pâturés des Hautes Plaines Algéro-Oranaises. Fonctionnement, évaluation et évolution des ressources végétales.* PhD Thesis. Université des Sciences et de la Technologie H. Boumediène. Alger.

Aidoud, A., Touffet, J. (1996). La régression de l'alfa (*Stipa tenacissima* L.), Graminée pérenne, un indicateur de désertification des steppes algériennes. *Sècheresse* 7, 187-193.

Aidoud-Lounis, F. (1998). *Le complexe à Alfa- Armoise-Sparte (Stipa tenacissima L, Artemisia herba alba ASSO, Lygeum spartum L) des steppes arides d'Algérie, Structure et dynamique des communautés végétales*. PhD Thesis. Université des Sciences et de la Technologie Houari Boumediene, Alger.

Amghar, F., Kadi Hanifi, H. (2008). Diagnostic de la diversité floristique de cinq stations steppiques du sud Algérois. *Cahiers d'Orphée,* 385-395.

Amghar, F., Forey E., Margerie P., Langlois E., Brouri L., Kadi-Hanifi H. (2012). Grazing exclosure and plantation: a synchronic study of two restoration techniques improving plant community and soil properties in arid degraded steppes (Algeria). *Revue d'Ecologie (Terre Vie)* 67, 257-269.

Bagnouls, F, Gaussen H. (1953). Saison sèche et indice xérothermique. *Bulletin Société d'Histoire Naturelle* 88, 193-240.

Belala, F., Meguenni, S. (2011). *Revue bibliographique des plantations à Atriplex Canescens dans la steppe Algérienne.* Msc Thesis. Université des Sciences et de la Technologie Houari Boumediene, Alger.

Bedrani, S. (1993). *Le secteur agricole et ses perspectives à l'horizon 2000: Algérie.* Luxembourg Office for Official Publications of the European Communities. Luxembourg.

Bessaoud, O. (2006). La stratégie de développement rural en Algérie. In: Chassany J.-P., Pellissier J.P. (eds.), *Politiques de développement rural durable en Méditerranée dans le cadre de la politique de voisinage de l'Union Européenne,* pp. 79-89. CIHEAM-IAMM, Montpellier.

Boulanouar, B., Chiriyaa, A. (2000). Browse foliage as a supplement to wheat straw for sheep. In: Gintzburger, G., Bounejmate, M., Nefzaoui, A. (eds.), *Fodder Shrub Development in Arid and Semi-arid Zones. Proceedings of the Workshop on Native and Exotic Fodder Shrubs in Arid and Semi-arid Zones, 27 October-2 November 1996, Hammamet, Tunisia,* Vol. 2, pp. 476-484. ICARDA, Aleppo, Syria.

Boughani A. (1995). *Contribution à l'étude de la flore et des formations végétales au sud des monts du Zab (Ouled Djellal, wilaya de Biskra): Phytomasse, application cartographique et aménagement.* Msc Thesis, Université des Sciences et de la Technologie Houari Boumediene. Alger.

Boukhobza, M. (1982). *L'agro-pastoralisme traditionnel en Algérie: de l'ordre tribal au désordre colonial.* Office de Publications Universitaires, Alger.

Bourbouze, A., Donadieu P. (1987). L'élevage sur parcours en régions méditerranéennes. *Option méditerranéennes*: *série Études Ciheam-Iamm* 11, 1-88.

Boutonnet, J., (1989). *La spéculation ovine en Algérie un produit clé de la céréaliculture.* Economie et Sociologie Rurale. ENSA, Montpellier.

Bouzid, A. Benabdelli, K. (2011). Contribution à l'évaluation de la biomasse verte d'une plantation d'atriplex halimus en zone aride de l'ouest algérien (région de naama). *Revue d'Ecologie* 66, 303-308.

Centre de Recherche sur les Ressources Biologiques Terrestres (C.R.B.T). (1978). *Rapport phytoécologique et pastoral sur les Hautes Plaines steppiques de la Wilaya de Saida.* CRBT, Alger.

Centre de Recherche sur les Ressources Biologiques Terrestres (C.R.B.T). (1995). *Rapport phytoécologique et pastoral sur les parcours d'Ouled Djellal (Wilaya de Biskra).* CRBT, Alger.

Cousins, S., Lavorel S., Davies, I.D. (2003) Modelling the effects of landscape pattern and grazing regimes on the persistence of plant species with high conservation value in grasslands in south-eastern Sweden. *Landscape Ecology* 18, 315-332.

Djebaili, S. (1978). *Recherche phytoécologique et phytosociologique sur la végétation des hautes plaines steppiques et de l'atlas saharien algérien.* PhD Thesis. Université de Montpellier, Montpellier.

FAOSTAT. (2009). Statistics of the Food Alimentation Organization of the United Nations. http://faostat.fao.org/

Fraser, L.H., Madson, E.B. (2008). The interacting effects of herbivore exclosures and seed addition in a wet meadow. *Oikos* 117, 1057-1063.

H.C.D.S. (2011). *Les potentialités agropastorales de la steppe algérienne.* Atelier de restitution. Djelfa, Argelia.

Hirche A. (1987). *Etude diachronique d'un transect au Sud Oranais.* Mém. D.E.S. U.S.T.H.B. Alger.

Hirche, A., Boughani, M., Salamani, M. (2007). Evolution de la pluviosité annuelle dans quelques stations arides algériennes. *Sécheresse* 18, 314-20.

Hirche, A. (2010). *Contribution à l'évaluation de l'apport de la télédétection spatiale dans la dynamique de écosystèmes en zones arides: Cas du Sud oranais.* PhD Thesis. Université des Sciences et de la Techologie H. Boumediene. Alger.

Ibáñez, J., Martínez Valderrama, J., Puigdefábregas, J.(2008). Assessing desertification risk using system stability condition analysis. *Ecological Modelling 213*, 180–190.

Kadi-Hanifi Achour, H. (1998). *L'alfa en Algérie. Syntaxonomie, relation milieu-végétation, Dynamique et perspectives d'avenir.* PhD Thesis. Université des Sciences et de la Techologie H. Boumediene. Alger.

Le Houerou, H.N. (1969). La végétation de la Tunisie steppique (avec référence au Maroc, à l'Algérie et à la Libye). *Annales de l' Institute National de Recherche Agronomique de Tunis* 42, 624.

Le Houerou, H.N. (1985). Aspects météorologiques de la croissance et du développement végétal. Diversité biologique, développement durable et désertisation, *Options méditerranéennes série B: recherches et études*, 1-396.

Le Houerou, H.N. (1992). An overview of vegetation in world arid lands. In: Dregne, H. E. (ed.), *Degradation and restauration of arid and semi arid lands*, pp. 127-163. Center For Arid and semi arid Lands Studies, Texas Technical University, Lubbock (TX), USA.

Le Houérou, H.N., Franclet, A. (1971). *Les Atriplex en Tunisie et en Afrique du Nord.* FAO, Forestry Production Division (Report n° 7), Rome.

Le Houerou, H.N., Hoste C.H.(1977). Rangeland production and annual rainfall in the Mediterranean Basin and in the African Sahel, Sudanian zones. *Journal of Range Management* 30, 181-189.

Levick, S., Rogers, K. (2008). Patch and species specific responses of savanna woody vegetation to browser exclusion. *Biological Conservation* 141, 489-498.

Nedjraoui, D. (1981). *Evolution des éléments biogènes et valeurs nutritives dans les principaux faciès de végétation (Armoise blanche, sparte* et al*fa) des hauts plateaux.*

Steppe de la wilaya de Saida. Msc Thesis, Université des Sciences et de la Technologie Houari Boumediene, Alger.

Nedjraoui, D. (2006). *Profil fourrager de l'Algérie.* FAO. http://www.fao.org/ag/AGP/agpc/doc/Counprof/Algeria/Algerie.htm

Nedjraoui, D., Bedrani, S. (2008). La désertification dans les steppes algériennes: causes, impacts et actions de lutte. *Vertigo* 8. http://vertigo.revues.org/5375.

Nefzaoui, A., Ben Salem, H., Ben Salem, L. (2000). Nitrogen supplementation of cactus based diets fed to Barbarine yearlings. In: Gintzburger, G., Bounejmate, M., Nefzaoui, A. (eds.), *Fodder Shrub Development in Arid and Semi-arid Zones. Proceedings of the Workshop on Native and Exotic Fodder Shrubs in Arid and Semi-arid Zones, 27 October-2 November 1996, Hammamet, Tunisia,* Vol. 2, pp. 512-517. ICARDA, Aleppo, Syria.

Nefzaoui A., Ben Salem, H. (2002). Forage, fodder, and animal nutrition. Chapter 12. In: Nobel, P.S. (ed.), *Cacti, biology and uses*. University of California Press.

Nefzaoui, A., El Mourid, M. (2010). Coping with climate change and risk management strategies for a sustainable livestock and rangeland systems in the WANA region. In: El-Beltagy, A., Saxena, M.S. (eds.), *Proceedings of the Ninth International Conference on Development of Drylands*, pp. 630-640.

Negussie, A., Aerts, R., Gebrehiwot, K., Muys, B. (2008). Seedling mortality causes recruitment limitation of Boswellia papyrifera in northern Ethiopia. *Journal of Arid Environments* 72, 378-383.

Redjel, N., Boukheloua, J. (2000). Algerian Experience with Fodder Shrubs Plantation. In: Gintzburger, G., Bounejmate, M., Nefzaoui, A. Fodder Shrub Development in Arid and Semi-arid Zones. Proceedings of the Workshop on Native and Exotic Fodder Shrubs in Arid and Semi-arid Zones, 27 October-2 November 1996, Hammamet, Tunisia (Vol. 1: pp. 203-209). ICARDA, Aleppo, Syria.

Ramos, M. C., Mulligan, M. (2004). Spatial modelling of the impact of climate variability on the annual soilmoisture regime in a mechanizedMediterranean vineyard. *Journal of Hydrology 306*, 287–301.

ROSELT, 2010. *Rapport final de l'observatoire des hautes plaines steppiques du sud oranais*. Projet OSS/ROSELT/CRSTRA Algérie.

Slimani, H. (1998). *Effet du pâturage sur la végétation et le sol et désertification. Cas de la steppe à alfa (Stipa tenacissima L.) de Rogassa des Hautes plaines occidentales algériennes.* Msc Thesis, Université des Sciences et de la Technologie Houari Boumediene. Alger.

Slimani, H., Aidoud, A., Rozeslimani, F. (2010). 30 Years of protection and monitoring of a steppic rangeland undergoing desertification. *Journal of Arid Environments* 7, 685– 691.

Van Delden, H., Luja, P., Engelen, G. (2007). Integration of multi-scale dynamic spatial models of socio-economic and physical processes for river basin management. *Environmental Modelling and Software, 22*, 223–238.

Wang, Y., Dawson, R., Han, D. (2001). Landscape ecological planning and design of degraded mining land. *Land Degradation and Development* 12, 449-459.

Zucca, C., Julitta, F., Previtali, F. (2011). Land restoration by fodder shrubs in a semi-arid agro-pastoral area of Morocco. Effects on soils. *Catena* 87, 306-312.

In: Steppe Ecosystems
Editors: M. B. Morales Prieto and J. Traba Diaz
ISBN: 978-1-62808-298-2

Chapter 15

SUSTAINABLE DEVELOPMENT IN THE CALDENAL (ARGENTINA)

Hugo Laborde [1]***, Osvaldo A. Fernández*** [1,2]***, María N. Fioretti*** [1]***,***
Sandra S. Baioni [1] ***and Roberto E. Brevedan*** [1*]
[1]Departamento de Agronomía, Universidad Nacional del Sur (UNS)
[2]CERZOS (CONICET), Bahía Blanca, Argentina

ABSTRACT

The Caldenal is a temperate semiarid region of about 40.000 km2 in central Argentina (32° to 40° latitude), a transition between the phytogeographical province of the Monte to the west and the humid Pampa to the east. It has a highly variable distribution of rainfall (350 to 550 mm) concentrated in the spring-summer season and very high (42°C) and very low (-14°C) temperatures. Soils have a petrocalcic horizon to a depth of 0.5 to 2.0 m.

The grazing lands of the Caldenal are an important asset. This natural resource is, however, generally both neglected and abused with the resulting low productivity of pasture, widespread erosion and a low level of livestock production.

In this work we propose two ways to improve the situation in the Caldenal: i) to increase the productivity of native grasslands. It is proposed the use of pasture legumes that can lead to a steady rise of soil nitrogen and organic matter. This has resulted in the development of a farming system in which the increased fertility associated with legumes is used by wheat crops or to increase livestock production through the quantity and quality of the legumes themselves and through the effect of improved fertility on the production of the pasture plants. ii) To minimize the adverse effects of grazing year-long, with the animals moving freely on a range unit, we propose a discontinuous grazing system. Under this system, the field would be divided into subunits and would hold all the livestock in one of them until total forage was consumed, then the livestock move to other subunit and so on. This system provides each subunit with intermittent rest periods from grazing. This

* Corresponding Author adress: Departamento de Agronomía, Universidad Nacional del Sur, Altos del Palihue, 8000 Bahía Blanca, Argentina. Email: ebreveda@criba.edu.ar.

system allowed so far the reintroduction of two valuable native species that had disappeared from certain locations.

Keywords: Caldenal, rangeland degradation, livestock production, Argentina, vegetation change, perennial grasses, legumes

INTRODUCTION

The Caldenal is a temperate semiarid region located in the central part of Argentina covering an area of about 40,000 km^2 of rangeland between approximately 32° and 40° S latitude. Most of the Caldenal is located in the central and eastern part of the province of La Pampa and has a width of approximately 100 km, running NW-SE, between the meridians 63° and 66° W longitude. Its oriental border is sinuous with various ecotones where the natural rangeland intermixes with the marginal zone of the humid Pampa. Its western border limits with the phytogeographical province of Monte (Cabrera, 1976) (Figure 1).

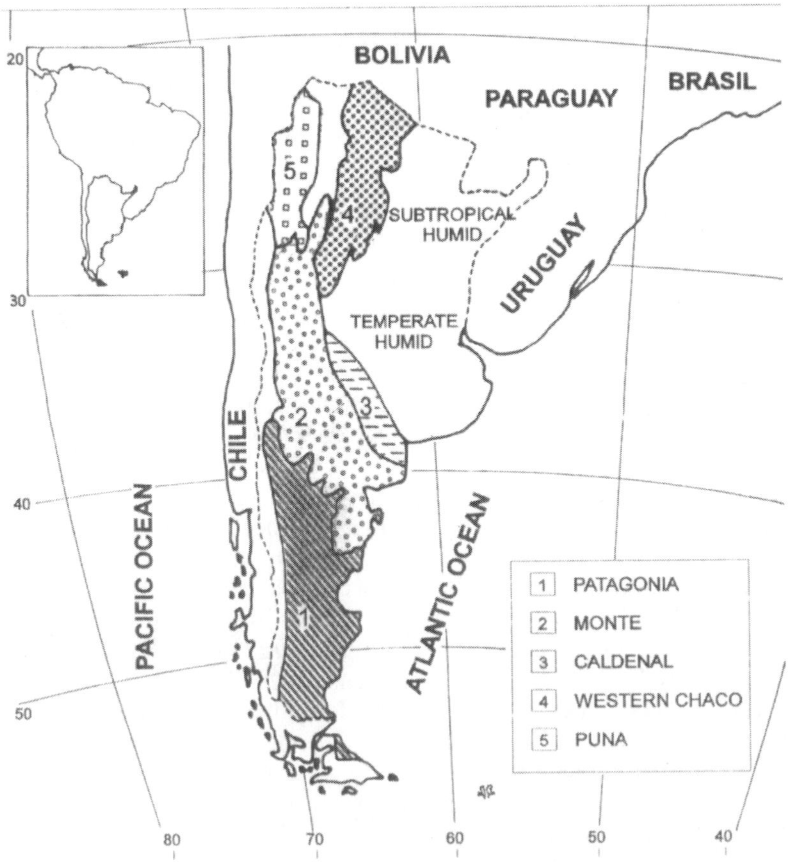

Figure 1. Major arid and semiarid rangelands of Argentina. The Caldenal appears as an ecotone in between the Monte desert to the west and the cultivated humid Pampa to the east. (Adapted from Cabrera, 1976 and Fernández and Busso, 1999).

The climate is temperate semiarid, and aridity increases westward and southward. The annual rainfall (350 to 550 mm) diminishes from east to west. In the northern part of the region there is a rainy summer season (from October to March) with 77% of the annual rainfall and a dry winter season (from April to September) giving the remaining 23%. In the southern region rainfall is more frequent and intense during spring and summer, with a dry winter. Annual potential evapotranspiration is 800 mm approximately. Mean annual temperature is 15.3 °C.

The most significant limitations are not only low rainfall and high temperatures but also low temperatures that can reach -14 °C, as well as the large variability in the occurrence of the first and last frost (Kent, 2010).

Due to the lack of ground and surface water in both quality and quantity, the main source of water for livestock is the rain. This water is collected by natural or artificial basins and then stored in ponds. One disadvantage of this technique is its low storage efficiency, with a use of around 50% owing to seepage and evaporation (Kent, 2010).

There are no rivers or streams in the area occupied by the Caldenal, but salt marshes are present. There are also freshwater lagoons generated by rainfall infiltration in dune areas, the quality is generally good for human consumption, livestock and irrigation.

In general in the eastern part of Caldenal, water is a few feet below the surface while in the western one is as deep as 100 m but still with the same poor quality for human and livestock (Lell, 2004).

The topography is characteristic of a well drained plain, undulating with gentle slopes, caused by wind action and ancient fluvial action, ranging from 50-100 m.a.s.l. in the southeast to 600 m.a.s.l. in the northern area. There is a loess mantle of variable thickness of up to 2 m on a rough hard and thick rock that can appear in surface. In many sectors the soil presents loose material, therefore forming important sand dune systems (Russo et al., 1979; INTA, 1980; Chiozza and Figueira, 1981).

Those soils have excessive drainage, very fast permeability, low moisture retention and lack of structure, except for the bulk surface. The most important limitations are climatic, as well as excessive drainage and greater susceptibility to wind erosion (Kent, 2005).

Soils are almost exclusively mollisols, with small inclusions of entisols and aridisols (INTA, 1980). They can be distinguished according to their organic matter content and mineral particle size (Quiroga et al., 1998). Both texture and organic matter content are believed to be the main causes of the differences between their physical and chemical properties (Quiroga, 1994).

The main different types of physiognomic vegetation are a dense shrubland covering a surface of 30.8% (Roberto et al., 2005, Figure 2). Different kinds of forest cover a surface of 16.1% and grassland 22.2%. The cultivated area (28.2%) is located mainly in the east of the region.

During the first decades of the last century, sheep rearing was the main cause of changes in the floristic composition and structure of the Caldenal, while in the last years the overgrazing by cattle, clearing and burning were the transforming main transforming factors (Roberto et al., 2005) (Figure 3).

Figure 2. Under pristine conditions original physiognomy of the vegetation consisted in a graminous steppe, with isolated shrubs and trees.

Figure 3. A large part of the Caldenal is now in regular or poor conditiondegraded.

The current vegetation is dominated by a *Prosopis* tree layer, especifically *P. caldenia* forming forest associated with *P. nigra, P. flexuosa* var *flexuosa*, and accompanied by *Goeffroea decorticans* and *Jodina rhombifolia*. Edaphophilous communities are common on dunes, along with psammophyte vegetation and brackish soils with halophyte shrublands (Cano and Movia, 1967; Anderson et al., 1970).

Shrubs can reach 1 m high. They are represented by *Condalia microphylla, Prosopis alpataco, Lycium chilense* var *chilense, L. chilense* var *confertifolium, Atamisquea emarginata, Ephedra triandra* and *Aloysia gratissima.*

The herbaceous canopy consisted of a grassland matrix, with prevailing high value forage grasses. There are is a considerable number of bunchgrasses, including: *Stipa tenuis, S. clarazii, S. longiglumis, Piptochaetium napostaense, Poa ligularis, Setaria leucopila, Aristida subulata, Digitaria californica, Trichloris crinita* and *Soghastrum pellitum.* Anderson (1982) proposed the domestication of *Sorghastrum pellitum* to facilitate its incorporation into degraded pastures, but its low seed yield and overlapping functional role with weeping lovegrass (*Eragrostis curvula*) discouraged initial expectations (Veneciano, 2006).

Grassland communities fulfill numerous functions in the community. They improve soil structure, increase the infiltration rate of water, protect the soil surface from rain-drop impact, capture solar energy and produce plant litter that eventually is incorporated into soil organic matter.

Perennial cool season grasses dominate the herbaceous layer in the rangelands of the southern Caldenal but in the northern area warm season grasses including weeping lovegrass, the main perennial forage crop in the region, prevail over cool-season grasses. During summer, warm season grasses produce an average concentration of 44-54% of the dry matter (DM) in the northern Caldenal. In spring there is great variability in the production of 21 to 39% of its annual total. There is a high correlation between the production rate and rainfall in the summer but not in the spring and fall seasons, when rainfall is necessary but not enough to determine optimal growth, and in which the temperature would also have a determinant role (Veneciano et al., 1996).

The C4 photosynthetic system of the warm season grasses have has general advantages, overcomes photorespiration, uses water more efficiently than C3 grasses requiring about one-third to one-half as much water to produce a unit of dry matter as C3 grasses do. Intermittent aridity may have contributed to the evolution of C4 grasses, although this does not mean that they are necessarily drought tolerant. Drought tolerance is not correlated directly with water use efficiency (Moser et al., 2004).

The Past Situation

In the early stages of field occupation by European immigrants, in the late nineteenth century, Argentina maintained a strong agricultural bias in the use of arable soils. Indeed, the use of land for the production of wheat was very profitable while soil high natural fertility lasted. This produced a long-lasting impression on the farmers' culture, since they always considered wheat production as the main source of income compared to other alternatives such as livestock production of meat or milk. In addition, wheat cultivation was a much simpler practice in relation to other management alternatives. Accordingly, and in the early part of the twentieth century, settlers entered the Caldenal from the border with the humid Pampa always with the greatest interest in cereal crops, wheat, oats or barley. After a laborious clearing, they once again took advantage of the relatively good natural fertility of their soils, which became evident immediately after the removal of shrubs and trees. In this case however, soils with lesser development had a lower potential fertility. Therefore, the rapid depletion of the fertility of these soils under cereal cultivation caused frequent crop failures, particularly in years of recurrent drought, which are unpredictable.

The call to expand further the agricultural frontier caused by the technification associated with the green revolution of the last quarter of last century promoted the cultivation of wheat and the subsequent development of corn and sunflower crops in the Pampas. This generated a new pressure to cultivate the Caldenal also with the advantage of agricultural mechanization that allowed a less laborious clearing. It also promoted the use of machinery that was expected to allow a better use of ground water for crops.

This scenario in which the average farmer made reasonably good earnings lasted until about 2004. Then a period of severe drought began, during which grass production dropped and cattle were sold to farmers from other areas or died from starvation. The cleared areas of Caldenal suffered a severe wind erosion process that had its peak in the southeastern portion, with important loss of soil by wind action. In bare soils remained only *Solanum elaegnifolium* weeds, which have no forage value.

There were no conservation measures in the Caldenal, while sheep and cattle pressures were high, so that the range experienced a strong deterioration over the last century (Defosse and Robberecht, 1987). Cattle numbers were usually too high in an area where on average only the equivalent of one cow can be carried on five hectares (Morris and Ubici, 1996).

The effects of heavy grazing are detrimental mainly to valuable species. Animals graze heavily on the most palatable plants, thereby depleting their food reserves and causing the loss of their ability to produce seeds; hence they are rapidly replaced by less palatable plants, perennials are replaced by annuals. The soil becomes progressively denuded, soil structure is degraded, and organic matter and infiltration rates are reduced. The original cover of predominantely gramineous plants is replaced by shrubs, which also results in reduced grazing.

During the first decades of the last century sheep rearing and logging were the main causes of changes in the species composition and structure of Caldenal while in more recent years cattle overgrazing, poor fire management and clearing were the main factors (Roberto et al.. 2005). That has promoted greater *Prosopis caldenia* density with shoots and appearance of woody shrubs, a reduction of valuable forage species and the consequent increase in larger size grasses of little or null forage value.

THE CURRENT SITUATION

Throughout the area, most of the available forage dry matter for the winter season is provided by the cumulative vegetation growth in the hot and humid season. In winter this resource is in an advanced stage of maturity, low in protein, high in lignified fiber and therefore of poor quality, which limits its intake by calves. Moreover, the lack of use of the pasture during the production cycle prevents activation of new growth points and increases the proportion of stems in mature forages that are less digestible than leaves (Veneciano, 2006).

Currently, the main use of this semiarid region is the breeding of domestic animals with the sole food source of natural vegetation. Most of the surface is undergoing some degree of degradation, i.e. the present degradation of plants modified by one or more vegetation disturbing factors, which may prevent the system to reach its climax (Anderson, 1980). In large areas, the overgrowth of shrub species has resulted in a decline in productivity. There is a replacement of valuable forage species by shrubs and lower quality grasses (Cerqueira et al., 2004). The overgrazing by domestic animals is the main factor determining the explosive spread of woody species in natural grazed ecosystems.

Overgrazing results in a gradual deterioration of vegetation, causes a reduction of the number of plants, the disappearance of valuable fodder species, the invasion of undesirable species and finally the decrease of available forage.

Fire had a very important role in the origin and evolution of the grassland, especially where woody vegetation could compete with the grass. Regular fire occurrence allowed controlling shrub density and prevented its spread within the natural grassland. The influence of fire today is markedly lower due to more intensive use of pastures, which reduces the accumulation of dry biomass.

Another structural modification in the vegetal community is the declining coverage pasture grasses (Jasic and Fuentes, 1991). A great part of the Caldenal has diminished its livestock capacity due to the loss of valuable grass species (Estelrich and Cano, 1985; Cano, 1990; Morici et al., 1996).

The agricultural frontier expansion has moved livestock to marginal areas and at the same time has led to an increase of meat production systems, for both intensive and semi-intensive systems to continue to supply the growing demand (Roberto et al., 2008). This movement of livestock is intensified because the productive efficiency index indices in more marginal areas are significantly lower than those of the higher potential areas, which affects considerably the economic sustainability of the production system (Roberto et al., 2008).

The management model is based on continuous grazing by animals with excessive charges resulting in a general deterioration of soils and vegetation. It is possible to find mismanaged paddocks where the proportion of grasses with little or no forage value and undesirable shrubs is very high (Cano, 1975).

Semiarid agroecosystems have in common their complex stability, risk of desertification and limited production (see also Chapter 14 in this volume), which makes it difficult to guess the combination of appropriate uses for these areas, which thet garantees the sustainability of the system (Boza et al., 1985; Boza, 1993). The limited structure of the soil and, therefore, its tendency to undergo erosion, makes recommendable for extensive livestock systems a shift to perennial grass-based production.

An overload situation leads to a decrease of grazing generating reducing livestock carrying capacity. This is produced by a reduction in vegetation cover, an increase in soil temperature, the oxidation of organic matter, the risk of wind and water erosion, and a decrease in the water holding capacity (Roberto et al., 2008).

The increase of livestock in the region is not accompanied by increases in the supply of fodder. Water deficit that characterizes the region limited forage production. Only an increase of fodder production accompanied by an appropriate management of rodeos will achieve, with the same stock, an increase in the stock removal rate and thus an increase in meat production (Roberto et al., 2008).

Livestock in the last hundred years experienced important changes, from an absolute predominance of sheep over cattle in the early stages to a reverse situation where cattle outweighs sheep, having increased its stock about 7 times (Roberto et al., 2008). The region is becoming increasingly important in the country in competitive calf production, compared to others where livestock has been replaced by more profitable crops.

In these conditions it is fundamental to consider the stocking capacity and responsiveness of the system. It is necessary to consider the adjustment between load and responsiveness, especially in marginal areas, where degradation of rangelands increases the negative consequences for extensive livestock, mainly reductions in animal numbers, and health deterioration (González Rebollar et al., 1993).

In summary, the Caldenal is a fragile ecosystem that human actionwhich has undergone a process of human-induced degradation throughout the twentieth century, which has not stopped yet. The effects of overgrazing of grasses are also evident in soil erosion. As loss of vegetation occurs, the land is exposed to drying out, to wind and rain. In this way the soil keeps on losing its organic content, fertility and ability to absorb and retain humidity.

In addition, low investment in infrastructures is characteristic of the region, so that establishments with a small number of paddocks, poor provision of water and low maintenance of fences are common. Therefore the design of each pasture and the distribution of watering places are aspects that determine the intensity and location of the disturbance caused by livestock. The use of vast paddocks may force the animals to make long journeys from the watering places to the opposite end of the farm (Kent, 2010).

Cerqueira et al. (2004*)* studied the nutritive value of the most important forage species of the Caldenal. The nutritional value was determined by samples that were collected in the fall, winter and spring under grazing conditions, for ranges of good and fair conditions, during two successive years. The crude protein (CP) concentration of short-winter grasses (*Piptochaetium napostaense*, *Poa ligularis*, *Stipa clarazii* and *Hordeum stenostachys*) was about 10%, mid-winter grasses (*S. tenuissima* and *S. gynerioides*) never reached 6% CP, summer grasses (*Digitaria californica* and *Trichloris crinita*) ranged from 7% to 9% CP. *In vitro* dry matter digestibility (IVDMD) was similar among short-winter and summer grasses (40-50%). Mid-winter grasses had the lowest IVDMD for all seasons (<40%). Good condition ranges provide a more acceptable forage supply than fair condition ones.

Butti et al. (2005) evaluated the production and quality of fertilized prevailing cool season grasses in the ecotone Caldenal-Monte. Studied species were *Poa ligularis*, *Stipa tenuis* and *Piptochaetium napostaense*. Accumulated rain from fertilization (winter) to the end of the vegetative cycle was 111.5 mm (2003) and 605 mm (2004). In 2003 accumulated phytomass averaged 400 kg ha^{-1}, whereas in 2004 was 5,340 kg ha^{-1}. Both years showed greater protein content in the fertilized treatment versus the control, 11.0% and 13.5% (fertilized), and 8.8% and 10.4% (control), respectively.

In order to provide an economic winter forage it would be useful to use forage produced during the previous growing season by warm season grasses. Stritzler et al. (1996) compared the winter forage quality produced during the previous growing season of 4 warm season grasses. *Tetrachne degreii* had generally, the highest dry matter production (10,132 kg ha^{-1}) CP 5.3%, IVDMD 50-40% and the lowest content of neutral and acid detergent fiber and lignin; *Panicum coloratum* was the second most productive and nutritious species followed by *Eragrostis curvula* cv. Tanganyka. *Panicum virgatum* cv. Pathfinder was generally the least productive species and its nutritive value was low.

In the northern area of the Caldenal grasses protein content in the whole plant, reached highest values in October (10.2% on average) when plants were preflowering, declining in November. From December to April these plants were kept between 7.0 and 8.3% with the lowest value in February (Veneciano, 2006).

SUSTAINABLE APPROACHES TO REGIONAL PROBLEMS

Livestock performance is often better on grass/legume mixtures than on grasses alone (Burns and Standaert, 1985). Legumes are an important component of grasslands providing inputs of nitrogen without which productivity would be low (Heichel and Henhum (1991). *Medicago minima* var *minima* is a high quality forage species during spring in both overgrazed and properly managed areas in the Caldenal. Under certain conditions, alfalfa (*Medicago sativa)* supplies nitrogen for growth of the associated grass by excretion, sloughing of nodules, or decay of the root system. Livestock grazed grasses may benefit from associated legumes because of the increase in forage quality and the biologically fixed nitrogen.

Certain legumes fix more nitrogen than is needed for their own growth and they build up soil nitrogen level as a result. This is important in regions where the use of nitrogen fertilizer is not profitable. The high nutritive value of the legumes is due to their high protein concentration, which in many cases is less degradable in the rumen. Nitrogen fertilizer is too expensive to be used in rangeland. Simbiotic fixation via legumes remains the only low-input method of increasing nitrogen. Legumes typically compete poorly with grasses for soil mineral nitrogen and therefore sometimes must rely almost exclusively on symbiotic nitrogen fixation for their nitrogen supply. The contribution of the symbiotic nitrogen fixation to legume nitrogen usually exceeds 80% (Thomas, 1995).

Environmental interactions make predicting the behavior of components of the association difficult. However, management can influence them, allowing some control over the components. Measures used to alter grass/legume balance will obviously depend on the direction in which the balance is to be shifted (Fales et al., 1996). Grass content generally increased as cutting interval is shortened and cutting height lowered. Control of grass to reduce competitiveness in grass/clover swards can be achieved by severe defoliation in late autumn or early spring (Davies and Evans, 1990). Due to the complexity of the grass/legume interaction, it is not possible to generalize about the direction in which dominance will be achieved for a given set of conditions (Fales et al., 1996).

The lack of stability of grass/legume mixtures is a serious limitation to their use and exploitation. Grassland management for the maintenance of a desired balance between grasses and legumes is very difficult compared with the management of grasses only.

Mixtures of weeping lovegrass (*Eragrostis curvula*) with legumes have great potential, and the best results have been obtained using alfalfa (*Medicago sativa*). When cut for hay, the mixture dried more rapidly and had lower leaf shattering losses than pure alfalfa. When grazed, bloating (a major problem for cattle grazing pure alfalfa), is minimized, while livestock performance and productivity were comparable to those obtained in well fertilized grass pasture (150 to 200 kg N ha^{-1}) (Rethman et al., 1986; Voight et al., 2004).

Rethman and de Witt (1984) indicated that mixtures of weeping lovegrass and alfalfa that were fertilized to meet the requirements of alfalfa, had similar DM yields and higher quality than weeping lovegrass pasture receiving 150 kg N ha^{-1} $year^{-1}$. A better distribution of the species in the stand was obtained from alternating row than from broadcast planting, and animal use was also improved (Rethman, 1987). Optimum seeding rates were 1.5 and 15 kg ha^{-1} of weeping lovegrass and alfalfa respectively (Rethman and de Witt, 1984). Over 5 years the proportion of alfalfa increased from 43 to 54% and 43 to 66% in hayed and grazed ecosystems, respectively. Weeping lovegrass and alfalfa mixtures were compatible with management requirements and use by grazing animals (Voigth et al., 2004).

Weeping lovegrass and *Digitaria eriantha* have been introduced in the Caldenal to improve forage production, in mixtures with alfalfa (Laborde et al., 2006, see Table 1). Mixtures of weeping lovegrass and alfalfa have great potential and are favored with the development of alfalfa cultivars with increased grazing tolerance (Brummer and Bouton, 1991). Medics *(Medicago minima)* showed a fairly good DM production mostly in winter time. IVDMD is over 70% in winter time (Gargano et al., 1986).

Table1. Dry matter yield of *Digitaria eriantha*, weeping lovegrass and alfalfa as pure or associated stands. Three-year average (Laborde et al., 2006)

Treatment	Dry matter (Mg ha^{-1})
Alfalfa	4.880 ab*
Digitaria	0.787 c
Weeping lovegrass	3.047 b
Alfalfa + Digitaria	5.042 ab
Alfalfa + Weeping lovegrass	7.180 a

* Means between treatments followed by the same letters are not statistically different (P>0.05).

Legumes have increased the productivity of native grasslands and using them results in a steady rise of soil nitrogen and organic matter. This has resulted in the development of farming systems in which the increased fertility associated with legumes is used by crops of wheat or to increase livestock production through the quantity and quality of the legumes themselves and through the effect of improved fertility on the production of other pasture plants. Also grassland productivity may increase when market conditions allows it, since responses to fertilization with nitrogen and phosphorus are significant.

In the southwestern area of the Caldenal all zones are considered submarginal for arable farming and best retained as pasture lands for cattle. The farms are commonly units between 1,000 and 10,000 ha in size, allowing an extensive form of stockraising mostly of store cattle that then is sent on for fattening to the more humid Pampa. The productivity of the land is lower than 30 kg live weight ha^{-1} yr^{-1}.

However, the rest of the land resulting from the absence of livestock allowed for the re-establishment of regional native species like *Stipa tenuis* and *Papophorum vaginatum* in a markedly hostile environment, particularly for the forage species that are most commonly sown in the Caldenal due to their greater productivity. After several years, the soil cover was almost complete and *S. tenuis* and *P. vaginatum* competed favorably with introduced weeds succesfully adapted to the region such as *Diplotaxis tenuifolia*. Since ground clearance is not required as in sowing of introduced species, the spread of *D. tenuifolia* occupying new areas is minimal.

In such conditions fodder production of reintroduced species has been 2,500 kg DM ha^{-1} yr^{-1} (Larreguy, V. personal communication), which is not negligible especially considering their cost of production was zero.

Free ranging livestock will tend to concentrate on areas that are within easy walking distance of to watering-places, especially on areas that are more attractive or easly grazed than others, such as the moister bottom-lands of flat topography. As a result, overgrazing is intensified on the most valuable parts of the range, whilst the production of large areas may go to waste and the areas eventually revert to worthless brush.

We propose to divide the farm into subunits using nonexpensive electric wiring inside the farm and have all the livestock in one of them till almost total forage consumption. Then livestock should move to another subunit and so on. This discontinuous grazing provides each subunit with intermittent rest from grazing, and ensures that grasses of the same subunit are not grazed year after year at the same growth stage.

The advantage of this discontinuous grazing system are: rest from grazing enables the herbage to gain in vigour and allow desirable species to produce seed. Livestock distribution over the range is better than otherwise, and grazing is more uniform; re-seeding is possible on resting range without interference from livestock. All these advantages should lead to greater livestock production.

It should be mentioned that the Caldenal management favoring native species may be accompanied in most favored soils by the introduction of species such as rye or weeping lovegrass. This is done to alleviate the pressure over the native stand.

No doubt that this type of land use of the Caldenal is now a change in the traditional extensive farming, which has been characteristic of the region. For example, the calculation of the daily allowance of food and maintenance of the electrical wiring necessary, require the permanent presence in the farm of the producer. Rodeo permanent observation leads to other advantages on livestock production, like detection of disease or parturition difficulties, which tend to be more efficient.

On the other hand, it must be recognized that the start of production in the Caldenal long and largely degraded areas, necessarily requires much more investment, which is not always taken into account. Native species mentioned have excellent ability to implant in these environments but its subsequent development up to a certain stage of defoliation by livestock, although indispensable, is very slow. There are examples already in the region where this system achieves calving percentages above 90%, without requiring nutrient inputs for livestock that are not produced *in situ* (Larreguy, V. personal communication).

The Caldenal, even in its current state, offers productive and sustainable answers for ruminant livestock, which can be achieved by using simple techniques, proven tested in other regions. The role of legumes may be mainly for legume-based pastures to increase livestock production and improve soil fertility.

Livestock productivity in the region is low, estimated at less than 30 kg liveweight ha^{-1} yr^{-1}. In such conditions, reinvestment opportunities in the area are also modest and must be programmed at a long term. No substantial changes can be expected in a short time but the key is to maintain an improvement strategy based on the knowledge gained from research in the area.

However, the whole region is in regression, with owners migrating to cities. The absence of settlers depresses economic and social activity in the region. Large areas are exposed to depopulation and actions of vandalism and theft of cattle and facilities, which makes up a vicious cycle. The region has not such extreme agroclimatic characteristics to deserve such state of things. If the region must increase its involvement in the production of calves, measures should be taken to change the *status quo* that prevails today. This situation is being maintained unchangedhas persisted for at least the second half of the 20th century and still continues.

REFERENCES

Anderson, D.L. (1980). La recuperación y mejoramiento de los pastizales naturales. *Ecología* 4, 9-12.

Anderson, D.L. (1982). Una reliquia del pastizal natural. *Informativo Rural* VII (17), 3-5.

Anderson, D.L., Del Águila, J.A. and Bernardón, A.E. (1970). Las formaciones vegetales en la provincia de San Luis. INTA, *Revista de Investigaciones Agropecuarias,* Serie 2, 153-183.

Boza, J. (1993). *Planificación ganadera del sudeste ibérico. Nutrición de ruminates en zonas áridas y de montaña y su relación con la conservación del medio natural*, pp. 59-66. Consejería de Agricultura y Pesca de Andalucía, Sevilla.

Boza, J., Silva, J.H., Azocar, P. (1985). *Recursos alimenticios en zonas áridas,* pp. 191-226. Simposio Internacional sobre explotación caprina en zonas áridas. Publ. Excmo. Cabildo Fuerteventura. Canarias, España.

Brummer, E.C., Bouton, J.H. (1991). Plant traits associated with grazing-tolerant alfalfa. *Agronomy Journal* 83, 996-1000.

Burns, J.C., Standaert, J.E. (1985). Productivity and economics of legume-based vs nitrogen-fertilized grass-based pastures in the United States. In: R.F. Barnes et al. (ed.), Proceedings Trilateral Workshop. *Forage legumes for energy-efficient animal production*,pp. 56-71. USDA-ARS, CSIRO, and DSIR. Nationall Technology Information Service, Springfield (VA), USA.

Butti L.R., Adema, E.O., Babinec, F.J. (2005). Producción y calidad de un pastizal fertilizado en el ecotono Caldenal-Monte Occidental, La Pampa. *Revista Científica Agropecuaria* 9, 153-162.

Cabrera, A.L. (1976). Regiones fitogeográficas argentinas. In: E.F. Ferreira Sobral (ed.), *Enciclopedia Argentina de Agricultura y Jardinería*, pp. 1-85. Acme, Buenos Aires.

Cano E., Movia, C. (1967). Utilidad de la fotointerpretación en la cartografía de comunidades vegetales del bosque de caldén (*Prosopis caldenia*). *La Vegetación de la República Argentina* 8, 1-44.

Cano, A.E. (1975). Pastizales en la región central de la provincia de La Pampa. *IDIA* 331-333, 1-5.

Cano, E., Morici, E., Fernández, B., Chirino,C. (1990). Características de tres pastizales bajos con distinto período de descanso. *Revista de la Facultad de Agronomía, UNLPam.* 5, 83-108.

Cerqueira, E.D., Sáenz, A.M., Rabotnikof, C.M. (2004). Seasonal nutritive value of native grasses of Argentine Calden Forest Range. *Journal of Arid Environments* 59, 645-656.

Chiozza, E., Figueira, R. (1981). Atlas físico de la República Argentina. In: *Atlas Total de la República Argentina.* Centro Editor de América Latina, Buenos Aires.

Davies, A.G., Evans, M.E. (1990). Effects of spring defoliation and fertilizer nitrogen on the growth of white clover in ryegrass/clover swards. *Grass Forage Science* 45, 345-356.

Defosse, G.E., Robberecht, R. (1987). Patagonia: range management at the end of the world. *Rangelands* 9/3, 106-109.

Estelrich, H.D., Cano, A.E. (1985). Disponibilidad forrajera y determinación de la capacidad de carga de un bosque de *Prosopis caldenia. Actas I Jornadas de Biología and II Jornadas de Geología, La Pampa. UNLPam* 1, 30-35.

Fales, S.L., Laidlaw, A.S., Albert, M.G. (1996). Cool-season grass ecosystems. In: L.E. Moser, Buxton, D.R., Casler, M.D. (eds.), *Cool-season Forage Grasses*, pp. 267-296.. American Society of Agronomy, Madison (WI), USA.

Fernández, O. A., Busso, C.A. (1999). Arid and semiarid rangelands: Two thirds of Argentina. *Rala Report 200. Proceedings Rangeland Desertification*, pp. 41-60. International Workshop of Agricultural Research Institute, Reykjavik.

Gargano, A.O, Möckel, F.E., Adúriz, M.A., Cantamutto, M.A. (1986). Evaluación de leguminosas anuales del género *Medicago* comparadas con *Vicia sativa. Revista Argentina de Producción Animal* 6, 573-580.

González Rebollar, J.L., Passera, C.B., de la Cruz, R. (1993). La intensificación de pastoreo intensivo y consecuencias no deseadas de la PAC. Algunos ejemplos y propuestas. *Paralelo 37°* 16, 141-145. Heichel, G.H., Henhum, K.I. (1991). Dinitrogen fixation, nitrogen transfer and productivity of forage legume-grass communities. *Crop Science* 31, 202-208.

INTA, Gobierno de La Pampa y Universidad Nacional de La Pampa. (1980). *Inventario Integrado de los Recursos Naturales de la Provincia de La Pampa.* Universidad Nacional de La Pampa, Buenos Aires.

Jasic, F.M., Fuentes, R.E. (1991). Why are native herbs in the chilean matorral more abundant beneath bushes: microclimate or grazing? *Journal of Ecology* 68, 665-669.

Kent, F. (2010). Planificación de tecnologías apropiadas para productores de la región del Caldenal, en el departamento de Utracán, Provincia de La Pampa. *XV Jornadas Nacionales de Extensión Rural y VII del MERCOSUR.*

Laborde, H.E., Brevedan, R.E., Fioretti, M.N. (2006). Technology and sustainability for the development of the Caldenal. In: Brebbia, C.A., Conti, M.E., Tiezzi, E. (eds.), *Management of Natural Resources, Sustainable Development and Ecological Hazards*, pp. 297-304. Wit Press, Southampton.

Lell, J.D. (2004). El Caldenal: una visión panorámica del mismo enfatizando en su uso. In: Arturi, M.F., Frangi, J.L., Goya, J.F. *Ecología y Manejo de los Bosques de Argentina*, pp. 1-18. Editorial Universitaria La Plata, Buenos Aires.

Morici, E., Chirino, C.C., Fernández, B.C., Estelrich, H.D. (1996). Aplicación del modelo de estados y transiciones en los pastizales de la región semiárida pampeana. *VI Jornadas Pampeanas de Ciencias Naturales (La Pampa)*, 167-172.

Morris, A., Ubici, S. (1996). Range management and production on the fringe: the Caldenal, Argentina. *Journal of Rural Studies* 12, 413-425.

Moser, L.E., Burson, B.L., Sollenberger, L.E. (2004). Warm-season (C4) grass overview. In: Moser, L.E., Burson, B.L. and Sollenberger, L.E. (eds.), *Warm-season (C4) Grasses*, pp. 1-14. American Society of Agronomy, Madison (WI), USA.

Quiroga, A.R. (1994). Influencia del manejo sobre propiedades físicas del suelo. Su relación con la granulometría y contenido de material organica. MSC Thesis. Universidad Nacional del Sur, Bahía Blanca, Argentina.

Quiroga, A.R., Buschiazzo, D.E., Peinemann, N. (1998). Management discriminant properties in semiarid soils. *Soil Science* 163, 591-597.

Rethman, N.F.G. (1987). Seedbed preparation and seeding techniques in southern Africa. In: Frasier, G.W., Evans, R.A. (eds.), *Seed and seedbed ecology of rangeland plants,* pp. 169-172. USDA-ARS.

Rethman, N.F.G., de Witt, C.C. (1984). The production potential of pure and mixed stands of *E. curvula* and *M. sativa* on the Eastern Transvaal Highveld. *Journal of the Grassland Society of South Africa* 1, 30-32.

Rethman, N.F.G., Odendaal, J.J., de Witt, C.C. (1986). Performance of *Medicago sativa* under dryland conditions on the eastern Highveld. *Journal of the Grassland Society of South Africa* 3, 122-127.

Roberto, Z., Frasier, E., Goyeneche, P., González, F., Adema, E. (2008). Evolución de la carga animal en la provincia de La Pampa. Período 2002-2008. *Publicación Técnica* 74. INTA EEA, Anguil, Argentina.

Roberto, Z.E., Adema, E.O., Rucci, T.E. (2005). Relevamiento fisonómico de la vegetación en el área del Caldenal. INTA *Publicación Técnica* 60. INTA EEA, Anguil, Argentina.

Russo, A., Ferello, R., Chebli, G. 1979. Llanura chaco pampeana. In: *Simposio de Geología Regional y Argentina*, 139-183.

Stritzler, N.P., Pagella, J.H., Jouve, V.V., Ferri, C.M. (1996). Semi-arid warm-season grass yield and nutritive value in Argentina. *Journal of Range Management* 49, 121-125.

Thomas, R.J. (1995). Role of legumes in providing N for sustainable tropical pastures systems. *Plant and Soil* 174, 103-118.

Veneciano, J.H. (2006). Gramíneas estisvales perennes para ambientes semiáridos: características y productividad. *Información Técnica* 171. INTA EEA San Luis. 84 pp

Veneciano, J.H., Terenti, O.A., Sager, R.L., Bertón, J.A. (1996). Variación estacional de rendimiento, proteína bruta y minerales en *Sorghastrum pellitum* (Hackel) Parodi (pasto de vaca). *Información Técnica* 139. INTA EEA, San Luis, Argentina.

Voigt, P.W., Rethman, N.F.G., Poverene, M.M. (2004). Lovegrasses. In: Moser, L.E., Burson, B.L., Sollenberger, L.E. (eds.), *Warm-season (C4) Grasses*, pp. 1027-1056. American Society of Agronomy, Madison (WI), USA.

INDEX

#

20th century, 167, 332
21st century, 290

A

absorption spectroscopy, 174
access, 299
accessibility, 236
accounting, 98
acid, 255, 256, 261, 276, 283, 328
acidic, 282
acidity, 292
acquisition of knowledge, 289
adaptation(s), vii, viii, ix, 9, 66, 76, 80
ADC, 228
adjustment, 328
adverse effects, 321
Africa, 9, 24, 243, 303, 311, 334
age, 1, 2, 5, 18, 40, 116, 165, 170, 177, 178, 179, 180, 183, 190, 301
agencies, 313
aggregation, 79, 221, 222, 223, 232, 236, 237, 244, 245, 249
agriculturally suitable soil, viii
agriculture, vii, 14, 38, 39, 166, 184, 186, 205, 206, 211, 285, 303, 315, 316
air temperature, 155, 159, 161
alfalfa, 329, 330, 332
algae, 130, 199
Algeria, 303, 304, 307, 309, 312, 314, 315, 317, 319
algorithm, 56, 65, 263
alien species, 184
alkalinity, 168, 180
Altai mountain steppes, 98
ambient air, 228, 281
amino, 275, 276
amino acid, 275, 276
ammonia, 276, 283
ammonium, 185, 280
animal disease(s), 306
animal husbandry, 114, 176
annuals, 21, 42, 203, 204, 298, 326
ANOVA, 228, 262, 292
anthropic action, x
aquifers, 168
Argentina, vi, 8, 20, 22, 71, 73, 81, 88, 92, 93, 95, 250, 321, 322, 325, 332, 333, 334
arithmetic, 78
ARS, 332, 334
arthropods, 47, 48
Artificial Neural Networks, 121
Asia, ix, 7, 24, 47, 137
aspiration, 291
assessment, viii, ix, 49, 78, 87, 92, 94, 98, 104, 148, 154, 197, 201, 202, 250, 251, 297
assimilation, 255, 256
atmosphere, 154, 161, 247, 279
attribution, 292
Austria, 71, 228
awareness, 289

B

bacteria, x, 253, 255, 257, 262, 267, 268, 270, 271, 272, 276, 280, 282, 283
bacterium, 267, 271, 274, 279, 282
ban, 315
banks, 12, 212
barriers, 56
base, 114, 162, 176, 199, 202, 204, 247
BD, 223, 225, 227, 228, 229, 230, 231, 232, 233, 237, 239, 240, 241, 242, 243, 244
bedding, 213

beetles, viii, ix, 47, 48, 49, 50, 55, 56, 58, 60, 61, 63, 64, 65, 66, 67, 69, 70, 71, 72, 90, 94, 95, 300
Beijing, 146, 147, 151, 253, 261, 282, 283
beneficial effect, 222, 315
beneficiaries, 311
bias, 166, 175, 325
bicarbonate, 5, 156
biodiversity, vii, viii, x, 47, 49, 69, 70, 74, 75, 81, 86, 87, 88, 91, 93, 94, 95, 98, 147, 150, 212, 283, 287, 288, 289, 290, 297, 301, 304, 312
biogeography, 44, 147
biological activity, 5, 155
biological control, 87
biological processes, 160
biomass, x, 47, 76, 80, 133, 138, 148, 161, 231, 232, 233, 243, 246, 247, 250, 254, 258, 259, 260, 262, 263, 265, 267, 268, 277, 278, 280, 285, 304, 309, 314, 316, 327
biosphere, 154, 212
biotic, vii, 76, 95, 115
biotic factor, vii
birds, vii, viii, 2, 11, 23, 25, 29, 30, 31, 32, 33, 37, 38, 39, 40, 43, 44, 45, 66, 72, 311
bison, 2
blood, 103
body size, 44
boreal forest, 94, 108, 183, 283
boreholes, 171, 173
BOS, 43
Botswana, 19
breeding, viii, 31, 39, 43, 44, 290, 297, 315, 316, 327
brevis, 58, 60
browser, 318
browsing, 3

C

Ca2+, 158, 167
calcium, 5, 157, 199, 225, 290, 292
calcium carbonate, 5, 157, 225, 290, 292
calibration, 190
CAM, 43
campaigns, 170, 184
canonical correspondence analysis, 36, 39, 78
capillary, 168, 169, 233
carbon, ix, 78, 79, 95, 153, 154, 155, 156, 157, 159, 161, 163, 222, 223, 226, 227, 229, 236, 246, 247, 249, 250, 254, 255, 256, 282, 283, 285, 286, 291, 293
carbon dioxide, 153, 154, 155, 156, 161, 163, 236, 256, 285
case study, 69, 151
catchments, 10, 12, 20
cation, 159
cattle, 21, 82, 91, 212, 282, 310, 315, 323, 326, 328, 329, 330, 332
causal relationship, 30
CCA, 29, 33, 36, 37, 38, 39, 40, 78, 79, 81, 83
CDC, 299
cell size, 31
Central Asia, viii, ix, 1, 5, 6, 9, 18, 40, 47, 139, 151, 183, 224
Central Europe, 137, 166, 184, 190
ceramic, 260
changing environment, 70
chemical, 153, 154, 162, 168, 199, 205, 206, 221, 222, 223, 230, 241, 243, 244, 249, 258, 278, 284, 290, 297, 298, 311, 323
chemical characteristics, 199, 258, 278
chemical properties, 205, 311, 323
Chicago, 150, 228
Chile, 20, 22, 92
China, vi, x, 51, 70, 111, 115, 116, 120, 147, 148, 149, 151, 183, 221, 222, 223, 224, 230, 231, 243, 244, 246, 247, 248, 249, 250, 251, 253, 257, 261, 281, 282, 283, 284, 285
chlorine, 159
chlorophyll, 122
circulation, 111, 189
cities, 287, 288, 332
civilization, 147
classes, 32, 42, 133, 134, 226, 227
classification, 30, 39, 78, 83, 84, 87, 89, 92, 100, 112, 120, 121, 122, 126, 134, 135, 150, 154, 155, 178, 282
climate change, ix, 40, 70, 79, 81, 91, 94, 150, 246, 319
climates, 2, 3, 6, 11, 22, 80
climatic factors, 29, 37, 39, 40, 79
clone, 253, 254, 255, 263, 270, 271, 272, 277, 281
closure, 180, 259
cluster analysis, 33, 78, 79
clustering, 33, 121
clusters, 89, 270, 271
CNS, 226
CO2, ix, 5, 153, 160, 161, 228, 235, 236, 237, 239, 256, 282
coding, 284
coefficient of variation, 228
Coefficient of Variation, 41, 42, 43
coleoptera, 288, 290
collaboration, 48
collection date, 87
colonisation, 298
colonization, 9, 65, 205, 208, 300
combined effect, xi

combustion, 226, 227
commercial, 86, 87, 291
common agricultural policy, 45
compaction, 221, 223, 227, 230, 231, 232, 236, 237, 239, 244, 247
comparative analysis, 44, 147, 155
compensatory effect, 246, 281
competition, vii, 3, 6, 18, 87, 208
competitiveness, 329
complement, 279
complementarity, 45
complex interactions, 56
complexity, 7, 15, 74, 81, 91, 179, 329
composites, 5, 8, 18, 20
compression, 227, 231
computer, 93
computer simulations, 93
computing, 71, 187
conceptual model, 135, 237
condensation, 9
conditioning, 1, 4
configuration, 39, 66, 122, 181
Congo, 24
Congress, 92, 149, 187, 188
congruence, 78, 79, 243
conifer, 3
consensus, 166, 167, 291
conservation, vii, viii, x, xi, 2, 23, 30, 39, 44, 45, 69, 71, 74, 75, 84, 86, 91, 93, 94, 98, 124, 147, 166, 188, 197, 198, 199, 201, 202, 205, 206, 208, 211, 212, 289, 292, 301, 311, 318, 326
conserving, x, 72
constituents, 161, 162, 168
construction, 56, 129, 132, 176, 289, 292, 300, 316
consumption, 254, 257, 259, 272, 273, 279, 280, 281, 285, 323, 331
consumption rates, 254, 259, 272, 273, 280, 281
contamination, 172, 175
Continental, 95, 188
contradiction, 237
control measures, 169, 172, 185
controversial, 2, 112, 125, 167
convergence, 8, 9
cooling, 131, 148, 178
correlation, 56, 58, 59, 61, 64, 65, 69, 81, 122, 123, 135, 174, 228, 231, 232, 262, 267, 325
correlation coefficient, 56, 228
correlations, 56, 60, 65, 163, 228, 243
cost, 298, 310, 311, 331
Council of Europe, 208
covering, viii, 20, 74, 135, 171, 298, 322, 323
cracks, 130
creosote, 18
Croatia, 95
crop, 38, 160, 161, 182, 183, 283, 285, 325, 326
crop production, 283
crop residue, 285
crops, 2, 18, 29, 30, 32, 36, 37, 38, 40, 285, 286, 289, 290, 321, 325, 326, 328, 330
crust, 159, 160, 186, 206
crystals, 198, 199
cues, x, xi
cultivars, 330
cultivation, 2, 13, 16, 199, 202, 203, 205, 206, 246, 249, 276, 281, 290, 292, 297, 298, 299, 301, 304, 325, 326
cultural values, 98
culture, 176, 192, 222, 255, 285, 298, 325
cycles, 176, 261
cycling, 67, 253, 254, 255, 277
Czech Republic, 187

D

damages, 222, 313
danger, 51
data set, 53, 66, 77, 84, 88, 135, 137, 228
database, 31, 32, 41, 42, 48, 69, 262, 263, 270, 308
DCA, 84, 88, 89, 112, 120, 123, 125, 126, 132
decay, 60, 66, 71, 72, 167, 329
decomposition, 67, 94, 232, 236
deficiency, 297
deficit, 21, 327
deflation, 130, 131, 139
deforestation, 21, 86, 169
degenerate, 280
degradation, viii, ix, xi, 20, 73, 74, 83, 114, 115, 120, 124, 126, 132, 139, 187, 193, 199, 221, 222, 231, 232, 243, 246, 250, 254, 257, 277, 288, 290, 299, 303, 307, 315, 316, 322, 327, 328
degradation process, 73, 74, 139
degraded area, 289, 290, 298, 313, 331
degraded soil properties, x
Delta, 272
demography, 289
dendrogram, 36, 262, 275
denitrification, 278
Denmark, 260
dependent variable, 30
deposition, 153, 221, 222, 247, 277, 280
deposits, 12, 116, 117, 155, 167, 168, 170, 174, 182, 212, 224
depression, 12, 153, 156
depth, 78, 160, 168, 175, 225, 260, 290, 291, 292, 321
desiccation, 80, 130, 139

destruction, 114, 118, 132, 139, 198, 221, 231, 237, 239, 244, 289, 291, 298
detectable, 215
detection, 261, 281, 286, 331
developed countries, 316
developing countries, 299
deviation, 42, 55, 310
digestibility, 328
digestion, 185, 260
discharges, 185
discontinuity, ix
discriminant analysis, 148, 151
discrimination, 246
dispersion, 161, 249, 298
displacement, 227
distilled water, 226
DNA, 253, 254, 260, 261, 262, 263, 264, 266, 269, 273, 284, 285, 286
DNA polymerase, 261
DNAs, 262
DOI, 285
domestication, 147, 149, 325
dominance, ix, 21, 38, 180, 182, 280, 329
downsizing, 315
drainage, 6, 12, 115, 166, 183, 192, 323
drought, vii, xi, 21, 74, 114, 199, 303, 307, 313, 325, 326
dry matter, 75, 277, 309, 325, 326, 328
drying, vii, 9, 17, 18, 260, 328

E

earnings, 326
Eastern Europe, 5
ecological data, 94
ecological processes, 199, 200
ecological requirements, 308
ecological restoration, 287, 289
ecology, viii, 44, 49, 53, 71, 98, 108, 149, 162, 208, 279, 281, 288, 289, 290, 316, 334
economic status, 245
economics, 332
ecosystem, viii, ix, x, xi, 3, 47, 50, 69, 70, 73, 74, 75, 76, 81, 86, 91, 98, 101, 114, 120, 129, 149, 153, 154, 155, 161, 165, 208, 222, 223, 224, 242, 243, 244, 245, 246, 247, 249, 250, 284, 285, 287, 289, 290, 291, 297, 298, 299, 311, 328
ectothermic, 65
EEA, 334
electrophoresis, 253, 254, 262, 282, 284
elongation, 261
emission, 153, 154, 155, 156, 160, 161, 278, 285
empirical studies, 174
endangered, viii, x, 75, 112, 207, 212, 215
endangered species, 75, 212, 215
endothermic, 65
endowments, 168, 169, 183
endurance, 149
energy, 66, 69, 72, 111, 226, 236, 255, 307, 325, 332
energy input, 226, 236
engineering, x, 148, 151, 288, 289, 299, 300, 301
England, 184, 228, 277
environment, viii, 4, 9, 36, 40, 56, 57, 66, 71, 73, 81, 86, 91, 138, 176, 190, 191, 202, 207, 222, 233, 245, 248, 267, 285, 286, 316, 330
environmental change, 76, 111, 114, 149
environmental conditions, 7, 8, 87, 132, 170, 199, 242, 312
environmental factors, 30, 39, 65, 75, 76, 78, 278, 285
environmental quality, 198
environmental variables, ix, x, 29, 31, 32, 33, 36, 39, 53, 54, 56, 58, 59, 76, 78, 81, 253, 257, 263, 267, 268, 277, 280
environments, vii, viii, ix, 2, 3, 5, 13, 21, 39, 65, 66, 72, 74, 104, 121, 186, 199, 205, 206, 222, 223, 237, 243, 331
enzyme, 255, 257
equilibrium, 76, 95, 114, 115, 135, 136, 139, 150
erosion, 116, 131, 132, 183, 187, 203, 206, 208, 241, 243, 248, 291, 308, 321, 323, 326, 327
ethanol, 261
EU, 32, 42, 197
eukaryotic, 199
Eurasia, 191
Europe, 7, 9, 11, 25, 30, 32, 44, 45, 67, 82, 171, 186, 187, 189, 208, 288, 300, 301
European Commission, 39
European Union, x, 29, 199, 207, 219
evaporation, 81, 153, 154, 155, 156, 159, 160, 161, 168, 169, 198, 199, 200, 202, 206, 323
evapotranspiration, 21, 65, 80, 112, 200, 323
evidence, viii, 113, 114, 116, 124, 139, 147, 166, 169, 175, 180, 192, 199, 230, 236, 239, 242, 244, 285
evolution, ix, 4, 24, 25, 162, 166, 167, 169, 170, 172, 175, 176, 181, 182, 188, 208, 304, 306, 307, 325, 327
exclusion, x, xi, 54, 221, 222, 223, 225, 228, 229, 230, 231, 233, 236, 237, 240, 243, 244, 245, 249, 250, 254, 257, 259, 263, 275, 278, 281, 286, 291, 295, 300, 318
excretion, 329
exile, 145
experimental design, 257, 290, 291
exploitation, vii, xi, 16, 293, 297, 329

exposure, 112, 120
extinction, viii, 43, 66, 75, 86, 87
extraction, ix, 139, 173, 185, 254, 260, 261, 262, 263, 264, 266, 269, 273, 286
extracts, 156, 158, 159, 261
extreme cold, 66, 179, 183
extreme weather events, 81

F

facies, 176, 191
families, 8, 79, 131, 203, 245, 282
farmers, 165, 315, 325, 326
farmland, 44, 45, 300
farms, 330
fauna, 2, 48, 57, 67, 69, 70, 165, 184, 187, 188, 287
fencing, 118, 137, 233, 313
fertility, 5, 150, 221, 240, 243, 244, 245, 248, 297, 298, 299, 321, 325, 328, 330, 331
fertilization, 292, 328, 330
fertilizers, 284
fiber, 326, 328
fidelity, 120
filtration, 226
financial, 313
fine wool, 258
Finland, 94
fires, 20, 51, 83, 114, 293
fixation, 308, 329, 333
flame, 174, 259
flooding, 181
floods, 169, 175, 181
flora, x, 8, 10, 13, 14, 24, 75, 79, 82, 94, 114, 131, 170, 175, 182, 183, 192, 197, 198, 201, 206, 208, 212, 213, 215, 220, 288, 289, 292, 311
flora and fauna, 170, 288, 289
flowering period, 77
flowers, 5
fluctuations, 168, 169, 176, 177, 180, 183, 307
FMC, 262
food, 65, 82, 86, 176, 304, 307, 310, 326, 327, 331
food chain, 86
forbs, 76
force, 227, 328
forest ecosystem, 297
forest fire, 50
forest formations, ix, 8
forest fragments, 93
forest management, 95
forested regions, vii
formaldehyde, 256
formamide, 262
formation, ix, 18, 20, 82, 100, 101, 156, 166, 198, 199, 200, 206, 221, 236, 237, 238, 239, 240, 242, 244, 245, 246, 247, 290
formula, 57
foundations, 291
fractures, 168
fragility, 297
fragments, x, 101, 102, 105, 106, 112, 130, 198, 261, 263
France, vi, x, 287, 288, 289, 293, 294, 295, 296, 300, 301
freshwater, 53, 281, 323
frost, vii, 130, 131, 323
functional changes, 154
funding, xi, 299, 315
funds, 310
fungi, 199, 215, 219, 220

G

gas diffusion, 227
gel, 170, 253, 254, 261, 262, 263, 266, 267, 274, 275, 282, 284
genes, 254, 257, 261, 284
genotyping, x
genus, 8, 18, 20, 21, 23, 82, 88, 178, 255, 303
geography, 191
geology, 18, 169, 182, 183, 201
Germany, 47, 108, 111, 221, 226, 227, 300
germination, 199
GIS, 32, 177, 183, 187, 308
global climate change, 149, 162
global scale, 76, 80, 114
global warming, 255
governments, 13
GPS, 120
gracilis, 6
granules, 121
graph, 39, 59, 61
graphite, 174
grass, viii, 3, 4, 10, 14, 15, 16, 20, 21, 22, 30, 38, 40, 51, 53, 79, 81, 82, 91, 101, 102, 103, 104, 105, 106, 107, 108, 160, 171, 177, 179, 182, 212, 240, 243, 245, 246, 257, 284, 305, 326, 327, 329, 332, 333, 334
grasses, 2, 3, 5, 6, 7, 8, 9, 20, 21, 22, 40, 51, 53, 73, 74, 75, 76, 79, 82, 91, 101, 102, 124, 131, 137, 138, 139, 178, 202, 203, 204, 212, 242, 259, 322, 325, 326, 327, 328, 329, 331, 333
gravity, 6, 168
grazers, vii
Grazing exclusion, 230, 233
Greece, 69

green revolution, 326
greenhouse, 255
greenhouse gas(es), 255
grids, 57, 60, 225, 229, 239
groundwater, 155, 167, 168, 169, 185, 189, 192, 291, 312
grouping, 37
growth, vii, 3, 5, 6, 7, 8, 10, 12, 21, 94, 112, 131, 137, 139, 155, 160, 180, 181, 243, 279, 307, 315, 325, 326, 329, 331, 333
guardian, 311

H

habitat(s), x, 2, 8, 9, 10, 12, 23, 29, 30, 31, 32, 36, 37, 38, 39,40, 42, 45, 44, 47, 50, 56, 64, 65, 66, 69, 76, 86, 87, 94, 102, 105, 120, 139, 167, 169, 171, 172, 180, 182, 183, 184, 198, 199, 200, 201, 202, 203, 204, 205, 206, 207, 213, 215
halophyte, 13, 104, 156, 159, 325
hardwood forest, 95
harvesting, 291, 309
hay meadows, 289
health, ix, 98, 135, 328
height, 8, 74, 87, 123, 125, 131, 139, 168, 202, 222, 233, 262, 304, 329
hemisphere, 20
heterogeneity, 71, 79, 84, 241, 243, 244, 245, 248
highlands, 70, 100, 112, 113, 114, 115, 120, 121, 126, 132, 136, 137, 138, 139, 304, 311
histogram, 57, 270
history, x, 11, 65, 148, 166, 170, 186, 187, 188, 193, 301
Holocene, x, 2, 13, 23, 112, 114, 116, 126, 131, 147, 148, 149, 150, 151, 170, 176, 177, 179, 180, 181, 182, 183, 184, 187, 188, 189, 190, 191, 301
homogeneity, 84, 89, 292, 309
horses, 2, 118
hotspots, 45, 69, 288
House, 148, 162, 219
human, vii, ix, x, xi, 2, 48, 76, 82, 86, 97, 98, 104, 113, 114, 121, 124, 149, 150, 165, 169, 182, 187, 198, 290, 291, 293, 300, 301, 306, 323, 328
human-impacted ecosystems, vii
humidity, ix, 5, 18, 22, 125, 131, 156, 328
humus, 5, 6, 21, 132
Hungary, 165, 166, 170, 171, 174, 182, 184, 185, 186, 187, 188, 189, 190, 191, 192, 193
hunting, 2
husbandry, 114
hybrid, 78
hypothesis, 79, 135, 199, 291, 297

I

Iberian steppe birds, 25
ideal, 39, 175
identification, 30, 40, 69, 77, 101, 104, 120, 122, 154, 180, 184, 281, 290
identity, 276
illumination, 262
image, 122, 134, 313
imagery, 98, 100, 121, 134
images, 54, 121, 135
immigrants, 325
impact assessment, 120
imports, 315
in transition, 19, 160
incidence, 78, 92, 142
income, 118, 325
incubation period, 235
independence, 304, 306
independent variable, 33
individuals, 79, 215
industry, 316
inertia, 80, 81, 84, 89, 91, 123
inferences, 174, 175
initiation, 168, 258
inoculum, 228
insects, 5, 53, 67, 72, 74, 86, 87, 91, 92, 93, 290
institutions, 313
Instron, 227
integration, 162
integrity, 100, 288, 298
interface, 93
interference, 331
international standards, 166
intervention, 291
invariants, 103
investment, 310, 328, 331
ionization, 259
ions, 79, 156, 167, 199
Iran, 71
irrigation, 13, 289, 323
Islam, 95
islands, 74, 87, 212, 221, 240, 242, 243, 246
isolation, 5, 15, 137, 285
isotope, 168, 185, 186, 189, 246
Israel, 49, 70
issues, viii, ix
Italy, 184

J

Japan, 261

K

Kazakhstan, 3, 4, 5, 50

L

lakes, 9, 12, 16, 48, 53, 130, 153, 155, 162, 166, 167, 168, 175, 177, 185, 188
land abandonment, 290
land tenure, 245
Late Pleistocene, 113, 192
latitudinal band, ix
leaching, 278
lead, 8, 39, 79, 108, 114, 118, 120, 222, 223, 244, 254, 257, 298, 316, 321, 331
leakage, 298
learning, 121, 122
learning process, 121, 122
legend, 103
legislation, 197, 200, 207
legume, 8, 13, 285, 329, 331, 332, 333
Lepidoptera, 95
lichen, 105, 208
life sciences, 112
light, 76, 114, 161, 190, 225, 233, 258, 278
lignin, 328
limestone, 116, 213
lipids, 282
livestock, vii, x, xi, 48, 51, 73, 74, 76, 79, 91, 98, 112, 114, 115, 118, 119, 126, 131, 132, 135, 138, 139, 150, 199, 206, 249, 277, 282, 303, 304, 306, 307, 309, 311, 312, 313, 315, 316, 319, 321, 322, 323, 325, 327, 328, 329, 330, 331
local community, 124
localization, 6
logging, 21, 326
low temperatures, 64, 65, 323
LSD, 228, 293
lying, 8, 180

M

machinery, 198, 202, 205, 206, 326
magnesium, 5, 79, 157, 159
magnitude, 239
majority, 2, 5, 6, 21, 126, 132, 169, 236, 254, 272, 279
mammal(s), ix, 112, 114, 126, 132, 311
man, viii, 2, 5, 20, 21, 162, 172, 288
manipulation, 57
mantle, 323
manure, 284
mapping, 97, 98, 104, 121, 151, 250
marsh, 173, 177, 180
Maryland, 69, 72
MAS, 108, 246
mass, 6, 22, 154, 226, 233
materials, viii, 13, 15, 16, 22, 202, 219, 220, 223, 232, 233, 241, 244, 245
matrix, 33, 37, 53, 56, 78, 80, 81, 83, 93, 122, 198, 262, 290, 297, 325
matter, 44, 79, 93, 153, 167, 208, 221, 222, 234, 236, 249, 323, 330
measurement, 33, 93, 95, 153, 155, 156, 160, 161, 259, 296, 309
meat, 304, 312, 315, 325, 327
mechanical loadings, 239
mechanical properties, 249
mechanical stress, 223, 236, 237, 238, 244
media, 175, 182
median, 53, 65, 228
Mediterranean, vi, vii, viii, x, 1, 2, 3, 4, 6, 8, 9, 10, 11, 18, 19, 20, 21, 23, 24, 25, 31, 39, 42, 44, 45, 49, 69, 199, 203, 204, 205, 207, 208, 242, 248, 287, 288, 293, 297, 299, 300, 301, 307, 318
Mediterranean climate, vii, 8, 9, 19, 293
melt, 5
memory, 29, 43, 47, 71, 122
MERCOSUR, 333
messages, xi
meta analysis, 247
Metabolic, 71
metabolism, 80
meter, 21, 54, 228, 290
methanol, 256
methodology, 57, 83
Mexico, 8, 208, 243, 246, 308
Mg2+, 158, 167
microbial cells, 281
microbial communities, 255, 257, 278, 282, 283, 285
microbial community, 254, 255, 278, 282, 283, 284
microclimate, 161, 333
microcosms, 283
microenvironments, 12
microhabitats, 184
microorganisms, 161, 254, 255, 279, 283
Middle East, 114, 308, 309, 311
migration, 154
military, 287, 289
mineralization, 5, 223, 227, 231, 232, 236, 237, 239, 244, 254, 277, 278, 285
Miocene, 9, 10, 12, 16, 150
mission, 150
Missouri, 24, 94, 146, 162, 301
mitochondrial DNA, 148

mixing, 255, 281
modelling, 70, 174, 248, 319
models, viii, xi, 2, 32, 36, 43, 44, 56, 150, 161, 167, 239, 319
modifications, 212
MODIS, 54, 69
moisture, x, 79, 80, 100, 159, 161, 231, 237, 246, 254, 265, 267, 268, 278, 280, 285, 323
moisture content, 254, 265, 267, 268, 278, 280
MOM, 226, 233, 236
Mongolia, viii, x, 3, 5, 6, 47, 48, 49, 50, 51, 52, 53, 54, 55, 58, 60, 61, 62, 64, 65, 68, 69, 70, 71, 72, 97, 98, 100, 101, 108, 109, 131, 148, 150, 178, 191, 221, 222, 224, 230, 231, 243, 244, 245, 246, 247, 248, 249, 250, 253, 254, 257, 258, 259, 265, 273, 276, 279, 280, 281, 282, 283, 284, 285, 286
Morocco, 6, 313, 319
morphology, 168
mortality, 114, 319
mosaic, 6, 13, 15, 18, 30, 39, 40, 45, 48, 100, 112, 122, 126, 128, 130, 133, 135, 136, 165, 179, 180, 182, 183, 184, 203
Moscow, 97, 108, 109, 162, 163, 219
mountain ranges, 3, 8, 12, 15, 48
multidimensional, 62, 78
multiple regression, 33
multivariate analysis, 72

N

Na+, 158, 167
Namibia, 19
NAP, 177, 178, 180, 181, 183
National Academy of Sciences, 71, 285
native species, 199, 300, 322, 330, 331
natural disturbance, 139
natural evolution, 2, 165
natural habitats, 137, 138
natural resources, xi, 97, 98, 304
nature conservation, 108
negative consequences, 328
negative effects, 247, 278
negative relation, 65
neglect, 315
nested PCR, 261, 282
Netherlands, 162
neural network(s), 121, 149
neutral, 237, 328
New Zealand, 7, 19, 20, 249, 253, 281
NIR, 122
nitrification, 203, 277
nitrogen, 78, 79, 199, 223, 246, 247, 250, 254, 282, 283, 285, 286, 321, 329, 330, 332, 333
nitrogen fixation, 285, 329
nitrous oxide, 284
NMR, 246
noble gases, 189
nodules, 329
normal distribution, 33
North Africa, 8, 9, 10, 14, 197, 200, 287, 288, 297, 298, 308, 309, 311
North America, 7, 9, 71, 243
Norway, 178
Norway spruce, 178
nuclei, 10
nucleic acid, 261, 262
nucleus, 298
null, 326
nutrient(s), 12, 67, 242, 248, 249, 253, 277, 279, 331
nutrition, 319

O

oceanic currents, 9
oil, 263, 283, 298
operations, 287, 289, 290, 291, 298, 299
opportunities, 331
organic matter, x, 5, 79, 161, 221, 222, 223, 226, 231, 232, 234, 236, 247, 248, 249, 250, 279, 290, 321, 323, 325, 326, 327, 330
organic soils, 6
organism, 313
organs, 161, 215
originality, 4
overgrazing, xi, 48, 74, 131, 139, 221, 222, 244, 250, 304, 307, 312, 323, 326, 327, 328, 331
overlap, 125, 313
overtraining, 134
ox, 80
oxidation, 255, 256, 260, 279, 281, 282, 283, 327
oxidation rate, 279
oxygen, 186, 255

P

Pacific, 18, 22
paleoecology, 24, 166, 297
paleoenvironmental changes, 172, 190
Paraguay, 20
parallel, vii, 6, 22, 131, 180, 181, 227, 289
parasites, 87
partition, 78
pasture(s), viii, ix, x, xi, 5, 6, 11, 16, 17, 20, 21, 45, 98, 111, 112, 114, 115, 118, 119, 120, 123, 124, 125, 126, 127, 129, 130, 132, 133, 135, 136, 137,

138, 139, 148, 149, 211, 245, 247, 250, 284, 287, 321, 325, 326, 327, 328, 329, 330 331, 332, 334
Patagonian steppe ecological health, ix
Patagonian steppes, 9, 20, 22, 92
pathways, 256
PCA, 122, 134, 262, 266, 267, 269, 273, 274
PCR, 253, 254, 257, 260, 261, 262, 264, 266, 269, 273, 274, 282, 286
peat, 82, 190
permafrost, 117
permeability, 6, 323
permission, xi, 23, 299
permit, 2
personal communication, 331
PET, 227
pH, x, 78, 156, 168, 225, 226, 228, 254, 260, 262, 265, 267, 268, 278, 280, 290, 292
Philadelphia, 149
phosphor(o)us, 199, 292, 297, 330
photographs, xi, 23, 174, 290
photosynthesis, 156, 161
phylogenetic tree, 275, 276, 280
physical properties, 161, 247, 249, 284
Physiological, 284
physiological factors, 1
physiology, 72
Pinus halepensis, 13, 16
pipeline, 298, 300
plant growth, 78, 223, 244, 278
plant type, 79
plasmid, 262
plasmid DNA, 262
platform, 16, 292, 299, 308
playing, 79
Pliocene, 187
Poland, 174
polar, 22
policy, 45, 118, 212
pollen, 2, 24, 124, 148, 151, 166, 171, 172, 174, 175, 176, 177, 178, 179, 180, 181, 182, 184, 185, 187, 188, 189, 191
pollinators, 86
polymerase, 261, 284
polymerase chain reaction, 284
ponds, 323
pools, 9, 12, 249, 278, 286
population, viii, 18, 25, 31, 39, 45, 97, 98, 197, 205, 206, 211, 215, 222, 290, 304, 306, 313, 316
population growth, 97, 98, 304
population size, 215
porosity, 237
Portugal, 207, 288
positive correlation, 60, 66, 267
positive relationship, 267
potassium, 199, 292, 297, 308
precipitation, ix, 7, 32, 33, 35, 36, 41, 42, 43, 49, 57, 58, 59, 62, 63, 65, 67, 75, 79, 80, 91, 100, 114, 115, 117, 118, 120, 129, 151, 153, 154, 159, 162, 168, 169, 202, 224, 244, 257, 273, 285
predation, 68, 87
predators, 65, 87, 311
predictor variables, 56, 58, 65, 121
preparation, 334
preservation, 98, 176, 178, 183, 197, 198, 199, 205, 290, 311, 312
President, 281
prevention, 236
principal component analysis, 148, 151
principles, 154, 170, 313
privatization, 118, 137
probability, 40, 59
producers, 248
productive efficiency, 327
profitability, 311
programming, 57
project, 43, 70, 71, 97, 222
proliferation, 255
property rights, 245
prophylaxis, 306
protected areas, x, 45, 211, 212, 213, 215, 218, 300, 313, 316
protection, 182, 198, 213, 215, 216, 218, 222, 223, 236, 237, 238, 243, 244, 250, 289, 308, 311, 314, 315, 319

Q

quantitative reconstruction, 189

R

radar, 150
radiation, 78, 112, 156, 223, 243
Radiation, 78
radius, 57, 176, 225
rainfall, viii, 6, 7, 18, 19, 20, 21, 22, 29, 35, 37, 39, 40, 44, 49, 51, 74, 76, 77, 81, 82, 125, 130, 168, 181, 206, 223, 248, 257, 288, 304, 305, 306, 308, 311, 313, 314, 318, 321, 323, 325
rainforest, 94
rangeland, xi, 75, 76, 115, 118, 135, 137, 138, 301, 307, 309, 311, 315, 316, 319, 322, 329, 334
reactions, 154
reality, 39, 312, 313, 316
recommendations, 202

reconstruction, 175
recovery, vii, x, 139, 205, 207, 221, 222, 230, 244, 276, 279, 280, 314
recurrence, 298
Red List, 197, 200, 219
redistribution, 223
redundancy, 253, 268
reference system, 71
regeneration, 203, 204, 245, 277, 297, 311, 312
regenerative capacity, 112
regionalization, 151
regions of the world, viii, 2
regression, 32, 56, 58, 65, 117, 156, 263, 332
regression analysis, 58, 117
regression model, 32, 65, 156
regulations, 165, 169
rehabilitation, 287, 290, 292, 298, 299
reinforcement, 316
reintroduction, 297, 298, 322
relevance, 29, 40, 111, 148, 304
reliability, 30, 156, 182
relief, 11, 101, 202, 206
remote sensing, ix, 112, 115, 121, 147, 148, 151, 247, 313
repellent, 112, 131
repetitions, 290, 291
representativeness, 213, 219
reproduction, 44, 65
requirements, 39, 76, 80, 309, 330
researchers, xi, 170
reserves, x, 75, 161, 197, 201, 205, 208, 282, 326
residuals, 32, 56
residues, 161, 236
resilience, vii, viii, ix, xi, 111, 112, 131, 135, 139, 290, 291, 297, 298, 301
resistance, 112
resolution, 32, 54, 70, 93, 121, 122, 184
resources, vii, viii, 65, 118, 121, 139, 162, 245, 247, 248, 303, 307
respiration, 153, 155, 160, 161, 162, 163, 225, 227, 236, 247, 282
response, viii, 7, 9, 17, 31, 32, 56, 71, 76, 84, 88, 91, 118, 135, 138, 147, 149, 150, 263, 283, 285, 301, 303
responsiveness, 328
restitution, 318
restoration, 98, 184, 208, 246, 255, 283, 289, 290, 291, 292, 293, 294, 295, 296, 298, 299, 300, 301, 308, 317, 319
rhizome, 259
ribosomal RNA, 257
risk, 38, 120, 153, 162, 318, 319, 327
risk factors, 153, 162
risk management, 319
risks, 86
river basins, 51
RNA, 254, 257
rodents, 2, 5, 66
room temperature, 226
root, 6, 103, 161, 260, 278, 329
root system, 6, 161, 329
roots, 6, 7, 8, 117, 130, 132, 161, 237, 238, 283
roughness, 112, 222
routes, 100
Royal Society, 92
rules, 122
runoff, 53, 168, 247, 308
rural areas, 312
Russia, vi, 4, 5, 18, 97, 108, 153, 162, 163, 211, 212

S

salinity, 155, 213
salt concentration, ix
salts, 4, 153, 157, 159, 162, 166, 167, 168, 169
saltwater, 53
sampling error, 53
samplings, viii
saturation, 81, 249
savannah, 3, 10, 20
scaling, 62, 78
scarcity, 199
scattering, 13
school, 182
science, 95, 162, 187
scientific knowledge, vii, xi
scope, 87
Scots pine, 178
sea level, 48, 100, 224, 257
seasonal flu, 65
seasonality, 32, 37, 58, 59, 65, 87
sedimentation, 226
sediment(s), 14, 15, 112, 116, 117, 130, 148, 151, 155, 172, 174, 185, 187, 190, 223, 241, 244, 247, 281, 282
seed, 199, 293, 297, 299, 300, 301, 318, 325, 331
seeding, 330, 331, 334
seedlings, 123, 199
segregation, 44
sensing, 121
sensitivity, x, 81, 222, 238, 239, 245, 261
sequencing, 283
services, 47, 316
settlements, ix, 48, 114, 120, 125
shallow lakes, ix, 21

sheep, ix, 74, 76, 77, 102, 103, 104, 114, 118, 130, 138, 203, 224, 225, 231, 232, 236, 249, 258, 259, 286, 287, 288, 290, 291, 297, 299, 309, 312, 315, 317, 323, 326, 328
shoots, 326
shoreline, 116
shorelines, 116, 147
shores, 53
shortage, 162
showing, 19, 38, 112, 136, 197, 254, 271, 276, 281, 283, 293
shrubland, 79, 83, 85, 89, 90, 208, 246, 323
shrubs, 2, 3, 6, 7, 12, 13, 14, 16, 18, 19, 20, 22, 51, 53, 73, 74, 75, 79, 81, 82, 91, 124, 125, 132, 148, 202, 207, 212, 242, 243, 309, 319, 324, 326, 327
Siberia, ix, 48, 109, 153, 183, 185
signals, 135
signs, 170
silica, 170, 260
simulation(s), 78, 149
skewness, 228
sludge, 255
smoothing, 59, 61
social sciences, 112
sodium, 21, 157, 159, 166, 167, 168, 169, 308
software, 32, 56, 84, 88, 94, 120, 150, 174, 228, 262, 263
soil erosion, 75, 81, 130, 207, 221, 231, 233, 312, 328
soil particles, 226
soil type, 6, 132, 157, 206, 246, 285
solution, 168, 226, 228, 315, 316
South Africa, 19, 149, 248, 334
South America, 22, 65, 66, 67
Soviet Union, 109, 177, 191
sowing, 161, 291, 295, 298, 299, 300, 331
SP, 282
Spain, v, viii, 1, 10, 12, 15, 17, 23, 24, 25, 29, 30, 31, 32, 38, 40, 41, 43, 45, 197, 199, 205, 206, 207, 208, 209, 243, 247, 288
specialists, 199
speciation, 66
species richness, viii, 30, 32, 33, 35, 37, 38, 39, 44, 47, 48, 54, 55, 56, 57, 58, 59, 61, 63, 64, 65, 66, 67, 68, 69, 70, 71, 74, 78, 79, 83, 86, 87, 91, 93, 131, 132, 215, 293, 295, 297, 311
specific tax, vii
spore, 174
stability, 1, 5, 12, 70, 114, 154, 223, 225, 227, 235, 237, 238, 239, 245, 246, 248, 250, 281, 318, 327, 329
stabilization, 221, 222, 223, 232, 237, 244, 247, 309
standard deviation, 42, 43, 53, 56, 77, 228, 230, 232, 234, 235, 309
standard error, 59, 265, 268
starvation, 326
state(s), ix, xi, 20, 31, 76, 92, 98, 105, 109, 135, 136, 154, 155, 166, 167, 201, 202, 205, 215, 290, 313, 331, 332
statistics, 292, 308, 313
steel, 156, 225
steppe ecosystems, vii, viii, ix, x, xi, 3, 5, 31, 48, 49, 67, 98, 100, 104, 154, 155, 161, 221, 244
stimulation, 278
stock, 75, 299, 327, 328
stoma, 20
storage, 250, 277, 323
stress, 60, 62, 66, 81, 223, 225, 227, 297
stretching, 111
structuring, 295, 298, 300
style, 245
subgroups, 213, 219
substitution, 5, 206
substrate(s), 3, 4, 10, 11, 14, 17, 18, 38, 52, 116, 126, 132, 135, 199, 200, 203, 208, 282
success rate, 304, 309
succession, 131, 203, 205, 206, 207, 208, 300
Sudan, 3
sugarcane, 282
sulfate, 198, 199, 202
sulphur, 199
Sun, 254, 284
supervision, 245
supplementation, 319
surface area, 38, 39, 288
surface layer, 5, 6, 292
surface properties, 112
surface tension, 87
survival, 65, 215, 309, 310
survival rate, 309, 310
susceptibility, 112, 323
sustainability, x, 327, 333
Sustainable Development, vi, 321, 333
Sweden, 174, 260, 318
Switzerland, 227
synthesis, 151, 309
Syria, 317, 319

T

target, 87
tax reform, 316
taxa, 9, 10, 40, 48, 64, 78, 79, 82, 120, 121, 126, 174, 175, 176, 177, 178, 179, 180, 181, 182, 183
taxes, 316

taxonomy, 83, 92
techniques, x, 78, 112, 115, 254, 283, 284, 288, 289, 300, 301, 307, 309, 316, 317, 331, 334
temperature, 7, 8, 22, 30, 32, 37, 40, 41, 42, 43, 48, 49, 52, 55, 57, 58, 59, 62, 63, 64, 65, 67, 71, 76, 77, 80, 81, 100, 115, 131, 151, 155, 156, 159, 160, 161, 181, 224, 227, 257, 261, 262, 278, 323, 325, 327
temporal variation, 70
tensile strength, 227, 235, 237, 239, 246
terraces, 116, 126
terrestrial ecosystems, 280
territorial, 100, 114
territory, 18, 20, 52, 57, 154, 155, 212, 215, 218, 307
tertiary sector, 316
testing, 122
texture, 225, 226, 233, 258, 259, 323
theft, 332
thoughts, 182
threats, xi, 81, 86, 198, 202, 289
Tibet, 2, 3, 6, 7, 111, 112, 113, 114, 115, 120, 124, 125, 126, 132, 147, 148, 149, 150, 151
time periods, 230
TIR, 121
tones, 51
topology, 276
total costs, 310
tourism, 291, 316
trace elements, 174
trade, 66
trade-off, 66
traditions, 98
training, 122, 133, 134
traits, vii, 9, 21, 114, 147, 301, 332
trajectory, 290, 298
transducer, 227
transformation, 13, 32, 33, 35, 86, 97, 98, 109, 118, 120, 148, 151, 162, 180, 181, 183, 278
transformations, 33, 153, 169, 181
translation, 169
transpiration, 22
transport, 167, 181, 298, 310, 316
transportation, 310
treatment, 265, 270, 277, 278, 280, 282, 291, 295, 301, 328
trial, 122
triggers, 283
tropical forests, 87
tuff, 167
tundra, 22, 94, 101, 102, 105, 178, 179, 183, 282
tungsten, 78
Turkey, 8, 9, 10
turnover, 48, 56, 57, 59, 60, 61, 62, 63, 66, 67, 69, 71, 74, 75, 81, 91, 123, 285
twist, 74

U

UK, 92, 262
Ukraine, 4, 5
UNESCO, 165, 171, 182
uniform, 174, 176, 258, 331
United, 95, 243, 283, 285, 308, 318, 332
United Kingdom, 283
United Nations, 318
United States, 95, 243, 285, 308, 332
urban, 39
urbanization, 86
urea, 262
urine, 277, 280
Uruguay, 20
USA, 163, 226, 228, 247, 261, 262, 263, 300, 318, 332, 333, 334
USDA, 332, 334
USGS, 54, 55, 72, 121
USSR, 71, 162
UV, 262

V

Valencia, 45
validation, 36, 122, 299
vandalism, 332
variables, 29, 30, 31, 32, 33, 35, 36, 37, 38, 39, 40, 41, 42, 54, 55, 56, 58, 59, 65, 67, 75, 78, 80, 81, 122, 243, 263, 267, 315
variations, 3, 7, 8, 37, 79, 86, 244, 273, 277
vector, 67, 262
vertebrates, 86
vertigo, 319
visualization, 57
vulnerability, 111, 112, 115, 135, 206, 215

W

walking, 87, 331
Washington, 146, 300
waste, 331
wastewater, 282
water evaporation, 161
water resources, 6, 22
watershed, 115, 120, 175, 181
web, 212
weeping, 325, 329, 330, 331

welfare, 98
Western Europe, 5, 186, 287
Western Siberia, 153, 154, 162
wetlands, 79, 80, 82, 182, 288
wildlife, 114
Wisconsin, 23
withdrawal, 291
wood, 16, 84, 123, 125, 290, 291, 301, 308
woodland, 3, 16, 126, 178, 182
worldwide, 1, 223, 288
worms, 5
wrestling, 304

X

X-axis, 64

Y

Yale University, 301
yearlings, 319
yield, 170, 263, 312, 325, 330, 334